PERIODIC CHART OF THE ELEMENTS

IA																	O
1 1.0079 **H** Hydrogen	IIA											IIIA	IVA	VA	VIA	VIIA	2 4.00260 **He** Helium
3 6.94 **Li** Lithium	4 9.0122 **Be** Beryllium											5 10.81 **B** Boron	6 12.011 **C** Carbon	7 14.0067 **N** Nitrogen	8 15.999 **O** Oxygen	9 18.9984 **F** Fluorine	10 20.18 **Ne** Neon
11 22.9898 **Na** Sodium	12 24.305 **Mg** Magnesium	IIIB	IVB	VB	VIB	VIIB		VIII		IB	IIB	13 26.9815 **Al** Aluminum	14 28.08 **Si** Silicon	15 30.9738 **P** Phosphorus	16 32.06 **S** Sulfur	17 35.453 **Cl** Chlorine	18 39.95 **Ar** Argon
19 39.09 **K** Potassium	20 40.08 **Ca** Calcium	21 44.9559 **Sc** Scandium	22 47.9 **Ti** Titanium	23 50.941 **V** Vanadium	24 51.996 **Cr** Chromium	25 54.9380 **Mn** Manganese	26 55.84 **Fe** Iron	27 58.9332 **Co** Cobalt	28 58.70 **Ni** Nickel	29 63.54 **Cu** Copper	30 65.38 **Zn** Zinc	31 69.72 **Ga** Gallium	32 72.6 **Ge** Germanium	33 74.9216 **As** Arsenic	34 78.9 **Se** Selenium	35 79.904 **Br** Bromine	36 83.80 **Kr** Krypton
37 85.47 **Rb** Rubidium	38 87.62 **Sr** Strontium	39 88.9059 **Y** Yttrium	40 91.22 **Zr** Zirconium	41 92.9064 **Nb** Niobium	42 95.9 **Mo** Molybdenum	43 **Tc** Technetium	44 101.0 **Ru** Ruthenium	45 102.906 **Rh** Rhodium	46 106.4 **Pd** Palladium	47 107.868 **Ag** Silver	48 112.40 **Cd** Cadmium	49 114.82 **In** Indium	50 118.7 **Sn** Tin	51 121.7 **Sb** Antimony	52 127.6 **Te** Tellurium	53 126.905 **I** Iodine	54 131.30 **Xe** Xenon
55 132.905 **Cs** Cesium	56 137.3 **Ba** Barium	57 138.905 **La** Lanthanum	72 178.4 **Hf** Hafnium	73 180.947 **Ta** Tantalum	74 183.8 **W** Tungsten	75 186.207 **Re** Rhenium	76 190.2 **Os** Osmium	77 192.2 **Ir** Iridium	78 195.1 **Pt** Platinum	79 196.967 **Au** Gold	80 200.6 **Hg** Mercury	81 204.4 **Tl** Thallium	82 207.2 **Pb** Lead	83 208.980 **Bi** Bismuth	84 **Po** Polonium	85 **At** Astatine	86 **Rn** Radon
87 **Fr** Francium	88 226.025 **Ra** Radium	89 **Ac** Actinium															

★

6	58 140.12 **Ce** Cerium	59 140.908 **Pr** Praseodymium	60 144.2 **Nd** Neodymium	61 **Pm** Promethium	62 150.4 **Sm** Samarium	63 151.96 **Eu** Europium	64 157.2 **Gd** Gadolinium	65 158.25 **Tb** Terbium	66 162.5 **Dy** Dysprosium	67 164.930 **Ho** Holmium	68 167.2 **Er** Erbium	69 168.934 **Tm** Thulium	70 173.0 **Yb** Ytterbium	71 174.97 **Lu** Lutetium
★★ 7	90 232.038 **Th** Thorium	91 231.036 **Pa** Protactinium	92 238.029 **U** Uranium	93 237.048 **Np** Neptunium	94 **Pu** Plutonium	95 **Am** Americium	96 **Cm** Curium	97 **Bk** Berkelium	98 **Cf** Californium	99 **Es** Einsteinium	100 **Fm** Fermium	101 **Md** Mendelevium	102 **No** Nobelium	103 **Lr** Lawrencium

Physical

SECOND EDITION

Science

With Modern Applications

MELVIN MERKEN

Worcester State College

 SAUNDERS GOLDEN SUNBURST SERIES

1980

SAUNDERS COLLEGE • Philadelphia

Saunders College
West Washington Square
Philadelphia, PA   19105

Front and back covers: This demonstration was made using the components of the Klinger
Blackboard Optics TM (Trademark) set.   Klinger Educational Products Corp., Jamaica, New York.

Physical Science With Modern Applications   ISBN   0-03-056793-9

0123   006   9 8 7 6 5 4 3 2 1

To
SHIRLEY

with all my love

PREFACE

This book, like the first edition, is addressed to non-science majors in four-year colleges and universities and two-year community and junior colleges, and to students in programs of continuing education, who live in a culture that is centered on science and technology and is growing more so with every passing day. My objective in writing it has been to present physical science in an interesting, intelligible, and enjoyable way to the future (or present) business manager, teacher, artist, journalist, lawyer, government worker, and to persons in many other fields for whom this may be the last formal glimpse into physical science. If one may judge from events since the publication of the first edition, many of the topics considered will be increasingly important in their lives in the remaining decades of the twentieth century.

I have presented some of the major ideas of physics, chemistry, astronomy, geology, and meteorology as they are presently accepted, indicating how they came to be, and applied them to topics of current interest. The emphasis throughout is on the nature of science as a creative human enterprise; the key role that it plays in modern society; its relationship to technology and thereby to the environment; its open-ended character as reflected in the dynamic nature of scientific concepts; and the human qualities of scientists and their social responsibility. The reader, I hope, will similarly concentrate upon attaining a broad view of physical science and show less concern over the mastery of minute details. Thus, when a natural law or phenomenon is discussed in this book, one learns also about the discoverer and what led to the discovery, as well as some of the applications and limitations. This sort of understanding and appreciation, I believe, promotes scientific literacy.

The usually limited background in science and mathematics that the non-science major brings to physical science should pose no problem. I assume no prior work in science, and the mathematical level is no more advanced than arithmetic and simple algebra. I have made a particular effort to keep the language as well as the mathematics as simple as possible. Previous users of the text may applaud the more conversational writing style in the present edition.

Scientific laws are stated first in verbal form and then in mathematical form. Many worked examples set off from the body of the text are included as models to aid the student in understanding how the principles are used, and mathematical operations are discussed fully as needed. The examples themselves frequently illustrate familiar applications of the principles discussed. Dimensional analysis is presented early as a useful technique and is applied often. Each chapter concludes with a set of exercises designed to test the student's grasp of the material, and with annotated selected references to books and articles from several generally accessible and appropriate journals: *Journal of Chemical Education, Chemistry, American Journal of Physics, Physics Today, Scientific American,* and *Smithsonian.* The updated references may serve as a basis for a student research report, or to extend a student's knowledge of a field.

The number of chapters in the present edition is essentially unchanged from the first. However, some chapters have been combined, many retitled, several new chapters added, and one placed in the Appendix. Chapters 7 and 8 in the first edition have been consolidated as Chapter 6: Heat—A Form of Energy; Chapters 10 and 11 have become Chapter 8: Wave Motion and Sound; Chapter 14 has been moved to Chapter 9: Light and Other Electromagnetic Waves; Chapters 12 and 13 are now Chapter 10: Electric and Magnetic Fields; Chapter 21 is Chapter 16: Water, Solutions, and Pollution, preceding Chapter 17: An Introduction to Organic Chemistry (which may be omitted if the design of a course precludes it). Following Chapter 18: Nuclear Science, are Chapter 19: Astronomy Today and Chapter 20: The Solar System. Chapter 25: The Planet Earth has been expanded to Chapter 21: The Earth's Crust and Interior—Invitation to Earth Science; Chapter 22: The Earth's Hydrosphere—Oceanography; and Chapter 23: The Earth's Atmosphere and Beyond—Meteorology. The text concludes with Chapter 24: The Chemical Basis of Life. Chapter 2: Measurement has been shortened and placed in the Appendix. Instructors who prefer a quantitative approach will find the essential material there. A Glossary is included in the second edition by popular request, and a set of multiple-choice questions has been added to every chapter. The number of topics covered allows the instructor considerable latitude in the design of courses of varying length and emphasis.

Revising a text is only slightly less arduous than writing a first edition, but a venture that is just as intensely cooperative. I am especially grateful to John J. Vondeling, Vice-President and Senior Editor, and Joan Garbutt, Physics Editor, at W. B. Saunders Company, who helped make this book possible. Mrs. Garbutt's ability, inspiration, enthusiasm, support, attention to detail, patience, warmth, and understanding have been everything that an author could desire in an editor. Although I owe sincere thanks to the entire staff of the publisher, I would like to express my appreciation to George Laurie, Patricia Morrison, and John Hackmaster of the Art Department; Lorraine Battista, Leo Peirce, and Betty Richter, Design; Lloyd Black, Copy Editor, and Thomas O'Connor, Production Manager for College. I am deeply indebted to the many individuals, publishers, institutions, and agencies who have made figures and photographs available to me.

It is a sincere pleasure to acknowledge the considerable assistance of colleagues around the country who read the entire manuscript and provided thoughtful critiques. Many of their comments and suggestions have found their way into the text. They are: Hans S. Plendl, The Florida State University; Jerry D. Wilson, Lander College; Lowell O. Christensen, American River College; Carl Naegele, Michigan State University; Malcolm Hults, Ball State University; P. Joseph Garcia, Bloomsburg State College; Keith R. Honey, West Virginia Institute of Technology; and Gene F. Craven, Oregon State University.

I wish to express my appreciation to President Joseph J. Orze, Worcester State College, my students, and the members of the faculty and administrative staff for their interest and encouragement during the revision of the text. Last but not least, I am deeply grateful to my family—Shirley, Stephen, Naomi, and Aaron—for their cooperation and understanding, without which I could not have completed this book.

MELVIN MERKEN

CONTENTS

Chapter 22

THE EARTH'S HYDROSPHERE—OCEANOGRAPHY 649

Chapter 23

THE EARTH'S ATMOSPHERE AND BEYOND—METEOROLOGY 665

Chapter 24

THE CHEMICAL BASIS OF LIFE 692

APPENDIX A—MEASUREMENT 709

CONTENTS

1

THE SCIENTIFIC ENTERPRISE

1–1 INTRODUCTION

In the three centuries that the scientific revolution has been in progress, science and technology have transformed the face of the earth.

New agricultural methods have made it possible for nations to raise nearly enough food for their own needs. Antibiotics and vaccines have eliminated many diseases, reduced infant mortality, and increased the average life span by many years. Television by satellite provides information and entertainment "live" as it occurs anywhere on earth, and even from outer space. Nylon, orlon, and other synthetic fibers lend added interest, comfort, and color to clothing and furniture. Computers, transistors and integrated circuits microwave ovens, "instant" cameras, and photocopiers have affected the lifestyles of many. Research in alternative energy sources has opened up the possibility of providing us with our energy needs long after the supplies of fossil fuels—coal, oil, and natural gas— have been exhausted. Science has made it possible to walk on the moon, live for months in skylabs, investigate the depths of the universe, and launch probes into space in a quest for intelligent life (Fig. 1–1).

In this chapter we'll learn what makes science tick and why it plays a dominant role in our lives. Do not be concerned about minute details. Instead, try to get an over-all view of science from this introduction. In later chapters we'll explore specific topics of science and note some of their modern applications.

Figure 1–1 The earth as seen from Apollo 8, with the moon in the foreground. (Courtesy of NASA.)

1

1–2   WHAT IS SCIENCE?

The scientific approach today is essentially that of the founders of the Royal Society of London, the oldest organization for the advancement of science in the world, chartered in 1662. Christopher Wren, Robert Boyle, and the other founders were by and large professional men who were interested in the "new philosophy," or natural science, that was then emerging.

They admired in particular the experimentation and observations made by Nicolaus Copernicus, Galileo Galilei, Sir William Gilbert, and Johannes Kepler. The Fellows, as they were known, conducted their own experiments and observations, which covered many fields, such as the circulation of the blood, the Copernican hypothesis, and the force exerted by ignited gunpowder. Believing that each problem was unique, they searched for an appropriate method to attack it rather than insisting upon following a particular pattern of experimentation.

One of the greatest scientists who ever lived, Albert Einstein, described science as nothing more than a refinement of everyday thinking. Percy Bridgman, another recipient of a Nobel Prize in physics, also maintained that there is no distinct scientific method as such. A scientist probing the unknown is like a composer facing a blank page or an artist facing an empty canvas. You cannot work through a code of rules to the discovery of new knowledge or to the invention of new theories. No such code has ever been found. According to this view, science is a process, a method, a quest for understanding. This aspect of science is science as search. Its objective is to describe and to understand nature.

Science differs from the arts, philosophy, and other fields in these important respects: Science is *self-testing, self-correcting* and *objective*. Scientists publish their work and expose their data to the scrutiny of other scientists, who may repeat the experiments with essentially the same results or quite different results. Errors in fact or interpretation are discussed openly. The final arbiter is experimentation or observation, not authority. Self-testing and self-correction in science, however, do not necessarily result in universal agreement. It is impossible to achieve complete certainty and universal agreement in science, since the discovery of a single fact may invalidate a theory. The most that can be said for a scientific truth is that it has a high degree of probability. Science has another aspect, which is a product of its first: It is a body of knowledge. Yet it is more than a mere compendium of facts; it is a search for relationships among them. Understanding is often followed by the ability to predict natural phenomena and ultimately to control them. For example, the discovery of the electrical nature of matter led eventually to practical methods of harnessing electrical energy, such as the battery and the generator.

Understanding in science is achieved through its law and theories, which create order out of the chaos of raw facts. *Laws* are statements of relationships. Newton's second law of motion, for example, describes the behavior of a particle when it is subjected to a force. Although laws are based upon past experiences, their utility lies in their ability to predict the unknown. Nature is reliable. Given the same set of circumstances, nature can be relied on to operate in the same manner. The "laws" of nature are human perceptions of natural phenomena and as such only approximate nature.

Laws, in turn, are explained by means of *theories*. A theory that is introduced to explain one law and is found to explain others as well is particularly valuable. Scientific theories are regarded as tentative. They can be replaced by simpler, more comprehensive, or more rational theories that provide better explanations of the laws of nature. For example, the highly successful caloric theory of heat eventually gave way to the theory of heat as a mode of motion.

The laws of one branch of science have never been found to conflict with the laws of another. There seems to be an underlying unity in nature that applies to both living things and inanimate objects. The laws and theories in one field of science often turn out to be applicable to other fields.

Theories are often made easier to grasp by means of *models*. A model is the product of a scientist's creative imagination, to be retained as long as it remains useful. When it has served its purpose, it may be replaced by one that is more adequate. In this manner, structures or processes that cannot be observed directly may be visualized. Occasionally, as in the case of theories of nuclear structure, a variety of models may exist side by side—the liquid-drop model, the shell model, and others—since each is uniquely suited to describe one aspect of nuclear structure, but no one model alone can account for all of the observed properties.

1–3 THE HUMANIST AND THE SCIENTIST

During the 19th century, according to C. P. Snow, a gap developed between the traditional culture and the explosive new sciences. It is marked by two contrasting approaches to understanding the world. By the beginning of the 20th century, this gap had become almost unbridgeable. The underlying causes for the polarization of society into two cultures and ways of closing the gap have been the subject of long and often heated debates.

The humanities are a group of studies consisting of the languages and literature, the fine arts (art, music, drama, dance), philosophy, religion, and logic (Table 1–1). They are presented in terms of qualitative arguments and concepts such as the good, the true, and the beautiful. The sciences include mathematics, the natural sciences (physics, chemistry, biology, astron-

Table 1–1 THE REALMS OF KNOWLEDGE AND EXPERIENCE

NATURAL SCIENCES
Mathematics
Astronomy
Physics
Chemistry
Biology
Geology
Meteorology
Oceanography

SOCIAL SCIENCES
Psychology
Sociology
Anthropology
Economics
Government
History

HUMANITIES
Literature
Fine Arts
Music
Drama
Philosophy
Religion

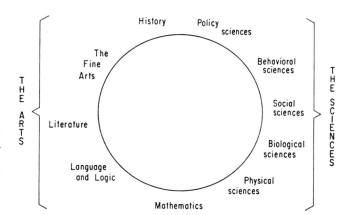

Figure 1–2 The arts (humanities) and sciences defined with reference to university disciplines. (From Cassidy, H. G., The Problem of the Sciences and the Humanities. *Journal of Chemical Education,* Vol. 38, No. 9. September, 1961, p. 444.)

omy, geology), and the social and behavioral sciences (psychology, sociology, anthropology, economics, government, and history). Scientific laws, theories, and models must be supported by empirical evidence to achieve acceptance. (See Fig. 1–2.)

Although the humanist and the scientist are concerned with the same world, they look at different aspects of it and see it with different eyes. For example, compare how a poet and a physicist—the extreme representatives of humanist and scientist—might describe a symphony or a sunset. The artist's approach is to express a personal vision through poetry, music, painting, or some other medium, whereas the scientist attempts to achieve objectivity by eliminating personal bias and subjective views (although the models scientists use are often chosen subjectively).

The humanist and scientist use the same conceptual tools, analysis and synthesis, but they differ in their concept of truth and how it may be attained. The humanist communicates the *qualitative* aspects of experience. In expressing an individual view of things, an appeal is made to the common humanity of the audience. The "truths" do not involve tests or corrections and may be accepted as certain in themselves. The scientist, on the other hand, is concerned with those aspects of phenomena that are verifiable and, at least in theory, subject to universal agreement. He or she aims for *quantitative* precision; their data are replicable, and open to test and correction. Scientific "truths" are tentative rather than eternal and absolute.

Science more than art is *cumulative*, the product of many individuals. The physics of Galileo and Newton is more comprehensive than that of Aristotle and Archimedes but less so than that of Planck and Einstein. The edifice of science grows from the contributions of scientists through the ages, and hence is "many-brained." The humanities, on the other hand, are "single-brained." Their products are unique. A painting by Rembrandt is not less beautiful today than one by Picasso. Einstein said that even if he himself had never lived we would still have had some form of the theory of relativity, but if Beethoven had never lived we would not have had an "Eroica Symphony."

Even though the humanities and sciences are so different, it is interesting to note that the great ages of science are the great ages of the arts. Because both depend on the exercise of imagination, art and science flourish, or languish, together. For example, Galileo and Shakespeare, who were born in the same year, grew into greatness in the same age.

1–4   THE LIMITATIONS OF SCIENCE

Anything beyond the limits of our senses and the instruments we build to extend them is outside the limits of science. Within these broad limits, the scope of science includes everything known to exist or to happen in the universe.

In principle, the arts and other humanities can be studied as psychological, anthropological, and biological phenomena, but something significant is lost in the process. A Mozart sonata or a rock-and-roll hit can be described by periodic oscillations and nerve impulses, but the scientific description cannot do full justice to the direct experience of the music.

Science has transformed the world by its selection and abstraction from nature of the measurable and quantifiable, but this does not mean that science deals with all of reality. Other human experiences, such as love, loyalty, beauty, and courage, are as real as any measurable experiences of our five senses. Most scientists do not subscribe to the doctrine of *scientism*, the belief that only scientific knowledge is worthwhile and that all other knowledge is nonsense.

Decisions as to how scientific knowledge shall be used are based on judgments that lie beyond the province of science. Scientific knowledge is ethically neutral, but it can be used for good or evil purposes.

Polio vaccine and napalm were both created in the laboratory, but society determined how each would be used. Scientists can help society to reach decisions, but in their capacity as responsible citizens rather than as scientists.

1–5 SCIENCE AND SOCIETY

Modern society faces difficult problems: a constant threat of war; an energy crisis; air and water pollution; depletion of natural resources; urban blight; poverty; famine; and the population explosion (Fig. 1–3). Many scientists hold that some of society's problems can be solved only by more research—pure and applied—and more technology. Take pollution as an example. This is a problem that has obvious scientific and technological components. It is difficult to think of a solution to photochemical smog or oil spills without an understanding of the fundamental aspects of the problem. A knowledge of more science, not less, is required.

Frequently the political, economic, and social aspects of a problem outweigh other considerations. Scientific solutions may be available which society is

Figure 1–3 The city epitomizes the problems of modern life: congestion, pollution, transportation, urban blight. (Photograph by Ernest Baxter, Black Star.)

not ready to accept. An example is our current transport systems, in which science and technology are not employed to their full potential. Such measures as staggered work schedules, exclusion of automobiles from a city's central business district during certain hours, a greater reliance on mass transportation, and the requirement of anti-pollution devices on vehicles all encounter various degrees of resistance from the public.

Assuming that it is possible to apply the methods of natural science to social or political problems, more collaboration will be needed than has been evident in the past among basic and applied scientists, engineers, social scientists, lawyers, and politicians. One proposal is the creation of national institutes that would be broadly interdisciplinary and team-oriented, and would apply the methods of science to our difficult social problems.

1–6　SCIENCE AND TECHNOLOGY

Through technology, the findings of science are translated into new or improved products or processes: high-speed computers, fertilizers that can double the size of a crop, transistors, antibiotics, organ transplants, jet travel (Fig. 1–4). The advances of technology have yielded benefits that on the whole vastly outweigh the injuries they have caused.

The phenomenal rate of change that a technology may introduce is evident in the field of transportation. The horse was the most rapid means of locomotion until the invention of the railroad, but we have moved from the propeller to the jet-propelled supersonic aircraft in one generation. The products and processes of science are the world's most powerful agents of change.

Until recently, technology was considered innocent until proved guilty. For example, DDT, a chemical compound first synthesized in 1874, was used during World War II to protect American soldiers against disease-carrying insects. Later it was used to protect civilian populations as well. Those who worked with DDT did not think of applying it to control insects infesting crops, livestock, or forests. But farmers and foresters, believing that what killed insects on people would also kill insects on plants,

Figure 1–4 Jet travel can be nearly emission-free with installation of controls. Photos show jet aircraft of similar types taking off with and without emission-reducing devices. (Courtesy of United Airlines.)

soon turned DDT into a massive assault on the environment. If the use of DDT had been restricted to the protection of people, it would not have become an environmental hazard.

Now the burden of proof is shifting to technology. The innovator is expected not only to show the impact that the innovation will have but also to prove that it will not produce harmful side effects. In order to test new products and methods for potentially dangerous effects, a new field has developed, known as *technology assessment*.

Technology assessment means trying to understand in advance some of the human, social, and environmental consequences of introducing new technologies or diffusing existing ones into new areas. It functions as an early-warning device, emphasizing the idea that technology exists to serve people. Even the products of technology now available—automobiles, pesticides, detergents—will have to prove their desirability (Fig. 1–5). Those that seem on balance to be beneficial will be encouraged, while those that seem to have risks that outweigh their benefits will be discouraged.

Although technology assessment may help avoid many unpleasant surprises, it will be difficult even

Figure 1–5　Traffic congestion in a major city. Are there better ways to move people? (H. Armstrong Roberts, Philadelphia, Pa.)

to guess at some of the future adverse effects of a new technology. On the other hand, failure to make assessments is almost sure to produce unpleasant surprises. To suggest that we turn our backs on further progress is somewhat like suggesting decapitation as a cure for headache. Once a society begins to move forward in technology, it must accept the accompanying responsibilities.

1–7　SCIENTISTS AND THEIR WORK

Unlike medicine or law, science is not a licensed profession. A person does not have to be a Ph.D. to be a scientist. The National Register of Scientific and Technical Personnel defines a scientist as anyone who is eligible for membership in a recognized professional society, such as the American Chemical Society. The scientific community includes a wide range of individuals: basic scientists in universities, research institutes, and government laboratories; engineers; workers in medicine and public health; graduate students and technicians working on scientific problems; and teachers of science in colleges and schools.

Scientists are ordinary human beings with essentially the same virtues and faults that most people have (Fig. 1–6). There seem to be no special qualities of personality, intelligence, background, or upbringing that enter into the making of a scientist. Science offers unusual opportunities for the exercise of the creative, imaginative, and manipulative talents while at the same time enabling one to contribute to the common good.

As recently as the 1930s, science was done by individuals pursuing the "art of the soluble," as the molecular biologist Peter Medawar described it, working on problems that they felt they could solve successfully with the very limited material means at their disposal. Samuel Goudsmit, a physicist, referred nostalgically to those days of "string and sealing wax," when basic science in the universities received little public or private support. Impelled by their own curiosity, scientists carried on their researches in basic science, very often with far-reaching results, working on problems that they themselves selected within a narrow field of knowledge. This is how penicillin, nuclear fission, and the genetic code were discovered.

Much of today's science, on the other hand, is conducted by teams of scientists. This "big science" is as characteristic of *basic* or "pure" science, which aims at extending knowledge, as it is of *applied science*, which is concerned with practical problems such as the development of radar or a moon rocket. It is centered around large laboratories that were first created around the turn of this century in industry and government.

Figure 1–6 Physicist Brian Josephson, who shared a Nobel Prize in Physics in 1973, crosses a glacier during the ascent of a mountain in the Austrian Alps. (Cover Photo of *Physics Today*, November, 1970.)

In dealing with a problem as complex as energy, a scientist soon finds that certain aspects of the solution lie beyond his or her immediate discipline. Such complex problems require teams of scientists representing various disciplines—physics, biology, chemistry, and so on—or different branches of the same discipline, such as physical chemistry, analytical chemistry, and biochemistry. The team, with its competencies in several fields, commands a greater amount of the knowledge necessary for progress than any individual could.

Important new scientific discoveries are made in many ways. Serendipity played a part in Wilhelm Roentgen's discovery of X rays. Precise measurements of the density of air led William Ramsay to discover the presence in air of the gases argon and krypton. On some occasions, systematic analysis is the key; at other times, a sudden thought, a flash of insight, even a vision may lead to discovery. The structure of the benzene molecule occurred to August Kekulé in a dream. To benefit from these flashes of genius, however, the scientist must be familiar with the knowledge of the past. As Louis Pasteur said, "Chance favors the prepared mind."

1–8 COMMUNICATION IN SCIENCE

More than the arts or literature, science is a cooperative activity. No scientist works alone, not even scientific geniuses. One of them, Sir Isaac Newton, freely admitted his debt to others when he wrote to Robert Hooke, "If I have seen further (than you and Descartes) it is by standing upon the shoulders of giants."

There were more connections between Newton and his predecessors and contemporaries than there were between Beethoven and Mozart in the field of music, or Picasso and Renoir in art. In our own day, two or three persons, frequently from different countries, may share a Nobel Prize for work for which they are jointly responsible.

Cooperation in science includes the publication of papers in scientific journals. Many appear under joint authorship, frequently bearing the names of three or more contributors. Books, letters, and oral presentations at scientific meetings are other means of communication among scientists. Through these means a scientist describes the results of experimental or theoretical work.

Scholarly journals frequently have expert referees to screen manuscripts, a system that helps sift out the good papers from the bad. Some of the more prestigous journals reject 80 to 90 percent of the articles submitted. The self-testing and self-correcting features of science are put into operation in this way. Scientists expect colleagues to be scrupulously honest. If the work is significant, others will soon attempt to replicate it. If other laboratories are unable to reproduce the original results, the scientist's reputation suffers. The basic difference between science and other endeavors is that inherent policing weeds out even the thought of fraud in science.

A new phenomenon is the use of the telephone as an instrument of scientific communication. Laboratories in Cambridge are in touch with those in New York, Chicago, and Pasadena all on the same day. By the time a paper is published in a journal, people working in that field in laboratories around the world seem to know about the work. Laboratories set at great distances from each other often collaborate as closely as if they were located on the same corridor.

Information travels almost with the speed of light. A bit of information picked up over lunch or in a corridor conversation in New York may be reported almost instantaneously in Houston. The system seems to function with amazing accuracy and with candor. It seems that scientists are telling each other everything they know, by telephone, as soon as they know it. The possibility of being scooped seems less of a problem today. Scientists realize there is so much to be learned that there can never be enough researchers. We have not reached the end of knowledge, but have only begun.

Since knowledge is the important outcome of science, great importance has been attached from Newton's day to the present to the issue of priority—to being the first to make a particular discovery and to be recognized as such. A frank account, based upon his own experience, of the extreme measures a scientist may take to assure priority, has been given by James Watson, who shared a Nobel Prize with Francis Crick and Maurice Wilkins for their work on the DNA molecule. In his book *The Double Helix,* Watson conveys the flavor of modern science as few others have done.

CHECKLIST

Applied science
Basic science
"Big science"
Experiment
Humanities
Interdisciplinary
"Little science"
Model

Objectivity
Replicable
Science
Scientific law
Scientism
Technology
Technology assessment
Theory

EXERCISES

1–1. Compare the concept of truth in science with truth in another realm of experience.

1–2. If ours is an "age of science," should scientists claim any particular privilege or be charged with special responsibilities not granted to or expected of other citizens?

1-3.   Discuss an instance of (a) an abuse of science; (b) an abuse of an art form.

1-4.   Is experimentation characteristic of all sciences? Explain.

1-5.   Why has the rate of scientific progress in the 20th century been so explosive?

1-6.   In your opinion, has the impact of science on the quality of life been on the whole positive or negative? Explain.

1-7.   Discuss how the humanities and the sciences have contributed to the problems of our age.

1-8.   Illustrate with a specific example the potential of a scientific discovery for good or evil.

1-9.   Discuss the merits of the proposal to call a moratorium upon scientific research until some of society's problems have been solved.

1-10.   Discuss some connections between science and the arts.

1-11.   Discuss the role of (a) cooperation and (b) competition in science.

1-12.   Cite evidence for and against the existence of "two cultures."

1-13.   *Multiple Choice:*

A. Scientific knowledge
   (a) starts anew with each generation
   (b) becomes dated quickly
   (c) is cumulative
   (d) is readily accessible

B. The methods that scientists use to solve problems in the laboratory
   (a) can contribute little to the solution of practical problems
   (b) are just as valid in dealing with all other problems
   (c) are too exacting for general use
   (d) probably cannot solve all human problems

C. The use of scientific discoveries
   (a) is controlled by the scientists who made them
   (b) is of no concern to non-scientists
   (c) is a matter for the people to decide
   (d) is authorized by the government

D. The future direction of science
   (a) can be predicted with great accuracy
   (b) cannot be predicted with great accuracy
   (c) is known by leading scientists
   (d) depends upon government funding

E. The fact that scientists and artists are often unable to communicate with each other
   (a) is the fault of the scientists
   (b) is the fault of the artists
   (c) is of no great concern to society
   (d) is a loss to both art and science

SUGGESTIONS FOR
FURTHER READING

Brush, Stephen G., "Should the History of Science Be Rated X?" *Science,* 22 March 1974.
   Asserts that the way scientists behave (according to historians) might not be a good model for students.
Fischer, Robert B., *Science, Man and Society.* 2nd ed. Philadelphia: W. B. Saunders Company, 1975.
   Discusses the nature of science and its relevance to man and society.
Holton, Gerald (ed.), *Science and Culture.* Boston: Houghton Mifflin, 1965.
   Humanists and scientists contribute to this outstanding collection of essays.
Klaw, Spencer, *The New Brahmins: Scientific Life in America.* New York: William Morrow & Co., 1968.
   Conveys a sense of what it is like to be a natural scientist in America.
Kuhn, Thomas S., *The Structure of Scientific Revolutions.* Chicago: University of Chicago Press, 1962.
   Analyzes the nature, causes, and effects of revolutions in scientific concepts.
Revelle, Roger, "The Scientist and the Politician." *Science,* 21 March 1975.
   Calls for the closest kind of cooperation among politicians and scientists in the search for ways out of our present dilemmas.
Snow, C. P., *The Two Cultures: A Second Look.* Cambridge, England: Cambridge University Press, 1964.
   Discusses the division of intellectuals into two groups that no longer communicate.
Watson, James, *The Double Helix.* New York: Atheneum, 1968.
   Candidly portrays the race for scientific priority, by one who was there.
Weisskopf, Victor F., "The Significance of Science." *Science,* 14 April 1972.
   Discusses the cultural and social aspects of science and its relation to society.

2

MOTION

2-1 INTRODUCTION

When an object changes its position relative to some other object or objects, we say it has moved. Motion fascinated even our prehistoric ancestors, who captured the beauty of wild beasts in motion in works of art (Fig. 2–1). Down through the ages some

Figure 2–1 Prehistoric paintings of beasts in motion in the Hall of Bulls, Lascaux Caves. (Courtesy of the French Government Tourist Office.)

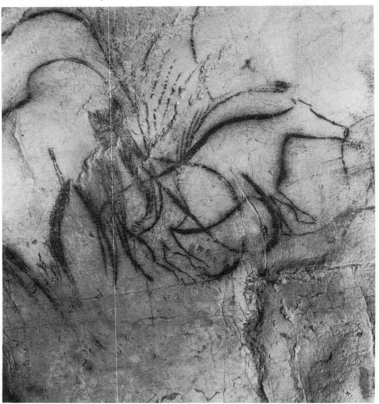

Figure 2–2 Marcel Duchamp's painting, *Nude Descending a Staircase*, conveys a sense of motion. (Courtesy of Philadelphia Museum of Art.)

of the best minds have been drawn to this subject. The development of the laws of motion in the 16th and 17th centuries signaled the birth of the modern scientific revolution. As we explore the realm of motion in this chapter, we will also be tracing the origins of modern science.

2–2 THE RESTLESS UNIVERSE

The universe is filled with things in motion, from majestic galaxies, stars, and planets, to atoms. The motions may be fast or slow, smooth or erratic, simple or complex (Figs. 2–2 and 2–3). The words that describe motion suggest an awareness of its variety: soaring, flying, leaping, running, galloping (Fig. 2–4). Challenged by the limitations imposed upon us by nature, we invent things to move faster, farther, higher, and deeper. The approximate speeds of some objects are listed in Table 2–1.

Some motions are simple. An athlete races down a track, going from start to finish in the shortest possible time. He or she is undergoing motion in a straight line, the simplest kind of motion. A record-player turntable

Figure 2–3 Stroboscopic photograph of a golfer's swing shows successive positions of the golf club. (Courtesy of Dr. Harold Edgerton, Massachusetts Institute of Technology, Cambridge, Mass.)

Figure 2–4 Physicist J. Robert Oppenheimer jumping. Might jumping in a laboratory identify a potentially great scientist? (Photograph by Philippe Halsman.)

undergoes circular motion, repeating the same path in a cycle in the same time. This motion, too, has an element of simplicity. The motion of a pendulum is repetitive, like that of the turntable. The pendulum bob swings to and fro, but never completely retraces its path. The swings become shorter and shorter and the bob is finally at rest again. Yet if we measure the time required for each swing, we find that it is almost exactly the same. How would you describe the motion of a ball thrown horizontally through the air, or straight up, or dropped straight down from a high spot? Before we discuss such questions, let's discuss *vector analysis*, a very useful tool.

2–3   VECTOR ANALYSIS

Certain physical quantities are described in terms of magnitude or size and dimension. They are known as *scalar quantities*. Examples are mass (the amount of matter in an object), time, and volume (the amount of space taken up by an object). Scalars are governed by the ordinary mathematical processes of addition, subtraction, multiplication, and division. For example, two objects with masses of 10 kg and 5 kg have a total

Table 2-1   SOME APPROXIMATE SPEEDS

OBJECT	SPEED	
	Miles/Hour	Meters/Second
Earth in orbit	67 000	30 000
Sun in galaxy	43 000	19 400
Earth satellite	18 000	8 300
Moon in orbit	2 200	1 000
Supersonic airplane	2 000	900
Racing car	200	180
Bird flying	60 to 180	27 to 82
Running elephant	25	11
Person running	6 to 15	2.7 to 6.8
Person walking briskly	3 to 4	1.4 to 1.8
Snail's pace	0.0025	0.0011

mass equivalent to an object with a mass of 15 kg. (See Appendix A for a discussion of units.)

Other physical quantities, however, are described completely only when direction is specified as well as magnitude. These are known as *vector quantities*. *Displacement*, the shortest (straight line) distance between two points, is an example.

Let us suppose that you commute by automobile from your home, H, to class, C (Fig. 2–5). The distance you actually travel on a given trip can be calculated from the difference in odometer readings of your automobile between the beginning and the end.

It is expressed in the number of miles traveled upon the particular road chosen for the trip. However, perhaps several alternate routes exist that you could take, with varying distances. Suppose, further, that we ignore the actual route you traveled and consider only the net change in position or displacement from your home to class, represented by a straight arrow extending from H to C. Note that your displacement will be the same no matter which route you take. The simplest way to represent your displacement is to draw an arrow from your home to class on a map. This arrow, a vector representing 5.0 miles, is characterized by its length (magnitude), direction, and dimension.

Imagine that on your way to class you stop to pick up a classmate. The displacement from your home to your classmate's, represented by a in Figure 2–6, is 4.0 miles, and the displacement from your classmate's home to class, represented by b, is 3.0 miles. The sum of these two displacements is equal to your net

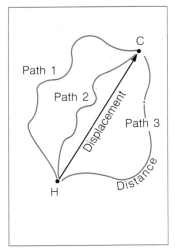

Figure 2–5 Illustration of the distinction between distance traveled over three alternate routes from H to C, and the net displacement, the latter represented by the arrow.

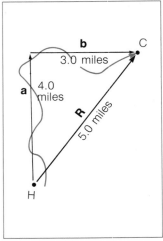

Figure 2–6 The net displacement, **R**, is equal to the sum of the displacements **a** and **b**.

Figure 2–7 Polygon rule of vector addition. The tail of vector **b** is placed at the head of vector **a** in *A* and *B*. The resultant vector **c** is drawn from the tail of **a** to the head of **b** and is equal to **a** plus **b**. Placing the tail of vector **a** at the head of vector **b**, as in *C*, gives the same resultant.

displacement of 5.0 miles, although the distance you actually traveled is greater than this. Thus, a new method of dealing with quantities having both direction and magnitude is needed. This method is called *vector algebra.*

We denote a vector by an arrow representing the magnitude at the same time that it specifies direction in space. If we adopt a suitable scale, such as 1 cm = 1 mile, then a displacement of 3.0 miles is represented by an arrow 3.0 cm long. A vector may be symbolized conveniently in handwriting by a letter with an arrow above it, such as $\vec{a}$, and its magnitude by the letter a without the arrow. In printing, vectors are symbolized by a boldface symbol, such as **a**. If we want to add vector **a** to vector **b**, the rule is to place the tail of one of the vectors at the head of the other (it does not matter whether the tail of **b** is placed at the head of **a,** or vice versa) (Fig. 2–7). The resultant vector, **c**, is drawn from the tail of the first to the head of the second and stands for the vector sum of the two vectors **a** and **b**. To add more than two vectors, the same process is continued, the tail of each successive vector being placed at the head of the last. The resultant or sum vector starts at the tail of the first and ends at the head of the last vector. This particular method of vector addition is called the polygon or "head-to-tail" method.

$$\mathbf{R} = \mathbf{a} + \mathbf{b} + \mathbf{c}$$

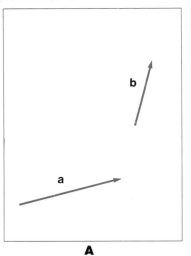

A

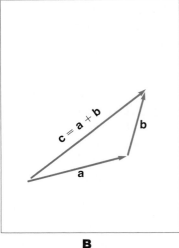

B

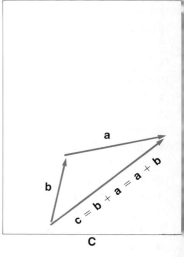

C

EXAMPLE 2–1

A helicopter travels 30 miles due east, then changes direction and travels 40 miles due north. How far and in what direction is it from its point of origin?

SOLUTION

The problem is to find the displacement—the straight-line distance and direction between the origin and destination—and to do this we'll use vector addition. We begin by setting up a coordinate system that enables us to specify direction; a useful one consists of horizontal (x) and vertical (y) axes. Along the x-axis, the direction to the right of the origin (the intersection of the x- and y-axes) corresponds to east, or 0°; the direction along the y-axis above the origin is north, or 90°; to the left, west, or 180°; and below the origin, south, or 270°.

Using a scale that is neither too small nor too large, such as 1 cm = 10 mi, construct a vector 3 cm long to the right of the origin along the x-axis, corresponding to a displacement of 30 miles due east (Fig. 2–8). From this arrowhead, construct a vector 4 cm long in the vertical direction, corresponding to the displacement of 40 mi due north, and place the arrowhead at the end of the line. The resultant displacement is the vector from the origin to the last arrowhead. Measure its length, 5 cm, corresponding to 50 miles. Determine the direction by measuring the angle with a protractor; it is 36.9°. The resultant displacement of the helicopter, then, is 50 miles in a direction 36.9° east of north from its point of origin, or 53.1° north of east.

2–4　RESOLUTION OF VECTORS

At times it is useful to replace a given vector, **R,** by two other vectors, **a** and **b,** whose sum is **R.** The vectors **a** and **b** are called the components of **R,** and this process

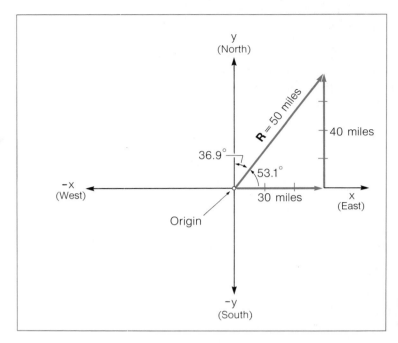

Figure 2–8 Determining a resultant displacement, **R**, by vector addition.

is called the resolution of **R** into its components. Although a vector may be resolved into its components in any two directions, most often the components in the x- and y-directions are desired.

EXAMPLE 2–2

A ship sails a straight course of 100 miles northeast (45°) from its origin. How far is it (a) to the east of its origin; (b) to the north of its origin?

SOLUTION

Construct an arrow 10 cm long (scale: 1 cm = 10 miles) in a direction of 45° (or 45° east of north) (Fig. 2–9). From the arrowhead of **R** drop a vertical line to the x-axis, construct an arrowhead there, and measure the length **a**. Next, construct a line parallel to the x-axis, and measure the length **b**. The length **a** represents the distance east the ship is from its origin, 71 miles, and the length **b** the distance north from its origin, also 71 miles.

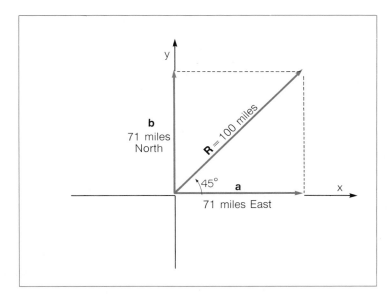

Figure 2-9 Resolving a vector, **R**, into two components: horizontal and vertical.

2-5 SPEED AND VELOCITY

Since motion involves a change in position, we are interested in such questions as the following in dealing with a moving body, whether it is a bowling ball, an auto or a spacecraft: Where is it? How fast is it traveling? Where is it going? Is is slowing down or speeding up? Is is moving steadily? Is it changing direction? For the sake of convenience we treat the moving body as if it were a point particle moving in space.

The key concepts of *kinematics*—the scientific description of motion, without regard to its cause—are position, speed, velocity, and acceleration. They, in turn, are derived from two of the fundamental quantities, length and time, discussed in Appendix A. The concepts that are presented here and in the following chapters evolved slowly over a period of centuries, and eluded even the greatest minds before their full significance was grasped.

We know that an automobile is moving along a highway from observing changes in its position (displacement) over a period of time. *Speed* is a measure of how fast a body travels in a given time, that is, the rate of change of position. If we know the distance traveled, and the period of time required, we can calculate the average speed from the formula:

$$\text{Average speed} = \frac{\text{Distance traveled}}{\text{Elapsed time}}$$

Equation 2–1   or, in symbols,

$$\bar{v} = \frac{s}{t}$$

(In physics, a bar placed over a symbol, such as $\bar{v}$, usually indicates an average value of that quantity.)

The greater the distance traveled in a given time, the greater the speed. An automobile traveling a distance of 50 miles in one hour has a lower speed than a jet airplane covering a distance of 500 miles in the same time. In either case, however, it is unlikely that a uniform speed was maintained for a period of one hour. The automobile may have had to slow down on curves or hills, or to stop for red lights or cattle crossings; conversely, it may have picked up speed on open stretches and while passing. In other words, we should say that the average speed is 50 miles/hour. In the course of the trip, the speedometer may have fluctuated between 0 and 60. A particular reading, such as 50, means that the speed at a given instant is 50 miles/hour. If this speed were maintained for one hour, the distance traveled would be 50 miles.

EXAMPLE 2–3

A racer in the famous "Indianapolis 500" travels 500 miles in 3 hours and 40 minutes. What is its average speed in miles/hour?

SOLUTION

It is good to tabulate the information given and the question asked in a column on the left, using appropriate symbols and units. Then search for some relation connecting the concepts involved, in this case, speed, distance, and time, and express that relation as an equation in symbols on the right. If necessary, rearrange the equation so that the unknown is to the left and other quantities are to the right of the equal sign. Next, substitute the specific values given, keeping the units, and

solve. Express the result by both a number and a
unit.

$s = 500$ mi $\bar{v} = \dfrac{s}{t}$

$t = 3$ hr 40 min $= 3.67$ hr $= \dfrac{500 \text{ mi}}{3.67 \text{ hr}}$

$\bar{v} = ?$ $= 136 \dfrac{\text{mi}}{\text{hr}}$

The simplest type of motion is that in which a body
moves with uniform speed in a straight line, that is, it
travels equal distances in equal intervals of time (Fig.
2–10). The average speed of such a body is given by
Equation 2–1. When a direction is associated with a
speed, we have a new quantity: velocity. The average
velocity of a body in uniform motion in a straight line is
defined as the displacement divided by the time
during which the displacement occurred:

$$\text{Average velocity} = \frac{\text{Displacement}}{\text{Elapsed time}}$$

Equation 2–2

$$\bar{v} = \frac{s}{t}$$

Since displacement, **s**, is a vector quantity, velocity is a
vector quantity, symbolized by **v**.

For uniform motion in a straight line, the magnitude
of the displacement is the same as the distance traveled

Figure 2–10 An example of
uniform motion in a straight line.
The car traverses equal distances in equal intervals of time.

POSITION:	A	B	C	D	E
DISTANCE:	0	30ft	60ft	90ft	120ft
TIME:	0	1 sec	2 sec	3 sec	4 sec

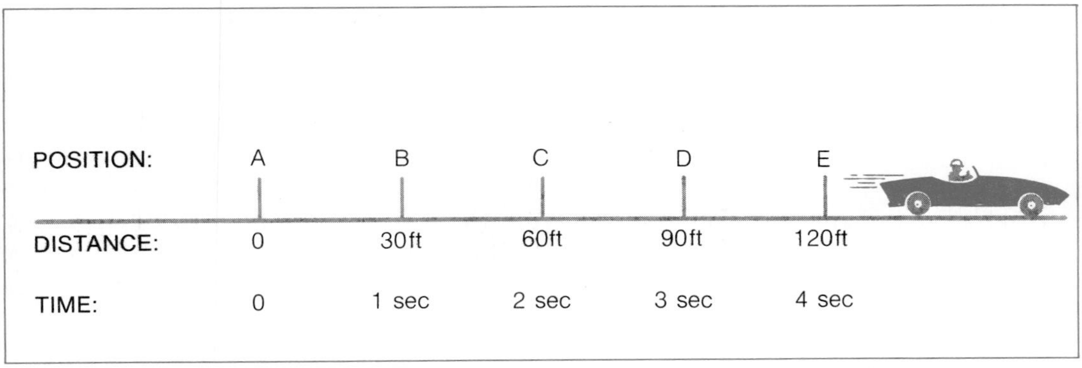

in a given time interval, and the magnitude of the velocity—for example, 30 ft/sec—is the same as the speed. The difference between velocity and speed is that velocity has associated with it the idea of direction; that is, velocity is a vector quantity, while speed is a scalar quantity. If the line of motion is horizontal, we can treat displacements to the right as positive and displacements to the left as negative; therefore, velocities to the right are positive and velocities to the left are negative. In Figure 2–10, the velocity can be expressed as 30 ft/sec to the right.

Instantaneous velocity refers to how fast a body is moving at a given instant and in what direction. It is defined as the average velocity over a very short time interval, one so short that for all practical purposes the velocity can be considered constant during that time interval. Measurements might indicate, for example, that a car traveled 44 ft in 1 second. Therefore,

$$\bar{v} = \frac{s}{t}$$

$$= \frac{44 \text{ ft}}{1 \text{ sec}}$$

$$= 44 \frac{\text{ft}}{\text{sec}}$$

If the car traveled 0.44 ft in 0.01 sec, we would calculate the average velocity to be

$$\bar{v} = \frac{0.44 \text{ ft}}{0.01 \text{ sec}}$$

$$= 44 \frac{\text{ft}}{\text{sec}}$$

This process, continued for shorter and shorter intervals of time, gives the instantaneous velocity.

Few highways are constructed with 50-mile stretches in one direction. On a real highway, an automobile undergoes more or less frequent changes in direction. Although the speedometer reading may remain constant for a period of time, the changes in direction mean that the velocity of the automobile is changing (Fig. 2–11). The velocity may be 50 miles/ hour north at one time and 50 miles/hour northeast at another. The numerical values—the magnitudes—of

Figure 2–11 Although the speed of the car is constant— 50 mi/hr—the velocity is varying because its direction is changing.

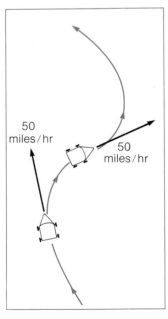

50 miles/hr

50 miles/hr

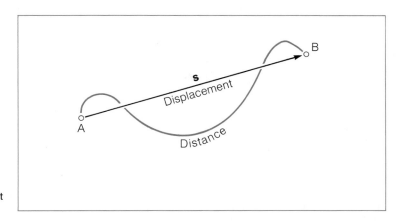

Figure 2–12 The displacement from A to B.

speed and velocity are identical: the speed is 50 miles/hour regardless of the direction.

Most bodies, in fact, move along paths that are not straight, and at rates that vary during the motion (Fig. 2–12). At one instant the body may be at position A, and at a later time at position B. During the time interval the body has traveled along the path a distance s. Its average speed is found by applying Equation 2–1, since it is the distance traveled divided by the time. In this motion, the displacement is given by the vector s, which is different in magnitude from the distance traveled along the path. The average velocity is given by Equation 2–2, the displacement divided by the time, in which t is the time required to travel from A to B. The direction of the average velocity is the same as the direction of the displacement.

2–6 ACCELERATED MOTION

For the motion of a body in which the velocity changes either in magnitude or direction, or both, we define a new quantity: *acceleration* is the rate of change of velocity.

Acceleration is an aspect of motion that we can often perceive. When riding in a car, even with our eyes shut, we can feel the velocity change. When the car is initially moving forward we may sink back into our seats; when it stops, we have the sensation of being pushed forward; when it rounds a curve, we experience a sideward push. We can feel the accelerations of a fast elevator at the start and stop of a trip. Each situation in which our velocity is changing rapidly involves a physiological awareness of the acceleration.

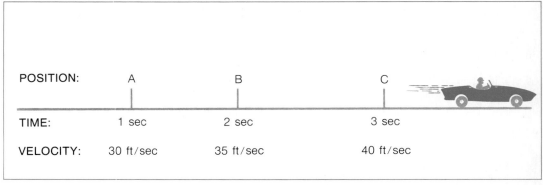

POSITION:	A	B	C
TIME:	1 sec	2 sec	3 sec
VELOCITY:	30 ft/sec	35 ft/sec	40 ft/sec

Figure 2–13   The velocity of the car is changing. The car is said to be accelerating.

An automobile is accelerating during periods when it "picks up" speed, slows down (although we usually refer to this as deceleration), or changes direction (Fig. 2–13). If it moves with a velocity, v_0, at point A, and at some time (t) later it is at point B with a velocity v, the average acceleration is the change in velocity divided by the time interval:

$$\text{Acceleration} = \frac{\text{Change in velocity}}{\text{Time taken}}$$

Equation 2–3

$$= \frac{\text{Final velocity} - \text{Initial velocity}}{\text{Time}}$$

$$\bar{a} = \frac{v - v_0}{t}$$

where v_0 is the initial velocity, which may or may not be zero, and v is the final velocity. Rearranging Equation 2–3, we obtain the useful expression:

Equation 2–4

$$v = \bar{a}t + v_0$$

If the initial velocity is zero, that is, if the body starts from rest, $v_0 = 0$, Equation 2–4 becomes:

Equation 2–5

$$v = \bar{a}t$$

By considering only motion in which the acceleration is constant in magnitude and direction, and the body is moving in a straight line, we shall not have to make a distinction between average and instantaneous acceleration. Therefore, the symbol **a** without the bar, a vector quantity, will represent the average acceleration from now on.

Except in problems such as those involving automobiles, where speedometer readings are available, it is often difficult to measure velocities. If the acceleration is constant, the distance traveled by an accelerated body has two contributions: (1) the distance that it would have traveled if the velocity v_0 had remained constant, $v_0 t$; and (2) the additional distance traveled as a result of the change in velocity:

Average

$$\bar{v} = \frac{v_0 + v}{2}$$ *Average*

$$\bar{v}t = \left(\frac{v}{2}\right)t = \left(\frac{at}{2}\right)t = \frac{1}{2}at^2$$

Equation 2–6

or:

$$s = v_0 t + \frac{1}{2}at^2$$

Some SI units (see discussion in Appendix A) for acceleration are meters per second per second (m/sec²); kilometers per second per second (km/sec²); and centimeters per second per second (cm/sec²). The kinematics equations are summarized in Table 2–2. The following examples will help you become familiar with those equations and the sometimes confusing units of acceleration.

Table 2–2 SUMMARY OF EQUATIONS OF KINEMATICS

STARTING FROM REST	INITIAL VELOCITY NOT ZERO
1. $s = \bar{v}\,t$	$s = \bar{v}\,t$
2. $s = \dfrac{v}{2}\,t$	$s = \left(\dfrac{v_0 + v}{2}\right)t$
3. $a = \dfrac{v}{t}$	$a = \dfrac{v - v_0}{t}$
4. $v = a\,t$	$v = a\,t + v_0$
5. $s = \dfrac{1}{2}a\,t^2$	$s = v_0 t + \dfrac{1}{2}at^2$

$t \sqrt{\dfrac{2s}{a}}$

$\bar{v} = \dfrac{v_0 + v}{2}$

$6t = \sqrt{\dfrac{2s}{a}}$

EXAMPLE 2–4

Starting from rest at the end of a runway, an airliner achieves its takeoff speed of 100 m/sec in 50 seconds. What is its acceleration?

SOLUTION

Tabulate the data given and the question asked in a column on the left. Do you see that these quantities are related by Equation 2–3? Write down this equation, then substitute and solve.

$$v = 100 \text{ m/sec} \quad a = \frac{v - v_0}{t}$$

$$v_0 = 0$$

$$t = 50 \text{ sec} \qquad = \frac{100 \frac{m}{sec} - 0}{50 \text{ sec}}$$

$$a = ?$$

$$= \frac{2 \text{ m}}{sec^2}$$

(read "2 meters per second per second," or "2 meters per second squared")

EXAMPLE 2–5

A rocket to the moon is given a constant acceleration of 2.0 m/sec². What is its velocity 90 seconds after liftoff?

SOLUTION

From the tabulated data, we find that we can apply Equation 2–5 to get the velocity.

$$a = 2.0 \text{ m/sec}^2 \qquad v = a\,t$$

$$v_0 = 0$$

$$t = 90 \text{ sec} \qquad = \left(2.0 \frac{m}{sec^2}\right)\left(90 \text{ sec}\right)$$

$$v = ?$$

$$= 180 \frac{m}{sec}$$

EXAMPLE 2–6

The driver of a car traveling 75 ft/sec slams on the brakes. If the car stops in a distance of 200 ft, (a) how long does it take to stop? (b) what is the acceleration?

SOLUTION

As a result of braking, the car is decelerated and comes to a stop. We can calculate the time from Equation 2–2, and the acceleration from Equation 2–3.

$v_0 = 75$ ft/sec   (a) $s = \bar{v}t$

$v = 0$ ft/sec

$$s = \left(\frac{v_0 + v}{2}\right) t$$

$s = 200$ ft

$a = ?$

$$\therefore t = \frac{s}{\left(\dfrac{v_0 + v}{2}\right)}$$

$t = ?$

$$= \frac{200 \text{ ft}}{\left(\dfrac{75\dfrac{\text{ft}}{\text{sec}} + 0}{2}\right)}$$

$$= \frac{200 \text{ ft}}{37.5 \dfrac{\text{ft}}{\text{sec}}}$$

$$= 5.33 \text{ sec}$$

(b) $a = \dfrac{v - v_0}{t}$

$$= \frac{0 \dfrac{\text{ft}}{\text{sec}} - 75 \dfrac{\text{ft}}{\text{sec}}}{5.33 \text{ sec}}$$

$$\frac{- 14 \dfrac{\text{ft}}{\text{sec}}}{\text{sec}}$$

$$= -14 \dfrac{\text{ft}}{\text{sec}^2}$$ (the minus sign here means deceleration, that is, slowing down)

2–7 A THEORY OF MOTION

For nearly 2000 years the ideas of Aristotle (384–322 B.C.) dominated the attempt to understand motion. Although the theory of motion was not his alone, Aristotle's exposition of it was the most convincing. Modern science began in large part when Aristotle's physics was successfully challenged.

Aristotle considered rest to be the natural state of the universe. Motion, therefore, involved a change in the natural state of order. He taught that all objects on or around the earth are made up of combinations of the four elements— earth, air, fire, and water (Fig. 2–14). Some objects appear to be light, and others heavy, depending upon the proportion in each body of the different elements.

According to Aristotle, earth was "naturally" heavy and fire "naturally" light; water and air fell between these two extremes. Each of the elements had a tendency to reach its "natural" place, where it remained at rest unless disturbed. If a body was heavy, its natural motion was downward. An apple being mostly of earth would fall downward when dropped. Smoke being mostly fire would rise upward since fire was above earth.

In Aristotle's system, heavy bodies fell because *they* possessed the property of gravity and the earth

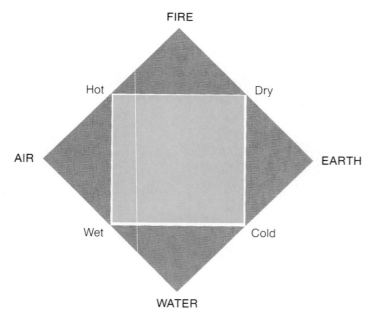

Figure 2–14 The "four elements" and "four qualities."

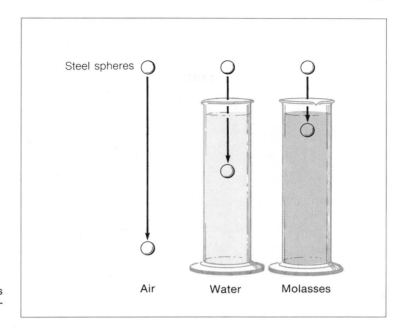

Steel spheres

Air   Water   Molasses

Figure 2–15 The more viscous
the medium, the slower the motion of the falling body.

was their "natural" place. Light bodies were thought to
possess the property of levity, which caused them to
rise upward to their "natural" place, the sky. Motion
was caused by the properties of gravity or levity which
the bodies themselves possessed, rather than by an
outside force such as the earth.

The medium through which a body moved played
an important role in Aristotle's theory. The denser the
medium through which a body moves, the slower the
motion (Fig. 2–15). (A rock falls more slowly through
water than through air.) In a vacuum, therefore, where
no medium exists, all bodies should fall with infinite
speed. Since this could not be, Aristotle considered it a
strong argument against the possibility of a vacuum.
(The first vacuum pumps were invented 2000 years
later.) Aristotle also believed that the heavier the body,
the greater its ability to overcome the resistance of the
medium. If two bodies, one twice as heavy as the other,
were dropped from the same height simultaneously,
the time for the heavier body to fall would be only half
as great as for the lighter body. (That such is *not* the
case was actually demonstrated many centuries later.)

In Aristotle's scheme of the universe, the earth was
assumed to be at rest and the stars, planets, and the sun
itself moved around the earth in circles. They
described circles because it was their "nature" to do so.
The celestial bodies were not made of the same four

elements as the earthly bodies but consisted of a fifth element ("quintessence," or "aether")—an unchangeable, perfect material unlike the four elements found on earth. It was thought that the natural motion of a body composed of aether had to be circular, the most perfect of all natural motions.

The earth had no "local motion," or movement from one place to another. It was not even supposed to rotate on its axis. Since the earth was composed of "imperfect," corruptible material, it was not "natural" for it to have a circular motion, such as moving about the sun or rotating upon its own axis.

Of course, objects very often move in "unnatural" ways; for example, an arrow shot from a bow or a rock thrown straight upward or straight across. Such motion, Aristotle believed, was "violent," or contrary to the nature of the body. It occurred only when some force acted to start and keep the body moving contrary to its nature. When the force was removed from the arrow or rock, the object would fall straight to the earth.

After the initial push supplied by the bow or hand, Aristotle imagined that the medium—the air, in this case—pushed the arrow or rock along by rushing in behind to prevent a vacuum (Fig. 2–16). Gradually, the push of the medium grew less and less, and eventually wore out. At this point, gravity took over and the body dropped under natural motion. Aristotle assumed that forced motion and natural motion did not mix. In a vacuum there could be no medium to serve as a "mover," and for this reason, also, Aristotle did not accept the reality of a vacuum.

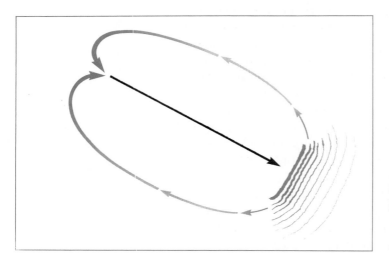

Figure 2–16   Forced or "violent" motion, according to Aristotle. The force is supplied by the medium—the air—rushing in behind the arrow to prevent a vacuum.

On the other hand, Aristotle himself had said that a medium, such as air, resists motion. How, then, could it be the cause of motion? If it is really the air which carries the rock along, why must the hand touch the rock at all? Again, a heavy rock can be thrown much farther than a very light one. This fact is difficult to explain if air is the cause of motion, since air would move light objects more easily than heavy ones. Questions such as these cast doubt on Aristotle's theory of motion.

2–8 GALILEO AND THE EXPERIMENTAL METHOD

In the centuries between Aristotle and Galileo Galilei (1564–1642), the study of motion made slow progress. Galileo is a transitional figure, standing at the end of the Middle Ages and at the beginning of modern science.

Galileo was not content with qualitative observations. He looked for *quantitative* observations that would describe a phenomenon with mathematical precision. In attacking scientific problems, he emphasized *how* things worked, not why. This approach was radically different from previous ones, and turned out to be a more fruitful one for answering scientific questions. In this respect, Galileo was the first modern scientist.

While a professor of mathematics at the University of Pisa, Galileo wrote *On Motion,* in which he discussed the problems of falling bodies and projectiles. Galileo broke with Aristotle's concepts of "heaviness" and "lightness." According to Galileo, all bodies are more or less heavy. Light bodies move upward because heavy bodies fall below them. Whether the medium is air or water, its resistance is a kind of buoyancy that supports lighter bodies more effectively than denser ones. Galileo at first supposed that in a vacuum, bodies would fall with speeds proportional to their densities. However, in a later book, *Dialogues Concerning Two New Sciences,* his ideas on motion had changed profoundly. This book laid the foundation for the science of mechanics and modern physical science.

Galileo argued against Aristotle's view that heavy objects falling freely fall faster than light ones. He

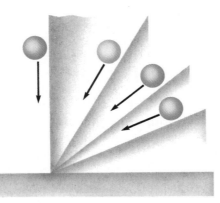

Figure 2-17 Spheres rolling down increasingly steep planes. The vertical plane, at 90°, matches free fall.

Figure 2-18 A stroboscopic photograph of a baseball and a golf ball released simultaneously. The time interval between photographs was 1/30 second. (From *PSSC Physics,* D. C. Heath and Company, Lexington, Mass., 1965.)

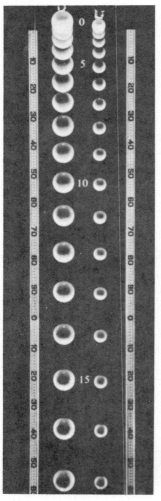

asked what would happen if a heavy body and a light body were tied together and were then dropped from a height. According to Aristotle's view, it could be maintained that the time would be the average of the times the two bodies would have taken to fall separately. It could also be argued that the time taken would be the same as that required by a body with their combined weights to fall from the same height. These results are contradictory; therefore, Galileo wrote that they showed Aristotle to be wrong.

There is an uncorroborated story that to find out what actually does happen when bodies fall, Galileo made a public demonstration from the Leaning Tower of Pisa before the students and faculty of the University. Today we have elaborate instruments and refined clocks to determine how the speed of fall varies with the time of fall or the distance. Since mechanical clocks were unknown in Galileo's day, he found the measurements on a body falling straight down beyond his means. The motion was too rapid to observe directly. If gravity could be "diluted," however, the motion would be slowed down, perhaps enough for measurements to be taken.

Galileo was convinced that smooth, metallic spheres rolling down a smooth, graduated inclined plane with measurable speeds were an example of diluted free fall and would follow the same laws as a body falling vertically. His argument was based on the fact that a ball resting on a horizontal surface does not move, whereas one falling parallel to a vertical surface moves as fast as if the surface were not there (Fig. 2-17). A ball on an inclined surface, Galileo reasoned, should therefore roll with an intermediate speed depending on the angle of inclination.

Galileo let spheres roll down planes tilted at

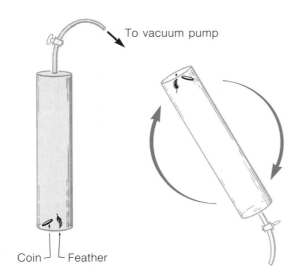

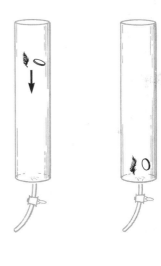

To vacuum pump

Coin — └ Feather

Figure 2–19   A coin and a feather fall at the same rate in the absence of air resistance.

various angles and determined the distances covered in different time intervals, which he measured with a water clock. He found that, regardless of their weights, the spheres all fell through the same distance in the same time. Extrapolating these results to bodies falling vertically, Galileo could say that, irrespective of their weights, all bodies fall with the same speed when dropped from the same height (Fig. 2–18). This result contradicted Aristotle's opinion that heavier bodies fall faster than light ones (Fig. 2–19).

In the account of his work, Galileo tells what apparatus he used, how it was set up, and how the experiment was performed (Fig. 2–20). He made many

Figure 2–20   Students performing Galileo's inclined plane experiment. (From Holton et al., *The Project Physics Course Text*, Unit 1, p. 55. Copyright © 1970 by Project Physics, Inc.)

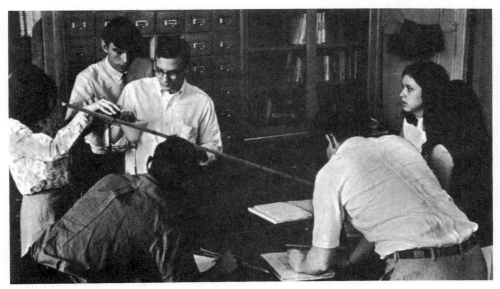

trials to insure that the results would be as accurate as possible. His experimental conditions were as perfect as he could make them; the plane was polished, and the metal spheres were smooth. To simplify his study and to get to the fundamentals of his problem, he ignored less important phenomena such as friction. He then analyzed his data mathematically and tested his hypothesis by further experiments. The information he obtained in this way would apply not only to the conditions of the particular experiment, but more generally to other bodies falling under gravity. The mathematical-experimental method, which character- izes science as we know it today, came to maturity in Galileo.

Galileo concluded that all bodies released from the same height simultaneously fall at the same rate, if air resistance is neglected. He also found that the rate changes with time. We now know that the velocity of a freely falling body increases by an increment of 9.80 meters per second (32 ft/sec) each second. The value 9.80 meters per second per second, or 9.8 m/sec^2 (32 ft/sec^2), is known as the acceleration due to gravity and is designated by the symbol g. The value of g varies slightly from place to place on the earth's surface. Extraordinarily precise measurements of g made by Robert H. Dicke and co-workers at Princeton University confirm Galileo's conclusion that all bodies fall at precisely the same rate.

2–9 PROJECTILE MOTION

Galileo applied his method to the problem of determining the path traced out by a projectile. He had learned through his experiments with rolling spheres that projectile motion could be considered to be made up of two components. A sphere rolled down one incline starts up a second incline with the same speed. As the angle of the second plane decreases, the sphere encounters less resistance to its motion until, when the plane is horizontal, it continues to move with uniform speed, slowed down only by friction (Fig. 2–21). Galileo concluded that in the absence of friction the sphere would move horizontally indefinitely at uniform speed, and this would be its natural motion. It is as if gravity were switched off when the sphere reached the

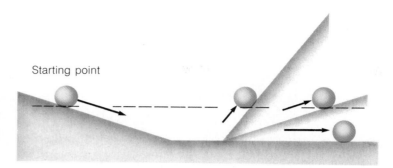

Starting point

Figure 2–21 In the absence of friction, a sphere rolling down one inclined plane would rise to the same height on another inclined plane. Galileo concluded that, if the second plane were horizontal, the sphere would travel indefinitely.

bottom of the incline, and the horizontal component of the ball's motion continued unaffected by gravity.

The conclusion that Galileo reached contradicted Aristotle's doctrine that a constant force is required to keep a body moving with uniform speed; no force that anyone could see kept the sphere moving horizontally. As far as Galileo was concerned, though he did not make the concept explicit, it is natural for a body in a state of uniform motion to remain in that state indefinitely, provided that it is not acted upon by an external force such as friction.

Galileo could approximate the motion of a projectile by considering a sphere rolling across a table with uniform speed until it comes to the edge. When it reaches the edge of the table the sphere, due to its own weight, acquires a vertical, accelerated motion while at the same time maintaining its uniform horizontal motion, and traces a curved path to the floor. In the horizontal direction, the constant uniform speed causes the body to cover equal distances in equal times. Vertically, the speed increases with time because of gravity, and the distances covered are proportional to the square of the time. Galileo determined that the form of the path followed by the sphere under these conditions would not be merely a curve of some sort, but a specific curve—a semiparabola (Fig. 2–22). The path of a projectile shot from a cannon would be a full parabola.

Galileo's theory of projectile motion—that it consists of horizontal and vertical components that do not interfere with each other—offered strong support to the hypothesis that the earth moves around the sun. Critics of this hypothesis argued that if the earth moved, a stone dropped from the top of a tower should land behind the tower—not at a point at the base

Figure 2–22 A ball rolling off a table traces a curved path—a semiparabola—to the floor. (From *PSSC Physics,* D. C. Heath and Company, Lexington, Mass., 1965.)

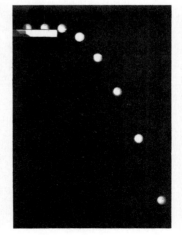

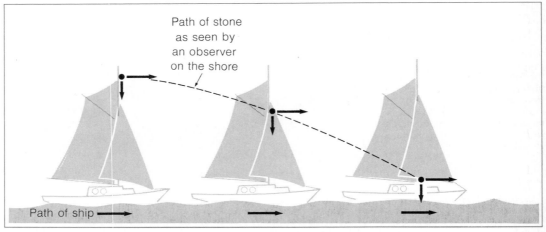

Path of stone
as seen by
an observer
on the shore

Path of ship ➙

Figure 2–23 A stone shares in the uniform horizontal motion of the ship. As the stone falls, it arrives at the same point simultaneously. The trajectory, as seen from the shore, is a semiparabola.

vertically below. The moving earth, it was supposed, would carry the base forward while the stone was falling.

To answer this criticism, Galileo drew the analogy of dropping a stone from the top of a mast of a ship moving with uniform speed (Fig. 2–23). When released, the stone has a horizontal motion equal to that of the ship and unaffected by its vertical motion. As the body falls while the ship moves, the horizontal motion of the stone carries it through the same horizontal distance that the ship travels. The stone and the mast, therefore, arrive at the same point simultaneously. To an observer on the ship who also shares its horizontal motion, the stone appears to fall straight down. In the same way, to an observer standing on the moving earth, a falling stone appears to drop vertically downward.

2–10   MOTION IS RELATIVE

To a passenger on a ship moving with uniform speed who shares the ship's horizontal motion, an object falling from the mast appears to fall vertically downward. The trajectory of the object appears curved, however, to an observer on land who does not share the ship's motion; the object appears to move forward with the ship as it drops. Which observer is correct?

Both are, if the frame of reference of the observer is considered. When the ship is standing still in the harbor, the trajectory appears to be a straight vertical line to both the passenger aboard ship and the observer on land. To the passenger, the trajectory is the same whether the ship is motionless or moving uniformly. The passenger cannot detect whether or not the ship is in motion from the trajectory alone. The observer on land sees both components, horizontal and vertical, which combine to give a semiparabola.

Such observations form the basis of the *Galilean relativity principle*, which states that the laws that describe physical experiments are the same in a reference frame at rest or in a reference frame moving with a constant velocity. Consequently, physical experiments will give the same results for different observers in reference frames moving at constant but different velocities. An object released from a ship's mast will land at the foot of the mast whether the ship is at rest or moving uniformly, and the laws of falling bodies are the same whether derived from experiments on a ship at rest or moving uniformly.

If the reference frames are moving uniformly with respect to each other (that is, at constant speed in a straight line), there is no preferred reference frame, according to Einstein's special theory of relativity. All reference frames are equally correct, although one may be more convenient to use than another. Without a frame of reference, the concept of motion has no meaning; there would be no way of determining whether or not motion is occurring.

We now know there is no absolute frame of reference by which you can say with confidence that whatever you see moving moves relative to you as you stay at rest. It was thought for centuries that the earth itself was an absolute frame of reference. However, experiments have shown that the earth rotates on its axis and undergoes other motions, although we are not aware of any of these motions.

CHECKLIST

Acceleration
Aether
Component
Displacement
Frame of reference
g
Kinematics
Polygon method

Relativity
Resultant
Scalar
Speed
Trajectory
Vector
Velocity
Violent motion

(See Appendix A for a discussion of scientific notation.)

2-1.   Contrast the vertical and horizontal components of the motion of a thrown football.

2-2.   From a physiological point of view, differentiate between constant velocity and acceleration as experienced in an elevator ride.

2-3.   Will a golf ball and a sheet of paper dropped from a bridge simultaneously strike the water at the same time? Explain.

2-4.   Discuss the features of Galileo's method of dealing with the problem of motion that are relevant to science today.

2-5.   Using examples from kinematics, justify the scientist's insistence upon precision in terminology.

2-6.   An automobile has a constant speed of 15 mi/hr. Is it possible for it to have an acceleration at the same time? Explain.

2-7.   A man walks 3 miles due east, then turns and walks 2 miles due north. How far is he from the starting point?

2-8.   A plane is climbing with a velocity of 150 m/sec at an angle of 30° to the horizontal. Determine the horizontal and vertical components of its velocity.

2-9.   A ship sails 75 miles due east, then sails in a direction 30° east of south for 60 miles, and finally heads due south for 40 miles. (a) How far has the ship traveled? (b) How far is it from its starting point?

2-10.   A hurricane is drifting at a velocity of 16 mi/hr in a direction 40° east of north. How fast is it moving in the (a) northerly and (b) easterly directions?

2–11.   An airplane travels with a velocity of 200 mi/hr in a direction 30° west of north for 1½ hours. How far north and how far west is it from its starting point?

2–12.   A driver travels 250 miles from New York to Washington in 5 hours. What is his average speed speed in miles per hour?

2–13.   A supersonic transport plane is traveling at a speed of 1200 mi/hr. How far does it travel in 1 minute?

2–14.   An earth satellite travels 2000 kilometers in 5 minutes. How far does the satellite travel in 24 hours?

2–15.   An automobile traveling at 22 ft/sec accelerates at a constant acceleration of 4 ft/sec² for 15 seconds. Determine its velocity at the end of this time.

2–16.   Starting from rest, how long must a car accelerate at 5 ft/sec² to reach a speed of 30 mi/hr? (Note: a useful conversion factor is 60 mi/hr = 88 ft/sec.)

2–17.   A truck traveling with an initial speed of 33 ft/sec is accelerated uniformly until it reaches a speed of 77 ft/sec. If the acceleration is 4 ft/sec², how much time is required for the truck to reach its final speed?

2–18.   Starting from rest, an electron is accelerated in a machine to a velocity of 10^4 m/sec in 0.01 seconds. (a) What is the acceleration? (b) How far does the electron travel in this time?

2–19.   A car traveling at 40 mi/hr skids to rest in 5 seconds. (a) What is its average speed during the stopping process? (b) How far does it go before stopping?

2–20.   (a) How long does it take sunlight to reach the earth? (The sun is 9.3×10^7 miles away; the velocity of light is approximately 186 000 mi/sec) (b) Moonlight? (The moon is 2.4×10^5 miles from the earth.)

2–21. What is the speed of a freely falling ball 3 seconds after it is released from rest?

2–22. A rock is thrown horizontally with a velocity of 20 m/sec from a bridge 30 meters high. (a) How long will the rock be in the air? (b) How far from the bridge will it land?

2–23. Multiple Choice

A. Which of the following is not a vector?
   (a) displacement   (c) velocity
   (b) speed   (d) acceleration

B. Acceleration is the rate of change of
   (a) displacement   (c) velocity
   (b) position   (d) time

C. Acceleration can be expressed by
   (a) m/sec   (c) m/sec^2
   (b) m^2/sec   (d) m^2/sec^2

D. If you cover 3000 miles in 5 days on a cross-country trip, 30 mi/hr could represent the
   (a) instantaneous speed   (c) velocity
   (b) average speed   (d) acceleration

E. According to Galileo,
   (a) rest is the natural state of an object
   (b) accelerated motion is the natural state of an object
   (c) if not acted upon by a force, a body continues in motion in a straight line
   (d) heavy objects fall faster than light objects

SUGGESTIONS FOR FURTHER READING

Adler, Carl G., and Byron L. Coulter, "Galileo and the Tower of Pisa Experiment." *American Journal of Physics,* March, 1978.
    The story of the Tower of Pisa experiment is probably more legend than fact. If the experiment had been done, would it have worked as well as the legend claims?

Casper, Barry M., "Galileo and the Fall of Aristotle: A Case of Historical Injustice?" *American Journal of Physics,* April, 1977.
    Presents Galileo's case against Aristotle's description of freely falling objects.

Chapman, Seville, "Catching a Baseball." *American Journal of Physics,* October, 1968.
    Shows that although baseball is not a scientific game, many principles of physics can be applied to it.

Drake, Stillman, "Galileo's Discovery of the Law of Free Fall." *Scientific American,* May, 1973.
    A recently discovered manuscript of Galileo's shows how he discovered the correct law of free fall.

Rogers, Terence A., "The Physiological Effects of Acceleration."
   Scientific American, February, 1962.
         At high accelerations, these effects are considerably mag-
         nified.
Seeger, Raymond J., "Galileo Yesterday and Today." *American
   Journal of Physics,* September, 1965.
         Considers Galileo's contributions to modern science.

PLANETARY MOTION

3-1 INTRODUCTION

Early peoples took an active interest in the skies. Those who were bound to the land learned from the rising and setting of constellations the times for sowing and harvesting (Fig. 3-1). Seafarers used the stars for planning their voyages and for navigation.

As astronomy developed, people mapped the heavens and charted the courses of the planets. The Egyptians and Babylonians were excellent observers of the periodic changes of sunrise and sunset during the year, the phases of the moon, the seasonal appearance and disappearance of constellations, and the more complex motions of the planets. They were aware of months and years and made calendars based on the moon cycle, the sun cycle, or combinations of the two. Until the Greeks, however, people were satisfied to describe the heavens and showed no interest in discovering an underlying regulatory mechanism.

3-2 PTOLEMY'S SYSTEM

From the very first, the Greeks tried to explain celestial phenomena. To do so they relied heavily upon the observations recorded by the Egyptians and Babylonians.

Long before Columbus, Pythagoras taught that the earth is a sphere. He may have been guided in this belief by several observations. The surface of the sea is curved, not flat. Because of this, we first see the tallest point of a distant ship as it approaches, then gradually

Figure 3–1 Stonehenge, in southern England, was probably an ancient observatory. It appears that it was used to keep track of certain days of the year and for predicting eclipses. (Gerald S. Hawkins, *Stonehenge Decoded.*)

the rest. In an eclipse of the moon, the circular edge of the earth's shadow also suggests that its shape is spherical. The fact that the sun and moon appear to us as circular disks is evidence of their spherical nature. Just as important to Pythagoras, however, was the belief that the sphere was incomparable in symmetry, beauty, and perfection, and was therefore the only proper shape for celestial bodies and the earth.

The belief that heavenly bodies move along circular paths was linked to the belief in spherical perfection. The circle, with neither beginning nor end, seemed the right path for them to follow. If some of the planetary motions seemed complicated, it was assumed that analysis would reveal that they could be reduced to uniform circular motions. The earth, however, was imagined to be at rest, and its center the center of the universe.

Pythagoras taught that the heavenly bodies were eternal, perfect, and unchangeable, and their motions uniform and circular. The earth, on the other hand, was subject to change and decay, and motions occurring on it were unpredictable and irregular. This dualism between the earth and the heavenly bodies had a lasting effect on astronomy.

The first attempt to explain astronomical phenomena in mathematical terms was made by Eudoxus, a pupil of Plato. His theory of homocentric spheres was the solution to a problem proposed by Plato to his pupils to "save the phenomena," that is, to account for the motions and positions of the heavenly bodies.

The peculiar motions of the planets baffled observers. At various times the planets seemed to stop, to retrograde, or to trace a curve such as a figure eight (Fig. 3–2). In his system, Eudoxus introduced 27 spheres to discover a combination of movements that a planet would seem to describe. The sun, moon, and each of the planets in this system were considered to be points on the surface of various interconnected spheres, all concentric ("homocentric"), with the earth at the center. The spheres were thought to turn at different rates, around different axes, and in different directions (Fig. 3–3). Eudoxus did not speculate on the real existence of these spheres or on the causes of their motions; his was a purely mathematical solution.

Admirable though it was, Eudoxus' system of homocentric spheres fell short of "saving the phenomena." It failed to account for variations in the brightness of planets, especially Venus and Mars, and for the difference in size between the sun and moon. However, considering the inaccuracies that were present in the data available to him, Eudoxus' system was remarkable. Others improved upon it by adding more spheres and brought the theory into closer

Figure 3–2 Retrograde motion of Mars for 1971. Mars proceeded eastward until July 13, retrograded westward, and resumed its eastward journey on September 11. (From Smith, E. V. P., and K. C. Jacobs, *Introductory Astronomy and Astrophysics*. Philadelphia: W. B. Saunders Company, 1973, p. 26.)

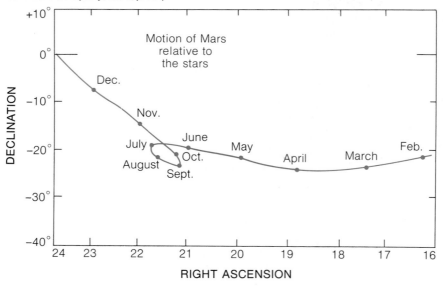

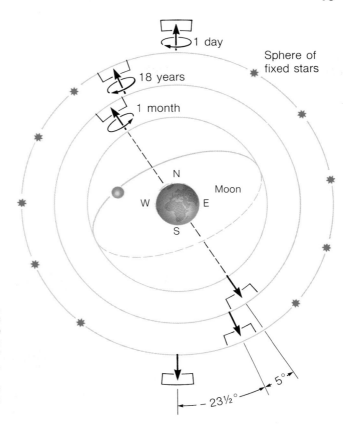

1 day

Sphere of
fixed stars

18 years

1 month

N

Moon

W   E

S

5°

— 23½°

Figure 3–3 With the earth at the center, Eudoxus employed three concentric ("homocentric") spheres to represent the moon's motion. Twenty-seven spheres were necessary for the entire planetary system. (Redrawn from Crawford, F. H., *Introduction to the Science of Physics.* Copyright © 1968 by Harcourt Brace Jovanovich, Inc., and reproduced with their permission.)

harmony with the observed motions. Aristotle converted the system from a purely geometrical to a mechanical structure that could be presumed to have a physical existence, although it is not certain that Aristotle believed in its physical reality. Even with the refinements and modifications, however, the theory of homocentric spheres still did not "save the phenomena."

The high point of Greek astronomy is found in the work of Hipparchus of Nicaea (c. 150 B.C.) and Claudius Ptolemy of Alexandria (c. 150 A.D.). Although Hipparchus' works are lost, he is known through the writings of Ptolemy and others. Ptolemy refers to Hipparchus so frequently in his writings that it takes some effort to recall that the two were separated by three centuries. Nor is it always clear from the context what was original with Hipparchus and what Ptolemy contributed. Both saw their tasks not so much as the taking and recording of observations as the mathematical explanation of the facts that the observations revealed. Their work culminated in Ptolemy's encyclo-

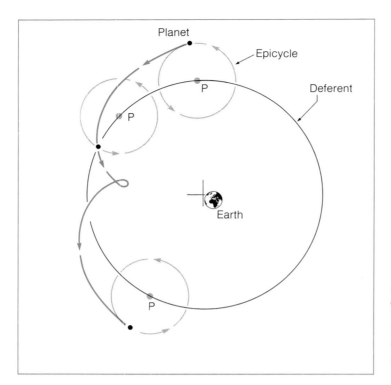

Figure 3–4 Ptolemy's basic devices: the *epicycle* and the *deferent.* A planet moved in a perfect small circle—the epicycle—around a point P, while P revolved in a perfect circle—the deferent. The earth is just off center *(eccentric).* Epicyclic motion saved Plato's perfect circles.

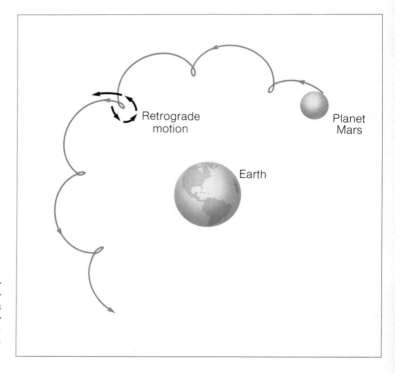

Figure 3–5 In Ptolemy's system, the path of a planet consisted of a series of loops as a consequence of epicyclic motion. A planet's retrograde motion occurs at the bottom of a loop.

pedic masterpiece, the *Almagest*. The *Almagest* displaced earlier works on astronomy and became the standard treatise on astronomy.

The *Almagest* developed the Ptolemaic system, a solar system that was completely earth-centered. The Ptolemaic system kept the Pythagorean concept of circular motions for the heavenly spheres. It met the deficiences of the theory of homocentric spheres with a mechanism consisting of eccentrics, epicycles, deferents, and equants (Figs. 3–4 and 3–5). The relative simplicity of Eudoxus' system gave way to one of complexity. The basic devices—epicycles and deferents—were used in various combinations to describe curves in which the planet was close to the center at times, distant at other times, stationary, or even retrograde. It was necessary only to choose the proper relative size of epicycle and deferent and speeds of rotation. Like Eudoxus, Ptolemy in all probability considered his system to be a mathematical model of the universe having no "real" existence.

For 1400 years Ptolemy's astronomical system encountered no serious challenge (Fig. 3–6). Its basis in the geocentric idea—an immobile earth at the center of a finite universe—was satisfying to most people. The Ptolemaic system was in complete harmony with the Aristotelian physics which it assumed—in particular,

Figure 3–6 Ptolemy's earth-centered model of the universe reigned supreme for 1400 years. It seemed to account for observations of planetary motions. The epicycles of Mercury and Venus always lie on the line joining the earth and the sun.

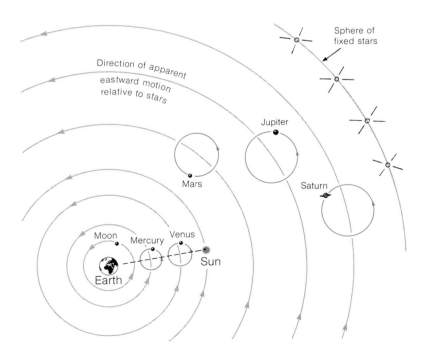

with the principle of uniform, circular motion for celestial bodies. Together, the two systems offered a coherent scientific world view that was, moreover, reconciled with the major religious systems of the West—Judaism, Christianity, and Islam.

3–3 THE COPERNICAN REVOLUTION

As astronomical observations improved, discrepancies showed up between observation and theory. When adjustments were made in certain parts of the Ptolemaic theory to account for these, other discrepancies appeared elsewhere. By the 15th century astronomy was in such a sorry state from the point of view of precision that Copernicus (1473–1543) spoke of a "monster" passing for a system of the world. He felt that no system as cumbersome and inaccurate as the Ptolemaic system had become could possibly be true of nature. The solution to the problem did not seem to involve further tinkering with the Ptolemaic system but a fundamental restructuring of astronomy.

The Copernican system was set forth in his work,

Figure 3-7 The sun-centered universe of Copernicus. The moon travels around the earth, but the earth and the other planets move about the sun. A less complex mechanism is required than in Ptolemy's system.

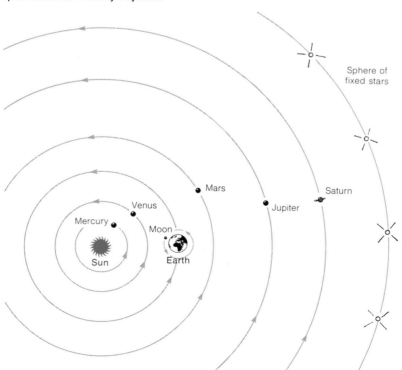

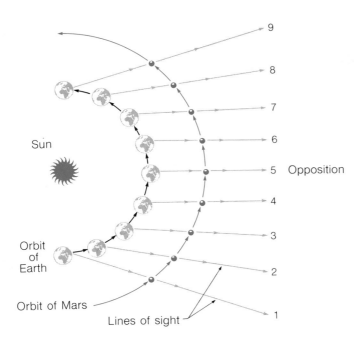

East

9

8

Opposition

West

Sphere of
fixed stars

Apparent orbit

Sun

Orbit
of
Earth

Orbit of Mars

Lines of sight

9

8

7

6

5

4

3

2

1

Figure 3–8 Retrograde motion of Mars explained in the Copernican system. The earth moves faster in its orbit than does Mars. Lines of sight drawn from a moving earth to a more slowly moving Mars illustrate how a retrograde loop in Mars' orbit is formed.

On the Revolutions of the Celestial Spheres, published in 1543. Copernicus received a copy of it on his deathbed. It reintroduced the idea of a sun-centered system, proposed 1700 years previously by the notable Alexandrian astronomer, Aristarchus, and replaced the notion of an earth at rest with one in motion (Fig. 3–7). The earth was described as having three motions: a daily rotation on its axis, an annual revolution in its orbit around the sun, and a wobbly motion to account for the precession of the equinoxes (Fig. 3–8).

In effect, Copernicus used the thought patterns of the Greeks and produced the simplest solution to the problem that Plato posed of "saving the phenomena." He kept the mechanism of concentric spheres, epicycles, and deferents, and the Pythagorean assumption that only circular, uniform motion is right for heavenly bodies. The number of spheres required, however, was reduced from Ptolemy's 80 or so to 34.

Copernicus believed firmly in the reality of the earth in motion and made this clear in his preface. In his final illness, the task of supervising the printing of the book was assumed by George Rheticus, a mathematician and student of Copernicus, and then by Andreas Osiander, a friend. Sensing the explosive contents of the book, Osiander wrote a second preface in which he warned the reader not to take the hypothesis of a moving earth seriously, but to view it only as a device for simplifying mathematical computation. The intent of this preface was to pacify those who would otherwise regard as blasphemous the dethronement of the earth as the center of the universe in favor of the sun. Martin Luther is quoted as saying: "This fool wishes to reverse the entire science of astronomy." To others, however, Copernicus was regarded as the "restorer of Astronomy" and the "Ptolemy of our age."

Astronomers freely used the Copernican system for calculation, and a few even supported the innovative ideas against which Osiander had warned. Popular writers, on the other hand, were hostile. It was difficult to give up the uniqueness of the earth in favor of a system in which the earth was merely another planet, and an insignificant one at that. The spread of Copernican ideas threatened to crack the intellectual structure of the medieval world, changing man's thinking about himself and the universe. It is in this sense that we speak of the Copernican revolution.

3–4 "GATEWAY TO THE SKIES": TYCHO BRAHE

Over the years tables were developed giving the changing positions of the planets. The Alphonsine planetary tables were based upon Ptolemy's geocentric (earth-centered) ideas. Following the appearance of Copernicus' heliocentric (sun-centered) system, the Prutenic tables were compiled and became the standard reference in the 16th century. Both tables gave the impression of considerable accuracy, stating planetary positions in degrees, minutes, and seconds of arc. However, a conjunction of the planets Jupiter and Saturn, when they appeared to be closest to each other, in 1563 indicated that both tables were in error in their predictions, the Alphonsine tables by a full month and the newer Prutenic tables by a number of days. To Tycho Brahe (1546–1601), making his first real observations in 1563 at the age of 16, these discrepancies meant that astronomy was in need of more accurate data. The sudden appearance in 1572 of a new star—what would now be called a nova or supernova—in the constellation Cassiopeia determined the direction of his career.

The appearance of a nova was not entirely without precedent. Hipparchus had witnessed one, but the fact that more were not recorded in the course of history is testimony to the low esteem in which observational astronomy was held. It was Tycho's opinion that the universe could never be understood until the positions of the stars and planets were known exactly. He was determined to compile the most accurate star catalogue possible, as well as data on the motions of the planets.

The new star of 1572 was as bright as the planet Jupiter at first and gradually faded away in 16 months. Tycho thought at first that it was a meteorological event, since Aristotle had taught that the heavens were perfect and unchangeable. But attempts to measure the parallax of the new star failed (Fig. 3–9). If it were below the moon, its relative nearness should have been revealed by its apparent shifts in position in relation to the backdrop of the stars. Tycho and others concluded, therefore, that the new star lay in the sphere of the fixed stars. Here was the first addition to the universe since the records of Hipparchus in the 2nd century B.C. It had to be admitted, then, that the heavens did change and therefore were not perfect.

Tycho wrote a book concerning this event, *On The New Star* (1573), which attracted the attention of King Frederick II of Denmark. The king persuaded Tycho to establish an observatory and conferred upon Tycho the feudal lordship of the island of Hveen on which to pursue his studies. He supported Tycho generously over the next 21 years.

On Hveen, Tycho built a castle-observatory exactly in the center of the small island. He called it Uraniborg, "Gateway to the Skies," in honor of the Greek muse of astronomy, Urania. It was a combination of residence, observatory, and school of astronomy, and an architectural as well as a scientific treasure. A short distance away he built a smaller observatory called Stjerneborg, the "Star Castle," sunk below ground level so that the wind would not affect the instruments.

Night after night Tycho peered at the sky, compiling the first massive body of data that astronomy had ever known. His observations were made without the aid of a telescope, which was then unknown. The

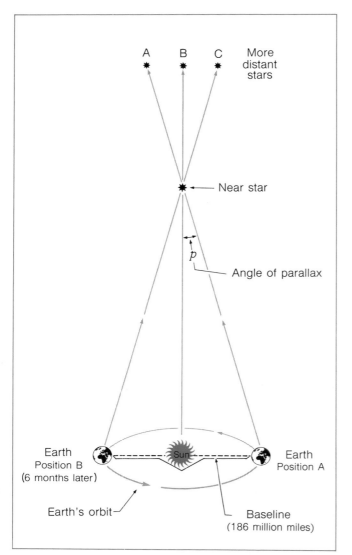

Figure 3–9 Parallax of a star *(p)* is the apparent shift in the position of a star against the backdrop of much more distant stars. As the earth revolves in its orbit, different distant stars are visible behind the near star. The angle of parallax is so small that the first was not measured until 1839 by F. W. Bessel. The parallax of the nearest star is less than one second of arc (0.75″).

instruments he used were traditional devices—quadrants, armillary spheres, parallactic rulers—which he improved and made larger than ever before. He also developed new instruments and techniques and with them attained an accuracy reaching the limits of vision with the human eye. Tycho's celestial catalogue eventually listed positions for more than 700 stars, more than three fourths of those that can be seen with the naked eye from the latitude of Denmark.

One of Tycho's major discoveries concerned the nature of comets, which Aristotle had maintained were meteorological phenomena. Based upon his observations of a comet in 1577, Tycho showed that it was not sublunar, but more distant than the moon. Like the new star, the comet had to lie in the sphere of the heavens with other true celestial bodies. These regions were thus capable of yet another change. The appearance of other comets confirmed this conclusion; no comet was ever found to be sublunar.

Tycho proved that the orbits of the comets penetrated the spheres of the planets. Many astronomers from Aristotle on had assumed that the spheres were "crystalline"—solid and transparent—and that the planets were embedded in them, for otherwise, what kept the planets in their positions? Tycho's conclusion that the heavens could not consist of real celestial spheres did much to sweep away ideas that had been held for centuries but which were not supported by scientific observations.

Tycho himself was not concerned with what kept the planets in their orbits. The use he made of his data in his own theoretical work was negligible, but the data were not wasted. When Frederick's successor, Christian IV, withdrew support from him, Tycho left Uraniborg and moved to Prague to serve as Imperial Mathematician to the Holy Roman Emperor, Rudolph II. There Johannes Kepler became an assistant to him in one of the most fortunate collaborations in the history of science. Tycho's years of patient observations and Kepler's genius at mathematics combined to complete the Copernican revolution.

3–5 HOW PLANETS MOVE: JOHANNES KEPLER

Johannes Kepler (1571–1630) won recognition with his book, *The Cosmographical Mystery* (1596). He sent copies of it to royalty and to distinguished scientists. Tycho received a copy and immediately recognized Kepler's mathematical ability. He thought that Kepler would make an ideal assistant at Uraniborg and offered him access to his vast collection of data if Kepler would join him. But Kepler did not accept.

Within three years, however, Kepler's life intersected again with Tycho's. Having lost his patron, Tycho left Denmark in 1599 and settled in Prague in the employ of the emperor, Rudolph II. Kepler, in

1600, was forced out of his position as district mathematician and teacher of mathematics, and went to Prague to see whether Tycho's offer was still open. The two reached agreement, and Kepler soon had at his disposal the data which Tycho had painstakingly gathered over many years. Their period of collaboration was brief, however; Tycho died in the following year, 1601. Kepler followed Tycho as Imperial Mathematician and retained access to Tycho's papers.

When Kepler arrived in Prague to work with Tycho, the first task he was assigned was to develop a theory of the orbit of Mars based upon Tycho's observations. Kepler brought to this task the belief of an ardent Copernican. He had no doubt that Mars, the earth, and the other planets revolved around the sun. Copernicus, however, had set the sun like a lamp in the middle of the solar system, with no other purpose than to provide heat and light. To Kepler, the sun was much more than just a lamp.

Kepler noted a correlation between a planet's speed and its distance from the sun. He observed that Venus moved more slowly than Mercury, the closest planet; the earth, at a greater distance than Venus, moved more slowly than Venus; and so on. These facts suggested to Kepler that the sun exerted a force that caused the planets to move and that this force fell off in strength as the distance to the planet increased. Whereas earlier theories had regarded the circular movements of the planets as natural, arising from some inherent property within the body, Kepler was guided by the hypothesis that the circular orbits were the results of forces from without.

Initially, Kepler worked out the elements of the orbit of Mars under the influence of the Pythagorean tradition, which required uniform, circular motion for planets. He did develop a circular orbit for Mars which agreed rather closely with Tycho's observations; the discrepancy was only 8 minutes of one degree. Kepler knew, however, that Tycho's data were accurate to about 2 minutes, and the discrepancy troubled him. His respect for the accuracy of Tycho's observations was so great that he abandoned the assumption of a circular orbit and searched for a new hypothesis.

The discrepancy between theory and observation meant to Kepler that the assumption of uniform motion for Mars was in error. The data indicated that Mars moved through its orbit not uniformly but erratically,

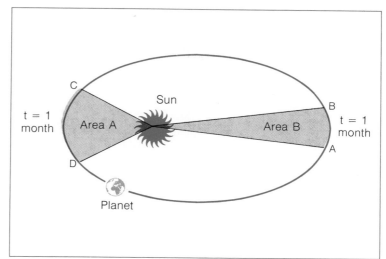

Figure 3-10 *Kepler's second law (law of equal areas)*: A line drawn between the sun and a planet sweeps out equal areas in equal intervals of time. Area A = area B. If the planet travels from point A to point B in the same time as from point C to point D, then it travels fastest in its orbit when nearest the sun.

its speed varying according to no obvious law. The planet moved faster when approaching the sun and more slowly when receding from it. Guided by the inverse relation between speed and distance through hundreds of pages of mathematical calculations, centuries before the introduction of the electronic computer, Kepler arrived at the law of planetary motion, which we know as his *second law*, or the *law of equal areas: A line drawn between the sun and a planet (the radius vector) sweeps out equal areas in equal intervals of time* (Fig. 3–10).

After finding the law of planetary motion, Kepler continued to work on the shape of the orbit of Mars. He became convinced by the data and by his calculations that the planet could not be traveling in a perfect circle eccentric to the sun. No curve except a perfect ellipse fit the data and the law of equal areas (Fig. 3–11). He was forced to conclude that the orbit was not a circle but an ellipse, with the sun at one focus (Fig. 3–12). He found that the orbits of all the planets, including the earth, are elliptical (Fig. 3–13). This discovery is known as Kepler's *first law,* or the *law of ellipses: Each planet travels in an elliptical orbit with the sun at one focus.*

Kepler announced his first and second laws in one of the great masterpieces of natural science, *The New Astronomy* (1609). Galileo praised it, but few astronomers took his ideas seriously. Kepler's *third law* of planetary motion appeared 10 years later, in *Harmony of the World* (1619): *The square of the period of*

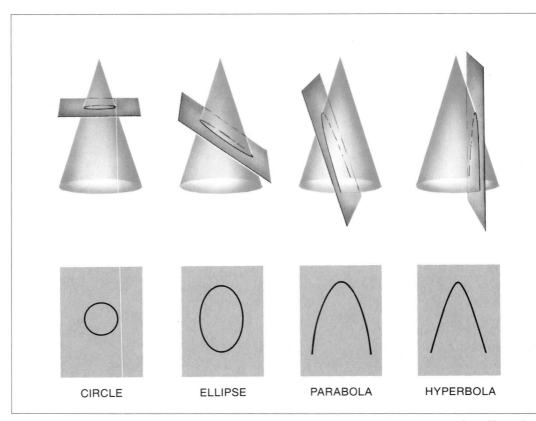

CIRCLE   ELLIPSE   PARABOLA   HYPERBOLA

Figure 3-11 An *ellipse* is a conic section—a curve produced but cutting a cone with a plane. If the cut is parallel to the base of the cone, the section is a *circle*. If the cut is made parallel to one side of the cone, the section is a *parabola.* An ellipse is obtained by an intermediate cut. A hyperbola is made by a cut perpendicular to the base.

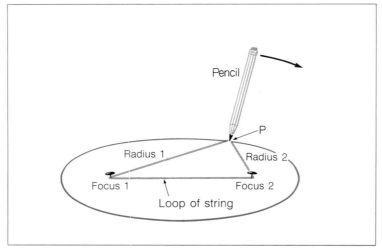

Figure 3-12 Constructing an ellipse. Stick two thumbtacks (the foci) in a piece of paper; place a loop of string loosely around them. Move a pencil within the loop of string, keeping the string taut. The sum of the distances of any point on the ellipse from the two foci (radius 1 and radius 2) is a constant.

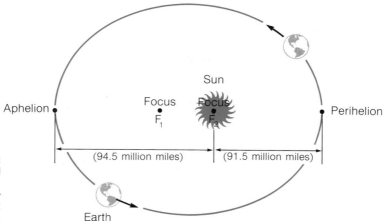

Figure 3–13 *Kepler's first law*:
The planets move in elliptical
paths, with the sun at one focus.
At *perihelion*, the planet is closest to the sun; at *aphelion*, it
is most distant from the sun.

Labels in figure: Aphelion, Focus F₁, Sun Focus F₂, Perihelion, (94.5 million miles), (91.5 million miles), Earth

*revolution of a planet (in years) is proportional to the
cube of the average distance of the planet from the sun
(in astronomical units, AU, the average sun-earth
distance).*

The third law is written symbolically as follows:

$$T^2 = kR^3$$ Equation 3–1

where T is the length of the planet's year—the time
required for it to complete one orbit around the sun; R
is the planet's average distance from the sun; and k is a
constant which has the same value for every planet,
1.00 when these units are used. Table 3–1 indicates
how Kepler's third law applies to the planets.

Table 3–1   SOME PLANETARY DATA AND KEPLER'S THIRD LAW

PLANET	DISTANCE FROM SUN (10^6 miles)	DISTANCE FROM SUN, R (AU)	REVOLUTION PERIOD, T (Earth-years)	T^2	R^3	T^2/R^3
Mercury	36	0.39	0.24	0.058	0.058	1.00
Venus	67	0.72	0.62	0.38	0.38	1.00
Earth	93	1.00	1.00	1.00	1.00	1.00
Mars	142	1.52	1.88	3.54	3.54	1.00
Jupiter	483	5.20	11.86	140.00	140.00	1.00
Saturn	886	9.54	29.46	868.00	868.00	1.00

EXAMPLE 3–1

Check Kepler's third law for the planet Mars.

SOLUTION

The period of revolution of Mars, T, in earth-years is 1.88 (see Table 3–1), and its distance from the sun is 1.52 AU. Using Equation 3–1, and setting K = 1,

$$T^2 = R^3$$

T = 1.88 earth-years

R = 1.52 astronomical units, AU

$$(1.88)^2 = (1.52)^3$$

$$(1.88)(1.88) = (1.52)(1.52)(1.52)$$

$$3.54 = 3.54$$

Hence, Kepler's third law checks for Mars.

Although Kepler had no more physical reasoning to support his three laws than Ptolemy had to support his theory, the advantage of Kepler's laws was that they gave a better description of planetary motions. Using Kepler's laws, the positions of the planets can be predicted more accurately than with Ptolemy's theory. Tycho, in fact, had retained Kepler primarily to complete the planetary tables and on his deathbed had obtained Kepler's promise that he would do this. Kepler kept his promise. His tables appeared in 1627 as the *Rudolphine Tables,* and they surpassed in accuracy the *Alphonsine Tables* and the *Prutenic Tables.* Kepler's three laws proved indispensable to Isaac Newton when he formulated his system of the world. After 1685, through Newton's work, Kepler's laws became well known and generally accepted.

3–6 GALILEO'S DISCOVERIES WITH THE TELESCOPE

The announcement of Kepler's first and second laws removed the mystery from planetary motion but exerted only a slight impact on the world. The

Copernican theory and the discoveries of Kepler had not yet become part of a coherent view of the world comparable to Aristotle's system upon which Ptolemy's astronomy was based. To people who thought at all about such matters, the quarrel between the supporters of Ptolemy and those of Copernicus concerned purely technical matters.

In 1609, word reached Galileo of the invention in Holland of an optical instrument that made distant objects look closer. Lacking further details, Galileo experimented with combinations of lenses and invented a telescope of the type still known as Galilean (Fig. 3–14). With improvements in the design, he constructed an instrument which made objects appear "a thousand times larger and thirty times closer." In the autumn and winter of 1609–1610, Galileo pointed his telescope toward the heavens. What he saw when he viewed the moon, the planets, the sun, and the stars he discussed in a book, *Sidereal*

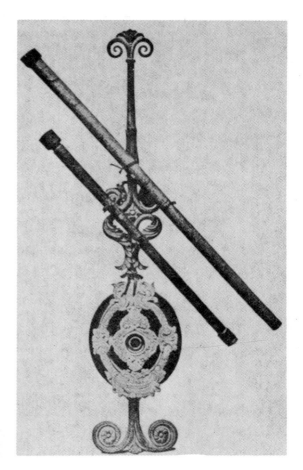

Figure 3–14 Two of Galileo's telescopes, on display in the Museum of the History of Science, Florence, Italy. (Yerkes Observatory photograph.)

Figure 3–15   Apollo 17 Astronaut Harrison H. Schmitt views a huge boulder on the Moon's surface. (Courtesy of NASA.)

Messenger (1610), which became an immediate success. A gifted writer as well as a brilliant mathematician and scientist, Galileo introduced astronomy to a wider public than ever before.

Wherever Galileo looked, he found evidence that either supported the Copernican system or weakened the Ptolemaic system. Thanks to the telescope, he found that the moon resembled the earth more closely than it did a perfect celestial body. Its surface was not smooth and polished, but rough and uneven, full of cavities, mountains, and deep valleys (Fig. 3–15). The moon's close resemblance to the earth refuted Aristotle's division of the universe into celestial and terrestial regions. Since the moon had a terrestial nature and revolved about the earth, why could not the earth revolve about the sun? Galileo attributed the moon's "secondary light" to the reflection of sunlight from the earth to the moon and back. From this he reasoned that if the earth shined, perhaps the other planets did also. (Prior to this time it had not been known whether the planets shined by their own light or by light reflected from the sun.)

Galileo observed a difference in the appearance of the planets and stars when viewed through the telescope. The planets appeared round and globular, like little moons with definite boundaries. The stars did not appear very different than they did when seen by the naked eye; they were not much enlarged, only brighter. Even in Galileo's day, the term "stars" applied both to the fixed stars and to the planets; now the telescope revealed differences between them. But he could see countless other stars, "so numerous as to be almost beyond belief." Galileo discovered that the Milky Way, about which there had been many disputes over the ages, consisted of a mass of innumerable stars that had never before been observed.

To Galileo, the most gratifying and important discovery was that of four satellites of Jupiter, "never seen from the very beginning of the world up to our own time." This discovery was offered as strong support of the Copernican system. Some objectors to the Copernican system would not accept the idea that the moon revolved around the earth while the earth revolved around the sun. If this were so, to their thinking the earth would lose the moon. But here were

Figure 3-16 The planet Venus exhibits phases similar to the Moon. Five phases of Venus are shown. Venus is a crescent only when it is in a part of its orbit that is relatively close to the Earth, and looks larger at those times. (Courtesy of Lowell Observatory.)

FIVE PHASES OF VENUS

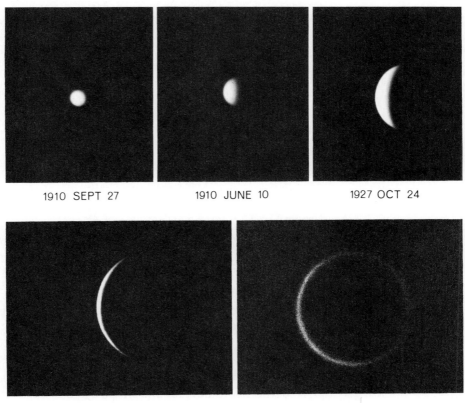

1910 SEPT 27   1910 JUNE 10   1927 OCT 24

1919 SEPT 25   1964 JUNE 19

four bodies moving around a planet while the planet moved around the sun. It was a model of the Copernican system. If Jupiter could move in an orbit and not lose its four moons, why would it be impossible for the earth to move and not lose its single moon? The fact that there were now two planets with satellites meant that the earth lost yet another claim to being unique; Galileo had proved that there was more than one center of rotation in the world.

Within the year, Galileo discovered yet another point of similarity between the planets and the earth. Through the telescope he observed that Venus exhibited phases like the moon (Fig. 3–16). The Copernican theory had predicted this, but no phases had ever been observed with the naked eye. Now there was proof that Venus shined by reflected light, and was similar in this respect to the earth and the moon. Once again a distinction between a celestial body and the earth was shown to be imaginary rather than real. As a result of Galileo's discoveries, more and more adherents were won over to Copernican ideas. By the end of the 17th century, the stage was set for a new physics, one suitable for a moving earth. That is the subject we examine in the next chapter.

CHECKLIST

Almagest	Orbit
Tycho Brahe	Parallax
Nicolaus Copernicus	Ptolemy
Ellipse	Pythagoras
Galileo	Retrograde motion
Geocentric hypothesis	Revolution
Heliocentric hypothesis	Rotation
Johannes Kepler	"Save the phenomena"
Kepler's laws	Telescope

EXERCISES

3–1. Account for the fact that the Ptolemaic system reigned unchallenged for 1400 years.

3–2. What effect did Copernicus' heliocentric (sun-centered) theory of the solar system have on humanity's perception of itself and its place in the universe?

3–3.   Explain the importance of Tycho's conclusions about the comet of 1577 with respect to the Ptolemaic system.

3–4.   Compare Tycho's observatory, Uraniborg, with a modern research institute with which you are familiar. (Name the institute.)

3–5.   Which discoveries of Tycho and Galileo posed serious difficulties for the Platonic view of heavenly perfection?

3–6.   Discuss some connections between the "old physics" and the "old astronomy."

3–7.   Why is the method of parallax impractical for very distant stars?

3–8.   Give an example of the interplay of theory and observation from the work of Tycho and Kepler.

3–9.   Why would Kepler's discovery of the ellipse as the preferred curve for planetary motion be objectionable to supporters of Ptolemy?

3–10.   A comet moves much faster when it is near the sun than when it is far away from it. Use Kepler's second law to explain this fact.

3–11.   The average distance of Jupiter from the sun is 5.20 AU. Determine the length of Jupiter's year, using Kepler's third law.

3–12.   If the earth is farther from the sun in summer than in winter, is it moving around the sun faster in summer or in winter? Explain.

3–13.   In what respects do Tycho's observatory and Galileo's application of the telescope to astronomy illustrate an aspect of scientific progress?

3–14.   What support did Galileo's discovery of four of Jupiter's moons lend to the Copernican theory?

3–15.   List several discoveries that Galileo made with the telescope which met the objections to a moving earth.

3–16.   How are "big science" and "little science" reflected in the careers of Tycho and Galileo? (Refer to Chapter 1.)

3–17.   Multiple Choice

    A. Ptolemy is known for his
      (a) invention of the telescope
      (b) study of free-fall
      (c) earth-centered model of the universe
      (d) sun-centered model of the universe

    B. Copernicus restated the
      (a) geocentric theory
      (b) heliocentric theory
      (c) theory of horizontal motion
      (d) theory of relativity

    C. Kepler discovered that planets travel in orbits that are
      (a) circular   (c) elliptical
      (b) parabolic   (d) hyperbolic

    D. Tycho Brahe
      (a) built an observatory   (c) invented the telescope
      (b) discovered Mars   (d) all of these

    E. Galileo discovered
      (a) the moons of Jupiter
      (b) the phases of Venus
      (c) the rough nature of the moon's surface
      (d) all of these

SUGGESTIONS FOR FURTHER READING

Gee, Brian, "400 Years: Johannes Kepler." *The Physics Teacher,* December, 1971.
    Emphasizes Kepler's collaboration with Tycho Brahe, and the development of Kepler's laws.

Gingerich, Owen, "Copernicus and Tycho." *Scientific American,* December, 1973.
    Relates the author's discovery of Tycho's personal copy of Copernicus' book, *De revolutionibus,* and its significance for modern science.

Postl, Anton, "Kepler's Anniversary." *American Journal of Physics,* May, 1972.

Reviews Kepler's life and work and his appreciation for the beauty and harmony of the universe.

Ravetz, Jerome R., "The Origins of the Copernican Revolution." *Scientific American,* October, 1966.

Suggests that Copernicus believed that as an astronomer he had to do more than merely "save the phenomena"; he felt that he should also exhibit the rationality and harmony of God's creation.

Wilson, Curtis, "How Did Kepler Discover His First Two Laws?" *Scientific American,* March, 1972.

Did Kepler begin with the hunch that the ellipse is the path described by planetary motions and then perceive that the calculated distances fitted into an ellipse?

4

LAWS OF MOTION
AND GRAVITATION

4-1 INTRODUCTION

Galileo's studies of falling bodies, motion on an inclined plane, and projectile motion did much to clarify the nature of motion. But later scientists were puzzled by the fact that although Galileo apparently understood clearly one of the laws of motion, the law of inertia (that a body in motion or at rest tends to continue in that state), there is no general statement of this law to be found in his writings. Moreover, although Galileo was the first to recognize the significance of acceleration, another law of motion that involves acceleration—the force law—also eluded him.

Kepler unraveled the mysteries surrounding the motions of the planets with the discovery of his three laws. Yet each seemed to be distinct from the other two. Although the ellipse replaced the circle as the curve favored for planetary orbits, the dualism between earthly motions and heavenly motions continued.

With the discoveries of Isaac Newton (1642–1727), we cross the threshold into the scientific age. As he himself freely acknowledged, however, without the work of Galileo, Kepler, and many others who had preceded him, he could not have achieved his synthesis of mechanics and astronomy.

4-2 ISAAC NEWTON'S "MARVELOUS YEAR"

On Christmas day of the year in which Galileo died, 1642, Isaac Newton was born at Woolsthorpe, in Lincolnshire, England. When he reached the age of 18, he entered Trinity College, Cambridge, where, except in mathematics, his record was not outstanding. In that field, however, he came under the tutorship of Isaac Barrow, who recognized his extraordinary talents. It was at Cambridge that Newton learned of the work of Copernicus, Kepler, and Galileo.

Newton took the Bachelor of Arts degree in January, 1665, just before the Great Plague forced the temporary closing of Cambridge. He returned to his home in Woolsthorpe and during the next 18 "golden months," between the ages of 23 and 25, made his greatest contributions to science. He made important contributions to algebra, invented the differential calculus, developed and applied the law of gravitation, carried out experiments with a prism that showed that white light contains all the colors of the spectrum, and invented a reflecting telescope. It may have been the most fruitful 18-month period in the history of creative thought. The rest of his scientific career consisted of an elaboration of the ideas he conceived at Woolsthorpe in that *annus mirabilis*, that "marvelous year."

4–3 THE *PRINCIPIA*

Nearly 20 years after the Woolsthorpe interlude, Newton found himself engaged in another 18-month period of intense concentration and productivity. This developed out of a visit by the astronomer Edmond Halley (1656–1742), of Halley's Comet fame, in August, 1684. Halley had a question concerning the attraction between the sun and the planets. Halley and Robert Hooke (1635–1703) had concluded from Kepler's study of planetary motions that the force of attraction between a planet and the sun must vary inversely with the square of the distance. But they had been unable to prove their hypothesis for elliptical orbits; it worked only for circular ones.

Halley asked Newton what curve would be described by the planets, assuming that gravity diminished as the square of the distance. Newton answered without hesitation, "an ellipse." Halley asked how he knew that. Newton replied that he had calculated it some five years previously. Halley wanted to see the calculations at once, but Newton could not find his notes, and promised that he would send the solution to Halley. Unable to find his earlier solution,

Newton reworked the theorems and proofs and sent them to Halley the following November. Urged by Halley to publish his results, Newton devoted the next 18 months to preparing a manuscript for the Royal Society. The publication of the *Principia* brought Newton almost immediate fame and made him the leader of science in England.

The *Principia* has been called the greatest scientific work ever published. It is divided into an introduction and three "books." In the introduction and the first book, Newton lays down his three laws of motion and considers the motion of bodies under various laws of force. In the second he treats the motion of bodies in various fluids. In the third book Newton discusses universal gravitation. Of himself and his achievements, Newton remarked:

> I do not know what I may appear to the world, but to myself I seem to have been only like a boy playing on the seashore, and diverting myself in now and then finding a smoother pebble or a prettier shell than ordinary, whilst the great ocean of truth lay all undiscovered before me.

4-4   NEWTON'S FIRST LAW OF MOTION: INERTIA

Newton's *first law* is an explicit statement of the *law of inertia:*

> *A body at rest or in uniform motion in a given direction remains at rest or in uniform motion in that direction unless acted upon by a net external force.*

Inertia is a property that becomes evident when an external force attempts to change the state of the body. Thus, if a net force acts upon the body, the body moves with changing speed, or in a changing direction, or both, but the body's inertia resists such changes.

The concept of inertia was borrowed from Galileo, who was aware of it but never stated it explicitly and generally. Newton applied it to both earthly objects, such as projectiles and tops, and to celestial ones— planets and comets—thus giving it universal significance.

Newton stated that motion is no less natural a state for a body than rest, provided that the body when placed in motion encounters no resistance (Fig. 4–1). Gone is

Figure 4–1 An illustration of *Newton's first law.* With no force acting on it in the direction of its motion, a ball on a horizontal plane moves along a straight line with constant speed.

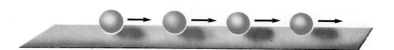

Figure 4–1 An illustration of *Newton's first law.* With no force acting on it in the direction of its motion, a ball on a horizontal plane moves along a straight line with constant speed.

Aristotle's physics of "natural" places for earth, air, fire, and water, and perfect circular motion for the "quintessence" making up celestial bodies. Gone, also, is the idea that the steady application of a force is required in "violent" motion, as in shooting an arrow or throwing a rock.

4–5 NEWTON'S SECOND LAW OF MOTION: FORCE

Newton's *second law,* the *law of force,* is stated as follows:

> *The acceleration of a body is directly proportional to the net force acting on the body and inversely proportional to the mass of the body, and is in the direction of the net force.*

If appropriate units are selected for force, mass, and acceleration, the proportionality can be expressed as an exact equation:

$$\text{Acceleration} = \frac{\text{Force}}{\text{Mass}}$$

Equation 4–1

$$\mathbf{a} = \frac{\mathbf{F}}{m}$$

This vector equation specifies that **a** is in the direction of **F** (Fig. 4–2). Rewriting this expression, we derive one of the most famous equations in science:

$$\mathbf{F} = m\mathbf{a}$$

Equation 4–2

Figure 4–2 *Newton's second law:* A force, **F,** acting on a block of mass, m, produces an acceleration, **a,** in the direction of the force, where **F** = m**a.**

Figure 4-3 A "frictionless" system: a dry-ice disk. (From *PSSC Physics,* D. C. Heath and Company, Lexington, Mass., 1965.)

The role of force in motion had long been puzzling. Aristotle's belief that a constant force was necessary to maintain a body in uniform "violent" motion seemed to be borne out by experience. If a spoon is placed on a table and given a push, it starts to move and its velocity is changed. An automobile traveling at a certain speed on a straight, level road gradually comes to a stop when its engine is disconnected by disengaging the clutch. It does seem that in the absence of a force, all bodies come to rest.

Recall, however, Galileo's investigations of motion on an inclined plane. He found that the smoother the surface, the greater the distance traveled by a polished metal sphere before slowing down and finally stopping. *Friction* between the plane and the sphere exerts a backward force on the sphere, retarding its motion and finally bringing it to a halt. If a perfectly smooth surface and sphere were available (a frictionless system), we would need only to set the sphere in motion and it

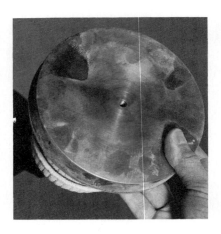

Figure 4-4 Carbon dioxide escapes through the hole in the bottom of the disk, forming a film of gas on which the disk glides. (From *PSSC Physics,* D. C. Heath and Company, Lexington, Mass., 1965.)

Figure 4–5 Forces are applied to the disk through springs mounted on the top and attached to strings. (From *PSSC Physics,* D. C. Heath and Company, Lexington, Mass., 1965.)

would continue forever without the application of additional force. Motion is as natural a state as rest.

It is possible to create approximate frictionless systems with which to study the effects of forces acting upon bodies. In one such arrangement, a heavy brass disk, highly polished on the bottom surface, supports a light metal tank carrying a supply of dry ice (solid carbon dioxide) (Fig. 4–3). The dry ice changes into a gas which escapes in all directions through a small hole in the bottom of the disk (Fig. 4–4). The disk and tank apparatus floats on a layer of escaping gas, which greatly reduces the friction between the disk and a glass surface. The film of gas supports the system and balances its weight. The only net force acting on the system is the one that we apply.

We can do this through a spring mounted on the top of the disk (Fig. 4–5). When the spring exerts no force on the moving disk, the disk moves with an approximately constant velocity, covering equal distances in equal times. If a constant force is applied, with the spring extended by a constant amount, the system experiences an acceleration (Fig. 4–6). If the same force is

Figure 4–6 A net force of one unit acting on a disk accelerates it to the right, as shown in this stroboscopic photograph. (From *PSSC Physics,* D. C. Heath and Company, Lexington, Mass., 1965.)

Figure 4–7 A net force of two units accelerates the disk to the right. Calculations show that the acceleration is doubled. (From *PSSC Physics,* D. C. Heath and Company, Lexington, Mass., 1965.)

applied to disks of different sizes, each disk will have a different acceleration inversely proportional to the mass. Doubling of the force on a given disk gives rise to a doubling of the acceleration (Fig. 4–7).

It is a tribute to Galileo, Newton, and other early physical scientists to whom frictionless systems were not available that they were so successful. They demonstrated that solving idealized problems, such as those involving a nonexistent, ideal, frictionless surface, can shed light on real problems in the physical world.

4–6 APPLICATIONS OF NEWTON'S SECOND LAW

The quantity m in $\mathbf{F} = m\mathbf{a}$ is called the inertial mass, or simply mass. It is a quantitative measure of the object's inertia, that is, its resistance to changes in its state of motion. A more massive object needs more force to move it, stop it, or change its direction. Mass is a scalar quantity and has a constant value for an object.

According to Newton's second law, force is anything that can change the state of motion of a body, that is, accelerate it. In its simplest sense, a force is a push or a pull. It may start a body moving, speed it up or slow it down, stop it, or change its direction. In each case, a force produces a change in the state of motion—an acceleration—and the body is accelerated in the same direction as the force is applied. When we throw a ball we accelerate it, and the acceleration is in the direction of the applied force. Force and acceleration are both vector quantities and are in the same direction.

In SI units, a force of one newton is defined as that force which accelerates a mass of 1 kilogram by 1 meter/sec^2 (Fig. 4–8).

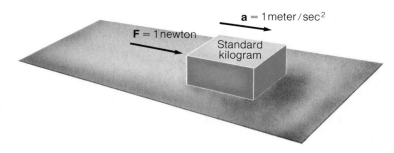

Figure 4-8 Definition of the SI
unit of force. A newton is the
force that gives a mass of one
kilogram an acceleration of one
meter per second squared.

$$1 \text{ newton} = 1 \text{ kilogram} \times 1 \frac{\text{meter}}{\text{sec}^2}$$

In cgs units (see Appendix A), a dyne is that force
which accelerates a mass of 1 gram by 1 centimeter/
sec^2.

$$1 \text{ dyne} = 1 \text{ gram} \times 1 \frac{\text{centimeter}}{\text{sec}^2}$$

The units newtons and dynes are related in the following way: Since 1 kilogram = 1000 or 10^3 grams, and
1 meter = 100 or 10^2 centimeters, then

$$1 \text{ newton} = 1 \text{ kilogram} \times 1 \frac{\text{meter}}{\text{sec}^2}$$

$$= 10^3 \text{ grams} \times 10^2 \frac{\text{centimeters}}{\text{sec}^2}$$

$$= 10^5 \frac{\text{gram-centimeters}}{\text{sec}^2}$$

$$= 10^5 \text{ dynes}$$

Table 4-1 SUMMARY OF MASS AND FORCE UNITS

MASS (m)	ACCELERATION (a)	FORCE (F = ma)	NAME OF FORCE UNIT
SI units: 1 kilogram	$1 \frac{\text{meter}}{\text{sec}^2}$	$1 \frac{\text{kilogram-meter}}{\text{sec}^2}$	newton (N)
cgs units: 1 gram	$1 \frac{\text{centimeter}}{\text{sec}^2}$	$1 \frac{\text{gram-centimeter}}{\text{sec}^2}$	dyne
	$\therefore 1 \text{ N} = 10^5 \text{ dynes}$		

Although other systems of force units are also in use, we shall confine our discussion to SI and cgs units in the present text. A summary is given in Table 4–1.

EXAMPLE   4–1

What constant net force will give a 1-kg disk an acceleration of 5 m/sec²?

SOLUTION

Analyzing the example, we see that the mass and acceleration of a disk are given, and that the force is the unknown. Set up the equation that relates the three quantities ($\mathbf{F} = m\mathbf{a}$), substitute, and solve.

$$m = 1 \text{ kg} \qquad \mathbf{F} = m\mathbf{a}$$
$$a = 5 \frac{m}{sec^2} \qquad = \left(1 \text{ kg}\right)\left(5 \frac{m}{sec^2}\right)$$
$$\mathbf{F} = ? \qquad = 5 \frac{kg\text{-}m}{sec^2}$$
$$= 5 \text{ newtons, or } 5 \text{ N}$$

The concept of weight is closely linked to Newton's second law. Not only is a body with a large mass resistant to changes in its motion; it is also hard to lift. This property is the gravitational force acting on a body, also called the weight of the body:

Equation   4–3 $$\mathbf{w} = m\mathbf{g}$$

in which $\mathbf{w}$ is the weight or gravitational force, m is the mass, and g is the acceleration due to gravity. The weight of a body, therefore, is directly proportional to its mass, and the factor of proportionality is numerically equal to g, which is nearly constant near the surface of the earth. The greater the mass of a body, the greater its weight and the harder it is to lift.

EXAMPLE 4-2

Calculate the weight of 1-kg mass when it is on the surface of the earth.

SOLUTION

The mass is given, and g, the acceleration due to gravity, may be considered a constant near the earth's surface. Knowing how weight is related to these quantitites ($w = mg$), set up the equation, substitute the appropriate quantities, and solve for **w**.

$$m = 1 \text{ kg} \qquad w = mg$$

$$g = 9.8 \frac{m}{sec^2} \qquad = \left(1 \text{ kg}\right)\left(9.8 \frac{m}{sec^2}\right)$$

$$w = ? \qquad = 9.8 \frac{kg\text{-}m}{sec^2}$$

$$= 9.8 \text{ newtons, or } 9.8 \text{ N}$$

Although mass is a constant, weight is a local property that depends upon the value of g for that locality. The value of g varies with location on the surface of the earth, increasing with latitude from 9.78 m/sec² at the equator to 9.83 m/sec² at the North and South Poles. It also varies with elevation, decreasing with elevation at a given latitude, as Table 4–2 shows. One's weight, therefore, would increase in traveling from the equator to the North Pole, and decrease in going from the surface

Table 4–2 ACCELERATION OF GRAVITY, g, AT VARIOUS HEIGHTS ABOVE THE SURFACE OF THE EARTH

HEIGHT (miles)	g (m/sec²)
0	9.80
1 000	6.27
2 000	4.35
3 000	3.26
10 000	1.57

Table 4–3   THE FOUR FORCES

FORCE	RELATIVE STRENGTH	PARTICLES AFFECTED
Nuclear	1	Nuclear particles
Electric	10^{-2}	Electrically charged particles
Weak	10^{-15}	Nuclear particles
Gravity	10^{-38}	All particles

of the earth at any location to some point above the surface. For example, you would weigh slightly more in Denver, Colorado, than at the top of Pike's Peak because, although these locations are at about the same latitude, Denver is 1600 meters above sea level as compared to 4290 meters for Pike's Peak.

A brick would have just as much mass as it has on the surface of the earth, and it would hurt your toes just as much to kick it in outer space as it would on earth. Yet the brick would not have much weight at all, since the value of g in outer space is low (see Table 4–2). If the brick were taken to the moon, its mass would not change, but since the gravitational field on the moon is only one-sixth as much as that on earth, it would weigh one-sixth as much. When you take your trip to the moon, your mass, too, will be unchanged; but if you weigh 150 pounds on earth, you will weigh only about 25 pounds on the moon.

All the forces in nature are of only four basic kinds. In addition to the gravitational forces just discussed, there are electromagnetic forces and two kinds of nuclear forces—weak nuclear and strong nuclear (Table 4–3). (Recent experiments indicate that there may be two more forces, the super-weak and the super-strong.) Regardless of its origin, each force plays the same role: it accelerates something that is sensitive to it. Since the number of forces is so small, scientists have wondered whether they can all be unified under one comprehensive theory. The idea of the unity of nature

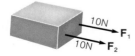

$R = 20$ N
(to the right)

Figure 4–9   Resultant of two forces F_1 and F_2 acting in the same direction.

strongly attracts the scientific mind and the search for
a unifying theory is continuing.

The vector properties of forces are further illustrated in the following examples.

EXAMPLE   4–3

A force (F_1) of 10 N, acting on a box toward the
right, combines with a similar force (F_2) of 10 N in
the same direction. Find the resultant force.

SOLUTION

The resultant force produces the same acceleration on the box as would a single force of 20 N
acting toward the right. (See Fig. 4–9.)

EXAMPLE   4–4

What is the net effect of a 10-N force (F_1)
pulling to the right, and a 3-N force (F_2) pulling to
the left?

SOLUTION

The net effect of these two forces is to produce
the same acceleration as a single force 7 N to the
right (See Fig. 4–10.)

Figure 4–10   Resultant of two forces F_1 and F_2
acting in opposite directions.

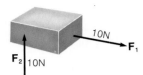

Figure 4–11   Resultant of two forces F_1 and F_2 acting at right angles.

EXAMPLE   4–5

A 10-N force (F_1) acts to the right, and another 10-N force (F_2) acts vertically upward on a body. Determine the resultant force. (See Fig. 4–11.)

SOLUTION

We apply the polygon method of vector addition discussed in Chapter 2. From the diagram we determine the magnitude of the resultant **R** and measure the angle α. We mean by this that a single force, **R**, acting upon the body at an angle of 45° east of north has the same effect as the two forces, F_1 and F_2, acting in their respective directions.

4–7   NEWTON'S THIRD LAW OF MOTION: ACTION-REACTION

Newton's *third law* is stated as follows:

> *Whenever one body exerts a force on another, the second body exerts a force equal in magnitude and opposite in direction on the first body; or, to every action force there is an equal and opposite reaction force.*

Newton's third law points out that forces always occur in pairs. For any force exerted on a body, there is a force exerted by the body; the two forces are an action-reaction pair and act on different bodies. Newton came to this conclusion from collision experiments between a moving pendulum and a stationary one. The motion of each pendulum after collision, he decided, resulted from equal forces exerted by each pendulum upon the other. Experience has shown that the third law applies to every known kind of force—gravitational, electromagnetic, and nuclear. In each instance, one of the

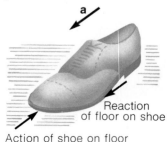

Figure 4–12   An action-reaction pair.

Figure 4-13 As a tire pushes against the road surface, the road reacts by pushing against the tire.

Reaction of road on tire

Action of tire on road

forces is arbitrarily designated the action force, and the other equal and opposite force, the reaction force.

We experience countless examples of action-reaction force pairs each day. Whenever a force is applied, a reaction force is created. Walking is possible because as we push against the floor, the floor pushes against us (Fig. 4-12). This reaction force accounts for our motion, and we depend upon friction to obtain it. Walking on ice is difficult because we cannot easily exert the necessary force on the ice, and therefore the ice cannot exert a reaction force to accelerate us. An ice skater pushes on a wall, and the wall exerts an equal and opposite force on the skater, causing the skater to accelerate away from the wall. The tires of a car push against the road surface, and the road reacts by pushing against the tires (Fig. 4-13). Some of the bruises we suffer arise from action-reaction pairs: when we accidentally kick a curbstone, the reaction force exerted by the curbstone upon our toes is unpleasant. The same action-reaction principle underlies rocket propulsion: a rocket pushes gases backward, and the equal and opposite recoil force exerted by the gases on the rocket pushes the rocket forward, even in the vacuum of space (Fig. 4-14). A spaceship maneuvers simply by firing rockets in the proper direction.

Figure 4-14 A "soda water" rocket. When the carbon dioxide cylinder is punctured, the rocket is accelerated along the wire guide.

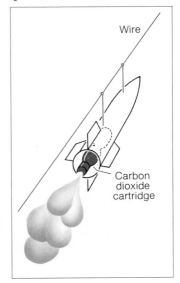

Wire

Carbon dioxide cartridge

4-8 THE "CENTER-SEEKING" FORCE

If you whirl a key at the end of a string, the key tends to fly off in a straight-line path, in accordance with the first law of motion. The fact that the key continues to move in a circle must mean that a force is acting on it that deflects it from its straight-line path (Fig. 4-15). This force, which is supplied by the hand and transmitted by the string and keeps pulling the key into the center, is called *centripetal force,* meaning "center-seeking." It is possible to show that centripetal force imparts an acceleration directed toward the center of the circle given by the expression

Equation 4–4

$$a = \frac{v^2}{r}$$

in which v is the constant speed, r is the radius of the circle, and **a** is the centripetal acceleration. The centripetal force **F** may thus be expressed as follows:

Equation 4–5

$$\mathbf{F} = \mathbf{ma} = \frac{mv^2}{r}$$

EXAMPLE 4–6

A racer takes a curve of radius 200 ft at a constant speed of 100 ft/sec. What is the centripetal acceleration?

SOLUTION

We may calculate the centripetal acceleration from Equation 4–4:

$$r = 200 \text{ ft} \qquad\qquad \mathbf{a} = \frac{v^2}{r}$$

$$v = 100 \frac{\text{ft}}{\text{sec}} \qquad\qquad = \frac{\left(100 \frac{\text{ft}}{\text{sec}}\right)^2}{200 \text{ ft}}$$

$$\mathbf{a} = ? \qquad\qquad = \frac{10^4 \frac{\text{ft}^2}{\text{sec}^2}}{2 \cdot 10^2 \text{ ft}}$$

$$= .5 \cdot 10^2 \frac{\text{ft}}{\text{sec}^2}$$

$$= 50 \frac{\text{ft}}{\text{sec}^2}$$

Figure 4–15 Centripetal force is needed to produce the inward acceleration that keeps the key moving in a circular path.

At every instant, the centripetal force is exerted at right angles to the direction of motion of the key and imparts an acceleration to it (Fig. 4–16). The centripetal force does not change the speed of the key, which remains constant; it changes only the direction of motion. If the key were not subject to the centripetal force, it would follow a straight path at constant speed along a tangent to the circle. In a sense, therefore, the key is always falling toward the center, veering away from the tangent direction that it would otherwise take (Fig. 4–17).

Centripetal force is not a new kind of force, merely a force that keeps a body moving in a curved path. In SI units, it is expressed in newtons, like any other force. It may be exerted by means of cords or springs, electricity, or gravitational attraction. If the string in the example above breaks, the centripetal force ceases to act and the key is free to travel in a straight line tangent to the path at that instant, following Newton's first law.

Like other kinds of force, a centripetal force is one member of an action-reaction pair. The reaction force is often called centrifugal force (for "center-fleeing"). The two forces, centripetal and centrifugal, are equal in magnitude and opposite in direction, in accordance with Newton's third law. They depend on the mass and the speed of the body moving in the circular path, and on the radius of curvature.

Only the centripetal force acts on the body. In the example of the string and key, the key exerts a reaction force—centrifugal force—upon the string. The two forces, therefore, act on different things: one acts on the key; the other on the string. The centrifugal force, ex-

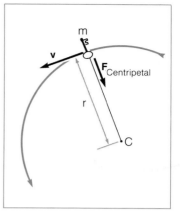

Figure 4–16 An acceleration **a** is imparted to the key, equal to v^2/r, and changes its direction. The centripetal force is exerted at right angles to the motion of the key.

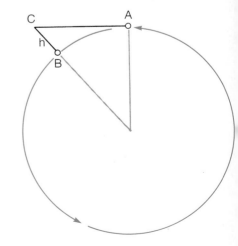

Figure 4–17 The key is always "falling" toward the center. When it moves from A to B, it falls through the distance h. If it had moved in a straight line, it would have reached point C.

erted *by* the key, and not *on* the key, has no effect upon the key's motion. The steady outward-pulling force felt by the hand—the centrifugal force as transmitted by the string—does not tend to pull the key away from the center of the circle. In the strict sense, there is no such applied force. Because of its inertia, the key is merely exhibiting its tendency to move in a straight line.

When an automobile goes around a curve, centripetal force arising from the friction between the tires and the surface acts on the car and keeps it in the curved path. The reaction force, or centrifugal force, is exerted by the car on the road. If you are a passenger not wearing a seat belt, you may feel pushed against the door—not because centrifugal force pushes you there, but because centripetal force, which would have been transmitted through the car's frame by a seat belt, is not keeping you in circular motion. You therefore tend to go in a straight line while the car travels in a curved path; consequently you and car door meet.

4–9   THE APPLE AND THE MOON

There is a famous story that while sitting in his garden and seeing an apple fall to the ground Newton came to wonder whether the force that the earth exerted on the apple might extend out to the moon. If the story is true, it may have been during the eventful period when Newton waited out the plague at Woolsthorpe and made so many remarkable discoveries. He wrote that ". . . in the same year I began to think of gravity extending to the orb of the moon. . . . "

When Newton returned to the problem years later, he found that the force of attraction between two spheres was the same as it would be if the mass of each sphere had been concentrated at the center. He therefore felt justified in considering a sphere as though its whole mass were concentrated at a point at the center. This result enabled him to treat the sun, moon and planets as point masses and to apply mathematical analysis with great precision to astronomical problems.

The moon in motion tends to follow a straight-line path, according to Newton's first law, the law of inertia. The fact that it follows a curved path meant to Newton that a force must be pulling it out of its straight-line path, just as the key described in Section 4–8 is deflected from its straight-line path.

If the earth, by its gravitational attraction, could cause the moon to "circle" the earth, perhaps it was the same force with which the earth attracted projectiles and caused an apple to fall. A similar force arising from the gravitational attraction of the sun might cause the planets to orbit around it. Newton made use of Galileo's data on falling bodies, Kepler's laws, and other findings, as well as his own experiments, intuition, and analysis, and arrived at the *law of universal gravitation:*

> *Every body in the universe attracts every other body with a force that is directly proportional to the product of their masses and inversely proportional to the square of the distance between them.*

Expressing this law symbolically, we obtain the following expressions ($\propto$ is a proportion sign):

$$\mathbf{F} \propto \frac{M_1 M_2}{r^2}$$

Equation 4–6

$$\text{or } \mathbf{F} = \frac{G M_1 M_2}{r_2}$$

in which $\mathbf{F}$ = force of attraction between two bodies
$\quad$ G = gravitational constant
$\quad$ M_1 = mass of one body
$\quad$ M_2 = mass of a second body
$\quad$ r = distance between the two bodies

Any object near the surface of the earth is approximately 4000 miles from the center (the earth's radius = 4000 miles). Since freely falling bodies near the surface have the same acceleration, 32 ft/sec², we may calculate the distance any body falls each second. In the first second of fall, the velocity is 0 at the beginning and 32 ft/sec at the end. The average velocity during the first second is therefore expressed as follows:

$$\bar{v} = \frac{\text{Initial velocity} + \text{Final velocity}}{2}$$

$$= \frac{0 + 32\frac{ft}{sec}}{2}$$

$$= 16\frac{ft}{sec}$$

and the distance traveled in the first second is given by:

$$\text{Distance} = \text{Average velocity} \times \text{Time}$$

$$s = \bar{v}t$$

$$s = 16\frac{\text{ft}}{\text{sec}} \times 1 \text{ sec}$$

$$= 16 \text{ ft}$$

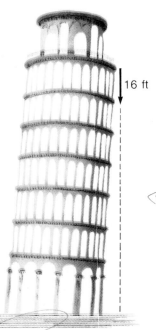

Figure 4–18   An apple falls 16 feet during the first second. How far does it fall in two seconds?

Thus, a freely falling body, such as an apple near the earth's surface, falls through a distance of 16 feet during the first second (Fig. 4–18).

Knowing that the moon's orbit is about 240 000 miles, or 60 earth-radii from the earth, and that the moon revolves around the earth once in 27.3 days, we can compute the moon's velocity in its orbit and its centripetal acceleration, v^2/r. The ratio of the moon's acceleration to the acceleration of an object near the earth's surface is found to be about 1:3600. The ratio of the moon's distance from the earth to the radius of the earth is 60:1, and $60^2 = 3600$. The acceleration, then, falls off with the square of the distance, and the force of gravitation is inversely proportional to the square of the distance.

The moon, like the apple, "falls" toward the earth, but in one second it falls 1/3600 of 16 feet, or about 1/20 of an inch. Yet, like the whirling key at the end of the string, it does not come any closer (Fig. 4–19). It falls away from the straight line it would travel if gravitational force were not acting. Newton felt justified in assigning the same cause—gravity—to the force that causes an apple to fall and that keeps the moon in its orbit.

Newton applied the law of universal gravitation to other heavenly bodies and to other phenomena. The motion of satellites about Jupiter and the motion of the planets around the sun appeared to be similar to the motion of the moon about the earth, and Newton assigned the same cause to them—gravitation. He showed that the orbits of comets were elliptical. He explained ocean tides as resulting from gravitational forces exerted by the sun and moon on the oceans. He showed how small irregularities in the motions of the planets could be explained by forces exerted on them by other planets. The law of universal gravitation and the laws of

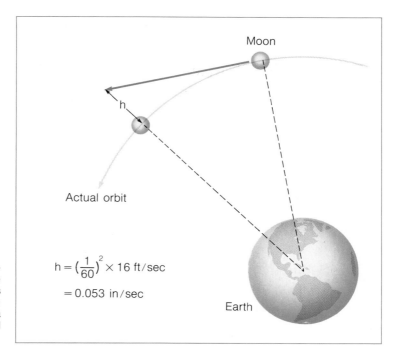

Moon

h

Actual orbit

$$h = \left(\frac{1}{60}\right)^2 \times 16 \text{ ft/sec}$$

$$= 0.053 \text{ in/sec}$$

Earth

Figure 4–19 The moon is continually "falling" toward the earth. For this reason it remains at the same distance from the earth. If the moon traveled in a straight line, it would be carried further from the earth.

motion seemed to apply to matter wherever it was found.

4–10 "WEIGHING THE EARTH"

Newton established the law of universal gravitation without being able to measure the small force of attraction between two bodies in the laboratory, but in the *Principia* he outlined a method that was followed successfully a century later. Once the value of G is known, the gravitational attraction between any two spherical bodies, whether bowling balls or planets, can be calculated, given their masses and the distance between their centers. Using this method, it is also possible to "weigh" the earth, the sun, and the other planets.

Henry Cavendish (1731–1810), during the period 1797–98, carried out the experiment to determine G so meticulously that his value differs by only about 1 per cent from the value accepted today. His experiment is as follows: Using a torsion balance, a wooden arm is suspended horizontally by a slender wire of known stiffness (Fig. 4–20). A small lead sphere of known mass and radius is mounted one each end of the arm. When large lead spheres, also of known mass and radius, are

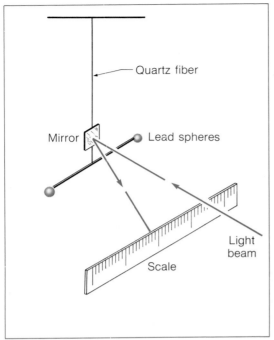
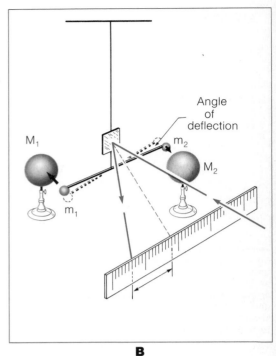

A

B

Figure 4–20   A diagram of the torsion balance used by Henry Cavendish. Knowing the masses of the spheres, the distance between their centers, and the force, one may determine the value of G.

positioned on either side of the small spheres, the suspension wire is twisted through a measurable angle by the gravitational forces between the spheres. In a separate experiment, the force necessary to twist the wire through a similar angle is determined. Knowing the masses of the spheres, and the force, the value of G—the only unknown—may be determined. The modern value, in SI units, is shown below:

$$G = 6.67 \times 10^{-11} \frac{\text{newton-meter}^2}{\text{kilogram}^2}$$

EXAMPLE 4–7

Apply the law of universal gravitation to "weigh the earth." Consider a mass M_1 of 1 kilogram on the surface of the earth, where it experiences a force of 9.8 newtons due to its weight. Assume that its distance from the center of the earth is 6.38×10^6 meters, the mean radius of the earth. Let the mass of the earth be M_2, and determine its value.

SOLUTION

Tabulate the data, then write the force equation which applies. Since the unknown is M_2 rewrite the equation with M_2 on the left and everything else on the right. Then substitute and solve.

$F \;= 9.8 \text{ N}$

$G \;= 6.670 \times 10^{-11} \dfrac{\text{N-m}^2}{\text{kg}^2}$

$M_1 = 1 \text{ kg}$

$r \;\;= 6.38 \times 10^6 \text{ m}$

$M_2 (\text{mass of earth}) = ?$

$$F \;= \frac{G M_1 M_2}{r^2}$$

$$M_2 = \frac{F r^2}{G M_1}$$

$$= \frac{9.8 \text{ N} \,(6.38 \times 10^6 \text{ m})^2}{\left(6.670 \times 10^{-11} \dfrac{\text{N-m}^2}{\text{kg}^2}\right)(1 \text{ kg})}$$

$$= 5.98 \times 10^{24} \text{ kg}$$

4-11 NEW NEIGHBORS IN SPACE

Newton's law of universal gravitation, Newton's laws of motion, and Kepler's laws form the basis of the branch of astronomy known as celestial mechanics. The motions of the planets can be calculated with such great precision that small discrepancies in their orbits have led astronomers to search for and to discover new planets.

Sir William Herschel (1738–1822) discovered a new planet, later named Uranus, in 1781. The details of its elliptical orbit were determined over the years. However, small but persistent discrepancies between the predicted orbit and actual observations appeared, for which no explanation could be found. There were even some who believed that the law of universal gravitation did not apply to the far reaches of the solar system.

In the 1840's, it occurred independently to two young astronomers, John Couch Adams (1819–1892) at Cambridge University and Urbain J. J. Leverrier (1811–1877) in Paris, that an unknown planet might be causing the perturbations of Uranus. In the solar system, every body disturbs every other body with "perturbing" forces, which affect the orbit of a planet as predicted from Kepler's laws; the departures from the predicted orbit are known as perturbations. Adams and Leverrier spent years at the difficult calculations and predicted the location of a new planet.

In October 1845, Adams wrote to the Astronomer-Royal at the Greenwich Observatory, advising him that

the perturbations of Uranus could be explained by assuming the existence of an outer planet situated then in
a specific latitude and longitude, and suggesting that a
search for the planet be conducted. Nothing came of it,
however. Later, Leverrier wrote to J. G. Galle at the
Berlin Observatory, also urging him to search for the
planet in a particular region of the sky. That very evening, September 23, 1846, after receiving Leverrier's
letter, Galle directed his telescope as Leverrier had
suggested and almost immediately found the new planet within a degree of the predicted position. The director of the observatory called it the most brilliant of all
planetary discoveries. The discovery of Neptune, as
the planet came to be known, was another triumph for
Newtonian physics, proving that Newton's laws are
valid for the entire solar system.

Within a century, small perturbations were discovered in the orbit of Neptune itself. Percival Lowell
(1855–1916) of the Lowell Observatory and, independently, William H. Pickering at the Mt. Wilson Observatory, ascribed them to yet another unknown planet.
Lowell tried unsuccessfully for years to find what he
called Planet X, and the search was continued after his
death. In 1930, Clyde W. Tombaugh at the Lowell Observatory detected an object that was identified as
Planet X and named Pluto (Fig. 4–21).

Are there other planets in the solar system? In 1972,

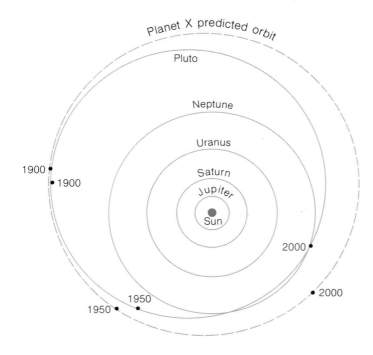

Figure 4–21   The orbit that Lowell
predicted for Planet X in 1905 compared with that observed for the new
planet, Pluto, discovered 25 years
later. The relative positions for 1900
were very close. (After Federer and
Tombaugh, *Sky and Telescope*, Sky
Publishing Corp.)

Joseph L. Brady at the University of California predicted the existence of a Planet Y based upon a computer analysis of comet perturbations, which obey the
same laws as planets. The new Planet Y is believed to
be three times the size of Saturn and could account for
perturbations in the orbit of Halley's Comet. Others
maintain that a planet of the presumed size would have
long since been discovered, and that there are other
reasons for the perturbations. Only time will tell which
view is correct.

4–12   EARTH SATELLITES

When launched at the proper angle and given
enough speed, objects may be projected for long distances. A missile may be launched from one spot on
earth and land at almost any other spot. Intercontinental
ballistic missiles (ICBM's) are designed to do just this.
Their trajectory may be measured by tracking with
radar or other means.

Theoretically, if an apple were given a suitable
horizontal speed from a high enough elevation, it could
"fall around the earth," as does the moon. Actually, a
projectile in an orbit this close to the surface of the earth
would experience so much atmospheric drag that it
would slow down and fall back to the ground. In principle, however, Newton was correct, and modern technology makes it possible to launch objects to altitudes
so high that atmospheric resistance is negligible. Under
these conditions they go into stable orbit around the
earth. In 1957, the earth satellite called Sputnik was
successfully launched. Since then, scores of other satellites have been placed in orbits around the earth.

While in orbit, even a natural satellite such as our
moon continually falls toward the earth. Its speed is
great enough, however, that it can continue to move in
orbit and not come any closer to the earth. The speed
required by a satellite to remain in orbit depends upon
its height above the earth.

The gravitational attraction of the earth holds satellites in various orbits. For circular orbits, the gravitational force is equal to the centripetal force:

$$\mathbf{F} = \underbrace{\frac{G M_1 M_2}{r^2}}_{\substack{\text{gravitational} \\ \text{force}}} = \underbrace{\frac{M_1 v^2}{r}}_{\substack{\text{centripetal} \\ \text{force}}}$$

where M_1 is the mass of the satellite, M_2 is the mass of the earth, and r is the radius of the orbit measuring from the center of the earth. Solving for the velocity of the satellite in an orbit,

$$v^2 = \frac{GM_2}{r}$$

Equation 4–7   and

$$v = \sqrt{\frac{GM_2}{r}}$$

Observe that the mass of the satellite cancels out of the equation, meaning that all satellites, regardless of mass, travel with the same velocity in a particular orbit.

The closer a satellite is to the earth, the faster it needs to travel to stay in orbit. As a spaceship moves from one orbit to another, its orbital speed undergoes a change, decreasing as it moves from a lower to a higher orbit. Near the earth's surface, a satellite must travel at more than 17 000 miles per hour to remain in orbit, while one 4000 miles above the surface orbits at 11 500 miles per hour. At the distance of the moon, the orbital speed is 2000 miles per hour. The time required to revolve about the earth, called the period of a satellite, is obtained from kinematics: time = distance/velocity, or $T = 2\pi r/v$. Satellites traveling in larger orbits have longer periods.

Hundreds of satellites have been circling the earth since the first Sputnik and Explorer ushered in the space age. Satellites have revolutionized global communications. Most television watchers have seen an athletic event or news reports beamed "via satellite" from the far corners of the world. A three-minute telephone call from New York to London that cost about $12 before the launching of the first commercial communications satellite, Early Bird, in 1965 then cost about $3.20. Satellites hovering in geosynchronous orbit relay messages over the Atlantic, Pacific, and Indian oceans to over 100 countries.

Weather pictures taken by meteorological satellites, or metsats, are displayed on almost every televised news broadcast. The GOES metsat (geostationary operational environmental satellite) beams back to ground stations detailed photographs both day and night of the entire Western hemisphere from a fixed altitude of 22 300 miles. The eye of a hurricane only 50 miles across can easily be seen in the pictures. Other satellites monitor the unfolding panorama of the weather on the earth below.

Landsats, still another type of satellite, can make fast, accurate surveys of forests and crops. Their cameras go over the same point on earth every nine days and radio data and pictures back to earth. Landsats have located deposits of copper and manganese as well as new sites for tuna and shrimp fishing.

Applied technology satellites, also known as ATS, beam educational and medical information. Others are used as navigational aids by ships and aircraft. Orbiting space laboratories and observatories are used to obtain knowledge about the universe that cannot be obtained in any other way.

CHECKLIST

Action-reaction pair
Centrifugal force
Centripetal force
Dyne
Force
G
g
Inertia
Mass
Newton (unit)

Newton's first law
Newton's second law
Newton's third law
Newton's law of universal
    gravitation
Perturbation
Planet X
Satellite
Weight

EXERCISES

4-1. Assume that you are standing in a bus. Explain why you tend to fall backward as the bus starts up.

4-2. Does an earth-orbiting space laboratory possess inertia? Explain.

4-3. The earth exerts a gravitational force on you. Identify the "reaction" force.

4-4. Explain the statement that although your mass is the same everywhere, your weight may vary from place to place.

4-5. With a diagram, show the force, F, exerted by a child pulling a sled, and indicate the horizontal and vertical components.

4–6.   (a) What is your weight? (b) What would your weight be on the moon where the acceleration due to gravity is one-sixth that on earth?

4–7.   (a) A force of 16 newtons is applied to a cart having a mass of 4 kg. What will the acceleration be? (b) If the cart started from rest, what is its speed at the end of 5 sec?

4–8.   A net force of 2500 newtons acts on a car, accelerating it 2.5 m/sec.^{2} What is the mass of the car?

4–9.   An astronaut pulls a 1500-gram camera across a spaceship. The camera is accelerated 25 cm/sec.^{2} What force is exerted?

4–10.   Once it has been in operation, why does it become easier to accelerate a rocket? (About 90 per cent of the mass of a rocket at the beginning of a launch is fuel.)

4–11.   A 1000-kg automobile accelerates from rest to 40 mi/hr (18 m/sec) in 5 sec. What force does the road exert on the car?

4–12.   A body having a mass of 50 kg moves with a constant acceleration of 2 m/sec.^{2} What force must be acting on it?

4–13.   A sprinter of mass 70 kg accelerates uniformly from a stationary start for the first 5 seconds of a race. His velocity is then 10 m/sec. Calculate the force provided by his legs.

4–14.   Explain the statement, "The moon is falling."

4–15.   Account for the fact that the moon does not travel in a straight line.

4–16.   Since the earth is attracted to the sun by gravitational force, what prevents the earth from falling into the sun?

4–17.   Does an earth satellite travel at constant speed in its orbit? Explain.

4–18.   How does the speed of an earth satellite in orbit depend upon its distance from the earth?

4–19.   Calculate the orbital velocity of an earth satellite in an orbit 400 miles above the earth.

4–20.   Multiple Choice

A. Newton's first law is a statement of the law of

(a) inertia   (c) action-reaction
(b) force   (d) gravitation

B. A measure of the inertia of a body is its

(a) speed   (c) mass
(b) friction   (d) weight

C. In Newton's third law, the action and reaction forces

(a) act on the same body
(b) act on different bodies
(c) are not necessarily equal
(d) are inversely proportional

D. A satellite is held in orbit around the earth because of

(a) a force pushing it around the orbit
(b) centrifugal force
(c) the sun's gravity
(d) the earth's gravity

E. A baseball thrown by a pitcher

(a) falls at a rate that depends on its horizontal speed
(b) begins to fall when it loses most of its speed
(c) falls just as fast as if it had no horizontal speed
(d) does not fall

SUGGESTIONS FOR FURTHER READING

Andrade, E. N. da C., *Sir Isaac Newton.* Garden City, N.Y.: Doubleday & Company, 1967.
    Discusses Newton's discoveries and the changes they made in man's view of the universe.
Franklin, Allan, "Principle of Inertia in the Middle Ages." *American Journal of Physics,* June, 1976.
    Shows that the revolution in mechanics in the seventeenth century did not occur out of thin air.
Freedman, Daniel Z., and Peter van Kieuwenhuizen, "Super-

gravity and the Unification of the Laws of Physics." Scientific American, February, 1978.

> A new theory of gravity that may lead to a unified theory of all the basic forces in nature.

Gamow, George, *Gravity.* Garden City, N.Y.: Doubleday & Company, 1962.

> Discusses the contributions of Galileo, Newton, and Einstein to our understanding of the still unresolved problem of the nature of gravity.

Thorne, Kip S., "Gravitational Collapse." *Scientific American,* November, 1967.

> The weakest force known, gravity, becomes the dominant force on an astrophysical scale and plays the role of "midwife and undertaker" in the birth and death of stars.

Weber, Joseph, "The Detection of Gravitational Waves." *Scientific American,* May, 1971.

> According to the general theory of relativity, accelerated masses should radiate gravitational waves, but the detection of these waves has been the subject of controversy.

MOMENTUM, WORK AND ENERGY

5-1 INTRODUCTION

Imagine rolling a bowling ball against one end of a row of bowling balls in contact. Only the ball at the opposite end moves; it rebounds from the far end as though it had been struck in the original impact. If two bowling balls strike the row, then two rebound at the far end. (An entertaining device based on this principle is shown in Figure 5–1.) Then imagine a bowling ball suspended like a pendulum and released from a position of rest. It rises no higher than its starting level when it completes a swing, as demonstrated in Figure 5–2. Two of the most powerful concepts in physical science are involved in such events: *momentum* and *energy*.

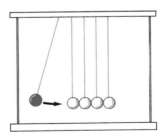
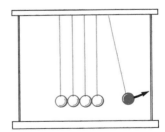
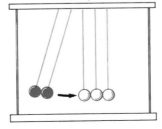
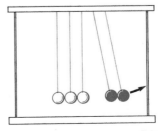

Figure 5–1 Collision spheres with a "memory." The release of one sphere on the left is followed by the ejection of one sphere on the right; two on the left, two on the right. The process then repeats from right to left.

99

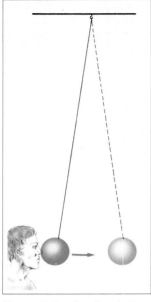

Figure 5–2 Bowling-ball pendulum. If released from the tip of one's nose with no initial velocity, the pendulum bob swings through an arc and returns just to the tip of the nose.

Equation 5–1

Table 5–1   UNITS FOR MOMENTUM

MOMENTUM	=	MASS	×	VELOCITY	UNIT
p	=	m	×	v	
SI System:		kilograms ×		$\dfrac{meters}{sec}$	$= \dfrac{kg\text{-}m}{sec}$
cgs System:		grams	×	$\dfrac{centimeters}{sec}$	$= \dfrac{g\text{-}cm}{sec}$

5–2 MOMENTUM

A bus is harder to stop than a compact car moving at the same speed. We can express this fact by saying that the bus has greater momentum than the car. *Momentum* is defined as the product of the mass of a moving body and its velocity. The greater the mass or velocity or both, the greater the momentum.

$$\text{Momentum} = \text{Mass} \times \text{Velocity}$$
$$\mathbf{p} = m\mathbf{v}$$

Like velocity, momentum is a vector quantity, and its direction is the same as the direction of the velocity. It is often convenient to choose velocities in one direction as positive and those in the opposite direction as negative; thus momentum may be either positive or negative. Some appropriate units for expressing momentum are shown in Table 5–1.

EXAMPLE 5–1

Compute the momentum of a baseball that has a mass of 0.2 kilogram and is moving at the rate of 30 m/sec.

SOLUTION

$$m = 0.2 \text{ kg} \qquad p = mv$$
$$v = 30 \text{ m/sec} \qquad\quad = (0.2 \text{ kg})(30 \text{ m/sec})$$
$$p = ? \qquad\qquad\quad = 6\frac{\text{kg-m}}{\text{sec}}$$

If the baseball in Example 5–1 were moving half as fast, at 15 m/sec, its momentum would be only half as great: 3 kg-m/sec.

5–3 IMPULSE

The concept of momentum is not recent. As a matter of fact, Newton referred to the product of mass and velocity, which we call momentum, as a "quantity of motion," and expressed his second law in terms of it. Since

$$\text{Force} = \text{Mass} \times \text{Acceleration}$$

and acceleration is equal to the rate of change of velocity, that is, the change of velocity per unit time, the second law may be stated as

$$\text{Force} = \text{Mass} \times \frac{\text{Change in velocity}}{\text{Time}}$$

or

$$\mathbf{F} = m\frac{(\mathbf{v} - \mathbf{v_o})}{t}$$

Performing the indicated multiplication by m, we obtain

$$\mathbf{F} = \frac{m\mathbf{v} - m\mathbf{v_o}}{t}$$

Therefore,

$$\text{Force} = \frac{\text{Change in momentum}}{\text{Time}}$$

and, multiplying by time,

$$\text{Force} \times \text{Time} = \text{Change in momentum} \qquad \text{Equation} \;\; 5\text{--}2$$

$$\mathbf{F} \times t = m\mathbf{v} - m\mathbf{v_o}$$

A force, then, is required to change the momentum of a body, and the force must act in an interval of time. The product of force and time is known as *impulse*. Thus,

$$\text{Impulse} = \text{Change in momentum} \qquad \text{Equation} \;\; 5\text{--}3$$

The change in momentum of the body on which the force is acting is a measure of the impulse of the force. If the body starts from rest, the impulse of the force is just the momentum the body acquires. Some appropriate units for impulse are the newton-second and the dyne-second.

In a sport such as tennis, we often want to impart as large a momentum as possible to an object. Since a large impulse produces a large change in momentum, the impulse is made as large as possible by extending the time of contact. A large force, for example, is applied for a maximum amount of time with a tennis racket, baseball bat, or golf club (Figs. 5–3 and 5–4). A player who "follows through" on a stroke is thus more likely to deliver a maximum impulse.

Figure 5–3 An example of an impulsive force. A large impulse delivered by the tennis racket sets the tennis ball in motion with a large momentum. Note how the racket and ball are distorted during contact in this high-speed photograph. (Courtesy of Dr. Harold Edgerton, M.I.T., Cambridge, Mass.)

Figure 5–4 Golf club and ball in contact during delivery of impulse. The ball recovers its normal shape as it moves away from the club. (Courtesy of Dr. Harold Edgerton, Massachusetts Institute of Technology, Cambridge, Mass.)

EXAMPLE 5–2

A 60-gram golf ball is subjected to an impulse that causes it to move with a velocity of 50 m/sec. If the golf club and the ball are in contact for 5×10^{-3} seconds, what average force acts on the ball?

SOLUTION

In order to maintain all units in the same system (SI in this case), express the mass of the golf ball in kilograms. Since the golf ball starts from rest, $v_0 = 0$. Then apply Equation 5–2:

$m = 60g = 0.060 kg$

$v = 50$ m/sec

$v_0 = 0$

$t = 5 \times 10^{-3}$ sec

$F = ?$

$$F \times t = mv - mv_0$$

$$F = \frac{mv - mv_0}{t}$$

$$= \frac{(0.060 \text{ kg})\left(50\frac{m}{sec}\right) - 0}{5 \times 10^{-3} \text{ sec}}$$

$$= 600\frac{\text{kg-m}}{\text{sec}^2}$$

$$= 600 \text{ newtons}$$

5-4 CONSERVATION OF MOMENTUM

Think of two bowling balls, A and B, that have equal masses, each suspended to form a simple pendulum (Fig. 5-5). If we allow A to swing and collide head on with B, then just before the collision, the momentum of the system ("system" in this case refers to the combination of A and B) is due to the motion of A and may be written:

$$\underbrace{\mathbf{p} = mv_A}_{\text{before collision}}$$

Following the collision, A comes to rest and B moves to the right with exactly the same velocity that ball A had just before the collision:

$$\underbrace{\mathbf{p} = mv_B}_{\text{after collision}}$$

In a perfectly elastic collision (which does not occur in nature but is often closely approximated) ball B would rise to exactly the same level from which ball A descended. In a real collision, air resistance and other factors cause B to deviate from the ideal situation, so that B does not quite reach the same velocity.

Figure 5-5 Collision of two bowling balls of equal mass, A and B, one moving and the other at rest, *(A)* before and *(B)* after impact. Ball B, which had been at rest, moves off with the same velocity that A had at the time of impact, while A comes to rest.

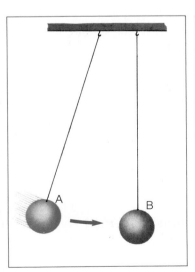

A

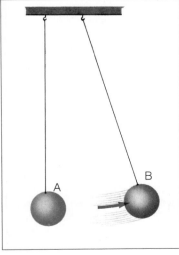

B

During the impact between A and B, the only forces acting are the force that A exerts on B and the force that B exerts on A. These two forces, by Newton's third law, are equal in magnitude but opposite in direction. Since they act for the same short time, they impart equal impulses to A and B. Equal impulses, accordingly, create equal changes in momentum. If one ball gains momentum, the other loses the same amount of momentum and the two changes cancel each other. The overall result is that the total momentum of the system remains constant. A quantity whose total amount does not change in some process is said to be conserved. Thus, momentum is conserved. In all collisions that have ever been studied, the *law of conservation of momentum* has been found to apply:

The total momentum of a system (isolated from outside forces) after interaction is equal to the total momentum of the system before interaction.

EXAMPLE 5–3

An astronaut and his equipment have a total mass of 100 kilograms in free space. If the astronaut removes his oxygen tank (mass = 8 kg) and throws it away with a velocity of 5 m/sec, what velocity does he acquire?

SOLUTION

Since the mass of the oxygen tank is 8 kg, the mass of the astronaut and remaining equipment must be 100 kg − 8 kg = 92 kg. Knowing the velocity which the oxygen tank acquires, we can apply the law of conservation of momentum to determine the recoil velocity of the astronaut. Since the total momentum of the system (the astronaut and oxygen tank) is constant, there can be no net gain or loss of momentum. The momentum of the astronaut must be equal to the momentum of the oxygen tank, but opposite in direction, or sign; then the total momentum after the event will equal the total momentum before the event, and momentum will be conserved. Thus, using the following equation:

Momentum of oxygen tank + Momentum of astronaut = Total momentum before event and after event

$$m_1v_1 + m_2v_2 = 0$$

we can substitute the values given, and solve:

$m_1 = 92 \text{ kg}$ $m_1v_1 = -m_2v_2$

$m_2 = 8 \text{ kg}$ $v_1 = -\dfrac{m_2v_2}{m_1}$

$v_2 = 5\dfrac{m}{sec}$

$v_1 = ?$ $= -\dfrac{(8 \text{ kg})\left(5\dfrac{m}{sec}\right)}{92 \text{ kg}}$

$$= -\dfrac{40\dfrac{m}{sec}}{92}$$

$$= -0.44\dfrac{m}{sec}$$

The minus sign means that the astronaut acquires a velocity opposite in direction to that of the oxygen tank.

For a collision between two bodies, the sum of the momenta of the two colliding bodies remains constant, that is:

Total momentum before collision = Total momentum after collision

or, in symbols:

$$m_1v_1 + m_2v_2 = m_1V_1 + m_2V_2 \qquad \text{Equation 5-4}$$

where m_1 and m_2 are the masses of the colliding bodies; v_1 and v_2 are their velocities before impact; and V_1 and V_2 are their velocities following impact. Although the momentum of each of the bodies may change, the sum of the two momenta is the same after the collision as it was before (Fig. 5-6).

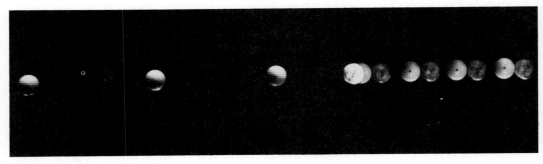

Figure 5–6 Photograph of a golf ball colliding with a putty ball. The golf ball moved in from the left and collided with the putty ball. The two balls moved off to the right, stuck together. (From *PSSC Physics,* D. C. Heath and Company, Lexington, Mass., 1965.)

Referring again to the two bowling balls, what would happen if we place some putty on bowling ball B at the point at which A strikes it, and repeat the experiment? A and B will move off together with a lower velocity than A possessed at the time of impact (Fig. 5–7). Knowing the velocity of A at the time of impact, we can calculate the velocity of the two bowling balls following impact from the law of conservation of momentum. For this sytem,

$$m_A v_A + m_B v_B = m_A V + m_B V$$

where V is the velocity of bowling balls A and B after they collide. Since B is stationary, $v_B = 0$, and $m_b v_B = 0$. Thus

$$m_A v_A = (m_A + m_B) V$$

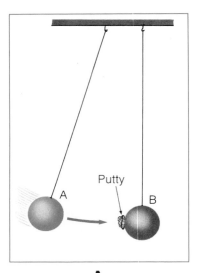

A

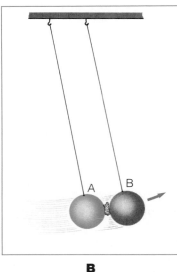

B

Figure 5–7 In this inelastic collision, following impact, A and B move off together with the same velocity (but lower than the velocity which A possessed at impact).

and

$$V = \frac{m_A v_A}{m_A + m_B}$$

EXAMPLE 5–4

A car traveling 30 km/hr strikes a parked car of equal size from the rear and the two cars lock bumpers. If the mass of each car is 500 kg, what is the common velocity of the two cars after the collision?

SOLUTION

The velocity of the parked car before the collision is 0. Apply the law of conservation of momentum to find the common velocity after collision.

$m_A = 500 \text{ kg}$ $m_A v_A + m_B v_B = m_A V + m_B V$

$m_B = 500 \text{ kg}$ $m_A v_A + m_B v_B = (m_A + m_B) V$

$v_A = 30 \text{ km/hr}$

$v_B = 0$ $V = \dfrac{m_A v_A + m_B v_B}{m_A + m_B}$

$V = ?$

$$= \frac{\left(500 \text{ kg} \cdot 30 \dfrac{\text{km}}{\text{hr}}\right) + (500 \text{ kg} \cdot 0)}{1000 \text{ kg}}$$

$$= \frac{15\,000 \dfrac{\text{km}}{\text{hr}}}{1000}$$

$$= 15 \frac{\text{km}}{\text{hr}}$$

Anything that travels in a curved path, such as a planet around the sun, an electron around a nucleus, or a whirling rock at the end of a string, has angular momentum. Just as for linear momentum, which applies to bodies moving in a straight line, angular momentum

Table 5–2 WORK UNITS

WORK W	= FORCE × DISTANCE = F × s	UNIT
SI System:	newtons × meters = newton-meters	= joules
cgs System:	dynes × centimeters = dyne-centimeters	= ergs

remains constant in a system in which no external forces are acting. That is, angular momentum is conserved.

5–5 WORK

There are two aspects of the effect of a force. We have previously discussed one of these, which is the impulse (force × time). The second, the product of a force and the distance through which it acts (force × distance), is a measure of the work done. The concept of work is one of the most useful in physical science.

Work is defined as the product of the net force exerted on a body and the distance through which the force acts. Work is a scalar quantity.

Work = Force applied × Distance through which the force acts

Equation 5–5

$$W = \mathbf{F} \cdot \mathbf{s}$$

Table 5–2 gives some units for work.

The unit joule (J) (rhymes with "rule") is named after James Prescott Joule (1818–1889). Erg is from *ergon*, the Greek word for work.

$$1 \text{ joule} = 1 \text{ newton} \times 1 \text{ meter}$$
$$1 \text{ joule} = 10^5 \text{ dynes} \times 10^2 \text{ cm}$$
$$= 10^7 \text{ dyne-cm}$$
$$= 10^7 \text{ ergs}$$

EXAMPLE 5–5

A boy pulls a sled. Assuming that the horizontal component of the force he applies is 15 newtons, how much work does he do in pulling the sled 300 meters?

SOLUTION

The work done on the sled depends upon the net force exerted in the horizontal direction and the distance in the same direction.

$$F = 15 \text{ N} \qquad W = F \cdot s$$
$$s = 300 \text{ m} \qquad = (15 \text{ N})(300 \text{ m})$$
$$W = ? \qquad = 4500 \text{ N-m}$$
$$= 4500 \text{ joules}$$

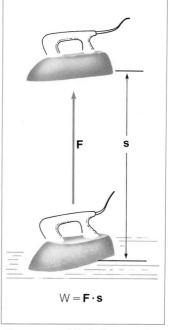

$$W = F \cdot s$$

Figure 5–8 Work done against gravity. The force **F** is applied vertically upward, moving the iron through a distance **s.** When the displacement is in the same direction as the applied force, the work done is W = **F · s.**

As defined in the physical sense, work is associated with both force and motion. Whenever work is done, a net force is exerted and something is moved by that force. When you push on a table and move it, you are performing work on it.

There are many cases in which you may exert a force and yet do no work. For example, you may push hard against a car, but if it does no move you have done no work on the car: the distance is zero; therefore the product of force × distance is zero. When you lift an iron, you do work against the earth's gravity (Fig. 5–8). But when you walk carrying the iron, you are doing no work against gravity. The force you exert on the iron is vertical and has nothing to do with the horizontal motion.

When you push a lawn mower you do work on it, even though you apply the force at an angle to the di-

Figure 5–9 Only the component of the applied force, **F,** in the direction of motion of the lawn mower, $F_{Horizontal}$, is used to do work on the lawn.

rection of motion (Fig. 5–9). The lawn mower moves horizontally because the applied force has a component parallel to the direction of motion. Only when the applied force is at right angles (90°) to the direction of motion will it lack a component parallel to the direction of motion; in such cases work, in the physical sense, is not done.

5–6 POWER

The total *work* done for a given purpose is often less important than the *rate* at which it is done. Given time, an automobile engine can do as much work as is done by the jet engine of an airliner during a New York-Chicago flight, but it could not do the work fast enough to get the airliner off the ground. *Power* is defined as the rate of doing work:

Equation 5–6

$$\text{Power} = \frac{\text{Work}}{\text{Time}}$$

$$P = \frac{W}{t}$$

The units for power were suggested by Thomas Savery (1650–1715), whose pumping engine was the first device to use steam power in industry. Since horses had been used to pump water in draining coal mines, Savery proposed as a standard of power the rate at which a horse could do work. James Watt (1736–1819) improved the steam engine, and in trying to sell his engines to mine owners was often asked, "If I buy one of your engines, how many horses will it replace?" To find out, Watt took Savery's suggestion and harnessed strong work horses to a load. He found that an average horse walking at the rate of 2½ miles per hour would steadily exert a 150-pound force for several hours. The rate at which a horse performed work became the standard *horsepower* (hp). It has a value of 33 000 foot-pounds per minute, or 550 foot-pounds per second. In SI units, the unit of power is defined as that capable of performing work at a rate of 1 joule/second, or 1 watt:

$$1 \frac{\text{joule}}{\text{second}} = 1 \text{ watt}$$

and   1 horsepower = 746 watts

A larger unit of power, the kilowatt (kw), is useful in rating engines and motors; it is equal to 1000 watts. A still larger unit, the megawatt (Mw), 10^6 watts, is applied to the rating of power plants. Although they are basically mechanical units, the watt and kilowatt can be used to express any power unit, such as electric power.

EXAMPLE   5–6

An 80-kilogram man runs up a flight of stairs 5 meters high in 10 seconds. What is the man's power output in watts and in horsepower?

SOLUTION

Since work = force × distance, the man's weight (force) must be considered. We can determine his weight, **w**, from the product of his mass and the acceleration due to gravity. Knowing the time, we can also determine his power and express it in appropriate units.

$$m = 80 \text{ kg} \qquad P = \frac{W_{ork}}{t}$$

$$s = 5 \text{ m}$$

$$= \frac{F \cdot s}{t}$$

$$g = 9.8\frac{m}{\sec^2}$$

$$= \frac{(mg)s}{t}$$

$$t = 10 \text{ sec}$$

$$\mathbf{F = w = mg} \qquad = \frac{(80 \text{ kg})\left(9.8\dfrac{m}{\sec^2}\right)(5 \text{ m})}{10 \text{ sec}}$$

$$= 392 \frac{\text{newton-meters}}{\sec}$$

$$= 392 \frac{\text{joules}}{\sec}$$

$$= 392 \text{ watts} \times \frac{1 \text{ hp}}{746 \text{ watts}}$$

$$= 0.525 \text{ hp}$$

$P = \dfrac{w}{t}$

$P = \dfrac{F \cdot S}{t}$

$P = \dfrac{(mg)s}{t}$

$W =$

$P = \dfrac{80kg\left(9.8m/sec\right)\left(5m\right)}{10 sec}$

$P = 392 \, N \cdot m$

$= 392 \, j$

$HP = \dfrac{392 \, j}{746}$

$= .525 HP.$

5–7 KINETIC ENERGY

Whenever work is done, a net force is exerted on a body over a distance and something is moved by that force. For example, when you throw a ball, your hand exerts a force on the ball and this force acts through a distance. The work you do on the ball causes it to leave your hand with a certain velocity. Because of its motion, the ball then has the capacity to do work on another object it strikes; for example, it may shatter a window. The ability to do work is energy. The energy of motion is known as *kinetic energy,* a term introduced by Lord Kelvin (1824–1907).

The connection between work and kinetic energy is illustrated as follows. When work is done on a body at rest, it is accelerated to some velocity and acquires KE. If the body is already moving, the work goes into increasing the KE. Since we know that work is going into increasing motion, might it be possible to express the energy acquired in terms of the amount of motion? The gain in the KE of the body is equal to the product of the force and the distance s through which the force acts, that is, the work done on the body:

$$KE = \mathbf{F} \cdot s$$

Starting from rest, the distance s for constant acceleration is $s = \frac{1}{2}at^2$. Therefore,

$$KE = \mathbf{F} \times \frac{1}{2}at^2$$

From Newton's second law, $\mathbf{F} = ma$; thus

$$KE = (ma)\left(\frac{1}{2}at^2\right) = \frac{1}{2}ma^2t^2$$

Substituting $v = at$ from kinematics, we get

Equation 5–7
$$KE = \frac{1}{2}mv^2$$

Kinetic Energy = one-half the mass times the velocity squared

Some appropriate units for kinetic energy are given in Table 5–3. So, kinetic energy is measured in the same

Table 5–3 SOME UNITS FOR KINETIC ENERGY

M	*V*²	½*MV*²	*UNIT*
SI System: kilograms $\times \left(\dfrac{\text{meters}}{\text{second}}\right)^2 = \dfrac{\text{kg-m}^2}{\text{sec}^2}$		$\dfrac{\text{kg-m}}{\text{sec}^2} \times$ m $=$ newton-meter $=$ joule	
cgs System: grams $\times \left(\dfrac{\text{centimeters}}{\text{second}}\right)^2 = \dfrac{\text{g-cm}^2}{\text{sec}^2}$		$\dfrac{\text{g-cm}}{\text{sec}^2} \times$ cm $=$ dyne-cm $=$ erg	

units as those for work. The distinction between energy and work is that energy is something a body *has*, whereas work is something a body *does*. A body can do work if it has energy, work being a measure of how much energy is transformed from one system to another.

EXAMPLE 5–7

A girl on skis reaches the bottom of a hill going 20 m/sec. If her total mass, including equipment, is 60 kilograms, what is her kinetic energy?

SOLUTION

$m = 60$ kg

$v = 20$ m/sec

KE = ?

$$KE = \frac{1}{2}mv^2$$

$$= \frac{1}{2}(60 \text{ kg})\left(20\frac{\text{m}}{\text{sec}}\right)^2$$

$$= 12\ 000\frac{\text{kg-m}}{\text{sec}^2} \cdot \text{m}$$

$$= 12\ 000 \text{ newton-meters}$$

$$= 12\ 000 \text{ joules}$$

5–8 POTENTIAL ENERGY

A stationary object may possess energy by reason of its position. In the process of raising an object—a pendulum, a sphere on an incline, a book—from one level to another, work is done on the body and its energy of position is increased (Fig. 5–10). In the process of

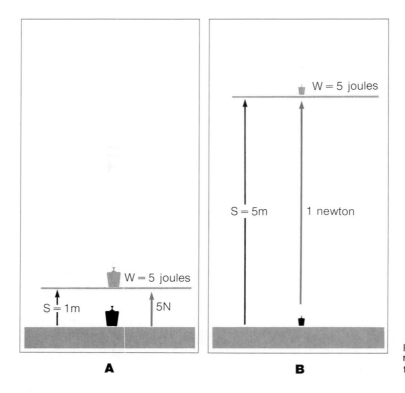

A   B

Figure 5–10 The work done in raising two different masses is the same in each case: 5 joules.

falling, this energy is converted to energy of motion (KE). William Rankine's (1820–1872) suggestion that such "stored" energy be called potential energy (PE) found general favor, and this term has been used ever since.

We see this principle applied in a pile driver (Fig. 5–11). There, a ram acquires potential energy when raised to an elevated position, which is converted to kinetic energy in the process of falling and does work upon a beam. When the ram strikes the beam, it moves.

This energy which is due to position is called gravitational potential energy. The body has PE by virtue of the earth's gravitational attraction. It takes work to separate it from the earth. The gravitational PE of a body is equal to the product of its weight (W) and its height (h) above the initial position. The reason for this is that you must exert a force on it at least equal to its weight and move it through a certain distance h.

Equation 5–8   Gravitational potential energy = Weight × Height

Since weight = mg,

$$PE = mgh$$

The work done against the gravitational force is stored by the body raised as potential energy. The body then has the ability to do work on something else.

5–9 CONSERVATION OF ENERGY

When a pendulum bob is raised to a certain height and released, its motion consists of a series of alternations between potential and kinetic energy (Fig. 5–12). The bob possesses maximum potential energy and zero kinetic energy at the two extremes of its swing. At the midpoint of its swing, the kinetic energy reaches its maximum value, and at that point the potential energy is at a minimum. As the bob travels up the arc, kinetic energy is converted into potential energy. In the absence of friction, the potential energy of the bob at the highest points in the arc is equal to the kinetic energy at the lowest point, and at all times the sum of the potential energy and kinetic energy is a constant. These energy relations are expressed in the statement of the *principle of conservation of mechanical energy:*

The sum of kinetic energy and potential energy is a constant if friction is not present.

$$KE + PE = \text{a constant}$$

or

$$KE + PE = E \, (\text{total energy})$$

Here are some other applications of energy transformations. The potential energy of water stored behind a dam, because of the elevated position of the water, is

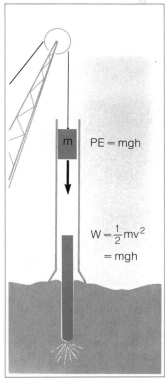

Figure 5–11 A pile driver does work. The potential energy of the block is transformed into kinetic energy by falling, and drives the post into the ground.

Equation 5–9

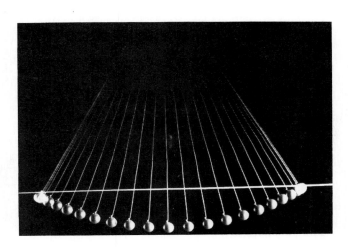

Figure 5–12 Stroboscopic photograph of a swinging pendulum. The bob has maximum potential energy at the extremes of each swing, and maximum kinetic energy at the midpoint. (Courtesy of Fundamental Photographs, from The Granger Collection.)

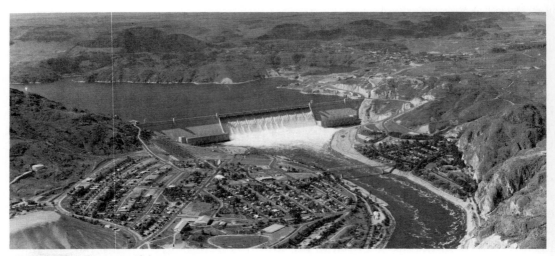

Figure 5–13 The potential energy of water at the top of the Grand Coulee Dam is converted to kinetic energy as it falls, which then does work on an electric generator. (Courtesy of U.S. Department of Agriculture.)

Figure 5–14 Some forms of energy and their conversion pathways.

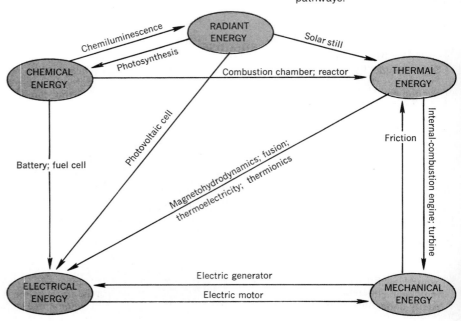

transformed into kinetic energy as the water falls (Fig. 5–13). An automobile perched at the top of a steep hill acquires kinetic energy if it rolls down the hill. A coiled spring contains potential energy owing to its configuration: when it is released, the spring can do work, that is, produce motion. In each of these examples, the potential energy is gravitational or mechanical in origin.

We have seen that an ideal pendulum would swing to precisely the same height when it reaches the end of its arc that it was released from initially. But a real pendulum does not; the swings become shorter and shorter and eventually stop. Where has the "missing" energy gone?

Whenever anything moves through a gas or a liquid, as a pendulum bob does in air, it encounters resistive or frictional forces. Whenever one surface slides or rolls over another, as a metal sphere does rolling down an incline, it also encounters resistive forces. When frictional forces work on a system, its total mechanical energy decreases. In performing work against these forces, the objects dissipate some of their mechanical energy. Some of the "missing" mechanical energy of the system is converted to an equivalent amount of energy in another form: heat energy.

Energy takes several different forms, all of which will be discussed in this text. Some of the forms of energy are mechanical energy, heat, sound, electromagnetic energy, chemical energy, and nuclear energy (Fig. 5–14). These various forms of energy can be transformed from one form to another. The kinetic energy of falling water can be transformed into electrical energy, which in turn can be transformed into light or heat energy. The chemical energy of gasoline can be converted into the kinetic energy of an automobile. Whenever energy is transformed from one form to another, no energy is destroyed: that is, the total amount of energy is conserved. It can very well be lost as far as the user is concerned, but it is around someplace in the universe. The result of considerable work on energy transformations is the single most powerful generalization in physical science, the *law of conservation of energy:*

> *The total energy content of a closed system is constant. Energy is never created or destroyed. Energy may be transformed from one kind into another, but its total magnitude remains the same.*

CHECKLIST   Conservation of energy   Impulse
Conservation of mechanical   Joule
    energy   Kinetic energy
Conservation of momentum   Momentum
Energy   Potential energy
Erg   Power
F × s   Watt
F × t   Work
Horsepower

EXERCISES   5–1.   What is the momentum of a 100-kilogram fullback moving at 4 m/sec?

5–2.   A 100-gram lump of clay moving eastward at 10 m/sec collides with a 150-gram lump of clay moving in the same direction at 5 m/sec. The two coalesce after impact. With what velocity does the combined mass move?

5–3.   Why does an automobile have to burn fuel continuously just to maintain a constant speed?

5–4.   Explain the spinning action of a lawn sprinkler.

5–5.   A heavy body and a light body have the same momentum. Which has the greater velocity?

5–6.   A man pushes a box across the floor by exerting a 10-newton force through a distance of 5 meters. How much work has he done?

5–7.   How much power does a 70-kilogram person develop by running up a 5-meter stairway in 10 seconds?

5–8.   An electron is traveling toward the screen of a TV set with a velocity of 2×10^6 m/sec. If its mass is 9.1×10^{-31} kilograms, what is its kinetic energy?

5–9.   You roll a bowling ball down the alley. If its mass is 6 kilograms, and it is moving with a velocity of 6 m/sec, what is its kinetic energy?

5–10.   A 60-gram ball is dropped 100 meters from rest. (a) How much potential energy is lost in the fall? (b) What speed does the ball have after it has fallen 100 meters?

5–11.   A helicopter lifts a load of supplies through a vertical distance of 500 meters. What potential energy does the load acquire if its mass is 100 kilograms?

5–12.   Why does a swinging pendulum eventually come to rest?

5–13.   A baseball pitcher throws a 140-gram ball at a speed of 50 m/sec. How much energy does the catcher absorb when he catches the ball?

5–14.   An athlete runs the 100-meter dash in 10 seconds. What is his kinetic energy while running if his mass is 75 kilograms?

5–15.   For what quantities are the following units appropriate? (a) kg-m/sec; (b) kg-m/sec²; (c) kg-m²/sec².

5–16.   A car (total mass = 1000 kilograms) is traveling at a speed of 50 m/sec. What is the car's KE?

$$\frac{1}{2}(1000)(50\ m/sec)^2$$
$$(1000)(2500\ m^2/sec^2)$$
$$1000 \times 1250\,000$$
$$1.25 \times 10^6\ kg\text{-}m^2/sec^2$$

5–17.   Is it possible for a body to have energy without having momentum?

5–18.   A 500-gram mass is dropped from a height of 10 meters. What is its kinetic energy just before it hits the ground?

5–19.   Is the energy that any mechanical device "wastes" destroyed? Explain.

5–20.   Multiple Choice

   A. Work is
      (a) energy times distance
      (b) force times distance
      (c) force times time
      (d) momentum times distance

B. You push against Plymouth Rock with a force of 100 newtons for 20 seconds. If the rock does not move, how much work have you done?
(a) 2000 joules   (c) 1 joule
(b) 5 joules   (d) 0 joule

C. When an automobile's speed is doubled, its kinetic energy is
(a) twice as large   (c) half as large
(b) four times as large (d) tripled

D. The rate at which work is done is called
(a) momentum   (c) kinetic energy
(b) potential energy   (d) power

E. Energy is
(a) lost if heat is produced
(b) a form of power
(c) conserved in a closed system
(d) equal to mv

SUGGESTIONS FOR FURTHER READING

The September 1971 issue of *Scientific American* is devoted entirely to articles on energy and power, among which are the following:

Dyson, Freeman J., "Energy in the Universe."
Shows that energy on the earth is part of the energy in the universe.
Starr, Chauncey, "Energy and Power."
Discusses the role of energy in human life: past, present, and future.
Summers, Claude M., "The Conversion of Energy."
Light bulbs, automobile engines, and home furnaces are a few of the devices that convert one kind of energy into another.

HEAT—A FORM OF ENERGY

6-1 INTRODUCTION

Although inventors have made great claims, a perpetual motion machine—a device that would put out more energy than it takes in—has never been built. Whenever a machine performs work a part of the energy put into the machine seems to disappear as friction converts it into heat. The useful work output, therefore, must be less than the original energy put into the machine. The efficiency of a machine, defined as the ratio of work output/work input, is thus always less than 100 per cent. Since this frictional loss in the form of heat always occurs, it is important to study the subject of heat in some detail. What is heat? How is it measured? How does it affect life?

6-2 TEMPERATURE MEASUREMENT

Our sense perceptions enable us to tell that an ice-cube tray is cold, while the sand on a sunny beach is hot. That such judgments can be ambiguous at times is indicated by a demonstration. Set up three pans of water—hot, cold, and lukewarm (Fig. 6–1). Place your right hand in the pan of hot water, A, and your left hand in the cold water, C, for the same length of time. Then place both in the lukewarm water, B. To the hand that had been in A, B seems cool, while to the hand that had been in C, B seems warm. Since subjective feelings of warmth and coldness can be misleading, an objective instrument is needed to measure temperatures.

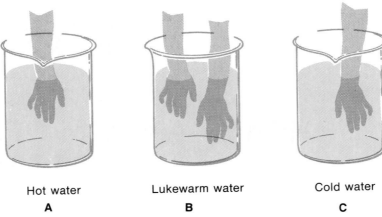

Hot water   Lukewarm water   Cold water

A   **B**   **C**

Figure 6–1   Your hand makes a poor thermometer. Subjective feelings of warmth and coldness can be misleading.

Figure 6–2   Galileo's thermoscope. When the bulb is warmed, the air expands and the water level drops; when the bulb cools, the air contracts and the water rises in the tube.

Air

Water

Thermometers depend on some property of matter that varies with temperature in a measurable way, such as expansion (gases, liquids, or solids), color change, or electrical resistance. Also, thermometers must have a reproducible scale, so that readings on different thermometers can be compared. For this purpose, two fixed points are usually calibrated, with the interval between them subdivided into equal degrees.

Probably the earliest thermometer was an instrument invented by Galileo, the air thermometer, or thermoscope. It has a glass bulb filled with air and a long stem with its end dipped into a reservoir of colored water (Fig. 6–2). When the bulb is warmed, the air expands and forces the water level to drop. Cooling the bulb causes the air to contract and the water level in the stem to rise. The instrument is surprisingly responsive to temperature changes, although Galileo apparently did not calibrate the stem. Santorio Santorio, a pupil of Galileo and a professor of medicine, used a variation of the thermoscope to indicate changes in body temperature during illness, thus making it the first clinical thermometer.

When Gabriel Daniel Fahrenheit (1686–1736) perfected a method of cleaning mercury, a substance that had been found to be particularly well suited to thermometers, he introduced the mercury thermometer into general use (Fig. 6–3). A quantity of mercury is enclosed within a bulb and extends into a narrow column in a glass tube. The volume of mercury responds uniformly to temperature changes. Even a small change in volume brought about by warming or cooling will cause a visible change in the length of the column of liquid in the tube. When the thermometer is placed in

a cup of hot water, both the mercury and the glass expand, but the mercury expands more than the glass and
rises in the stem of the thermometer. When the thermometer is cooled, the mercury contracts more than
the glass and the level of mercury drops. While a clinical thermometer contains mercury as the thermometric
substance, other household thermometers are more
likely to contain alcohol made more visible by a dye,
usually red or blue.

6–3   TEMPERATURE SCALES

While experimenting with a number of systems as
possible fixed points for the scale of his thermometers,
Fahrenheit happened to visit the astronomer Olaf
Römer (1644–1710) and found him calibrating thermometers. With some minor changes, Fahrenheit
adopted the same principles in his own work. Römer
had chosen the temperature of a mixture of ice, salt,
and water as the lower fixed point, and the boiling point
of water as the higher fixed point. On Fahrenheit's
scale, the freezing point of water, or ice point—the
temperature of a mixture of pure ice and water at standard atmospheric pressure—became 32 degrees (Fig.
6–4). The boiling point of water, or steam point—the

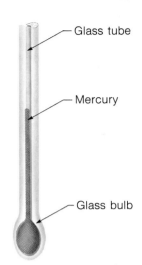

Figure 6–3 A mercury thermometer, a liquid-in-glass thermometer. As the mercury in the
bulb expands or contracts because its temperature changes,
the level in the narrow column
rises or falls.

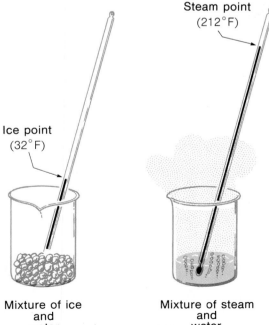

Figure 6–4 Calibrating a thermometer on the
Fahrenheit scale. By definition, a mixture of ice
and water is at 32°F, and a mixture of steam and
water at 212°F.

Ice point
(32°F)

Steam point
(212°F)

Mixture of ice
and
water

Mixture of steam
and
water

temperature of steam from pure boiling water at standard atmospheric pressure—was assigned the value of
212 degrees. There are, therefore, 212 − 32, or 180
degrees between the steam point and the ice point.

Another temperature scale has been even more
widely adopted. Anders Celsius (1701–1744) proposed
a scale based on 100 degrees between the steam point
and the ice point. Although Celsius suggested that the
boiling point of water be assigned the value 0 degrees,
and the freezing point 100 (possibly to avoid negative
temperatures below the melting point of ice), his col-
league, Märten Strömer, inverted the scale and identified 0 with the freezing point and 100 with the boiling
point of water. The Celsius-Strömer scale is called
simply the Celsius scale (Fig. 6–5).

Since there are 100 Celsius degrees between the
steam point and the ice point, and 180 Fahrenheit degrees between these two points, each Celsius degree
is nearly twice as large as a Fahrenheit degree or, more
exactly, 180/100 = 9/5 as large. To convert temperatures
from one scale to the other, this is one factor we must
consider. A second factor is the difference in the cali-
bration of the two scales. On the Celsius scale, the ice

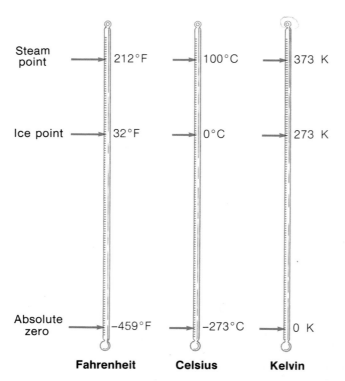

Figure 6–5 Comparison of the Fahrenheit, Celsius, and Kelvin scales.

point is 0°, but on the Fahrenheit scale it is 32°. The number 32, therefore, also enters into the conversion factor.

Suppose we want to convert a reading of 10°C to its equivalent in °F. We are 10 degrees above the freezing point of water. On the Fahrenheit scale, we are 9/5 (10) or 18 degrees above the freezing point of water, 32°F. The reading of 10°C corresponds, therefore, to 32 + 18 = 50°F. On the other hand, suppose we want to convert a reading of 95°F to its equivalent in °C. On the Fahrenheit scale we are (95 − 32) = 63 degrees above the ice point. On the Celsius scale, we are 5/9 (63) or 35 degrees above the ice point, 0°C. Therefore, 95°F is equivalent to 35°C. Thus by simple reasoning you can convert from one scale to the other without any use of formulas. You are encouraged to practice this until it is easy so you need not rely on the following conversion formulas:

$$°F = \frac{9}{5}°C + 32°$$ Equation 6–1

$$°C = \frac{5}{9}(°F − 32°)$$ Equation 6–2

EXAMPLE 6–1

A patient's temperature is 100°F, slightly above normal (98.6°F). What is the equivalent temperature on the Celsius scale?

SOLUTION

$$°C = \frac{5}{9}(°F − 32°)$$

$$= \frac{5}{9}(100° − 32°)$$

$$= \frac{5}{9}(68°)$$

$$= 37.8°C$$

EXAMPLE   6–2

The melting point of silver is 960.8°C. What is the equivalent Fahrenheit temperature?

SOLUTION

$$°F = \frac{9}{5}°C + 32°$$

$$= \frac{9}{5}(960.8°) + 32°$$

$$= 1729.4° + 32°$$

$$= 1761.4°F$$

6–4   THERMOGRAPHY

Although we think of normal human body temperature as about 98.6°F (37°C), that is only the oral temperature. The range of skin temperature is considerable. The temperature on the skin of the back may be 90°F, the legs 85°F, and the feet 50°F or even lower. The variations result from differences in the skin and in the tissues under it. Where disease or abnormal physiological states are present, there are often changes in skin temperature. More than 95 per cent of all breast cancers, for example, are associated with a skin temperature of at least 1°C (1.8°F) higher than the uninvolved part of the same breast. Fractures, infections, and other conditions also yield their thermal "fingerprints."

A study of temperature data of the body can yield valuable information concerning the diagnosis and treatment of certain disorders. In thermography, variations in body temperature are detected and transformed into visible signals that can be recorded on photographic film. The result, called a thermogram, shows the relatively warm parts of the body as a light gray and the relatively cool parts a darker gray. The progress of disease and the effectiveness of therapy can be monitored by studying the thermal patterns of the skin before and after treatment.

6–5 TEMPERATURE AND LIFE; THERMAL POLLUTION

Life is confined to a narrow range of temperatures, from about 0°C to 50°C (32°F to 122°F). Very few organisms can live long at temperatures above or below these limits. Among the exceptions to this rule are some bacteria that can survive for months at a temperature of −180°C (−300°F). At the other extreme, some bacteria thrive at 70°C (158°F) or above, and a blue-green alga can live in some of the pools of Yellowstone National Park at 85°C (185°F).

Life is sensitive to the temperature of the environment because of the effects of temperature upon enzymes, those biochemical catalysts that control the rates of life-sustaining reactions. If the organism's temperature decreases, the rates of such reactions decrease. At high temperatures, the large, complex enzyme molecules become unstable, and enzyme-catalyzed reactions cease. Changes of temperature at both ends of the scale influence the rate of reactions, limiting the range within which life is sustained.

The temperature of the earth when life first appeared is not known, but it was probably higher than it is today. Life is so nicely attuned to present temperatures that it is estimated that if the average daily temperature were suddenly raised or lowered by only 10°C most forms of life would perish. In spite of the large differences in environmental temperature—from the tundra to deserts and jungles, from season to season, and from day to night—the body temperature of most mammals remains close to 37°C (98.6°F).

The discharge of waste heat into natural bodies of water—thermal pollution—is, therefore, a matter of concern. The electric-power industry, which uses more than 80 per cent of the cooling water used by industry, is the principal source of this waste heat (Fig. 6–6). The production of waste heat is inevitable in generating plants, and water is the only practical medium for carrying the heat away. Water is pumped from a river or lake through steam condensers in generating plants. It is then discharged into the same body of water at a temperature that is from −12°C (10°F) to 10°C (50°F) higher than it was originally. The hot discharge water, being warmer and lighter than the main body of water, spreads over the surface and disperses its heat to the river or lake and to the atmosphere.

Figure 6–6   Cooling towers for an electric generating station. Such stations are normally used when a natural or artificial body of water is not available or to prevent thermal pollution. (Courtesy of Sacramento Municipal Utility District.)

The increased temperature of the water has a number of effects, most of them harmful to the ecology of the system. It decreases the amount of the dissolved oxygen while at the same time raising the metabolic rate of fish and therefore their need for oxygen. Above some maximum temperature, from about 24°C (75°F) to 33°C (92°F), thermal death follows from a breakdown of one or more vital processes: respiration, circulation, or nervous response.

The reproductive mechanisms of fish are stimulated by normal seasonal changes, such as the warming of waters in the spring, and can be upset by abnormal temperatures that simulate such changes. A desirable fish population may die out and be replaced by less desirable species as the waters warm up. The brook trout, for example, becomes more active as the temperature rises from 5°C (40°F) to 9°C (48°F). However, its swimming speed decreases as the temperature goes from 9°C to 16°C (60°F); death occurs at 25°C (77°F).

The dispersion of waste heat may be approached from two directions: as the disposal of a waste product or as an opportunity for the recovery of a valuable by-product. As an example of the latter approach, heated water between 32°C and 55°C (90°F and 130°F) from

a pulp and paper plant is used by farmers for such purposes as frost protection and extending the growing
season by heating soil in spring and fall. Thermal energy
has also been used to desalt sea water by evaporation,
and the desalinated water, enriched with nutrients, is
used to grow high-quality fruits and vegetables within
controlled-environment greenhouses. As the economic
feasibility of these and other uses is demonstrated, we
may expect to see other productive applications of waste
heat.

6–6   THE LOWEST TEMPERATURE

The highest temperatures in the universe probably
occur in the interiors of certain stars, which may be as
high as 4 billion degrees Celsius. There is no theoretical upper limit to temperature.

In contrast, there *is* a theoretical *lower* limit to
temperature. If a gas at 0°C is cooled at constant volume, its pressure decreases by 1/273 for every decrease
of temperature by 1°. (Pressure is the ratio of force/area;
in SI units, N/m².) At −273°C the gas pressure should
drop to zero (Fig. 6–7). The temperature −273°C is
known as the absolute zero of temperature. If no further
reduction in pressure is possible, no further reduction
in temperature is possible.

William Thomson, Lord Kelvin (1824–1907) proposed a scale based upon the absolute zero of tempera-

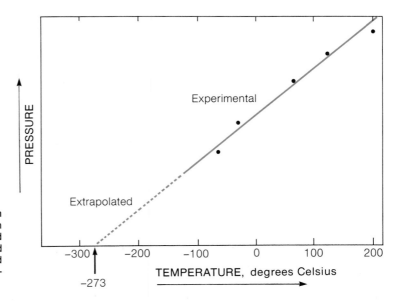

Figure 6–7 Plot of the variation
of the pressure of a gas with
temperature. When extrapolated
downward, the pressure would
become zero at −273°C, defined
as the absolute zero of temperature.

ture. The Kelvin or absolute scale keeps the Celsius degree, but the zero point is intended to be the lowest attainable temperature, 273° below zero on the Celsius scale (459.69° below zero on the Fahrenheit scale). Temperatures within 0.000001 K have been reached. The freezing point of water is $0° + 273° = 273$ K, and the boiling point of water is $100° + 273° = 373$ K. As on the Celsius scale, there are 100 Kelvin degrees between the ice point and the steam point. The Kelvin and Celsius scales are related in the following way:

Equation 6–3

$$K = °C + 273°$$

EXAMPLE 6–3

What is the equivalent of 113°F on the Kelvin or absolute scale?

SOLUTION

First convert the Fahrenheit temperature to Celsius, then to Kelvin.

$$°C = \frac{5}{9}(°F - 32°)$$

$$= \frac{5}{9}(113° - 32°)$$

$$= \frac{5}{9}(81°)$$

$$= 45°C$$

$$K = °C + 273°$$

$$= 45° + 273°$$

$$= 318 \text{ K}$$

6–7   THE KINETIC THEORY OF MATTER

According to the *kinetic theory of matter,* matter is made up of tiny particles called molecules, and these molecules are in a constant state of motion. The mole-

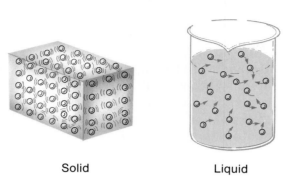

Solid   Liquid

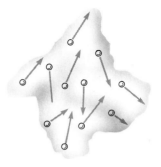

Gas

Figure 6–8 States of matter as described by the kinetic molecular theory.

cules of a gas are thought to be rather far apart and to move in constant, random, and rapid motion (Fig. 6–8). Because of the wide separation of the molecules, a gas is mostly empty space and can be compressed easily, forcing the molecules closer together. A gas has no fixed shape or volume; the forces between the molecules are so weak that they do not stay together. A gas thus fills any container; the molecules move randomly in all directions, from top to bottom to all sides (Fig. 6–9).

Gas pressure results from collisions between molecules and the wall of the container. The collisions among gas molecules are assumed to be perfectly elastic, and both kinetic energy and momentum are conserved. Except for exchanges of kinetic energy during collisions, there are no forces acting between gas molecules or between them and the wall.

In a liquid, the molecules are much closer together than in a gas, so compression is more difficult. Although the forces between the molecules are not strong enough to hold them in fixed positions, they are strong enough to keep the molecules fairly close together. Thus, a liquid maintains its volume but assumes the shape of its container. For example, when a liter of water is poured from a pitcher into several glasses, the water assumes the shapes of the glasses but the volume remains 1 liter. Since liquids and gases both have the ability to flow, they are referred to as fluids.

In a solid, the molecules are arranged in an orderly, fixed array, and are attracted to one another by relatively strong forces. Solids therefore have a fixed shape and size and are incompressible. The molecules cannot move very far, but vibrate about nearly fixed positions. An increase in temperature causes the molecules of a solid to vibrate faster around these positions.

Figure 6–9 The molecules of a gas have velocities of random magnitudes and directions. The dots represent molecules, and the arrows their direction of motion.

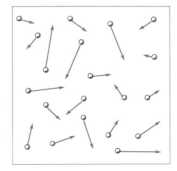

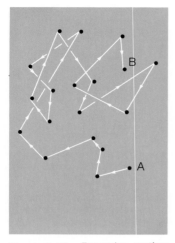

Figure 6–10   Brownian motion. Path of a pollen particle suspended in water. Positions determined at 2-minute intervals with a high-powered microscope.

The most convincing direct evidence for the kinetic theory occurs in the motion of fine particles suspended in air or water. If a sample of air with fine smoke in it is observed with a microscope, the particles of smoke suspended in the air are seen to move constantly in a random zigzag path. The smaller the particle, the faster its motion. The effect was first discovered by Robert Brown (1773–1858), a botanist, while observing pollen granules suspended in water, and is called Brownian motion (Fig. 6–10).

Brownian motion was not really understood until Albert Einstein (1879–1955) explained it by the kinetic theory. Molecules of the medium in which the particles are suspended are in perpetual motion and collide with the particles. Air or water molecules are too small to be seen, but a particle's motion resulting from these collisions can be followed under a microscope. Einstein used the kinetic theory and statistics to predict the average speed of a particle undergoing Brownian motion. Jean Perrin (1870–1942), noted for his quantitative studies of Brownian motion, experimentally determined the average velocity of a particle, measured the mass, and calculated its kinetic energy, and found the result to be in close agreement with the kinetic theory.

At the microscopic level, temperature is related to the average kinetic energy of molecules. Kinetic theory tells us that the Kelvin temperature of a gas is proportional to the average kinetic energy of its molecules.

Equation   6–4
$$T \simeq \frac{1}{2}m \overline{v^2}$$

If there are N gas molecules in a box moving at various speeds in different directions, and colliding with other gas molecules and with the molecules of the wall, the average kinetic energy of the gas molecules is given by the following expression:

$$\text{average kinetic energy of molecules} = \frac{3}{2}kT$$

Equation   6–5
$$\overline{KE} = \frac{1}{2}m \overline{v^2} = \frac{3}{2}kT$$

where T is the Kelvin temperature and k is 1.38×10^{-23} $\frac{\text{joule}}{\text{molecule} \cdot \text{K}}$. It follows, therefore, that a typical molecular speed at temperature T is given by

$$\bar{v} = \sqrt{\frac{3kT}{m}}$$

Equation 6–6

The pressure on the walls of a container containing a gas results from collisions of the gas molecules against the walls. When the temperature of a gas is increased, the molecules have more kinetic energy and strike the walls with greater force. Molecules may well have speeds of hundreds or thousands of miles per hour. As can be seen in the above formulas, the greater the mass (m) of the molecule, the lower the velocity ($\bar{v}$) for a given temperature. This concept can be extended to particles of enormous size compared to molecules. For example, in Brownian motion the particle velocity is still appreciable, but it is negligible for more massive bodies.

EXAMPLE 6–4

What is the average speed of oxygen molecules at 300 K? (Mass of an oxygen molecule = 5.32×10^{-26} kg.)

SOLUTION

Apply Equation 6–6. The constant k is known, and T and m are given. Solve for $\bar{v}$.

$$T = 300 \text{ K}$$

$$m = 5.32 \times 10^{-26} \text{kg}$$

$$\bar{v} = ?$$

$$k = 1.38 \times 10^{-23} \frac{\text{joule}}{\text{molecule} \cdot \text{K}}$$

$$\bar{v} = \sqrt{\frac{3kT}{m}}$$

$$= \sqrt{\frac{3\left(1.38 \times 10^{-23}\right)\frac{\text{joule} \cdot (300 \text{ K})}{\text{molecule} \cdot \text{K}}}{5.32 \times 10^{-26}\text{kg}}}$$

$$= \sqrt{2.33 \times 10^5 \frac{\text{m}^2}{\text{sec}^2}}$$

$$= 483 \frac{\text{m}}{\text{sec}}$$

6–8   TEMPERATURE AND HEAT

When you boil a kilogram of water on a hot plate (Fig. 6–11), it takes a certain amount of time to raise the temperature of the water from room temperature to the boiling point. If you boil 5 kilograms of water on the same hot plate, the time required is five times as great. Something apparently passes from the hot plate to the water, and more of this quantity is needed for the greater amount of water.

The first person to distinguish clearly between heat as a quantity distinct from although related to temperature was probably Joseph Black (1728–1799), a professor of medicine at the University of Glasgow. Heat is what is transferred from the hot plate to the water. It is a physical entity and we can measure its amount just as we measure an amount of water. (The science of heat measurements is known as *calorimetry*.) Temperature, on the other hand, refers to the *intensity* of heat, its "hotness" or "coldness." At the boiling point, both samples of water have the same temperature, but the beaker with 5 kilograms of water has the greater heat content.

According to the kinetic theory, heat is the transfer of molecular energy from one body to another because of a temperature difference. The direction of flow is from a high temperature to a low temperature. For example, the fact that the plate is hot means that its molecules are moving very rapidly. (Temperature is directly proportional to the average kinetic energy of the molecules.) When these high-speed molecules strike a beaker, they transfer some of their energy to the molecules of the beaker, thus raising its temperature. The faster-moving molecules of the beaker then collide with the cooler water molecules and increase their kinetic en-

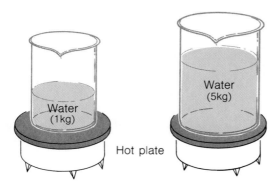

Figure 6–11   It takes five times as long to boil 5 kg of water as to boil 1 kg.

ergy so that the temperature of the water increases as well. Energy is thus transferred from the hot plate to the water.

Not all of the energy transferred to the water goes into increasing the translational kinetic energy of the water molecules; some of it increases the rotational and vibrational motion of the molecules (Fig. 6–12). The total energy of the water molecules—the sum of the kinetic energy, rotational energy, and vibrational energy—is referred to as *thermal energy*. The terms temperature and thermal energy thus refer to different properties. In 5 kilograms of water there is five times as much thermal energy as in 1 kilogram of water at the same temperature.

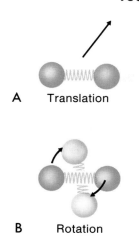

A   Translation

B   Rotation

6–9 SPECIFIC HEAT

When a kilogram of warm water at 75°C is mixed with a kilogram of water at 25°C, the temperature of the mixture is found to be 50°C (Fig. 6–13). The temperature of the warm water is lowered 25° while that of the cold water is raised 25°. It appears that the heat lost by the hot water is equal to the heat gained by the cold water.

Suppose that equal masses of two different liquids are mixed, ethyl alcohol at 75°C and water at 25°C. The temperature of the mixture is only 43.6°C. The alcohol cools by 31.4° while the water becomes warmer by only 18.6°. If we assume that heat is conserved, that is, that the heat lost by the warm alcohol is equal to the heat gained by the cold water, then a smaller quantity of heat is needed to increase the temperature of alcohol than is needed to increase the temperature of water by the same number of degrees. Alcohol has less *thermal capacity*, or capacity for heat, than water does. Every substance has its own thermal capacity, which can be determined by experiment.

The *specific heat* of a substance is defined as *the amount of heat required to raise the temperature of one gram of the substance one Celsius degree*. Let us take water as an example. The amount of heat required to raise the temperature of one gram of water one Celsius degree (from 14.5°C to 15.5°C) is defined as the *calorie* (lower-case "c"). Thus, the specific heat of water is 1 cal/g·C°. One kilocalorie, or "giant" calorie—*Calorie* (capital "C")—is the heat required to raise the temperature of one kilogram of water one Celsius de-

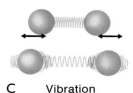

C   Vibration

Figure 6–12   Molecules can have *(A)* translational kinetic energy, moving from one place to another; *(B)* rotational energy; and *(C)* vibrational energy—the atoms vibrate back and forth in the molecule.

Figure 6–13   A kilogram of water at 75°C is mixed with 1 kg of water at 25°C. The mixture has a temperature of 50°C.

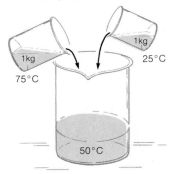

Table 6–1   SPECIFIC HEATS

SUBSTANCE		SPECIFIC HEAT (cal./g · C° or BTU/lb · F°)
Solids	Ice	0.49
	Wood	0.33
	Aluminum	0.21
	Asbestos	0.195
	Glass	0.19
	Copper	0.094
	Silver	0.056
	Tin	0.055
	Platinum	0.032
Liquids	Water	1.00
	Ethyl alcohol	0.6
	Benzene	0.41
	Mercury	0.033
Gases	Steam	0.48
	Air	0.17

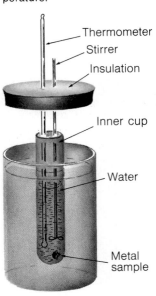

Figure 6–14 A simple, double-can calorimeter. The specific heat of a metal is determined by the method of mixtures. The metal sample, for example, copper, is pre-heated to 100°C and placed in the inner cup, which contains water at room temperature.

Thermometer
Stirrer
Insulation

Inner cup

Water

Metal sample

gree. It equals 1000 "small" calories and is the unit used to rate the heat value of foods and fuels.

In engineering work, one *British thermal unit* (BTU) is defined as *the amount of heat required to raise the temperature of one pound of water one Fahrenheit degree*. For water, the specific heat is 1 BTU/lb·F°, and for other substances the numerical value of specific heat in BTU's is identical to the value of cal/g·C°. One British thermal unit is equal to 252 calories. Table 6–1 gives the specific heats of a number of substances.

6–10   CALORIMETRY

Measurements on the quantity of heat transferred from one substance to another are carried out in a specially designed vessel called a *calorimeter,* in which heat exchanges with the environment are reduced to a minimum (Fig. 6–14). In the method of mixtures, the heat given up by a hot substance placed in the calorimeter is equal to the heat absorbed by the calorimeter and its contents.

Since the specific heat of liquid water is 1 cal/g·C°, supplying 5 calories of heat to 1 gram of water increases the temperature 5 C°. The same quantity of heat sup-

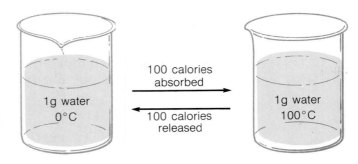

Figure 6–15 To raise the temperature of 1 g of water from 0°C to 100°C, 100 calories must be supplied. In turn, when the water cools from 100°C to 0°C, 100 calories are released.

plied to 1 gram of alcohol raises its temperature 8.3 C°, since the specific heat of alcohol is 0.6 cal/g·C°. To raise the temperature of 1 gram of water from the freezing point (0°C) to the boiling point (100°C), 100 calories must be supplied (Fig. 6–15). The amount of heat that water releases to the environment upon cooling through the same temperature difference is equal to the amount of heat supplied in these processes initially. One gram of water thus furnishes 100 calories to its surroundings in cooling from its boiling point to its freezing point. Thus, the amount of heat a substance loses or gains depends upon three factors: (a) the mass of the substance; (b) the nature of the substance; (c) the temperature change. This relationship is expressed as follows:

$$\text{Heat lost or gained} = \left(\begin{array}{c}\text{mass of}\\\text{substance}\end{array}\right) \cdot \left(\begin{array}{c}\text{specific}\\\text{heat}\end{array}\right) \cdot \left(\begin{array}{c}\text{change in}\\\text{temperature}\end{array}\right)$$

or, in symbols:

$$H = m \cdot s \cdot (T_2 - T_1)$$

Equation   6–7

in which H is the amount of heat lost or gained; m is the mass of the material; s is its specific heat; and $(T_2 - T_1)$ is the temperature change.

EXAMPLE   6–5

How much heat must be supplied to 20 grams of tin, originally at 25°C, to raise its temperature to 100°C?

SOLUTION

The mass of tin, the initial temperature, and the final temperature are given. The specific heat of tin can be obtained from Table 6–1.

$m = 20 \text{ g}$

$s_{tin} = 0.055 \text{ cal/g} \cdot \text{C}°$

$t_1 = 25°\text{C}$

$t_2 = 100°\text{C}$

$H = ?$

$H = m \cdot s \cdot (T_2 - T_1)$

$= (20\text{g})\left(0.055\dfrac{\text{cal}}{\text{g} \cdot \text{C}°}\right)$

$\qquad (100 - 25)°\text{C}$

$= (20)(0.055)(75)\,\text{cal}$

$= 82.5 \text{ cal}$

In a very real sense, foods are body fuels. The heat value of fuels and foods is determined from the *heat of combustion*—the heat produced per unit mass of substance burned in oxygen. Knowing their heats of combustion, it is possible to compare one kind of fuel with another or different grades of the same fuel with respect to their heat value. Kerosene, for example, yields 11 200 kcal/kg when burned, while alcohol furnishes 6400 kcal/kg. To get the same amount of heat from alcohol as from kerosene, the mass of alcohol burned must be nearly twice as great. For weight-watchers, the heat of combustion of foods is a factor in deciding upon their daily diet: a scrambled egg yields

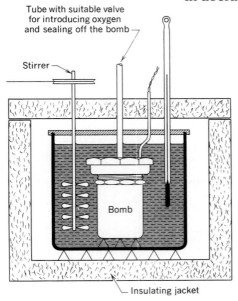

Tube with suitable valve
for introducing oxygen
and sealing off the bomb

Stirrer

Bomb

Insulating jacket

Figure 6–16   Bomb calorimeter for determining the heat of combustion of fuels or foods. (From Shortley, G. H., and D. E. Williams, *Principles of College Physics.* 2nd ed. Englewood Cliffs, N.J. Prentice-Hall, Inc., 1967, p. 305.)

more calories (2100 kcal/kg) than a boiled egg (1600 kcal/kg).

The heat of combustion of a fuel or food is measured in a bomb calorimeter (Fig. 6–16). The "bomb" is a strong steel cylinder fitted with a heavy cap that screws tightly into place. A crucible suspended from the cap contains a known mass of the material to be burned. A small platinum wire suspended from two insulated rods dips into the fuel. The cap is adjusted, and oxygen is admitted through a valve in the cap to a pressure of about 15 atmospheres. The bomb is then placed in a water calorimeter. An electric current sent through the wire heats it to incandescence and ignites the fuel, which burns violently in the oxygen atmosphere. The large amount of heat that is produced causes the temperature of the metal bomb and the surrounding water to increase. The heat of combustion is then computed by the method of mixtures:

$$\text{Heat produced} \atop \text{by fuel} = {\text{Heat gained by} \atop \text{calorimeter}} + {\text{Heat gained} \atop \text{by water}}$$

Table 6–2 gives the heats of combustion of some fuels and foods.

Table 6–2 HEATS OF COMBUSTION OF FUELS AND FOODS

	HEAT OF COMBUSTION (kcal/kg)
Fuels	
Natural gas	8000–12 000
Gasoline	11 300
Diesel oil	10 500
Fuel oil	10 300
Anthracite coal	7000–8000
Alcohol	6400
Wood (pine)	4500
Foods	
Bread (white)	2600
Butter	7950
Eggs (boiled)	1600
Eggs (scrambled)	2100
Ice cream	2100
Meat (lean)	1200
Milk	715
Peanuts	5640
Potatoes (white boiled)	970
Rice (cooked)	1120
Sugar (white)	4000

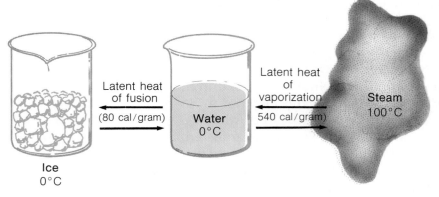

Figure 6–17 Phase changes. Ice absorbs 80 cal/g at 0°C to melt. When water freezes at 0°C, 80 cal/g are released. The heat of fusion of H_2O is 80 cal/g.

6-11   CHANGE OF STATE

The sensation of coldness that you feel when you touch a piece of ice is an indication that heat is quickly drawn from your hand to the ice. Surprisingly, however, the temperature of the dripping water is no greater than the temperature of the ice. During the melting process, a large amount of heat—80 calories per gram—is absorbed by the ice *without* raising its temperature (Fig. 6–17). The heat that the ice absorbs in this way without increasing its temperature is known as the *latent heat of fusion,* L_f. The latent heat of fusion is defined as the *quantity of heat that must be added to one gram of solid at its melting point to change it to liquid at the same temperature and pressure.*

As viewed by kinetic theory, the molecules of ice vibrate faster and faster about their normal positions as the ice is heated. At 0°C (when the pressure is 1 atmosphere), the molecular vibrations are so vigorous that the molecules break out of their fixed positions in the crystal lattice and start rolling over one another; that is, the ice begins to melt. During the melting process, the temperature does not change: the heat of fusion does not increase the kinetic energy of the molecules. Instead, work must be done to overcome the attractive forces between the ice molecules, in order to separate them into the more random motion of the molecules of liquid water, and this raises their potential energy. The quantity of heat required to melt ice at 0°C and 1 atmosphere depends upon the mass of ice and the heat of fusion of ice, and may be calculated from the following equation:

Equation   6–8 $$H = m \cdot L_f$$

in which H is the quantity of heat, m is the mass of ice, and L_f is the latent heat of fusion of ice.

EXAMPLE 6–6

How much heat is required to melt 50 grams of ice at 0°C?

SOLUTION

Use Equation 6–8 to determine the quantity of heat involved in the melting process of ice at 0°C, assuming that the temperature of the resulting water is 0°C. (See Table 6–3 for latent heat of fusion.)

$$m_{ice} = 50 \text{ g} \qquad H = m \cdot L_f$$

$$L_{f_{ice}} = 80 \text{ cal/g} \qquad = \left(50 \text{ g}\right)\left(80\frac{\text{cal}}{\text{g}}\right)$$

$$H = ? \qquad = 4000 \text{ cal}$$

Once the ice has melted, further heating increases the KE of the molecules and the temperature rises (Fig. 6–18). At the boiling point, the molecules are moving so rapidly that the attractive forces can no longer confine them to the limited volume of a liquid, and the molecules move out in all directions as a gas. The work done against the attractive forces raises the PE of the molecules. The energy required to do this work, called the *latent heat of vaporization*, L_v, is de-

---

Table 6–3 LATENT HEATS OF FUSION AND VAPORIZATION (AT ATMOSPHERIC PRESSURE)

SUBSTANCE	MELTING POINT (°C)	LATENT HEAT OF FUSION, L_f (cal/g)	BOILING POINT (°C)	LATENT HEAT OF VAPORIZATION, L_v (cal/g)
Alcohol, ethyl	−117.3	24.9	78.5	204
Ammonia	−75	108.1	−34	327
Copper	1083	42	2595	1760
Mercury	−38.7	2.8	357	70.6
Water	0	80	100	540

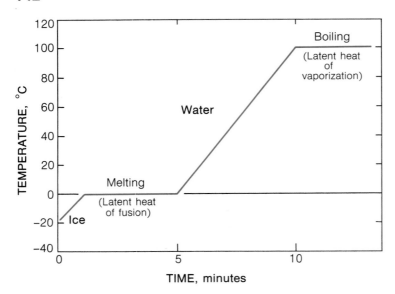

Figure 6–18 A graph showing the temperature of ice and water heated at a constant rate. Note that the temperature remains constant as the ice melts, and again as the water boils.

fined as *the quantity of heat that must be added to one gram of the liquid at its boiling point to change it to vapor at the same temperature and pressure.*

The value of L_v of water is 540 calories per gram, a value more than six times as large as its latent heat of fusion. The reason is that the energy needed to separate liquid molecules into the gas is greater than that needed for the smaller molecular separation required when a solid becomes a liquid. The corresponding equation for the process of vaporization is:

Equation   6–9

$$\dot{H} = m \cdot L_v$$

EXAMPLE 6–7

How much heat is required to vaporize 50 grams of water at 100°C?

SOLUTION

Apply Equation 6–9.

$$m_{water} = 50 \text{ g}$$
$$L_{v_{water}} = 540 \text{ cal/g}$$
$$H = ?$$

$$H = m \cdot L_v$$
$$= \left(50 \text{ g}\right)\left(540\frac{\text{cal}}{\text{g}}\right)$$
$$= 27\,000 \text{ cal}$$

Table 6–3 gives the latent heats of fusion and vaporization of several substances.

The phase changes of a substance can be reversed. For example, water vapor condenses on the outside of a glass containing a cold drink. Water molecules in the air near the cold glass lose KE when they collide with the molecules of the glass. As more and more water molecules slow down and approach one another, the attractive forces between them cause them to come together and form a drop of water on the glass. In this process, the latent heat of vaporization of water, 540 calories per gram, is released to the surroundings.

In a steam heating system, water is boiled in the furnace, absorbing 540 cal/g at the boiling point as it changes to steam. Entering a radiator at 100°C, steam is cooled and condenses to water, releasing the latent heat of vaporization to the surroundings. This is the principle behind keeping tubs of water near vegetables or plants to protect them on a cold night. The freezing of the water is accompanied by the release of heat to the air—the latent heat of fusion—and this keeps the surrounding temperature from falling too low.

The cooling effect of a dab of rubbing alcohol on the skin comes from the withdrawal of surface heat, causing the alcohol to evaporate. For the same reason, you feel cool after a swim, particularly if there is a breeze; the latent heat of vaporization of water is large, and as the water evaporates it withdraws body heat. The cooling effect of evaporation may be used to keep liquids cool; evaporation from a moist, porous, canvas container helps to keep the contents cool. A burn from "live" steam is more harmful than one caused by an equal mass of hot water. As it comes into contact with the cooler skin, each gram of steam releases the latent heat of vaporization of steam—540 cal/g. This is a much greater value than the specific heat of water—1 cal/g $\cdot$ C° —released by each gram of water for each degree it is cooled.

When you place moth flakes in a dresser drawer, you may find that they disappear eventually without leaving a trace. The moth flakes, a chemical substance called p-dichlorobenzene, pass directly from the solid phase to the gas phase, without passing through the liquid phase. This process is called *sublimation*. Besides moth flakes, solid iodine and "dry ice" are familiar examples of substances that sublime. Ordinary ice also sublimes to some extent. For example, when wet cloth-

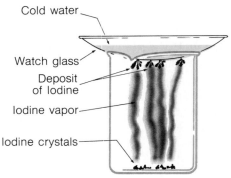

Cold water

Watch glass

Deposit
of Iodine

Iodine vapor

Iodine crystals

Figure 6–19 Sublimation of iodine crystals and their deposition on a watch glass.

ing is put out to dry in zero-degree weather, the water in it freezes, but after a period of time much of the ice sublimes.

We do not usually observe sublimation, because the more common vapors are usually colorless. Iodine crystals, however, give off a purple vapor when they sublime. If a few crystals are placed in a beaker covered with a watch glass, iodine vapor fills the beaker (Fig. 6–19). The presence of cold water in the watch glass causes iodine crystals to be deposited on the underside of the watch glass. The blackening inside a tungsten incandescent bulb is another example. When the tungsten filament is hot, some of the tungsten goes directly into the vapor state. As the vapor encounters the cooler glass of the bulb, tungsten is deposited on it and darkens it.

6–12 THERMAL EXPANSION

The volume of a gas at constant pressure decreases by 1/273 of its value at 0°C (273 K) for each degree decrease in temperature. The nature of the gas does not matter; all gases undergo approximately the same fractional change in volume with the same change in temperature. This discovery was made independently by Jacques Charles (1746–1823) and Joseph Gay-Lussac (1778–1850), and is known as the *law of Charles and Gay-Lussac.* Using absolute temperatures, it tells us that *the volume of a gas varies directly with its absolute temperature (if the pressure is held constant):*

Equation 6–10

$$\frac{V}{V'} = \frac{T}{T'}$$

in which V is the initial volume of the gas, V′ the final volume, T the initial absolute temperature, and T′ the final absolute temperature. This is an observation that led to the suggestion that molecules at higher temperatures are moving faster and on the average are farther apart from one another, one of the cornerstones of the kinetic theory. So, the gas will take up more space as its temperature is increased.

EXAMPLE 6–8

One liter of helium is heated from 25°C to 100°C. Assuming that the pressure is constant, what is the final volume?

SOLUTION

A key point in this example is that Celsius temperatures must be converted to their equivalents on the absolute scale. Since the temperature is increased, we predict that the volume of helium will increase.

V = 1 liter

T = 25°C + 273° = 298 K

T' = 100°C + 273° = 373 K

V' = ?

$$\frac{V}{V'} = \frac{T}{T'}$$

$$V' = \frac{VT'}{T}$$

$$= \frac{(1 \text{ liter})(373\,\text{K})}{(298\,\text{K})}$$

$$= 1.25 \text{ liters}$$

When the temperature of a solid is increased, generally the solid expands. Unlike gases, however, different solids expand by different amounts for each degree change in temperature. An aluminum rod, for example, expands in length twice as much as an iron rod, and nearly six times as much as tungsten for the same change in temperature. Experiments show that the amount of expansion in length depends upon three factors: (1) the increase in temperature—expansion is greater if the

Table 6–4   COEFFICIENTS OF LINEAR EXPANSION
(Change in length per unit length per degree change in temperature)

MATERIAL	α/°C
Aluminum	24×10^{-6}
Copper	17×10^{-6}
Steel	13×10^{-6}
Iron	12×10^{-6}
Concrete	12×10^{-6}
Ordinary glass	8.5×10^{-6}
Tungsten	4.3×10^{-6}
Pyrex glass	3.3×10^{-6}
Invar (a nickel-steel alloy)	0.9×10^{-6}
Fused quartz	0.5×10^{-6}

temperature is increased 50 degrees than if it is increased 10 degrees; (2) the original length—a long rod expands more than a short one because of the greater number of crystals, each of which undergoes the same expansion; and (3) the nature of the metal. These facts are expressed in the equation for the linear expansion of a solid as follows:

Equation   6–11

$$e = \alpha \cdot \ell \cdot (T_2 - T_1)$$

in which e is the change in length; α is the coefficient of linear expansion for the material; ℓ is the original length; and $(T_2 - T_1)$ is the change in temperature. Table 6–4 gives the coefficient of linear expansion of several materials, that is the fractional expansion per degree change in temperature.

EXAMPLE   6–9

An aluminum rod is 50.0 centimeters long at 25°C. How long will the rod be at 100°C?

SOLUTION

We expect the rod to elongate with an increase in temperature. We solve for the expansion e, and add that amount to the original length to find the length at 100°C. (Find the coefficient of linear expansion of aluminum from Table 6–4.)

$a_{Al} = 24 \times 10^{-6}/°C$

$\ell = 50.0$ cm

$T_1 = 25°C$

$T_2 = 100°C$

$e = ?$

$e = \alpha \cdot \ell \cdot (T_2 - T_1)$

$= \dfrac{(24 \times 10^{-6})}{°C}(50.0 \text{ cm})(100 - 25)°C$

$= 0.1$ cm

Length at $100°C = 50.0$ cm $+ 0.1$ cm

$= 50.1$ cm

There are many familiar examples of expansion and contraction of solids caused by changes in temperature. A thick-walled glass tumbler may crack if hot water is poured into it, because stresses are created as the inner surface layers expand before the outer layers are heated. Concrete sidewalks and highways are laid with gaps between sections to accommodate expansion and prevent buckling. Expansion joints are provided between sections of a steel bridge to allow for daily and seasonal changes in temperature. Power lines are more taut in winter than in summer. The design of certain types of thermostats is based upon the

Figure 6–20 Differential expansion. A bimetallic strip of brass and iron bends, as shown, because brass expands (or contracts) more when heated (or cooled) than does iron.

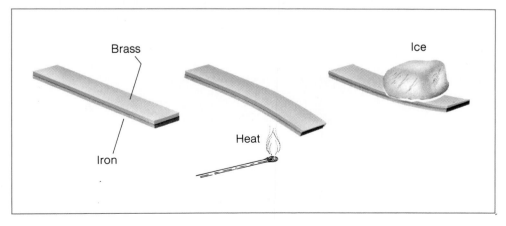

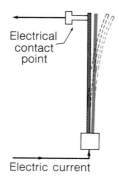

Figure 6–21 The operation of a bimetallic thermostat depends upon the unequal expansion of the two metals.

thermal expansion of two different metals, such as brass and iron, welded or riveted together (Fig. 6–20). When the temperature increases, the metals expand unequally, causing the bimetallic strip to bend. Many thermostats that control furnaces or air conditioners use bimetallic strips to open and close electrical contacts (Fig. 6–21).

The behavior of liquid water with temperature changes is unusual. When a beaker of water at 4°C is heated, the water expands; but when it is cooled below this temperature, instead of contracting, the water expands again. A given mass of water thus has the smallest volume when it is at 4°C. Since density is the ratio of mass to volume, water has the greatest density at this temperature.

Aquatic life in colder climates is affected by the thermal behavior of water. In a body of fresh water, as the surface layer of water cools to 4°C, it becomes more dense than the water below and sinks, pushing up warmer water to the surface where it, in turn, is chilled. Considerable mixing occurs as successive layers at the surface cool to 4°C and sink, until the entire body of water is 4°C. When the surface layer is cooled below 4°C, its density becomes less than that of underlying layers and it remains at the surface while freezing (Fig. 6–22). Lakes and streams thus freeze from the top down, and this enables plant and animal life to survive the winter in an above-freezing environment (0°C to 4°C; 32°F to 39°F). The lower density of the ice (ice is 10 per cent less dense than water) keeps it on top where it acts as an insulating blanket to retard the cooling of the underlying water. When water freezes in a confined space it can exert a tremendous force and break even very strong metal. For this reason antifreeze must be used in automobiles in cold climates and water pipes must be buried deep underground.

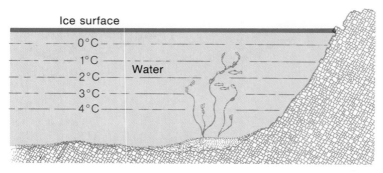

Figure 6–22 The temperature profile of water in a pond in winter.

6–13 PRESSURE EFFECTS ON FREEZING AND BOILING POINTS OF WATER

The statement that water freezes at 0°C is valid only at normal atmospheric pressure. Since water expands upon freezing and cannot freeze without expanding, an increase in pressure lowers the freezing point of water by increasing the resistance to expansion.

One effect of the unusual behavior of water at the freezing point (most substances contract upon freezing) is that you can skate or ski on ice. When you do so, the ice or snow melts under two influences: (1) the pressure exerted by the skate blade or ski (a large force—weight —distributed over a small area), and (2) friction. A thin, slippery film of water is formed which acts as a lubricant between the blade and ice. As soon as the pressure returns to normal, the water refreezes.

Regelation—the process of ice melting at a temperature lower than 0°C because of increased pressure, then refreezing when the pressure is removed—can be demonstrated by suspending a heavy weight from a fine wire looped over a block of ice (Fig. 6–23). The increased pressure under the wire lowers the freezing point of the ice, causing some of it to melt and allowing the wire to drop down. The water refreezes above the wire, where the pressure is less and the freezing point is normal. In this way the wire passes through the ice, leaving a path of fine air bubbles, but the ice is still frozen solidly.

The boiling point of a liquid is the temperature at which its vapor pressure is equal to the surrounding atmospheric pressure. As the boiling point is approached, the molecules are moving so rapidly that

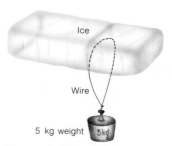

Figure 6–23 Regelation. A weighted wire loop cuts through a block of ice. The increased pressure under the wire lowers the freezing point of the ice. The water refreezes above the wire.

Table 6–5 VAPOR PRESSURE OF WATER

TEMPERATURE (°C)	PRESSURE (torr = mm Hg)
0	4.58
20	17.53
40	55.32
60	149.4
80	355.1
100	760 (1 atm)
120	1489

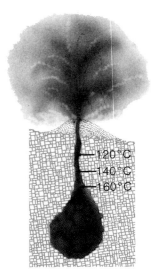

Figure 6–24 A geyser is a natural "pressure cooker." When superheated water under pressure boils, an eruption of steam and water soon follows.

—120°C
—140°C
—160°C

Figure 6–25 Water boiling at reduced pressure. Pouring cold water on a flask of boiling water causes steam to condense, reducing the pressure and lowering the boiling point.

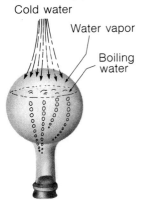

Cold water

Water vapor

Boiling water

some of them may enter the gaseous state and form a bubble. With continued heating, the vapor pressure within the bubble builds up and the bubble rises to the top. At the boiling point, the vapor pressure within the bubble is equal to the pressure exerted on the liquid. If the external pressure is increased, it becomes more difficult for bubbles of vapor to form, and the boiling point is raised. In a pressure cooker, the confined steam increases the pressure upon the surface of the liquid, thereby raising the boiling point of water from 100°C to 110°C or more. This reduces the time required for cooking food. Table 6–5 gives the vapor pressure of water at various temperatures.

Geysers are natural pressure cookers. Old Faithful, a geyser in Yellowstone National Park, is so named because on the average it erupts at regular intervals (about 65 minutes). One model of geyser action assumes that water from a hot spring collects at the bottom of a fissure—a vertical, pipelike opening (Fig. 6–24). The water near the base of the fissure is under considerable pressure and may reach a temperature of from 120°C to 175°C—considerably above its normal boiling point and hotter than the water higher up the fissure. When the water begins to boil and changes to steam, the expanding steam pushes some water out of the pipe in a brief, quiet fountain approximately 6 feet high. The pressure in the entire column is relieved by the overflow, but the water is now superheated at a temperature above its new boiling point. Within 60 seconds some of the water boils, and the expanding steam causes an eruption of steam and water 150 feet into the air that continues for about 4 minutes. The pressure and temperature at the base then return to normal, water seeps in again, and the cycle is repeated.

A decrease in the external pressure on the surface of a liquid has the opposite effect, that is, lowering the boiling point. Consider a round-bottomed flask in which water is boiled for a few minutes to expel the air. When steam comes from the mouth of the flask, it is tightly sealed with a stopper containing a thermometer (Fig. 6–25). Pouring cold water over the flask causes some of the steam above the surface of the liquid to condense to water, creating a partial vacuum, and the vapor pressure above the water is reduced. The boiling point of the water is thus lowered, and the water boils vigorously at a lower and lower temperature. Water can even be made to boil at room temperature in this way.

6–14 METHODS OF HEAT TRANSFER

When there is a temperature difference between two objects in close proximity, or between two parts of the same object, the warmer region loses heat and the cooler region gains heat. The transfer of heat may occur by one or by a combination of processes: *conduction, convection,* and *radiation.*

Conduction.   Stirring boiling water with a silver spoon causes the spoon to heat up so quickly that it soon becomes too hot to handle. Silver conducts heat readily; the transfer of heat along the spoon is called *conduction.* Spoons made of other metals—aluminum, iron, lead, brass—also conduct heat, but at different (and generally lower) rates. Conduction involves the transfer of thermal energy through the interaction of neighboring molecules or atoms *without* the conducting material itself being moved. With an increase in temperature, the molecules vibrate faster and faster and transmit this energy through collisions with adjacent molecules in the material.

Metals as a group are better conductors of heat than non-metals, but metals differ in their conductivity. Silver is excellent, aluminum is good, and iron and lead are relatively poor metallic conductors. Glass is such a poor conductor of heat that one end of a 5-inch rod may be held comfortably in the hand while the other end is melting in a flame. Other non-metals, such as wood, paper, and cotton are poor conductors of heat and are therefore good insulators. Air and gases generally are quite poor conductors of heat. The warmth-retaining quality of feathers, fur, cork, and other materials is frequently due to the insulating property of air trapped in small spaces between fibers or other structures. Storm windows are effective heat retainers for this reason; two layers of glass are separated by a poorly conducting air space. The thermal conductivities of some materials are listed in Table 6–6.

Convection.   Fluids transfer heat by *convection,* a process that causes mixing of the warmer with the cooler regions of the liquid or gas. The bottom-most layer of water in a kettle is heated by contact with the hot kettle and expands (Fig. 6–26). The increase in volume lowers the density of the warm water; being lighter, it is pushed to the top by the buoyant force of the cooler, denser, surrounding water. The cooler water that moved in to replace it is itself heated and is then

Table 6-6 THERMAL CONDUCTIVITIES OF COMMON MATERIALS

MATERIAL	THERMAL CONDUCTIVITY (calories/cm · sec · C°)
Metals	
Silver	0.99
Copper	0.92
Gold	0.72
Aluminum	0.48
Iron and steel	0.11
Non-Metallic Solids	
Concrete	0.0022
Brick	0.0015
Asbestos sheet	0.0004
Wood	0.0003
Snow	0.0003
Glass wool	0.0001
Liquids	
Mercury	0.016
Water	0.0013
Gases	
Hydrogen	0.00041
Air	0.000057

also pushed upward. The convection currents set up a circulation, and the mixing of the warm and cold water soon establishes a more or less uniform temperature throughout.

Central heating systems in buildings operate by convection of hot air, hot water, or steam. Some systems rely on natural convection as just described. Other systems use supplementary blowers or pumps to circulate the heated fluid. When a room is warmed by a radiator, heat is transferred in part by convection currents that are set up in the air. Most of the heat in an open fireplace is lost by convection up the chimney.

Radiation.   The sun is the source of most of the energy on the earth. Since the space between these two

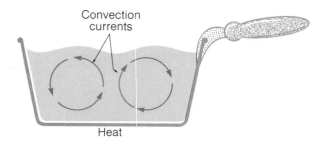

Convection currents

Heat

Figure 6-26 Convection currents circulate heated water.

bodies is largely empty, it cannot transfer heat by conduction or convection, both of which require a material medium. Instead, energy is transferred by a third process—*electromagnetic radiation.*

Any body above absolute zero radiates heat to its surroundings. The hotter the body, the greater the rate at which energy is radiated. All bodies radiate energy as electromagnetic waves at a rate proportional to the fourth power of the absolute temperature of the body, T^4. Radiant energy may pass through air or glass without warming it, because only a small part of it is absorbed. But when radiant energy falls on a rock, pavement, or leaf, which absorbs more of it, there is an increase in the internal energy of the object, followed by a rise in temperature. Experiments show that a rough, unpolished surface radiates heat better than a polished surface, and a black surface radiates heat better than a white surface at the same temperature. Good radiators of heat are also good absorbers of heat.

The methods of heat transfer have been considered in the design of the thermos bottle, which is intended to keep its contents at a given temperature: hot liquids hot or cold liquids cold (Fig. 6–27). The space between the walls of the double-walled thermos bottle is evacuated, reducing heat transfer by conduction and convection to a minimum, and the walls themselves are coated with silver to reduce radiation. Little heat is conducted along the glass or by the cork. Thus, there is a minimum amount of heat transferred into or out of the flask.

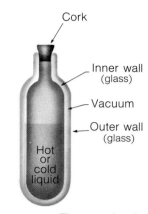

Figure 6–27 Thermos bottle keeps contents hot or cold by minimizing heat transfer by conduction, convection, and radiation.

6–15 HEAT AND LIFE

Justus von Liebig (1803–1873) proved that heat is produced by the combustion of food in body tissues. Later it was shown that heat production and heat loss are controlled by the nervous system. The body reacts to cold with exercise, shivering, and the unconscious tensing of muscles. Heat loss, meanwhile, is reduced to a minimum by cutting down the rate of circulation of blood to the skin. In hot weather, body heat is dissipated faster by the increasing the circulation to the skin, by sweating, and by faster respiration.

In a very cold environment, the body's metabolic production of heat may not keep up with the loss of heat by radiation, conduction, and convection. Normally, the body's thermostat in the hypothalamus monitors and adjusts the internal temperature of the body from the inside. Even after a hot meal or a cold drink, the body's internal temperature is maintained at a practically constant value. Fever has long been recognized as a sign of disease, but why sickness is so often accompanied by a rise in body temperature is not yet known.

Animals living in cold environments have ways of conserving body heat. The fur of arctic animals is considerably thicker than that of tropical animals, and its insulating ability is many times greater. An arctic fox can be comfortable at −50°C without increasing its rate of metabolism, although the feet, legs, and nose are uncovered. If they allowed the escape of body heat, mammals and birds could not survive in cold climates. The warm feet of an animal standing on snow or ice would melt it and would soon freeze in place. A gull or duck swimming in icy water would lose heat through its webbed feet faster than it could produce it.

To reduce the loss of heat from unprotected extremities, the warm outgoing blood in the arteries heats the cool blood returning in the veins in a network of veins, the *rete mirabile,* near the junction between trunk and extremity. Therefore, the extremities can become much colder than the body without draining off the body heat. This mechanism also serves as a means of enabling thickly furred animals to release excess heat: the flow of blood to the extremities is increased and heat is radiated. Heat is also dissipated by evaporation from the mouth and tongue.

CHECKLIST

Absolute zero of temperature	Kelvin scale
Boiling point	Kinetic theory
Brownian motion	Latent heat of fusion
BTU	Latent heat of vaporization
Calorie	Law of Charles and
Calorimetry	Gay-Lussac
Celsius scale	Linear expansion
Conduction	Pressure
Convection	Radiation
Evaporation	Specific heat
Fahrenheit scale	Sublimation
Fixed points	Temperature
Freezing point	Thermal energy
Heat	Thermal pollution
Heat of combustion	Thermography
	Thermometric property

6-1. Determine the equivalent of 0°F on (a) the Celsius
scale; (b) the Kelvin scale.

6-2. Express the following temperatures in °F: (a) −40°C; (b)
218 K; (c) 6000°C; (d) 360°C.

6-3. The material that has the lowest temperature most of
us are likely to encounter is dry ice, −79°C. What would its
temperature be on the Fahrenheit scale?

6-4. (a) What is the difference between 70°F and 70°C in
Celsius degrees? (b) In Fahrenheit degrees?

6-5. Liquid oxygen rocket propellant is kept at its boiling
point of 90 K. What would the temperature be on the Fahrenheit scale?

6-6. When a thermometer is placed in a beaker of hot water,
the mercury level drops, then rises. Explain.

6-7. Room temperature is often identified as 68°F. What
temperature is this on (a) the Celsius scale; (b) the Kelvin
scale?

6-8. Can you think of any advantage to using the Celsius
scale over the Fahrenheit scale?

6-9. Explain the meaning of the absolute zero of temperature.

6-10. Why are high temperatures resulting from thermal
pollution often fatal to fish?

6-11. What is the basis for the belief that matter is made up
of small particles in motion?

6-12. What prevents the earth's atmospheric gases from
leaking off into space?

6-13. How is the concept of temperature interpreted by the kinetic theory?

6-14. How is the concept of gas pressure interpreted by the kinetic theory?

6-15. The gas pressure on a piston is 7.5×10^4 newtons per square meter. If the area is 2.0×10^{-3} square meter, what force does the gas exert?

6-16. A smoke particle undergoing Brownian motion in air has a mass of 2.0×10^{-5} kg. (a) What is its average velocity if the temperature is 300 K? (b) What is its average kinetic energy?

6-17. How much heat is required to raise the temperature of 10 kilograms of aluminum by 80°C?

6-18. A hot water bottle containing 1 kilogram of water cools from 70°C to 20°C. How much heat was given off by the water?

6-19. A kilogram of copper gains 1 kilocalorie of heat. By how many degrees is the temperature of the copper changed?

6-20. If 500 kilograms of water at 200°F is poured into a lake that is at 50°F, how much heat is added to the lake?

6-21. To make iced coffee, 1 kilogram of ice at 0°C is added to 2 kilograms of coffee at 95°C. Assuming that the specific heat of coffee is the same as that of water, what is the temperature of the mixture?

6-22. A 65-kilogram woman is on a 2500-kcal/day diet. If a corresponding amount of heat were added to 65 kilograms of water at 37°C, to what temperature would the water be raised?

6-23. Why does the temperature of boiling water remain the same as long as the boiling continues?

6–24.   How much more heat is required to vaporize 100 grams of water than to melt 100 grams of ice?

6–25.   Explain why evaporation is a cooling process.

6–26.   Explain why water condenses on the inside of a window on a cold day.

6–27.   Distinguish between the temperature and the thermal energy of a cup of hot coffee.

6–28.   In a steam heating system, water leaves a radiator at the same temperature as steam entered, 100°C. How is the room heated?

6–29.   What characteristic of water makes it useful as an automobile engine coolant?

6–30.   Why does an oak floor feel colder to the bare foot than a rug at the same temperature?

6–31.   How much heat does 50 grams of steam at 125°C release in the process of condensing to water at 100°C?

6–32.   Why is an ice cube at 0°C more effective in cooling a drink than the same quantity of water at 0°C?

6–33.   When the pressure on water is increased, will its boiling point be higher or lower than 100°C?

6–34.   Why might an astronaut's blood boil if he left his capsule without his pressurized spacesuit?

6–35.   A 20-gram ice cube at 0°C melts, and the temperature of the water rises to 25°C. How much heat is involved?

6–36.   What water temperature would you expect to find near the bottom of an ice-covered lake?

6–37.   What is the seasonal variation in the height of a 300-foot steel-frame building if the temperature range is −20°F to 95°F?

6–38.   If a house is insulated to reduce heat loss in winter, how comfortable will it be in summer?

6–39.   Explain why perspiration occurs when the body is hot, but produces a cooling effect.

6–40.   Multiple Choice

A. Temperature is a measure of
   (a) the amount of heat in a body
   (b) the average total energy of molecules
   (c) the average kinetic energy of molecules
   (d) all of these

B. As water freezes, its temperature
   (a) increases
   (b) decreases
   (c) first increases, then decreases
   (d) does not change

C. The motion of small particles due to collisions with molecules in a fluid is called
   (a) evaporation   (c) Brownian motion
   (b) sublimation   (d) pressure

D. The lowest possible temperature is
   (a) unknown   (c) infinitely low
   (b) absolute zero   (d) 273K

E. If the temperature of a gas is increased, its volume is
   (a) reduced by half
   (b) reduced by one fourth
   (c) doubled
   (d) increased

SUGGESTIONS FOR
FURTHER READING

Armstrong, George T., "The Calorimeter and Its Influence on the Development of Chemistry." *Journal of Chemical Education,* June, 1964.
   The calorimeter made combustion and heat among the most intensively investigated areas of chemistry, and it still contributes valuable data.

Chalmers, Bruce, "How Water Freezes." *Scientific American,* February, 1959.
   The freezing of water depends upon a delicate interdependence of temperature, pressure, and geometry.

Gershon-Cohen, Jacob, "Medical Thermography." *Scientific American,* February, 1967.
  The measurement of heat emitted through the skin is useful in the diagnosis and treatment of various disorders.
Sproull, Robert L., "The Conduction of Heat in Solids." *Scientific American,* December, 1962.
  The measurement of heat conduction by metals and nonmetals relates it to the conduction of both sound and electricity.

ENERGY CONVERSION

7-1 INTRODUCTION

The fact that you can produce fire from the friction of rubbing together two sticks of wood, and from flint and iron, suggests a connection between motion and heat. Count Rumford (1753–1814), an American scientist and soldier who had emigrated to Europe, demonstrated through cannon-boring experiments that mechanical friction could produce large quantities of heat, enough to boil a massive amount of water without any fire. On the basis of his experiments, he asserted that work must be equivalent to heat. However, Rumford's experiments were not quantitative. That is, he could not say *how much* work was equivalent to a given amount of heat, and he admitted that he did not keep track of the heat absorbed in other ways than boiling the water. We now look at the connection between heat and work and of the conversion of the one into the other, called *thermodynamics*.

7-2 MECHANICAL EQUIVALENT OF HEAT

James Prescott Joule (1818–1889), a brewery owner and amateur scientist, established the equivalence of work and heat through experiment. Although he was not a trained scientist in the usual sense, Joule did scientific research throughout his life as an avocation. Realizing the need for precise instruments, he and a Manchester instrument maker developed the first accurate and reliable thermometers in England. For about 40 years, Joule experimented on the mechanical equivalent of heat, an interest he came to from studies on electric motors.

The electric motor, based upon Michael Faraday's discoveries in electromagnetism, had recently been invented. Joule saw in the electric motor the possibility of converting the family brewery from steam to electric power, and set out to improve it. But when he compared the work done by the motor with the cost of the zinc consumed in the batteries which operated the motor, he became convinced that the electric motor would be an expensive substitute for the steam power produced by burning coal.

In his investigations, Joule measured the heat developed by an electric current conducted in a wire. When he turned his attention to Faraday's dynamo —a machine in which a spool of wire is mechanically rotated between the poles of a magnet and an electric current is generated—he found that the current produced heat in the wire also. Thinking that the source of this heat was the work done in turning the handle that rotated the spool, he measured the work done on the handle and related it to the heat in the wire.

It then occurred to Joule that mechanical work might be converted directly to heat, avoiding the electrical step. For many years (from 1842 to 1878) he did experiments on such conversions: friction between two cast-iron plates moving over one another; liquids heated by the rotation of paddle wheels; fluid friction when liquid was forced through small tubes; compression of air with a hand pump. His goal was to determine the quantitative conversion factor between work and heat for every conceivable situation which lent itself to experimental study. In some of the experiments, the temperature changes in his calorimeter were less than 1°F, but the sensitive thermometers that he designed enabled him to obtain readings to 1/200°F.

In Joule's paddle-wheel experiments, a vessel filled with water contained a rotating shaft with several stirring paddles attached to it (Fig. 7–1). Vanes attached to the wall of the vessel prevented the water from rotating freely with the paddles, increasing the internal friction. The shaft and paddles were driven by a falling weight suspended from a pulley. The mechanical work, the product of the weight and its distance, was converted into heat through the friction of the rotating water, and the temperature of the water rose. The effect on the water was the same as if the water had been

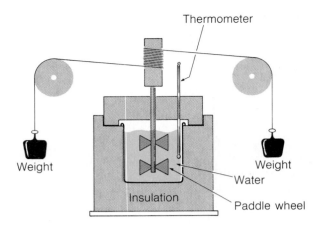

Figure 7–1 Joule's apparatus for measuring the mechanical equivalent of heat. As the weights descend, they cause a paddle wheel to turn in water. The potential energy of the weights is converted into kinetic energy of the paddle wheel, then into heat energy of the water.

heated. Knowing the amount of water, its specific heat, and the rise in temperature, Joule could calculate the amount of heat produced [$w \cdot s \cdot (T_2 - T_1)$] and relate it to the work done.

Joule found that the same amount of work, however it was done, yielded the same quantity of heat. He concluded, therefore, that heat was a form of energy. The equivalence between mechanical units of energy and heat is called the *mechanical equivalent of heat* and is a constant designated by the symbol *J* (in honor of Joule).

$$\text{Mechanical equivalent of heat} = \frac{\text{Work}}{\text{Heat}}$$

Equation   7–1

$$J = \frac{W}{H}$$

The value of J, as determined from the results of many experiments, is

$$J = 4.18\frac{\text{joules}}{\text{calorie}}$$

Since the mechanical equivalent of heat is a constant, calories and joules must be different units of measurement for the same quantity. In other words, 1 calorie is the same quantity of energy as 4.18 joules, just expressed in different units. Further, all of the energy in the systems measured is conserved: the input energy (measured in joules) ultimately appears as heat (measured in calories).

EXAMPLE 7–1

An automobile with a mass of 1000 kilograms (including passengers) is traveling at 36 km/hr. How many calories of heat are produced in bringing it to a stop?

SOLUTION

The kinetic energy of the automobile is equivalent to the work necessary to stop it. From the information given, we calculate the KE, then relate it by Equation 7–1 to the heat evolved in the brakes.

$m = 1000 \text{ kg}$

$$KE = \frac{1}{2}mv^2$$

$$v = 36\frac{km}{hr} = \frac{3.6 \times 10^1 \times 10^3 m}{3600 \text{ sec}}$$

$$= \frac{1}{2}(1000 \text{ kg})\left(10\frac{m}{sec}\right)^2$$

$$= \frac{3.6 \times 10^4 \text{ m}}{3.6 \times 10^3 \text{ sec}} = \frac{10 \text{ m}}{sec}$$

$$= \frac{1}{2}(1000 \text{ kg})\left(100\frac{m^2}{sec^2}\right)$$

$$J = \frac{4.18 \text{ joules}}{\text{calorie}}$$

$$= 50000\frac{kg\text{-}m}{sec^2} \cdot m$$

$$H = ?$$

$$= 5.0 \times 10^4 \text{ joules}$$

$$J = \frac{W}{H}$$

$$\therefore H = \frac{W}{J}$$

$$= \frac{5.0 \times 10^4 \text{ joules}}{4.18\dfrac{\text{joules}}{\text{calorie}}}$$

$$= 1.2 \times 10^4 \text{ calories}$$

Like Joule, the physician Julius Robert Mayer (1814–1878) acquired an interest in the idea of the conservation of energy. Neither man knew that the other had simultaneously arrived at the same ideas. After taking his M.D. degree, Mayer served for a time as ship's physician on a Dutch East Indies freighter, the *Java*. In 1840, he sailed for the East Indies and arrived there after a four-month voyage.

Mayer had taken along Antoine Laurent Lavoisier's (1743–1794) treatise on chemistry, the first modern chemistry textbook. He soon became fascinated by Lavoisier's suggestion that animal heat is generated by the slow combustion of food. This suggestion set him to thinking about the relation between oxidation and the production of energy.

When the *Java* reached the East Indies, many of the crewmen were ill with fever. Mayer resorted to the accepted treatment of the day—bleeding the patient—and made an interesting observation. In his medical studies, Mayer had learned that blood in the arteries is bright red and blood in the veins dark red. But he observed that the crewmen's venous blood was bright red, almost as red as blood in the arteries.

Lavoisier had stated in his treatise that in warm surroundings, less internal combustion is needed to keep the body warm than in cold surroundings. The reason that arterial blood is bright red is that it has a high oxygen content. Mayer concluded, therefore, that the venous blood of his patients was bright red like arterial blood because it, too, had a high oxygen content. This meant that in the tropical East Indies the body did not consume as much oxygen as in cooler places. Mayer went beyond Lavoisier and related the heat of metabolism to the heat the body loses to its surroundings and to the work the body performs. Although he guessed that there was an equivalence between heat and work, and that they were two manifestations of energy, he did not have any experimental proof. However, both Joule and Mayer received fame and recognition following one of the most intense disputes over priority in the history of science.

7–3   THE FIRST LAW OF THERMODYNAMICS

The equivalence of heat and work is incorporated in the *first law of thermodynamics*, which is the same as the *law of conservation of energy:*

Energy cannot be created or destroyed, but may be converted from one form to another, as from work into heat. However, the total energy in the universe is constant.

The statement, "You can't get something for nothing," that is, it is impossible to create energy, characterizes the first law. This truism is based on human experience down through the ages and means that it is futile to try to invent a perpetual motion machine that would perform work without absorbing energy. Al-

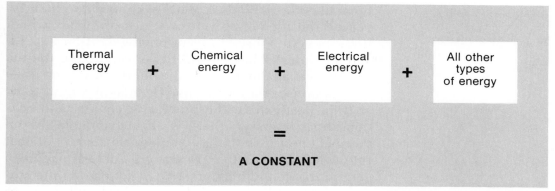

| Thermal energy | + | Chemical energy | + | Electrical energy | + | All other types of energy |

=

A CONSTANT

Figure 7-2 The total of all types of energy in an isolated system remains a constant. Transformations of one type into another may occur as the first law of thermodynamics states.

though claims for such machines have been offered, none has ever been substantiated.

The first law of thermodynamics means that the total energy, including heat, of a closed system isolated from its environment remains constant (Fig. 7–2). Hermann von Helmholtz (1821–1894), who was trained as a physician but made inquiries into physics, physiology, and meteorology as well, convinced the scientific community of the validity of the first law. Declaring that perpetual motion machines were impossible, he showed mathematically and experimentally that this led to the conclusion that energy is always conserved. Both heat and work, he pointed out, are forms of energy; what is conserved is the total of the two forms rather than either taken separately.

7–4 THE SECOND LAW OF THERMODYNAMICS

Although by the first law of thermodynamics, mechanical work can be converted entirely into an equivalent amount of heat, the reverse is not true: *heat cannot be completely converted into work*. In any operating heat engine, such as a steam engine, in which heat is transformed into work, some of the initial energy is wasted. Of the unavailable waste energy in a real engine, some is lost through overcoming friction, some through warming the engine and the atmosphere, some through leakage and other ways. The second law of thermodynamics means that "You can't break even," because in energy-exchange processes heat cannot be completely converted into work.

Nicolas Léonard Sadi Carnot (1796–1832), an army engineer, imagined an ideal engine that was completely insulated, frictionless, and leak-proof. He then invented the idea of a cycle, a series of events which brings a system back to its original condition. For example, water is heated until it is vaporized, the expanding steam forces a piston to move, the steam cools and condenses into water, and the piston returns to its starting position. In the process of cooling and condensation, however, there is an unavoidable loss of thermal energy, so that a complete conversion of heat into work is impossible. In any type of heat engine, a certain fraction of heat must be conducted to the surroundings at a lower temperature. This fact is the basis of another expression of the second law: _The natural direction of heat flow is from a reservoir of heat at a high temperature to a reservoir of heat at a low temperature_.

The second law imposes a limit on the efficiency with which heat can be converted to other forms of energy. Every heat engine makes use of a hot working fluid such as steam or combustion gases in converting heat to mechanical work. The converted energy is obtained from the hot working fluid at a temperature T_1. However, at the lowest temperature of the cycle, T_2, there is heat remaining in the fluid that is not converted to work but is dissipated—either discarded with the working fluid or carried away by the condenser cooling water. In a jet engine, heat is lost in the hot gases of the jet; in a steam turbine, heat is lost in the condenser cooling water.

Suppose Q_1 is the heat that flows into the engine, Q_2 is the heat that flows out of the engine, and W is the work that the engine does (Fig. 7–3). Since heat is equivalent to energy, Q_1 also represents the total energy

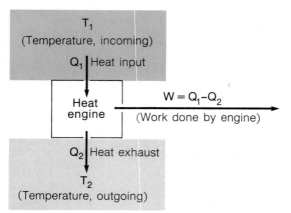

Figure 7–3 Principle of a heat engine. Heat (Q_1) flows into the engine at temperature T_1 and is exhausted (Q_2) at temperature T_2. The work (W) done by the engine = $Q_1 - Q_2$.

input. The heat expelled and the work, $Q_2 + W$, represent the total energy output. According to the first law of thermodynamics, energy is conserved; therefore,

Heat input = Heat output + Work done

$$Q_1 = Q_2 + W$$ Equation 7–2

from which

$$W = Q_1 - Q_2$$ Equation 7–3

From Equation 7–3, we see that the maximum amount of work is obtained from the quantity of heat Q_1 when Q_2 is zero. How can Q_2 be made equal to zero? For the ideal or Carnot engine, it can be shown that

$$\frac{Q_1}{Q_2} = \frac{T_1}{T_2}$$ Equation 7–4

from which

$$Q_2 = Q_1 \frac{T_2}{T_1}$$ Equation 7–5

in which T_1 and T_2 stand for the initial and final absolute temperatures, respectively, between which the engine is operating. Only when $T_2 = 0$ (absolute zero) is the heat expelled, Q_2, equal to zero:

$$Q_2 = \frac{0}{T_1} Q_1 = 0$$

If no heat is exhausted, no energy is lost. Thus all the heat input in an engine can be converted to useful work only if the final temperature is absolute zero.

The maximum possible efficiency of a Carnot engine, that is, the fraction of heat that can be transformed into work, is given by

$$\text{Efficiency} = \frac{\text{Useful work}}{\text{Energy input}}$$

$$= \frac{W}{Q_1}$$

$$= \frac{Q_1 - Q_2}{Q_1}$$

and by Equation 7–4

Equation   7–6
$$\text{Efficiency} = \frac{T_1 - T_2}{T_1}$$

The maximum efficiency of a steam engine depends, Carnot discovered, only upon the temperature difference between the boiler and condenser and the amount of heat that passes between them. All heat engines working between the same temperature levels would have the same maximum efficiency, whether they are turbine or cylinder engines, or whether they use steam, air, or any other working substance.

EXAMPLE   7–2

A highly efficient engine operates between 2000 K (T_1) and 700 K (T_2). What is its maximum possible efficiency?

SOLUTION

We can determine the efficiency of an ideal (Carnot) engine operating between these temperatures. Since a real engine can do no better than this, the efficiency represents the maximum possible.

$T_1 = 2000$ K

$T_2 = 700$ K $\text{Efficiency} = \dfrac{T_1 - T_2}{T_1}$

Efficiency = ?

$$= \frac{2000 \text{ K} - 700 \text{ K}}{2000 \text{ K}}$$

$$= \frac{1300 \text{ K}}{2000 \text{ K}}$$

$$= 0.65, \text{ or } 65\%$$

Since we cannot make $Q_2 = 0$ or avoid friction, some of the original energy Q_1 is not put to work and becomes unavailable, though not destroyed. Consequently, our supply of energy must be decreasing. This

principle of the degradation of energy leads to another
statement of the second law of thermodynamics:

*As a result of natural processes, the
energy in our world available for work is continually decreasing.*

Rudolph Clausius (1822–1888), a mathematical
physicist, combined the work of Joule, Mayer, and
Helmholtz on conservation of energy with the work
of Carnot and gave the second law still another formulation:

*As a result of natural processes the en-
tropy of our world is continually increasing.*

Entropy is the quantitative measure of the disorder of a system. Whenever there is a process of mixing, entropy is created. Entropy S is defined as the
amount of heat received or lost by the body divided
by the body's absolute temperature:

$$\text{Entropy} = \frac{\text{Heat for process}}{\text{Absolute temperature}}$$

$$S = \frac{Q}{T}$$

Equation 7–7

The second law of thermodynamics says that the
total entropy of any closed system must either remain
constant or increase with time; it can never decrease.
A mixture, for example, will not unmix itself spontaneously; this would reduce its entropy. Mixing is an ir-
reversible process.

7–5 "HEAT DEATH," "TIME'S ARROW," AND MAXWELL'S "DEMON"

Clausius predicted that a time will come when
the entire universe will be at the same temperature,
if the laws of thermodynamics apply to the universe as
a whole. This end is called the "heat death" of the universe, when there will be no available energy. Although
the total amount of energy will be the same as always,
there will be no way to use it. With no differences in
temperature—no hills of heat—there can be no work.
The universe will be incapable of further change and

will be in a state of final chaos. The entropy of the universe will have reached its maximum value.

There is no evidence to either confirm or contradict the applicability of thermodynamics to the universe as a whole. We know that conservation laws that are believed to be universally valid must be constantly reexamined and tested. But even if the "heat death" prediction is valid, that event will not come to pass for billions of years.

An interesting aspect of the second law of thermodynamics is that it differentiates between forward and backward in time. There is a preferred direction of natural processes called the "arrow of time." Imagine a motion picture of any scene of ordinary life, then run it backward in your mind. In the time-reversed view of diving off a board into a pool, the diver leaps backward from the pool to the diving board. The backward-in-time views are amusing because things do not happen that way, not because they are forbidden by energy considerations, that is, the first law, but because they violate the second law of thermodynamics.

We know that a sequence of events can occur in one order and not another, because a spontaneous process is associated with an entropy increase. We recognize the impossibility of a natural process in which entropy is decreasing. Of course, there are many examples of a seeming decrease in entropy—a picture is painted, a letter is written, a symphony is played, a baseball game is held—in each case, something becomes more highly organized than it was before. But to accomplish each such organization of matter, energy from food is being inefficiently converted—into sound waves, muscular motion, and so on—so that the total entropy of the universe is still increased.

James Clerk Maxwell (1831–1879), a professor of natural philosophy, imagined a "demon" that could violate the second law and create what Maxwell called a "perpetual motion machine of the second kind." Work would be obtained from a source at a uniform temperature by separating it into a region with molecules having greater than average energy (hotter) and a region with less than average velocities (colder) without expending any energy (Fig. 7–4). Once a temperature difference was established, it could be used to drive a heat engine that would perform useful work. However, if the energy expended by the demon is taken into account, as it must be, then there is no conceivable way a demon could

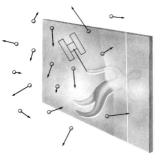

Figure 7–4 Maxwell's "demon" can allegedly establish a temperature difference by separating fast from slow molecules.

be constructed that could violate the second law, and none has been.

7–6   THE STEAM TURBINE

There are practical limits to the sizes of different types of engines. Watt's reciprocating steam engine, which inaugurated the "Age of Steam" and completely revolutionized world industry, was capable of developments on a larger scale than Newcomen's atmospheric engine. The turbine can be built on a still larger scale.

At about the time when reciprocating steam engines had been pushed to their limit, Charles A. Parsons (1854–1931), an engineer, demonstrated the reaction steam turbine. In principle, this turbine is a descendant of Hero of Alexandria's reaction sphere (Fig. 7–5). Some of the early Parsons turbines were built as small as 10 horsepower, but the large sizes were more common. The high speeds of the turbine (10 000 to 30 000 revolutions per minute for the primary shaft) are achieved with a minimum of vibration.

The steam turbine differs from the reciprocating steam engine in that the work expended by the steam in expanding and cooling is done on a set of rotating blades. In principle, a steam turbine is the essence of simplicity; it is a pinwheel driven by high-pressure steam rather than by air. However, it is subject to the same limitation on thermal efficiency as the steam engine.

In the steam turbine engine, jets of steam are directed against the blade or "buckets" of the turbine, exert forces on the blades, and keep the wheel in motion (Fig. 7–6). Fixed blades are mounted between adjacent moving wheels. The steam jet strikes against the first set of moving buckets and rebounds, giving up some of its energy. Then the jet, reversed by the fixed blades, strikes the next revolving wheel. The steam gradually gives up momentum and energy as it does work. The turbine, connected to an electric generator, does work in turning the shaft. The heat energy of the steam is converted in this way to the kinetic energy of rotary motion. Steam turbines account for more than 75 per cent of the electric power generated in the world (most of the rest is hydroelectric). When large unit output and maximum efficiency are essential, as for the propulsion of big ships and power generation, the steam turbine is unchallenged.

Figure 7–5 In this model of Hero of Alexandria's aeolipile, the escape of steam through the nozzels causes the sphere to rotate.

Figure 7–6 One of the largest steam turbines built with a single shaft. Its capacity is 400 000 horsepower. (Courtesy of Brown, Boveri & Company, Ltd., Baden, Switzerland.)

7–7   INTERNAL COMBUSTION ENGINES

The essential concept of internal combustion is that combustion can take place inside the working fluid without any need for a firebox and a boiler. Carnot had suggested that air in a heat engine can serve as the medium of energy conversion and at the same time, because of its oxygen content, can participate in the generation of heat. By successfully working out Carnot's idea, Nicolaus August Otto (1832–1891), an inventor, and Rudolf Diesel (1858–1913), a professor of thermodynamics, achieved a revolution in energy conversion.

More than 10 million reciprocating internal-combustion engines are built in the United States each year to provide power for automobiles, trucks, boats, locomotives, and small airplanes. More than 100 million such engines are in service. Nearly all are based on a principle first demonstrated in 1876 by Otto, who had found a way to burn a compressed mixture of illuminating gas and air in the cylinder of an engine without producing destructive explosions.

In an "Otto cycle," four functions must be per-

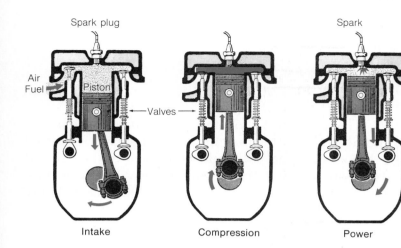

Intake   Compression   Power   Exhaust

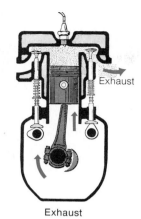

Figure 7–7 Conventional piston engine employs 4-stroke cycle first demonstrated by Nicolaus August Otto: intake, compression, power, exhaust. After compression, the mixture of fuel and air is ignited by a spark plug. The expanding gases produce the power stroke.

formed (although not necessarily in four distinct strokes): fuel-and-air intake, compression, power, and exhaust (Fig. 7–7). Two, four, or six strokes may be used to accomplish the four functions. An engine of the Otto type takes in a controlled mixture of fuel and air, compresses it, and ignites it by a spark plug. The original Otto engine had an efficiency three to four times that of a steam engine.

Although automobiles are usually powered by an Otto engine, the Diesel has revolutionized heavy transportation—heavy trucking, earth-moving equipment, and ships under 20 000 horsepower. Rudolf Diesel considered the engine that he patented in 1892 to be not just an improved heat engine, but a revolutionary one. He insisted that it was an engine built on scientific principles, a "rational" engine, and he expected it to displace the steam engine within a few years and to drive everything from battleships to sewing machines. He thought of it as a Carnot engine that would have a thermal efficiency of 73 per cent (compared to 7 per cent for the best steam engines of the time). As it turned out, the Diesel engine we know today is not a Carnot engine at all, and is quite different from Diesel's original idea.

The Diesel evolved over a period of 25 years. In World War I, most submarines and some ships were powered by Diesels. The modern Diesel appeared in the 1920's, when suitable fuel injectors were perfected (Fig. 7–8). It is a high-compression engine, with a compression ratio of about 16:1 (compared to Otto ratios of about 7 or 8 to 1). Air is the working fluid. The fuel is injected near the end of the compression stroke, and is

Figure 7–8 Modern Diesel engine. Near the end of the compression stroke, fuel is injected and ignites by the heat of compression.

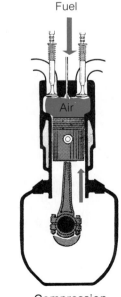

Compression

ignited by the heat of compression. The thermal efficiency is high—up to 40 per cent for the best Diesels.

7–8   THE ROTARY ENGINE

There are strong incentives to replacing the piston engine. Foremost among them are new laws requiring a reduction in the pollutants contained in the exhaust gases and a reduction in manufacturing costs. Several alternatives to the reciprocating piston engine are being explored: the gas turbine, electric propulsion, and a number of steam or vapor engines.

A proposed successor to the reciprocating piston engine is one invented by Felix Wankel (b. 1902) and known as the Wankel engine, the rotary engine, or the R. C. engine (rotating combustion engine) (Fig. 7–9). The apexes of the triangular rotor of a Wankel are always in contact with the housing, and create three separate variable volumes. There are no valves in the Wankel; the rotor itself opens and closes the fuel-air inlet port and the exhaust port at appropriate times in the combustion cycle. With only about half as many moving parts as a conventional engine, a Wankel is only about half the size and weight, and is almost completely vibrationless. It may be more reliable and serviceable than piston engines. Its fuel-octane requirement is less, and it can run well with unleaded fuel.

7–9   GAS TURBINE OR TURBOJET ENGINE

For powering high-speed aircraft, the gas turbine (also known as the turbojet, or jet engine) has replaced

Figure 7–9   The Wankel engine, a rotary engine, undergoes a 4-stroke Otto cycle in each of three small, variable chambers in one revolution of the rotor: intake, compression, ignition by a spark plug, exhaust.

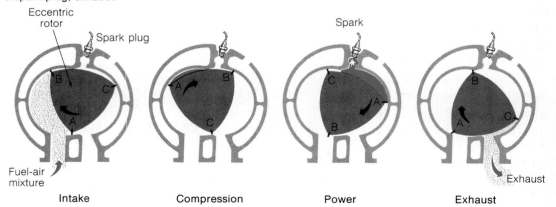

| Intake | Compression | Power | Exhaust |

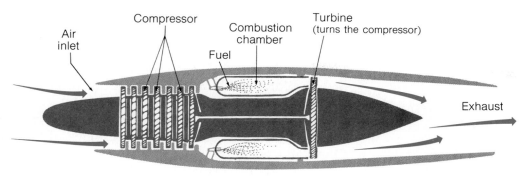

Intake   Compression   Power   Exhaust

Figure 7–10 Turbojet engine. In relation to the internal combustion engine, the basic 4-stroke functions are carried out simultaneously in different parts of the turbojet engine.

the piston engine. It is a non-reciprocating internal combustion engine which transforms heat energy directly into work, eliminating the intermediate steps of the piston, shaft, wheel, and propellor. The gas turbine liberates a large amount of energy in a short time by operating continuously and at higher speeds than a piston engine, since it does not have reciprocating parts.

The gas turbine works as follows: Air is drawn into a compressor in which its pressure is raised, then into a combustion chamber (Fig. 7–10). There it is mixed with fuel and produces an intense flame. The combustion gas passes through a turbine, its velocity increasing, and then through an exhaust nozzle and out to the atmosphere. A powerful thrust (net force) is developed at high speeds from the change in momentum of the gases between the inlet and the exhaust. The turbine drives the compressor.

The efficiency of jet engines, as with other heat engines, depends upon the temperature limits between which they operate. The most serious efficiency limitation is the temperature of the combustion gas, which must be kept low enough to prevent destruction of the turbine blades, usually below 1000°C. The efficiency of an aircraft gas turbine is about 20 per cent.

7–10 ROCKETS

The simplest type of jet propulsion, in principle, is the rocket. A gas is generated at high temperature and high pressure as it flows through the exhaust nozzle. In the combustion chamber, chemical energy is converted into heat with an efficiency on the order of 95 per cent, and in the exhaust nozzle heat is converted into the kinetic energy of the jet with an efficiency of about 50 per cent. The unconverted heat is expelled with the hot gas. Unlike the turbojet, the rocket engine carries its own oxidizer, usually liquid oxygen or concentrated nitric acid, and does not take in air. Liquid hydrogen and liquid oxygen powered the 200 000-lb thrust J-2 engines in the second and third stages of the Saturn V-Apollo system that sent astronauts to the moon.

The rocket principle itself is centuries old. The Chinese may have used rockets in the 13th century; Arabs of the period referred to rockets as "Chinese arrows." These rockets were fueled with black powder, a mixture of sodium nitrate, charcoal, and sulfur. For fireworks rockets, fine metal filings might be added to produce a shower of sparks. Their tendency to explode during manufacture led Alfred Nobel (1833–1896), a chemist, to develop smokeless powders from nitrocellulose, nitroglycerine, and stabilizers, which became the basis of modern solid-fuel rockets.

With a solid-fuel rocket, nothing can be done to control its flight after it has been ignited. Konstantin Tsiolkovsky (1857–1935), an inventor and rocket expert, in 1903 suggested a new approach—that rockets be designed for liquid fuels. With liquid fuels, the rate of burning can be adjusted by means of valves or injection nozzles, either by a radio signal or directly in a manned rocket.

Robert Hutchins Goddard (1882–1945), a physicist at Clark University, in 1919 presented the first modern paper on rocket theory, "A Method of Reaching Extreme Altitudes," in which he included a reference to liquid fuels. In his book, *The Rocket into Interplanetary Space* (1923), Hermann Oberth strongly recommended the use of liquid fuels, with liquid oxygen as the oxidizer. On March 16, 1926, Goddard launched the first liquid-fuel rocket, consisting of gasoline and liquid oxygen, on a farm in Auburn, Massachusetts. The rocket landed 184 feet away, the precursor to the flights to the moon of the 1960's and 1970's (Fig. 7–11).

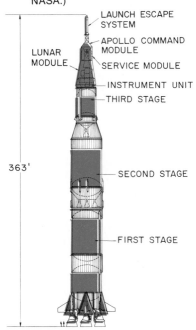

Figure 7–11 The Apollo/Saturn V rocket, which propelled astronauts to the moon. (Courtesy of NASA.)

LAUNCH ESCAPE SYSTEM

APOLLO COMMAND MODULE

LUNAR MODULE

SERVICE MODULE

INSTRUMENT UNIT

THIRD STAGE

363'

SECOND STAGE

FIRST STAGE

7–11   THE REFRIGERATOR

The refrigerator acts as a heat engine in reverse, a heat pump. It extracts heat from food and water at a low temperature, does work, and delivers heat to the air surrounding the refrigerator (the kitchen).

In the electric refrigerator, an electric motor operates a compressor (Fig. 7–12). The working substance, called the refrigerant, is a gas that is easily condensed by pressure. Ammonia (NH_3), sulfur dioxide (SO_2), or Freon (CCl_2F_2) is commonly used. The refrigerant is taken through a cycle that is repeated until the temperature is reduced to the point where a thermostat opens a switch and cuts off the motor. When the interior warms up enough, the thermostat closes the switch and the cooling cycle begins again.

A typical cycle may start with the refrigerant as a vapor. The vapor is taken to the compressor, where a pump compresses it and heats it to a temperature slightly above room temperature. In the pipes of the condenser, the compressed and liquefied fluid gives up heat to the circulating air and is cooled. Then it goes to the cooling unit where it evaporates and reduces the temperature of the surroundings. In this process heat is

Figure 7–12 Principle of the refrigerator. (From Lee, G. L., et al.: *General and Organic Chemistry.* Philadelphia: W. B. Saunders Company, 1971, p. 400.)

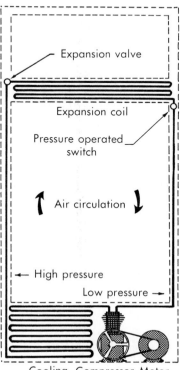

Expansion valve

Expansion coil

Pressure operated switch

Air circulation

High pressure

Low pressure

Cooling coil   Compressor   Motor

extracted from the contents of the cooling unit of the refrigerator and is used to vaporize the refrigerant. The vapor is returned to the compressor, and the cycle is repeated.

7–12   FUEL CELLS

Most of the electrical energy produced in the world is generated through some form of heat engine fueled by the combustion of coal, petroleum, or natural gas. The source of that energy is the chemical bonds in these fossil fuels. In the process of obtaining electrical energy in this way, most of the chemical energy is wasted in the three-step conversion process: The chemical energy is first converted by combustion of the fuel into heat; the heat is then transformed into mechanical energy in a heat engine; and, finally, the mechanical energy is converted into electrical energy in a generator. The overall efficiency of this process suffers from incomplete combustion, heat losses, and the limitations of mechanical devices. Even under ideal conditions, the thermodynamics of heat engines limits the efficiency of converting thermal energy into mechanical work. At the operating temperatures of heat engines, the second law of thermodynamics decrees that more than half of the original chemical energy must be lost as wasted heat. (Nuclear power plants are subject to this same limitation and are even less efficient than fossil-fuel plants because of their lower operating temperature.)

In a fuel cell, the heat cycle is bypassed and the chemical energy of fuels is converted directly into electricity. This idea is not new. Sir William Grove (1811–1896), a lawyer whose hobby was physics, in 1839 constructed a battery in which hydrogen reacted with oxygen to form water and an electric current was generated. The basic fuel is used directly in a true fuel cell, unlike zinc or lead in batteries which must be refined at great expense. With fuel cells, efficiencies of 75 per cent, more than twice that of the average steam power-station, are possible. Considerable development work will be required to realize such possibilities on a wide commercial scale, although fuel cells have already proved feasible in the Gemini and Apollo spacecrafts.

In order to send astronauts to the moon and back, efficient and lightweight energy sources had to be developed. Since batteries and solar cells were judged inadequate for such space flights, the fuel cell appeared to be the best solution. A fuel cell can produce up to six times as much electricity as the best alkaline battery. Moreover, a fuel cell operating on hydrogen and oxygen produces pure water as a by-product, which can be used by the astronauts for drinking, cooking, and for cooling equipment. Fuel cells provided on-board electrical power for Apollo spacecraft, each vehicle using three power plants of 31 cells each, connected in series. Many believe that the fuel cell has a bright future, although commercial fuel cells may rely on the conversion of hydrocarbons and air, rather than hydrogen and oxygen, into electricity. If they are right, in the 1990's cars and homes could be powered by fuel cells.

CHECKLIST

Carnot engine
Diesel engine
Entropy
First law of thermodynamics
Internal combustion
Liquid-fuel rocket
Maxwell's "demon"
Mechanical equivalent of heat
Otto engine
Reciprocating engine
Rotary engine
Second law of thermodynamics
Thermal efficiency
Turbine
Turbojet engine
Wankel engine

EXERCISES

7–1.   What did Joule's experiments prove about energy?

7–2.   Explain what is meant by the conservation of energy.

7–3.   Give an example of a device that can perform the conversion of (a) electrical energy to thermal energy; (b) mechanical energy to thermal energy; (c) chemical energy to thermal energy.

7–4.   Water at Niagara Falls undergoes a 160-foot drop. Would you expect a difference in the temperature of the water between the top and bottom of the Falls? Explain.

7–5.   A 75-kilogram man climbs a staircase 7 meters high. How much energy (in calories) does he expend?

$3000 kcal \left(\dfrac{1000 \, cal}{1 \, kcal}\right)\left(\dfrac{4.18 \, j}{1 \, cal}\right)$

$= 1.25 \times 10^7 \, j$

#7

$W = mgh$

$h = \dfrac{W}{mg} = \dfrac{3000 kcal}{75 kg (9.8 m/s^2)}$

$= \dfrac{1.25 \times 10^7 j}{(75 kg)(9.8 m/s^2)}$

$= 1.7 \times 10^4 \, m$

7-6. In highly developed countries, a person consumes approximately 3000 kilocalories per day as food. Explain what happens to this energy.

7-7. If the energy in the food which a man consumes in one day, 3000 kcal, were all converted into mechanical energy, how high would it lift a 75-kilogram mass?

7-8. Which law of thermodynamics can be paraphrased as follows: "You can't win; the best you can do is break even"? Discuss.

7-9. Could a heat engine be designed with a thermal efficiency of 100 per cent? Explain.   *NO*

$eff = \dfrac{W}{Q_1} = \dfrac{2000 j}{5000 cal}$

$= \dfrac{480 cal}{5000 cal}$

$= 9.6\%$

7-10. The input of a heat engine is 5000 calories per cycle, and the work output is 2000 joules per cycle. What is the efficiency of the engine?   *9.6%*

7-11. The efficiency of an engine that obtains heat at 260°C and exhausts heat at 60°C is 16 per cent. What is its maximum possible efficiency?   *must be K°*

$EFF = \dfrac{T_1 - T_2}{T_1}$

$= \dfrac{533°k - 333°k}{533°k}$

$= \dfrac{200}{533}$

$= 38\%$

7-12. A steam turbine operates between the temperatures of 800°C and 20°C with an efficiency of 33 per cent. How does its efficiency compare with the efficiency of a Carnot engine operating between the same temperatures?

7-13. What is the Carnot efficiency of a steam turbine working between 300°C and 40°C?

7-14. Describe what happens to the waste heat in an automobile engine, which has an efficiency of 25 per cent.

7-15. Apply the first law of thermodynamics to the motion of an automobile.

7-16. State some advantages and disadvantages of rotary engines over reciprocating engines.

7–17. How does a rocket engine differ from a turbojet engine?

7–18. How did James Watt influence the modern world?

7–19. Would you expect a kilogram of ice to have more or less entropy than a kilogram of water?

7–20. List some favorable and adverse effects of the internal combustion engine upon society.

7–21. How are the concepts of entropy and "time's arrow" related?

7–22. Multiple Choice

A. A heat engine is designed to
(a) heat houses
(b) cool houses
(c) change work into heat
(d) change heat into work

B. The law of conservation of energy is expressed by the
(a) first law of thermodynamics
(b) second law of thermodynamics
(c) first law of motion
(d) second law of motion

C. A mixture of salt and pepper separating into separate layers when shaken violates the
(a) conservation of energy
(b) conservation of momentum
(c) law of entropy
(d) principle of efficiency

D. The mechanical equivalent of heat is expresed by
(a) J   (c) H
(b) W   (d) C

E. The Carnot efficiency of a Diesel engine is
(a) easy to attain
(b) its maximum possible efficiency
(c) 100 per cent
(d) a measure of its thermal energy

SUGGESTIONS FOR
FURTHER READING

Angrist, Stanley W., "Perpetual Motion Machines." *Scientific American,* January, 1968.

> Proposals of getting something for nothing have foundered on either the first or second law of thermodynamics.

Bryant, Lynwood, "Rudolf Diesel and His Rational Engine." *Scientific American,* August, 1969.

> Although the Diesel engine failed to attain the ideal of the Carnot cycle, it has revolutionized heavy transportation.

Cole, David E., "The Wankel Engine." *Scientific American,* August, 1972.

> The aim of this rotary engine is to combine the virtues of the piston engine and the gas turbine.

Fickett, Arnold P., "Fuel-Cell Power Plants." *Scientific American,* December, 1978.

> These devices convert the energy of fuel directly into electricity. They are being scaled up for possible use by electric utilities.

Hossli, Walter, "Steam Turbines." *Scientific American,* April, 1969.

> These machines efficiently convert heat energy into the kinetic energy needed to drive generators and large ships.

Pierce, John R., "The Fuel Consumption of Automobiles." *Scientific American,* January, 1975.

> Can the efficiency of American cars be increased by at least 40 per cent? The author suggests ways to do this.

Wilson, David Gordon, "Alternative Automobile Engines." *Scientific American,* July, 1978.

> The search for an alternative to the conventional Otto engine is stimulated by requirements for cleaner and more efficient engines.

WAVE MOTION AND SOUND

8-1 INTRODUCTION

As we have seen, a particle is an entity that has a definite mass associated with it. A baseball moving through the air may be viewed as a particle that by virtue of its motion carries energy and momentum. Although a wave, too, can carry energy and momentum, it does not necessarily carry any mass with it. Yet a wave can cause dishes to rattle, or the eardrum to oscillate, or a glass to shatter.

Waves are a direct connecting link between our senses and our physical environment. The colors of a sunset, the sounds of children at play, the surface of the sea, the flapping of a flag, the shock waves emitted by a jet plane flying at supersonic speeds—these are all wave phenomena. Waves of all kinds—mechanical, acoustical, electrical, and optical—behave fundamentally alike in most respects. They propagate, reflect, refract, superpose, interfere, and diffract.

8-2 THE WAVE CONCEPT

Think of a *wave* as a disturbance that is propagated through a medium. The disturbance—a change in the condition of the medium—may be a displacement of a material medium, as in the case of water waves and sound waves. Or it may be a change in an electromagnetic field in empty space, as in the case of light and radio waves. In all waves, energy is transferred from one point in space to a neighboring point.

For example, when you throw a rock into a quiet pond, circular ripples spread out from the point of impact and travel over the surface of the pond. If the wave passes a toy sailboat, the sailboat bobs up and down but does not move forward with the advancing wavefront (Fig. 8–1). Although there is no large scale movement of water from the point of impact of the rock to any other point, the wave causes physical effects some distance away. The features of wave motion are: a source of energy—the splashing rock; a medium—the water; and a moving pattern of displacement that we call the wave.

If the disturbance is a single event of short duration, it is called a *pulse* (Fig. 8–2). Supersonic jet aircraft set up pulses called *shock waves* in air, speedboats produce them in water, and earthquakes produce them in rocks. When the disturbance is repeated at a regular interval, called the period T, a *periodic wave* is propagated.

We'll demonstrate some wave properties with a ripple tank (Fig. 8–3). A shallow glass-bottomed tank containing water is illuminated from above. Ripples are produced by a rod that pushes up and down on the surface of the water, and their shadow is cast on a screen of white paper on the table under the glass. The highest points of waves are called *crests,* the lowest points *troughs.* The crests and troughs of the waves appear on the screen as alternating bright and dark circular bands moving outward from the rod (Fig. 8–4). The *amplitude* is the height of a crest or the depth of a trough, as measured from the water level when there is no disturbance (the equilibrium position).

Figure 8–1 As the wave-front proceeds from left to right, the boat bobs up and down in place.

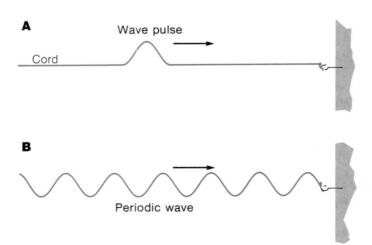

A

Wave pulse

Cord

B

Periodic wave

Figure 8–2 *A,* A single wave pulse sent out along a cord. *B,* A periodic wave.

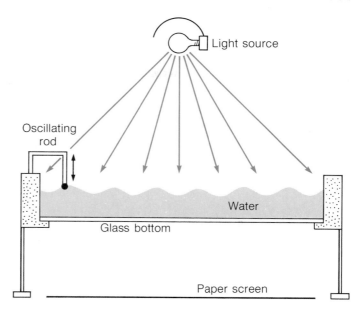

Figure 8-3 A ripple tank.

Figure 8-4 Alternate crests and troughs produced in a ripple tank. (Courtesy of Bill Jack Rodgers.)

As long as the rod oscillates at a constant rate, the disturbance is repeated at a regular interval called the *period,* T. The number of waves produced per unit time is the *frequency,* f (Fig. 8–5). If the frequency is 10 waves/second, the period is 0.1 second/wave and therefore:

$$\text{Period} = \frac{1}{\text{frequency}}$$

Equation 8–1

$$T = \frac{1}{f}$$

from which

Equation 8–2

$$f = \frac{1}{T}$$

We see then that T and f are two ways of expressing the same idea. If the period is short, the frequency is high. The SI unit for frequency, the cycle/second, is named the hertz (Hz) after Heinrich Hertz (1857–1894), the physicist:

$$1 \text{ hertz (Hz)} = 1\frac{\text{cycle}}{\text{second}}$$

Figure 8–5 Frequency of a wave, measured in cycles per second (hertz), is the number of crests or troughs that pass a given point in one second. The frequency of wave **B** is greater than that of wave **A.**

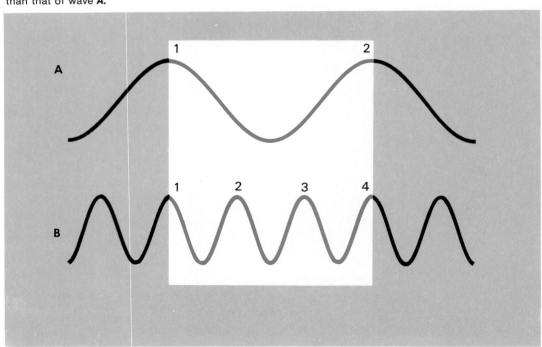

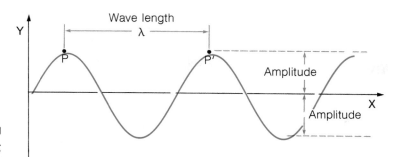

Figure 8-6 Wavelength and amplitude of a periodic wave. Points P and P′ are "in phase."

The *wavelength*, λ, is the distance between any two points in the medium that, at any instant, are in precisely the same state of disturbance (Fig. 8–6). The distance from crest to crest on two successive waves, therefore, or from trough to trough, is the wavelength. Points on a wavetrain that are a wavelength apart are said to be "in phase," since they are points of equal disturbance; they are also in phase if they are 2, 3, 4, ... wavelengths apart. On the other hand, points on a crest and a trough and similar points that differ by half a wavelength are completely "out of phase."

If we watch a particular crest of a wave, we will see it traveling with a constant velocity (Fig. 8–7). Initially, at point P, the crest moves to points P_1 and P_2 in successive small time intervals. During the time T, the crest has moved forward a whole wavelength to the position occupied by the preceding crest at the start of this time interval. The velocity v with which a periodic wave is propagated is, therefore, given by

$$v = \frac{s}{t}$$

$$= \frac{\lambda}{T} \text{ or } \frac{1}{T} \cdot \lambda$$

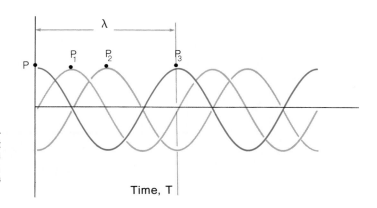

Figure 8-7 A periodic wave propagating to the right shown at different time intervals. The crest at P moves forward a whole wavelength in time, T. The velocity, v, therefore equals λ/T.

Since $1/T = f$, we have a relationship called the *wave equation:* velocity = frequency × wavelength, or

Equation 8–3 $$\mathbf{v} = \mathbf{f} \cdot \lambda$$

Since for many types of waves the velocity is well known, we may use the wave equation to calculate the wavelength when the frequency is known, or the frequency when the wavelength is known. For a wave of a particular velocity, if we generate waves at a faster rate, their crests will be closer together. So, the higher the frequency, the shorter the wavelength.

EXAMPLE 8–1

In a ripple tank, waves are generated at a rate of 10 per second. If the wavelength is 2.5 centimeters, (a) what is the wave velocity? (b) what is the period?

SOLUTION

Since the frequency and wavelength are known, the wave equation may be applied to find the velocity. We can get the period from Equation 8–1.

$$f = 10\frac{\text{waves}}{\text{sec}} \qquad \text{(a)} \;\; v = f \cdot \lambda$$

$$= \left(10\,\frac{\text{waves}}{\text{sec}}\right)\left(\frac{2.5\;\text{cm}}{\text{wave}}\right)$$

$$\lambda = 2.5\;\text{cm}$$

$$v = ? \qquad\qquad\qquad\qquad = 25\frac{\text{cm}}{\text{sec}}$$

$$T = ?$$

$$\text{(b)} \;\; \text{Period} = \frac{1}{\text{frequency}}$$

$$T = \frac{1}{10\,\dfrac{\text{waves}}{\text{sec}}}$$

$$= 0.1\frac{\text{sec}}{\text{wave}}$$

The simplest type of regular wavetrain is called *sinusoidal* (since it has the same shape as a graph of the

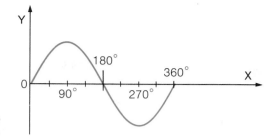

Figure 8–8 One wavelength of a sine wave divided into 360°. Waves of this type are referred to as sinusoidal.

sine function in trigonometry) (Fig. 8–8). A small cork on water undergoes sinusoidal motion as it is displaced to a maximum from its equilibrium position, first in one direction, then in the opposite direction, by a progression of wavecrests and wavetroughs. The motion of a swing and the oscillations of a pendulum are also sinusoidal, alternating between a maximum displacement in one direction, a return to the equilibrium or starting position, and a maximum displacement in the opposite direction.

8–3 TRANSVERSE WAVES

Imagine a long horizontal spring with one end attached to a wall and the other end held in the hand and pulled taut. By moving the near end up and down rapidly, you can generate a wave that travels along the spring in a horizontal direction. Each particle in the spring, however, oscillates up and down in a vertical plane at right angles to the direction in which the wave is traveling (Fig. 8–9). This kind of wave is called a *transverse wave*.

Light waves are transverse waves. They differ from transverse waves on a spring or on the surface of water; the latter are mechanical waves in which material particles are displaced. Light waves are electromagnetic waves in which the disturbance involves changing electric and magnetic fields traveling with a velocity of 3×10^8 m/sec (about 186000 miles/sec) in a vacuum, which is equivalent to traveling around the earth at the equator seven times in one second! Visible light ranges in wavelength from about 4000 Å at the violet end to about 8000 Å at the red end. Although our eyes cannot see beyond this range on either side, we know from measurements that there are electromagnetic waves both longer and shorter; radio waves are among the longest, and gamma rays the shortest (Table 8–1).

Figure 8–9 Transverse wave traveling along a coiled spring from right to left. The ribbon vibrates at right angles to the direction of motion of the wave. (From Physical Science Study Committee, *Physics.* Boston: D. C. Heath & Co., 1960, p. 250.)

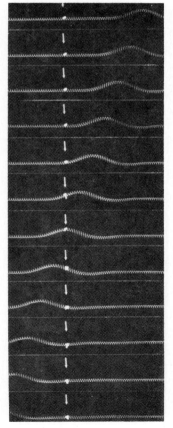

Table 8–1   WAVELENGTHS AND FREQUENCIES OF TYPICAL
ELECTROMAGNETIC RADIATIONS

TYPE OF WAVE	WAVELENGTH (cm)	RANGE (Å)*	FREQUENCY RANGE (Hz)
AM radio	6×10^4 to 1.5×10^3	6×10^{12} to 1.5×10^{11}	0.5×10^6 to 2×10^7
TV and FM radio	7.5×10^2 to 1.5×10^2	7.5×10^{10} to 1.5×10^{10}	4×10^7 to 2×10^8
Microwaves	30 to 0.1	3×10^9 to 10^7	10^9 to 3×10^{11}
Near infrared	0.03 to 1.0×10^{-4}	3×10^6 to 8 000	10^{11} to 3.0×10^{14}
Visible light			
Red	7.6×10^{-5}	6 300–8 000	3.9×10^{14}
Orange	6.1×10^{-5}	5 900–6 300	4.9×10^{14}
Yellow	5.9×10^{-5}	5 600–5 900	5.1×10^{14}
Green	5.4×10^{-5}	4 900–5 600	5.6×10^{14}
Blue	4.6×10^{-5}	4 000–4 900	6.5×10^{14}
Ultraviolet	4×10^{-5} to 3×10^{-6}	4×10^3 to 3×10^2	7.5×10^{14} to 10^{16}
X-rays	3×10^{-6} to 10^{-10}	3×10^2 to 10^{-2}	10^{16} to 3×10^{20}
Gamma rays	3×10^{-8} to 10^{-13}	3 to 10^{-5}	10^{18} to 3×10^{23}

*1 Å = angstrom = 10^{-10} m

EXAMPLE 8–2

A television channel broadcasts waves with a frequency of 55 MHz. What is their wavelength?

SOLUTION

Since you know the frequency and velocity of the waves, apply the wave equation to get the wavelength.

$$v = 3 \times 10^8 \text{ m/sec} \qquad v = f \cdot \lambda$$

$$f = 55 \text{ MHz}$$

$$\therefore \lambda = \frac{v}{f}$$

$$= 55 \times 10^6 \frac{\text{cycles}}{\text{sec}}$$

$$= 5.5 \times 10^7 \frac{\text{cycles}}{\text{sec}} \qquad = \frac{3 \times 10^8 \frac{\text{m}}{\text{sec}}}{5.5 \times 10^7 \frac{\text{cycles}}{\text{sec}}}$$

$$\lambda = ?$$

$$= 0.55 \times 10^1 \frac{\text{m}}{\text{cycle}}$$

$$= 5.5 \frac{\text{m}}{\text{cycle}}$$

8–4 LONGITUDINAL WAVES

The spring described at the beginning of Section 8–3 can oscillate in another mode—forward and backward in a horizontal direction. The spring remains straight, but a wave of alternating compressions and rarefactions travels along its length (Fig. 8–10). (A *compression* is a region in which the coils are closest together; a *rarefaction* is one in which they are spread out.) A particular coil in the spring undergoes periodic motion in a horizontal direction parallel to the direction in which the wave is traveling. Wave motion of this kind is called *longitudinal* (or *compressional*). Sound waves are our most familiar example of longitudinal waves: the disturbance of the medium occurs parallel to the direction in which the wave is proceeding.

The wavelength of a longitudinal wave is defined as the distance from a point on one compression to a point in the same phase of oscillation on an adjacent compression. Each wave is composed of a compression and a rarefaction. The frequency is the number of compressions or rarefactions that are produced in one second.

The energy transfer that can take place with compressional waves can be illustrated as follows. A number of metal or glass spheres are laid side by side between two meter-long sticks. When a sphere is rolled along the groove and strikes one end of the row, a sphere at the other end is kicked away some distance. The spheres in the middle of the row are not perceptibly disturbed, although they transmit energy, as is evident from the behavior of the sphere at the end of the row. These spheres transmit the disturbance while they themselves oscillate imperceptibly about their equilibrium positions in the direction in which the wave is traveling.

Figure 8–10 A longitudinal wave on a stretched spring. The particles of the spring oscillate in the direction of the wave. A wavelength is the distance between two adjacent points that are in phase.

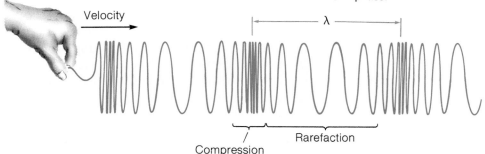

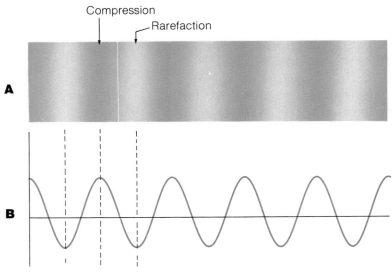

Compression

Rarefaction

A

B

Figure 8–11 Compressions
and rarefactions of longitudinal
waves *(A)* compared to crests
and troughs of transverse waves
(B).

A sound wave in air consists of a train of compressions (high pressure regions) following one another in rapid succession with the frequency of the sound, and separated by rarefactions (regions of lower pressure). For the sake of convenience, we represent sound waves graphically as sinusoidal waves (Fig. 8–11). A crest on a sine wave representing sound corresponds to a compression, and a trough a rarefaction. If a light pith ball is held by a thread close to a vibrating tuning fork, the ball is thrown violently outward and comes to rest only as the sound ceases. The air is compressed when the tine of the fork moves outward, and is rarefied when it moves inward (Fig. 8–12). The vibrating surfaces of the tuning fork thus become the center of compressions and rarefactions that spread out in all directions with the speed of sound. The individual molecules of air through which sound is being transmitted oscillate backward and forward in a direction parallel to the disturbance.

That sound can be heard only if there is a medium —solid, liquid, or gaseous—between the source and the ear to transmit it is shown graphically by the bell-in-vacuum demonstration (Fig. 8–13). A ringing alarm clock cannot be heard in a jar from which the air has been exhausted, but becomes audible when air is reintroduced. The ear is sensitive to an amplitude in air of around a thousand-millionth of a centimeter (even less amplitude in liquids and solids). The maximum

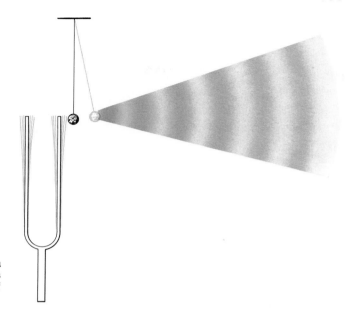

Figure 8–12 Vibrating tuning fork is a center of compressions and rarefactions that travel outward with the speed of sound.

displacement of the molecules in air from their original positions is seldom more than 0.5 millimeter.

8–5 VELOCITY OF SOUND WAVES

It is a matter of common observation that sound has a definite velocity in air. We hear thunder moments after seeing a flash of lightning. Sitting in a grandstand in a baseball park, we see a fly ball before hearing the crack of the bat. This means that sound takes a measurable period of time to reach us.

The speed of sound can be determined directly by time-interval measurements. A gun or cannon is fired while observers stationed at various distances measure the time interval between the flash of the gun and sound of the firing. Dayton C. Miller (1866–1941), a physicist at what is now Case-Western Reserve University, applying this method with coastal defense guns, made some of the most accurate measurements in modern times. The development of modern electronics has made possible even more precise measurements of small time intervals. In air at 0°C and 1 atmosphere of pressure, the velocity of sound is 331 m/sec (1087 ft/sec). The velocity varies with the temperature, changing approximately 0.6 meter per second per Celsius

To vacuum pump

Figure 8–13 The sound of the bell is not transmitted through a jar from which the air has been pumped out.

degree (2 ft/sec/C°). Therefore, over a wide range of temperature, T, the velocity of sound is given by the following expressions:

Equation 8–4
$$v_T(\text{m/sec}) = v_0 + 0.6\,T$$

Equation 8–5
$$v_T(\text{ft/sec}) = v_0 + 2\,T$$

EXAMPLE 8–3

Calculate the velocity of sound (m/sec) in air when the temperature is 35°C.

SOLUTION

Use Equation 8–4 for determining the velocity of sound in air at various temperatures.

$$T = 35°C$$

$$v_0 = 331 \text{ m/sec}$$

$$v_{35°C} = \,?$$

$$v_{35°C} = v_0 + \frac{0.6\dfrac{\text{m}}{\text{sec}}}{°C}(35)°C$$

$$= 331\frac{\text{m}}{\text{sec}} + 21\frac{\text{m}}{\text{sec}}$$

$$= 352\frac{\text{m}}{\text{sec}}$$

Table 8–2 VELOCITY OF SOUND IN VARIOUS MEDIA

	MEDIUM	*m/sec*	*ft/sec*
Gases	Air (0°C)	331.4	1 087
	Carbon dioxide (0°C)	258	846
	Helium (20°C)	927	3 040
	Hydrogen (0°C)	1 270	4 165
	Oxygen (0°C)	317	1 041
	Water vapor (35°C)	402	1 320
Liquids	Mercury	1 450	4 650
	Sea water	1 520	4 900
	Water (20°C)	1 461	4 794
Solids	Glass	5 500	18 033
	Granite	3 950	12 600
	Lead	1 230	3 950
	Pine	3 320	10 900
	Steel	5 000	16 400

The speed of sound waves is different for each gas. In liquids and solids, the speed of sound is considerably greater, as Table 8–2 shows. The difference is apparent if a steel rail is struck by a hammer. A distant observer will hear two sounds, the first coming through the rail, which transmits the sound readily; the second through the more slowly transmitting air.

8–6 THE DOPPLER EFFECT

The word *frequency,* as we have seen, is an objective property; that is, it is the number of cycles per second. When we hear a high-frequency sound, we say, subjectively, that the "pitch" is high. Likewise, for low-frequency sounds, we say we hear a low pitch. When there is relative motion between a source of waves and an observer, an apparent change occurs in the frequency of the sound; that is, there is a change in pitch (Fig. 8–14). A few familiar examples illustrate this. The pitch of the siren on an ambulance moving in traffic seems to be higher than normal as it approaches a pedestrian on a sidewalk, then suddenly drops to lower than normal as it passes. The pitch of an automobile horn or a jet plane undergoes a similar change as it approaches or recedes from us.

Christian Doppler (1803–1853), a physicist and mathematician, worked out the theory of such frequen-

Figure 8–14 Doppler effect. Waves are produced by a source moving to the right. The waves bunch up on the right, the wavelength is reduced, and the frequency is higher than if the source were at rest. (Courtesy of Education Development Center, 39 Chapel St., Newton, Mass.)

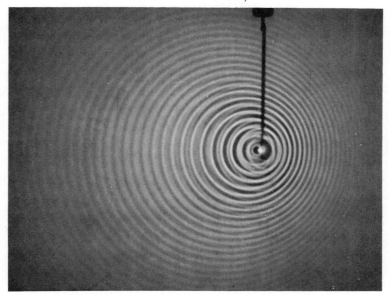

cy shifts, now called the *Doppler effect,* and experimental tests confirmed his equations. The sound was provided by several trumpeters, and musically trained observers evaluated it by ear. First, the trumpeters were positioned on a moving railroad flat car while the observers were stationed on the ground. Then they exchanged places, and the observers moved past the stationary trumpeters.

Although the pitch of the trumpets is unchanged as played, to an observer in relative motion with the source it seems to increase or decrease. When the source is moving toward the observer, with each new wave the wavetrain is closer to the observer than when the preceding wave started. The distance that each new wave has to travel is therefore less than that traveled by the preceding one, and more waves reach the observer per second than when the train is stationary. The effect is to raise the apparent pitch of the trumpet. When the source is receding, the pitch is lowered, since fewer waves reach the observer per second.

The Doppler effect applies to light waves as well as to sound waves. If a star is approaching the earth, its light seems more violet, indicating higher frequencies; its radiation is said to shift toward the violet. If it is receding, the star's radiation becomes redder, indicating lower frequencies, and there is a "red shift."

8–7 REFLECTION OF WAVES

When waves encounter a barrier or an abrupt change in the nature of the medium, some *reflection* occurs at the surface in a manner similar to the way a ball bounces after striking a wall. The waves turn back upon themselves; what is not reflected is absorbed or transmitted by the medium in which they are traveling. Light is reflected from mirrors, sound from buildings or canyons, certain radio waves from the ionosphere and satellites, and laser beams from the moon. It is found experimentally for all waves that the angle of incidence is equal to the angle of reflection (Fig. 8–15). The incident and reflected waves are measured from a line perpendicular to the plane of the reflecting surface called the normal. The two waves and the normal lie in the same plane.

A reflected sound that can be easily distinguished from the original is called an *echo.* To hear an echo, it

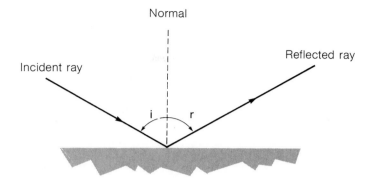

Figure 8–15 *Law of reflection.* The incident ray and the reflected ray make equal angles with the normal to the surface: i = r.

is necessary to stand 60 feet or more away from a large reflecting barrier and to produce a loud pulse of sound. The distance is critical, because the sensation of sound persists for about one tenth of a second. If the speed of sound is 1100 feet/second, sound travels 110 feet in 0.1 second. Therefore, if the distance between source and reflector is 60 feet or more, the pulse has time to travel to the barrier and back (120 feet) and to be heard as a sound distinct from the original. By then the sensation of the original sound will have decayed.

Sound can undergo rapid multiple reflections that cannot be analyzed into distinct echoes, an effect called *reverberation.* Such sound may persist for several seconds after the original sound has ceased. Wallace C. Sabine (1868–1919), who established the basic principles of modern architectural acoustics, found that reverberation time, an index of the "liveness" of sound, is one of the factors determining the acoustical qualities of auditoriums and concert halls. By using appropriate materials on interior surfaces—fabrics, acoustic tiles and panels, stone, brick, wood, and plaster—architects design music rooms and lecture halls that have optimum reverberation times. Boston Symphony Hall, the first structure designed upon the basis of Sabine's work, is still regarded as one of the few excellent concert halls in the world.

8–8   REFRACTION OF WAVES

When the speeds of the waves differ in two adjacent media, as in air and water, then in addition to reflection a wavetrain undergoes a change in direction and bends as it passes at an angle from one medium

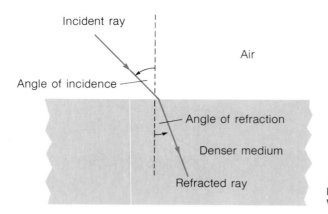

Figure 8–16 Refraction of waves. As they pass from one medium into another at an angle, waves change direction.

into the other. This bending of the waves upon entering a new medium is called *refraction* (Fig. 8–16). Investigation shows that an increase in the angle of incidence is followed by an increase in the angle of refraction. It has been determined that the ratio of the speeds of the waves in the two media is a constant, n, called the index of refraction of the second medium relative to the first. Values of n for various media relative to air are listed in tables.

The bending of the wavefront is toward the normal as the waves enter a medium in which their velocity is slower. As part of the wavefront enters the second medium in advance of the rest of the front, its velocity changes, and there is a change in the direction of the

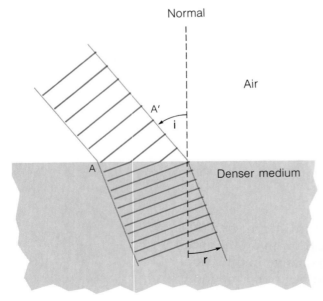

Figure 8–17 Refraction explained by a difference in the wave velocity in the two media. Point *A* enters the denser medium first and slows down, while *A'* is still traveling in the less dense medium.

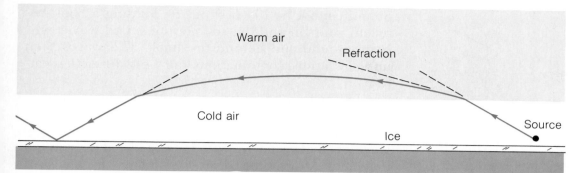

Warm air

Refraction

Cold air

Ice

Source

front (Fig. 8–17). If the wavefront is found to bend away from the normal, the waves must be traveling with a greater velocity in the second medium than in the first.

A layer of warm air acts as if it were an acoustically different medium than a layer of cold air, because of the difference in the velocity of sound waves in the two layers. Sound refraction occurs when sound passes obliquely from cold air into warm air. This explains why sound travels great distances over a frozen lake. As the sound waves travel upward from the ice, their speed is greater at the higher altitude, where the air is warmer (Fig. 8–18). The waves are refracted by the warmer air, and strike the surface of the lake again. There they are reflected, and the cycle of refraction and reflection is continued with the result that sound travels great distances.

Figure 8–18 Refraction of sound waves over a frozen lake. Sound waves are bent—*refracted*—downward by layer of warm air above cold, then reflected by the ice.

8–9 DIFFRACTION

Although we cannot see around corners, we can hear a person talking around a corner or in the next room. Low-pitched tones, in particular, carry well. In similar fashion, ocean waves bend around a breakwater and cause boats behind the breakwater to bob up and down. The bending of waves around barriers is not explained by reflection or refraction, since the medium is the same and we may assume that the temperature is the same. The bending can be explained, however, by diffraction.

Diffraction is that property which enables waves to bend around an obstacle or spread through an opening and not cast a completely sharp shadow (Fig. 8–19). To shut out noise from a room a door or window must be closed completely; leaving it open even slightly may admit nearly as much sound as when it is wide open, because of the tendency of the waves to spread.

Figure 8–19 Diffraction of water waves. The waves spread when they bend around obstacles or pass through openings.

Shadow zone

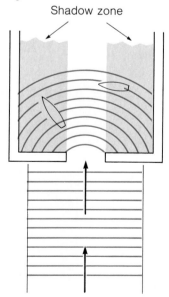

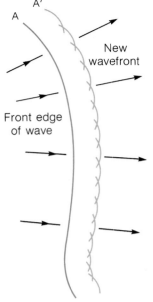

Figure 8-20 Huygens' theory of wave propagation. Every point on an advancing wavefront is considered a source of secondary wavelets. The envelope around the secondary wavelets marks the new position of the wave.

Sound cannot be blocked out by holding a book between yourself and the source; the sound waves will bend around and behind the book. The waves then unite in various combinations and yield diffraction patterns.

Diffraction spreading is explained by Huygens' theory of wave propagation (Fig. 8–20). Each point on a wavefront may be regarded as a source of secondary wavelets that spead out with the speed v of the wave at that point. After a time t, these secondary wavelets have radii v · t, and the envelope of these wavelets will be the new wavefront at the end of the time interval. The wavelets combine to form a replica of the original wavefront, and the wave has advanced. When waves pass around an obstacle or through a small opening, they can thus change direction and spread. The greater the wavelength and the smaller the opening, the more pronounced the diffraction (Fig. 8–21).

8–10   SUPERPOSITION OF WAVES; INTERFERENCE

When two or more waves move through a medium at the same time, each wave proceeds as if it were the

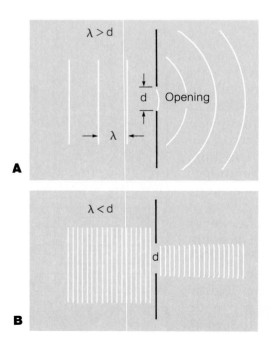

Figure 8-21 Diffraction spreading is less pronounced when wavelengths are short compared with the size of the opening.

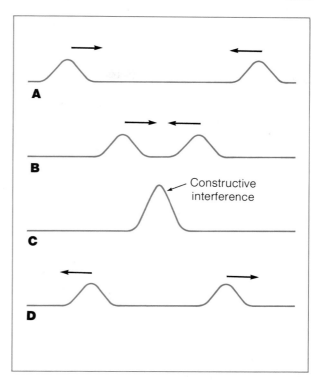

Figure 8–22 *Superposition.* Two waves traveling in opposite directions on a cord add where they meet and then pass through each other.

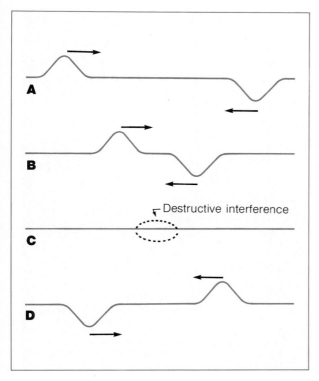

Figure 8–23 *Destructive interference.* A crest and a trough traveling in opposite directions cancel each other when they meet.

only one present, except where they meet. In those regions the resultant displacement of the medium is the vector sum of the displacements of the individual waves that would occur if each wave alone were present (Fig. 8–22). Then each wave passes on in its original direction without change, as though the other waves were not present. This behavior, unique to waves, is known as the *superposition principle*.

If two waves having the same amplitude arrive at a given point where each wave is at a crest, the two amplitudes combined will be exactly twice the amplitude of a single wave alone. This effect is known as *constructive interference*. But if one wave is at a crest and the other is at a trough, the two waves give a combined amplitude of zero, and there is *destructive interference* (Fig. 8–23). For other combinations, the resultant amplitude will be greater than zero but less than the maximum.

Thomas Young (1773–1829), a physician and physicist, performed experiments that did much to establish the wave nature of light. When a panel in which two thin slits had been cut was illuminated by sunlight, two

Figure 8–24 Illumination of two narrow slits produces interference pattern on screen. (From Huggins, Elisha R., *Physics One.* Menlo Park, Calif.: W. A. Benjamin, Inc., 1968, p. 369.)

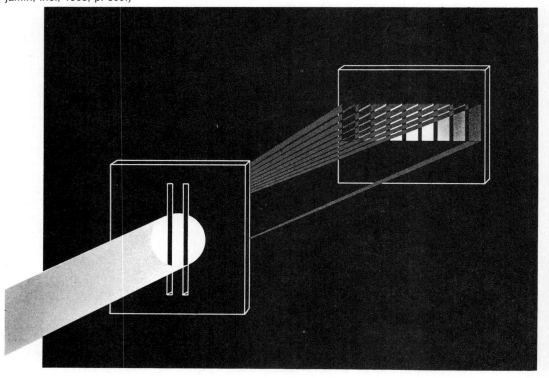

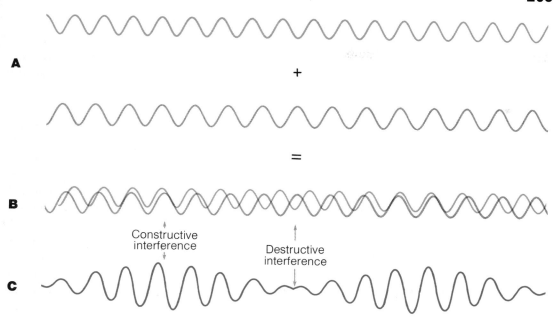

A

+

=

B

Constructive
interference

Destructive
interference

C

Resultant wave

Figure 8–25 Beats result from interference of two waves having slightly different frequencies. A periodic change in amplitude results in fluctuation in loudness of the sound waves.

light beams emerged from the other side of the panel, each slit itself acting as a source. On a screen some distance away, Young saw alternating areas of brightness and darkness (Fig. 8–24). The results can be explained only if light is considered to have a wave nature as opposed to a particle nature. Bright fringes occur where light waves from the two sources superpose in similar phases, as crest superposed upon crest, while the dark areas result from destructive interference, as when a trough superposes with a crest. The rainbow colors of soap films and oil slicks on a wet pavement are due to interference of light waves reflected from the two surfaces of the film. The combination of constructive and destructive interference causes some wavelengths to be reinforced while others are canceled.

Two wavetrains of sound of slightly differing frequencies but equal amplitudes interfere to produce a series of *beats:* points of maximum amplitude alternating with points of minimum amplitude in the vibration (Fig. 8–25). The beat frequency is given by the difference of the two wave frequencies. Thus, when two tones of slightly different frequency are sounded together, say 128 Hz and 136 Hz, the loudness fluctuates at a beat frequency of 8 beats per second.

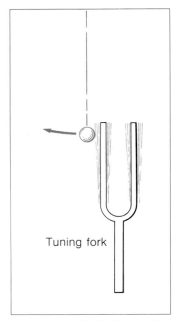

Figure 8–26   A vibrating tuning fork.

8–11   SOUND AS VIBRATION

When you strike a tuning fork with a rubber mallet, it emits its tone. Although the vibrations of the tuning fork are too rapid to be seen, the edges present a fuzzy outline while sound is emitted, and the vibrations can be detected by touching the fork with the finger (Fig. 8–26). The vibrating surface becomes the center of compressions and rarefactions in the air, which spread out in all directions. Any vibrating object surrounded by an elastic medium will produce a compressional wave in that medium—the buzzing wings of a fly, the vibration of the vocal cords, the speaker of a radio or stereo system.

Although a tuning fork emits a pure tone, the sound cannot be heard more than a short distance away. If the handle is pressed against a table top, though, the sound can be heard throughout the room. The tuning fork forces the table top into vibration, and the large vibrating area of the table radiates sound well. A vibrating string by itself also produces very little sound, since it is too narrow to disturb much air. But the sound of the string can be amplified by passing the string over a support which rests on a wide plate called the soundboard (Fig. 8–27). The vibrations of the string are conveyed to the soundboard, which in turn communicates them to a much greater mass of air. Without amplification, many sounds would be inaudible. The three stages in the production of sound are: the initial disturbance, the amplification of the disturbance, and the radiation of sound.

The use of a stretched string or wire as a source of sound is the basis of the violin, guitar, and piano. The tone it emits when twanged, bowed, or struck depends on the length, mass and tension of the string. The vibration frequencies of strings are inversely proportional to their lengths, directly proportional to the square roots

Figure 8–27   String and soundboard, which amplifies the sound produced by the vibrating string.

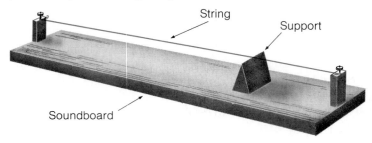

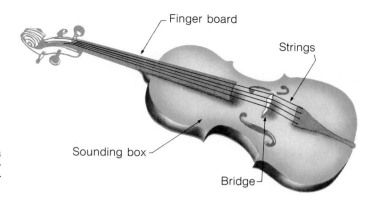

Figure 8–28 The four strings of a violin are tuned to different frequencies which can be compared to certain notes on a piano keyboard.

of their tensions, and inversely proportional to the square roots of their linear densities.

A string fixed at each end is set into vibration by plucking, bowing, or striking. In the violin and guitar, varying the length of a string by fingering causes the frequency to vary. Since the four strings of a violin are all of the same length but of different thickness, their fundamental frequencies—the lowest frequency the string can emit—differ. The wooden bridge communicates the vibrations of the strings to the body of the violin and the air inside it, which are then set into vibration and amplify the sound (Fig. 8–28). In the piano, the length, tension, and thickness of a wire vary from note to note. Like the body of the violin, the soundboard of the piano moves a large mass of air that radiates sound to the surroundings.

8–12   STANDING WAVES

If the frequencies are properly selected, a string may be made to vibrate in several natural vibrational modes. Each direct wave is reflected at the end of the string and upon its return meets an ongoing wave. Regions of constructive and destructive interference occur at fixed positions, and a *standing wave* is produced (Fig. 8–29). The stationary pattern is called a "standing" wave because it involves no motion along the length of the string; unlike a traveling wave, a crest does not move from one point to another. Each section appears, instead, to move transversely to the length. Vibrating at its fundamental frequency, a string forms a simple pattern with a region of maximum constructive interference, the *antinode,* in the middle, and a sta-

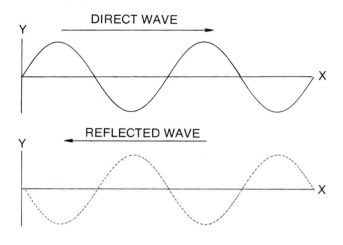

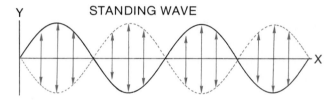

Figure 8–29 When two waves of the same frequency and amplitude travel in opposite directions, a standing wave is produced.

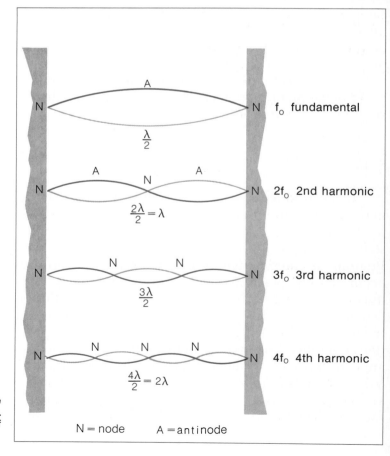

Figure 8–30 Standing wave patterns in a stretched string. The first four harmonic modes of vibration are shown.

tionary point, or *node*, at each end, the result of complete destructive interference.

Standing waves are useful because of the fixed wave pattern they form. To get a specific note from a musical instrument, a fixed wave pattern—that is, one that does not change its shape too rapidly—is necessary. Otherwise, a mixture of sounds is produced that is not identifiable as a note. The sound that comes from a violin string, an organ pipe, or air columns of wind instruments is the result of standing waves. Plates, rods, and diaphragms that are parts of percussion instruments also produce standing waves.

As the frequency of vibration is increased, standing wave patterns consisting of two, three, or more segments can be developed with nodes alternating with antinodes (Fig. 8–30). The distance between two consecutive nodes or antinodes is one-half wavelength, $\frac{1}{2}\lambda$. When the string vibrates as one segment, with a node at each end and only one antinode between them, the length L of the string equals one-half wavelength, $\frac{1}{2}\lambda$; the wavelength, therefore, is twice the length of the string: $\lambda = 2L$. This is the maximum wavelength of a standing wave for the string, and corresponds to the fundamental frequency, f_0—the lowest frequency of which the string is capable.

When the string is made to vibrate in two segments, the length of the string contains two half-wavelengths: $L = 2\frac{\lambda}{2}$. The wavelength thus equals the length of the string: $\lambda = L$. Since the frequency and wavelength are inversely proportional, according to the wave equation ($V = f\lambda$), the corresponding frequency is twice that of the fundamental, or $2f_0$. This frequency is called the first octave, or first overtone. Three segments in the standing wave pattern ($L = 3\frac{\lambda}{2}$) produce a frequency three times the fundamental, $3f_0$, called the second overtone. Four segments produce the frequency $4f_0$, the third overtone, and so on. The frequencies of these modes of vibration form the series f_0, $2f_0$, $3f_0$, $4f_0$, in which the frequencies are integral multiples of the frequency of the fundamental; the frequencies in such a series are called *harmonics*. The fundamental is called the first harmonic, the higher frequencies the second harmonic, third harmonic, and so on. Although it is difficult to visualize, a string vibrates in multiple modes at the same time, producing the fundamental frequency and various overtones.

EXAMPLE   8–4

Give three possible (a) wavelengths and (b) corresponding frequencies of standing waves in a wire stretched between two supports that are 1 meter apart. (Assume the velocity of sound to be 340 m/sec.)

SOLUTION

For the fundamental frequency, the length L of the wire equals $\frac{\lambda}{2}$, since there will be a node at each end. Therefore, $\lambda = 2L$. For the first overtone, $\lambda_2 = L$; and for the second overtone, $\lambda_3 = \frac{2L}{3}$. The frequencies are in the ratio 1:2:3.

$$L = 1 \text{ m}$$

$$V = 340 \text{ m/sec}$$

$$\lambda_0, \lambda_1, \lambda_2 = ?$$

$$f_0, f_1, f_2 = ?$$

$$(a)\, \lambda_0 = 2L$$
$$= 2(1\text{m})$$
$$= 2\text{m}$$

$$\lambda_1 = L = 1\text{m}$$

$$\lambda_2 = \frac{2L}{3} = \frac{2\text{m}}{3} = 0.67\text{m}$$

$$(b)\, V = f_0 \cdot \lambda_0$$

$$f_0 = \frac{V}{\lambda_0}$$

$$= \frac{340\frac{\text{m}}{\text{sec}}}{2\text{m}}$$

$$= 170\,\text{Hz}$$

$$f_1 = 2f_0$$
$$= 2(170\,\text{Hz})$$
$$= 340\,\text{Hz}$$

$$f_2 = 3f_0$$
$$= 3(170\,\text{Hz})$$
$$= 510\,\text{Hz}$$

8–13 VIBRATING AIR COLUMNS

Since wind instruments are silent unless someone blows upon them, the vibrations are driven by energy put in from the outside. The standing wave pattern set up in an air column is like the standing wave on a string except that the air has a longitudinal motion in a direction parallel to the wave. The simplest standing wave pattern has an antinode at the open end, where the air has maximum possible motion, and a node at the closed end, where the air cannot move. Many other standing wave patterns are possible, but in all there will always be an antinode at the open end and a node at the closed end.

For the simplest case mentioned, the length (L) of the tube is very close to one-quarter wavelength, $\frac{1}{4}\lambda$ (Fig. 8–31). The maximum wavelength is, therefore, 4L, corresponding to the fundamental, the lowest frequency of sound possible for the tube to emit. The ratio of the set of frequencies for such a tube is $1f_0 : 3f_0 : 5f_0 : \ldots$, or odd multiples of the fundamental. The frequency $3f_0$ corresponds to the first overtone or second harmonic; $5f_0$ to the second overtone or third harmonic, and so on.

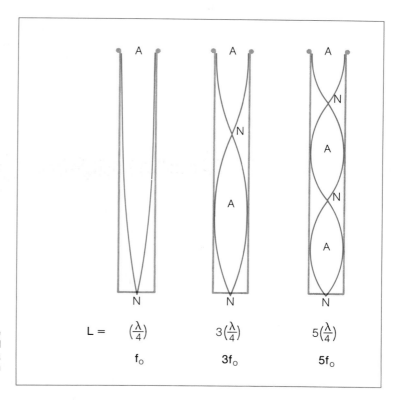

$$L = \quad \left(\frac{\lambda}{4}\right) \qquad 3\left(\frac{\lambda}{4}\right) \qquad 5\left(\frac{\lambda}{4}\right)$$

$$f_0 \qquad\qquad 3f_0 \qquad\qquad 5f_0$$

Standing wave patterns of an air column closed at one end. There is always a node at the closed end and an antinode at the open end.

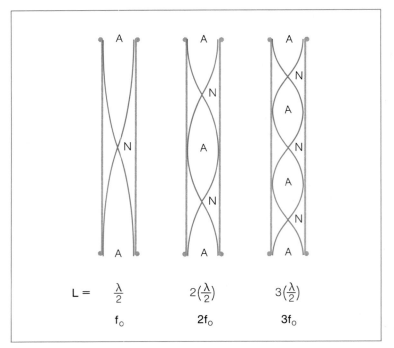

$$L = \quad \frac{\lambda}{2} \qquad 2\left(\frac{\lambda}{2}\right) \qquad 3\left(\frac{\lambda}{2}\right)$$

$$f_o \qquad\qquad 2f_o \qquad\qquad 3f_o$$

Figure 8–32 Standing wave patterns of an air column open at both ends.

In a tube open at both ends, the simplest standing wave has an antinode at each end with one node between them (Fig. 8–32). The length L of the tube is very close to one-half wavelength; therefore, the wavelength of the fundamental is approximately twice the length of the tube, or 2L. The possible frequencies emitted by an open tube are in the ratio 1f: 2f: 3f: . . . , or both odd and even multiples of the fundamental.

The same tone, say with a fundamental frequency of 100 Hz, sounds different depending on whether it is produced by a closed tube or an open tube, since the number and strength of overtones present differ. With the closed tube, only odd multiple overtones may be present in the tone: 300 Hz, 500 Hz, and so on. With the open tube, both the odd and even multiples are possible: 200 Hz, 300 Hz, 400 Hz, 500 Hz, and so on. While the lowest tone generally establishes the pitch, the remaining harmonics determine the timbre or quality of the sound.

8–14   TIMBRE OR TONE QUALITY

A musical note emitted by a vibrating string or air column is a combination of pure tones. The mathematician Jean Baptiste Joseph Fourier (1758–1830) showed that any wave motion can be represented as a sum of sinusoidal motions of different frequencies and amplitudes, each of which is known as a component (Fig. 8–33). One sine curve represents the fundamental, and there is a sine curve corresponding to each of the overtones, which are heard simultaneously. A sound may be regarded as a combination of many independent sinusoidal sound waves, each behaving as if it were the only one present.

The "vibration recipe" of a sound—that is, the presence in it of certain frequencies—explains our ability to perceive sounds as being different even when they appear to have the same pitch and loudness. Each of us speaks with a complex vibrating system that has a recognizable vibration recipe. A sound produced by a musical instrument is a complex of tones of definite pitch, detectable by the ear and called partial tones or partials. The partials give the particles of the medium a motion that is the resultant of a group of simple periodic motions.

Figure 8-33 Complex wave associated with a violin note. The sound is analyzed into four prominent harmonics, each of which is a pure sine wave. The musical note is a combination of pure tones.

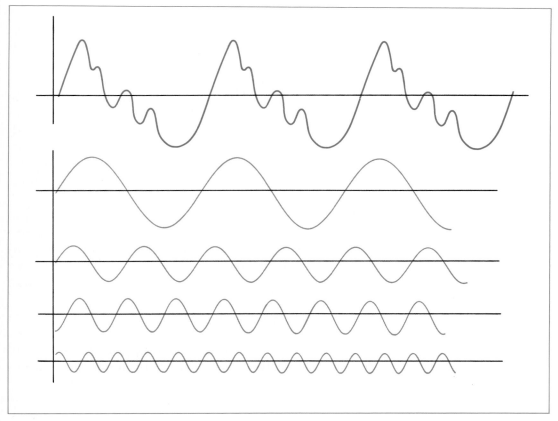

The distinctive tone quality of a sound depends upon the particular mixture of component frequencies that compose it. Most instruments give both the fundamental and higher harmonics. Without the latter, all instruments would produce rather lifeless sounds of the same quality, indistinguishable by the ear. The ear tends to hear each harmonic as a separate tone, and the brain combines them in some way. The fundamental normally has a greater amplitude than any of the overtones and is the frequency that we generally hear as the pitch. Different instruments emphasize different overtones. The difference in quality between a violin and a cello when sounding the same note depends upon the relative intensities of the partials or harmonics produced by each instrument. The fifth partial of a violin is given particular emphasis. The sound of a tuning fork consists almost entirely of the fundamental, and is musically uninteresting. The presence of harmonics in a sound gives it brilliance and brightness.

8–15 INTENSITY AND LOUDNESS

The *intensity* of a sound wave refers to the amount of energy the wave is carrying. It is defined as the amount of sound energy crossing a unit area in unit time and is commonly measured in watts per square meter. The intensity is directly related to the amplitude of the wave. The least intense sound detectable by the human ear is about 10^{-12} watt/m^2 corresponding to amplitudes in air that may be less than one millionth of a centimeter. The amplitude in air produced by rather intense sounds (the largest intensity that can be detected by the ear is 1 watt/m^2) may be about one hundredth of a centimeter.

Loudness is a physiological sensation that is experienced when sound waves strike the eardrum. It is related to the physical intensity of the compressions and rarefactions of the air. Loudness is a highly subjective perception, and individual differences can be quite great.

The smallest change in intensity to which the average ear is sensitive is an increased intensity 1.26 times the original intensity. The change in intensity of a sound resulting from 10 such increases in the original intensity, $(1.26)^{10}$ times, is called a *bel*, after Alexander Graham Bell (1847–1922), a teacher of speech for the

Table 8-3   INTENSITIES OF TYPICAL SOUNDS

SOUND	INTENSITY LEVEL IN DECIBELS (dB)
Rustle of leaves (faint)	10
Soft whisper at 3 feet	20
Library	30
Mosquito buzzing	40
Ordinary conversation	50
Noisy store, busy office	60
Vacuum cleaner/Average street traffic	70
Garbage disposal/Diesel truck/ Tabulating room	80
Motorcycle/Symphony music	90
Subway train/Jet fly-over	100
Hard-rock music/Outboard motor	110
Thunderclap/Construction noise	120
Jet aircraft taking off	140
Rifle	160
Moon rocket	200

deaf and inventor of the telephone. Each of the 10 equal steps, representing an increase by a factor of $10^{0.1}$ or 1.26, is one tenth of a bel, or a *decibel* (dB). Thus, 1 bel = 10 decibels (10 dB). The average individual can distinguish about 130 steps in intensity, from the threshold of hearing to the level just below that at which the ear interprets the intensity as pain rather than sound.

The relative intensity of two sound waves is measured in decibels, a unit that expresses a ratio between two sound intensities, I/I_0: the reference intensity I_0 and the intensity I of the source being measured, such as a dishwasher or a truck. A particular intensity is selected to represent zero decibels, the standard reference level: 10^{-12} watt/m² at a frequency of 1000 Hz. This is close to the faintest sound that can be heard by a young adult in a quiet location. A wave 10 times as intense is at 10 decibels; a wave 100 times as intense, representing another tenfold increase, is at 20 decibels, and so on. Table 8-3 lists some typical sound levels.

8-16   THE EAR AND HEARING

A sound wave striking the ear penetrates the outer opening and sets the eardrum into vibration (Fig. 8-34). Three small bones (ossicles) in the middle ear transmit these vibrations to a fluid in the cochlea of the inner

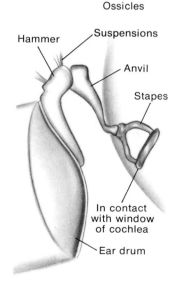

Figure 8-34 Structure of the ear, showing the three tiny bones (ossicles) that connect the ear drum to the oval window of the cochlea.

Ossicles

Suspensions

Hammer

Anvil

Stapes

In contact with window of cochlea

Ear drum

ear, forcing the thin basilar membrane into vibration. The latter transmits the stimulus to the organ of Corti, where the endings of the auditory nerves are located, and the organ of Corti converts the vibrations into nerve impulses. This organ can identify the component frequencies in a complicated sound. The combination of ear and brain distinguishes pitch, loudness, and quality.

To be audible, a sound must have a minimum intensity. This intensity, called the threshold of hearing, is different for different frequencies. A low-frequency sound at 30 Hz cannot normally be heard until its sound intensity reaches 60 decibels, while a 4000 Hz tone might be heard at a level of 5 decibels. An audiogram gives a profile of an individual's threshold of hearing for a range of frequencies (Fig. 8–35). Pure tones at various frequencies are presented, and the intensity that is necessary for audibility at each frequency is noted.

The ear is least sensitive to low frequencies, and most sensitive to the range of 500 to 4000 Hz. The insensitivity to the lower frequencies is fortunate, for otherwise we would hear the bodily movements, muscular contractions, and vibrations of our own

Figure 8–35 An audiogram. The ear is most sensitive to frequencies near 2000 Hz. Above or below this frequency the intensity must be increased considerably for the sound to be audible.

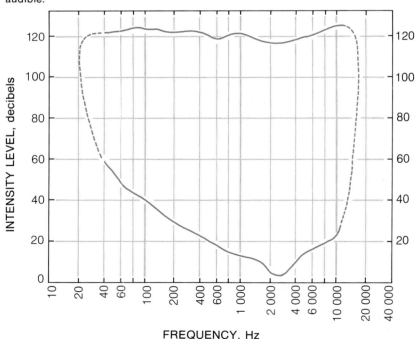

bodies. With advancing age, the sensitivity of the ear to high-frequency sound normally deteriorates. A person at age 40 may lose about 80 Hz sensitivity at the higher frequencies every six months.

Hearing also occurs via bone conduction through the skull. The sound of clicking teeth is heard mainly through vibrations of the bones in the skull; some of the vibrations bypass the middle ear and are transmitted directly to the inner ear. When we speak, some of the sound reaches the ears via the air and some through bone conduction. Humming with closed lips is transmitted almost entirely by bone conduction. Speaking involves two different sounds to the speaker, but only one—through air conduction—to another listener. We normally do not hear ourselves as others hear us—the low-frequency vibrations we hear through bone conduction are not heard by others—and a recording of ones's own voice may seem disappointingly thin for this reason.

8–17 NOISE POLLUTION

The range of sound quality extends from pure sine waves, through musical tones consisting of a set of harmonically related frequencies, to the chaos of unrelated sounds called noise. Beyond ordinary noise there is a pure confusion of sound called *gaussian noise*, or *white noise*, containing sound of *all* frequencies, just as white light is made up of all frequencies of light.

The British Committee on the Problems of Noise defines noise as "unwanted sound." In a noisy environment—near elevated railroads, trucking routes, airplane flight paths—about 25 per cent of the people say they are not disturbed by the noisy activities. Another 10 per cent seem to be annoyed by any noise, regardless how faint, that they themselves do not make.

Noise is a by-product of the conversion of energy. The basic problem is essentially incurable; it is a price we must pay for a machine civilization. Although the noise of modern technology cannot be eliminated, it can be controlled to minimize its adverse effects. There is legislation establishing industrial standards, building codes, and transportation requirements to limit the transmission of noise. The Noise Control Act of 1972 authorized the federal government to study controls and set standards to protect the public health and welfare. The Walsh-Healy Act allows an overall sound intensity level of 90 dB for eight hours' exposure of employees in a plant selling a product or providing a service to the government. Deafness, partial or total, has become a major category of industrial injury, and workers are now protected with compensation for work-associated hearing disabilities. If the noise intensity is too high, special ear muffs or ear plugs must be worn.

Noise is one of the most annoying and pervasive of all types of pollution. Certain types of noise and vibration can cause disease, dizziness, and nausea. The psychological effects are more difficult to measure, but through the persistence of annoyance, fatigue, and the inability to concentrate, efficiency may be reduced and accidents caused.

Every problem having to do with excessive noise or vibration originates in the periodic motion of a structure. The most widely used means of controlling vibrations is to isolate the mechanical energy of a vibrating structure from surrounding areas by installing soft, elastic separators along the transmission route. An example is the isolation of automobile-engine mount vibrations from the passenger compartment by means of coil springs and pads of elastomer.

Another approach to controlling noise and vibration is acoustical absorption, used mainly in architectural applications to make buildings quieter. Acoustic tiles and baffles constructed from porous materials are used; the air penetrates the structure and expends its energy through frictional flow. This interception of acoustical energy in the air protects the ears from exposure to intense pressure waves. Acoustical building codes are applied to new construction.

Basic research is needed in many areas if much noise-control is to be realized. Noise production by jet-engine exhausts is not well understood. The impact of one body on another is perhaps the least understood of all mechanical sources of noise. Yet the noise of industrial machinery is largely produced by impacts: pneumatic drills, hammers, saws, engine blocks, milling machines—all make high-speed impacts. Sustained public concern for a quieter environment can lead to results. For example, consumer demand for less noise in appliances resulted in reduction of the noise level of a portable dishwasher from 85 to 70 dB through a combination of vibration control mechanisms; and the more annoying high-frequency components were decreased by 40 dB.

8–18 ULTRASONICS

Sound waves with frequencies ranging from 20 000 Hz, the upper limit of human hearing, to a billion or more Hz are described by the term *ultrasonics*. Waves of these frequencies can be generated by the vibration of thin pieces of quartz or other crystals, set into motion by electrical means. Frequencies up to 20 billion Hz (20 gigahertz) has been achieved. The wavelengths of these ultra–high-frequency waves are about the same as those of visible light.

Ultrasonic waves travel through liquids and solids in the same way that audible sound waves do. They are used for such purposes as underwater echo-ranging by sonar (*so*und *n*avigation *a*nd *r*anging)—the location of the object and the measurement of the distance between it and the source; detecting flaws in the interior of solids; destroying microorganisms; and mapping underground structures for oil and mineral deposits. They are also used in medical diagnosis. In sounding out the abdomen, as an example, the sound waves pass through the different tissues at speeds that depend on the elasticity and density of the tissue. As they collide with different structures, they send back echoes, which are picked up by sensitive microphones and turned into electrical signals on a TV screen. From the pattern of the echoes, tumors and other abnormalities can be picked up.

More efficient and more sensitive than any sonar contrived by man, however, is the sonar system of certain bats that, like dogs, crickets, and other animals, perceive acoustic frequencies inaudible to us. Not all bats use sonar; the largest bats have prominent eyes and depend on their vision instead. Small bats rely largely on echolocation; they emit and hear sounds of frequencies at least up to 100 000 Hz, and the precision of their sonar echolocation is so great that they can get along perfectly well without eyesight. Insect-eating bats can easily catch hundreds of moths in an hour in the dark of night. Far from being a crude collision-warning device, the echolocation system of bats is very sharp and precise.

Example 8–5 illustrates a typical echo-ranging situation.

EXAMPLE 8–5

An acoustic signal is transmitted from the bottom of a ship and is reflected from the bottom of a channel in 0.034 sec. Assuming the velocity of sound in sea water to be 1520 m/sec, how deep is the channel?

SOLUTION

If the elapsed time is 0.034 sec, the time required for the signal to travel from the ship to the bottom is one half of this, or 0.017 sec. From kinematics, the depth can be determined.

$$\bar{v} = 1520 \text{ m/sec} \qquad s = \bar{v}t$$

$$t = 0.017 \text{ sec}$$
$$= 1520\frac{\text{m}}{\text{sec}}(0.017 \text{ sec})$$

$$s = ?$$
$$= 26.0 \text{ m}$$

8–19   RESONANCE

Resonance occurs when a system that can vibrate with a certain natural frequency is acted upon by a periodic disturbance that has the same frequency (Fig. 8–36).

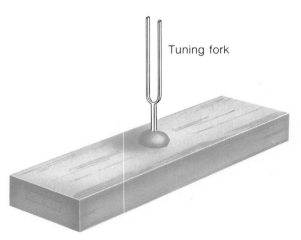

Tuning fork

Figure 8–36 Tuning fork mounted on a resonating box. The sound of the vibrating tuning fork is amplified by the resonance of the air in the box.

Figure 8–37 During a gale, the Tacoma bridge was set into resonance vibration to one of its natural frequencies; the bridge oscillated, and ultimately collapsed. (From Stevenson, R., and R. B. Moore, *Theory of Physics.* Philadelphia: W. B. Saunders Company, 1967, p. 391.)

A child's swing illustrates this principle. A small effort applied at the right moment produces the best effect, either by the child's own "pumping" of the swing, or by gentle pushes from another. A suspension bridge can swing to and fro like a pendulum; if excited into its natural resonant vibration by a wind, the amplitude can lead to its collapse. This happened to the Tacoma Narrows Bridge in 1940, which was set into such violent vibration during a gale that it collapsed only four months after its completion (Fig. 8–37). A musical instrument may set windows or other objects into vibration.

Many sources of sound, such as the vocal cords or a violin string, produce such faint sounds that they would be almost inaudible if there were no way to increase the sound. By means of resonance, a relatively small amount of energy can be amplified by the larger mass of air set into vibration. The resonant cavities in the head reinforce and amplify the sound produced by the vocal cords.

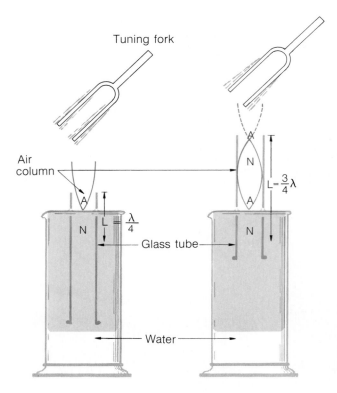

Tuning fork

Air
column

Glass tube

Water

Figure 8–38 Resonance between a tuning
fork and an air column. The tube is moved
up and down in the water to tune it to the
frequency of the fork. When the length of
the column = λ/4, the sound increases
greatly in loudness, and again at L = ¾λ.

When you hold a sea shell to your ear, you hear
sounds of certain frequencies that are reinforced by the
shell and stand out from other sounds. If you speak into a milk bottle, certain frequencies present in your
voice are emphasized by the resonance of the bottle.
The air is set strongly in motion by a sound wave that
has very nearly the same frequency as one of its resonant frequencies.

A striking demonstration of resonance involves
a tuning fork struck and held over an adjustable column
of air (Fig. 8–38). Several positions of the column can
be found that reinforce the sound of the tuning fork.
The shortest length L of the tube in which the air can
be in resonance with a wavelength λ is λ/4, since there
will be a node at the closed end and an antinode at the
open end. If the tube is long enough, resonance occurs
again when the length of the tube is 3 λ/4, since there
will again be a node at the closed end and an antinode
at the open end. Muscial instruments are essentially
resonators; the mouthpiece, string, or reed is the source
of excitation, and certain frequencies are emphasized
through resonance by the instrument.

Amplitude
Antinode
Audiogram
Beats
Bell-in-vacuum
Bone conduction
Closed tube
Constructive interference
Crest
Decibel (dB)
Destructive interference
Diffraction
Doppler effect
Frequency
Fundamental
Intensity
Interference
Law of reflection
Law of refraction
Longitudinal wave
Loudness

Node
Noise
Open tube
Ossicles
Overtone
Partial
Period
Quality
Reflection
Refraction
Resonance
Reverberation
Sine wave
Superposition principle
Threshold of hearing
Transverse wave
Trough
Vibrating air column
Vibrating string
Wave equation
Wavelength

EXERCISES

8–1.   What evidence can you give that sound is a form of energy?

8–2.   What evidence is there that light is a form of wave motion?

8–3.   Discuss (a) some ways in which sound waves resemble light waves; (b) some ways in which they differ.

8–4.   When you are watching a distant batter, why does the crack of the bat striking the ball seem to occur moments after the event?

8–5.   There may be a problem for the marchers in a long column to keep in step with a band up front. Explain.

8–6.   Six seconds after a lightning flash was seen, thunder was heard. How far away was the lightning if the temperature was 30°C?

8–7.   An echo returns from a building in 0.5 second. How far away is the building? (Assume the temperature of the air to be 20°C.)

8–8.   Assuming that your range of hearing is from 16 Hz to 20 000 Hz, what are the corresponding wavelengths (a) in air at 20°C? (b) in water at 20°C?

8–9.   Dolphins emit very short sound waves with a frequency of 2×10^5 Hz. What is their wavelength in water? (Consider the velocity of sound in sea water to be 1520 m/sec.)

8–10.   An ocean wave with a wavelength of 100 meters has a speed of 16.3 m/sec. How many waves would pass a given point in 1 second?

8–11.   Waves with a wavelength of 3 centimeters are generated in a ripple tank when the frequency is set for 10 Hz. What is the speed of propagation?

8–12.   A particular wavelength of orange light is 6.06×10^{-5} cm. (a) What is its frequency? (b) What is its period?

8–13.   A broadcasting station transmits radio waves at a frequency of 10^6 Hz. What is their wavelength?

8–14.   Arrange the following in the order of their decreasing frequency: (a) gamma rays; (b) orange light; (c) radio waves; (d) infrared rays.

8–15.   Distinguish between reflection and refraction.

8–16.   Waves traveling in air at 3×10^8 m/sec enter a medium in which their velocity is reduced to 2×10^8 m/sec. If the waves strike the interface at an angle of incidence of 35°, in which direction will they move in the second medium?

8–17.   A tone with a frequency of 512 Hz is sounded at the same time as a 520-Hz tone. How many beats per second are produced?

8–18.   Explain why sound travels well on a quiet night in summer.

8–19.   Why is a long reverberation time undesirable in a concert hall?

8–20.   Assume that you are in a moving car. Apply the Doppler effect to an ambulance siren approaching you from the opposite direction and passing you.

8–21.   An air column closed at one end resonates with a tuning fork which has a frequency of 128 Hz. What is the length of the air column?

8–22.   Two pipes, one open at both ends and the other closed at one end, are 0.5 meter in length. What is the fundamental frequency of each? (The velocity of sound is 340 m/sec.)

8–23.   What length of open organ pipe would give a frequency of 440 Hz at 20°C?

8–24.   Are the long pipes or short pipes of an organ the high-frequency pipes? Explain.

8–25.   Explain how a wire of a given length is capable of emitting notes of different frequencies.

8–26.   What is "stationary" about standing waves on vibrating strings?

8–27.   Use diagrams to show the fundamental and first two overtones of (a) a vibrating string; (b) a vibrating air column in a tube closed at one end.

8–28.   Assuming that the frequency of the fundamental waves in Exercise 11–7 is 100 Hz, what are the frequencies of the overtones in each case?

8–29. A string 2 meters long is fixed at both ends. What are the three longest wavelengths of standing waves established in the string? (Assume the velocity of sound to be 340 m/sec.)

8–30. What conditions must be satisfied for resonance to occur between two mechanical systems?

8–31. Determine the 5th overtone of middle A (440 Hz) on a piano.

8–32. The frequency of a tuning fork is 256 Hz. What is the frequency of a tuning fork one octave higher?

8–33. What are some possible wavelengths of standing waves in a cord stretched between two supports 50 centimeters apart?

8–34. Why does a recording of one's own voice often seem surprising when heard for the first time?

8–35. Why do identical notes emitted by a trumpet and a trombone sound different?

8–36. What is considered musical in one country may be considered noise in another. Discuss.

8–37. A quieter environment will not come cheaply. Discuss whether you would be willing to pay your share toward it in the form of higher prices for goods and services.

8–38. A state law requiring the sound level of snowmobiles not to exceed 73 dB was revoked for one year by the governor of that state when dealers claimed that the snowmobiles they had in stock could not meet this standard and they would suffer financially. Comment on the appropriateness of the governor's action.

8–39. Discuss whether, in your opinion, the quality of life in terms of noise pollution has improved during the past five years.

8–40. Intense ultrasound can be applied to destroy tumors. What is the wavelength in air if a frequency of 10^6 Hz is used?

8–41. *Multiple Choice*

   A. A wave transmits
      (a) matter   (c) energy
      (b) momentum   (d) velocity

   B. The distance from crest to crest of a wave is called its
      (a) frequency   (c) amplitude
      (b) wavelength   (d) velocity

   C. A sound wave is
      (a) longitudinal   (c) resonant
      (b) transverse   (d) harmonious

   D. The bending of a wave around a barrier is called
      (a) reflection   (c) interference
      (b) refraction   (d) diffraction

   E. The frequency of a sound wave is related to its
      (a) pitch   (c) quality
      (b) amplitude   (d) loudness

SUGGESTIONS FOR FURTHER READING

Beranek, Leo, "Acoustics." *Physics Today,* November, 1969.
    Modern acoustics is applied in obstetrics and gynecology, music, chemistry, oceanography, biology, and psychology.
Devey, Gilbert B., and Peter N. T. Wells, "Ultrasound in Medical Diagnosis." *Scientific American,* May, 1978.
    It is possible to explore structures within the human body painlessly, safely, and at relatively low cost with sound waves.
Koch, Winston E., *Sound Waves and Light Waves,* Garden City, N.Y.: Doubleday & Co., 1965.
    Discusses sound waves and light waves and their applications in such devices as microwave transmitters and electromagnetic lenses.
Schelleng, John C., "The Physics of the Bowed String." *Scientific American,* January, 1974.
    The physics of the behavior of strings can be of considerable importance to the player of a stringed instrument.
Shaw, Edgar A. G., "Noise Pollution—What Can Be Done?" *Physics Today,* January, 1974.
    Noise is associated with a vast increase in the use of energy for transportation and labor-saving machinery. The control of noise is likely to be of major concern for years to come.

LIGHT AND OTHER ELECTROMAGNETIC WAVES

9–1   INTRODUCTION

Can you imagine anything "purer" or more elementary than a beam of sunlight? The discovery of its complex nature, therefore, came as a surprise. Through a small hole in a window shutter, Newton admitted a beam of sunlight to a darkened room and directed the beam through a prism. A band of colors—a spectrum—appeared like a rainbow on the opposite wall (Fig. 9–1). Newton then used a second prism, inverted with respect to the first, and recombined the colors into white light (Fig. 9–2).

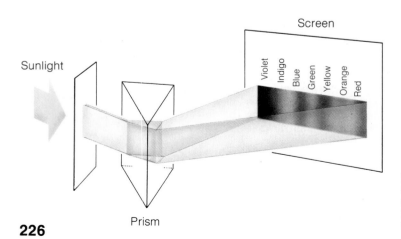

Sunlight

Screen

Violet Indigo Blue Green Yellow Orange Red

Prism

Figure 9–1   Isaac Newton discovered that a prism separates a beam of sunlight into its component colors and produces a spectrum.

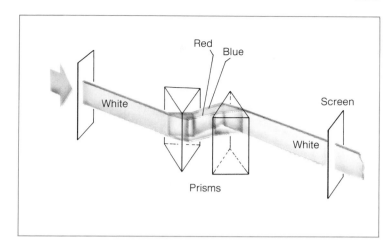

Figure 9–2 A second prism recombines the colors; a white spot appears on the screen.

Dispersing sunlight through a prism as Newton had done, the astronomer William Herschel (1738–1822) placed several thermometers in the region just beyond the red end of the spectrum and discovered invisible but heat-carrying radiation (Fig. 9–3). This region of the spectrum is now called the *infrared* (*infra*, below). Later, the physicists Wilhelm Ritter (1776–1810) and William Hyde Wollaston (1766–1828) detected *ultraviolet* rays (*ultra*, beyond) at the opposite end of the visible spectrum through the blackening effect of these rays on crystals of silver iodide. Gradually, other radiations from various souces were discovered. Although the concept of "light" is at times confined to stimuli that affect the sense of vision, a more inclusive definition includes radiation, which shares with visible light many of the same properties, such as reflection, refraction, polarization, and the same velocity.

Figure 9–3 William Herschel discovered infrared radiation while measuring the temperature of the spectrum. The temperature increased from the violet end to the red end and, surprisingly, remained high beyond the visible red.

9–2 THE VELOCITY OF LIGHT

While studying the eclipses of Io, the innermost of the satellites of the planet Jupiter, Olaf Römer (1644–1710), the astronomer, came upon an apparent discrepancy in the satellite's behavior. After determining that the average period of Io was about 42½ hours, Römer thought he could predict the times when Io emerged from Jupiter's shadow, and he prepared a long-term table of Io's appearances. However, Io appeared further

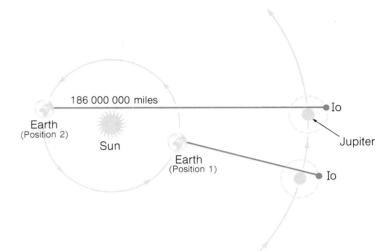

Figure 9–4 Olaf Römer's method of determining the speed of light. With the earth at position 2, light from Jupiter's satellite Io must travel an added distance, the diameter of the earth's orbit.

and further behind schedule as the earth receded more and more from Jupiter in its annual orbit around the sun. As the distance between the earth and Jupiter decreased on the earth's return, the discrepancy between the predicted and observed times also decreased until Io's appearances were back on schedule.

Römer reasoned that Io's appearances were on schedule and that the delay represented the transit time of light across the earth's orbit (Fig. 9–4). If the difference between the maximum and minimum separation between the earth and Jupiter is the diameter of the earth's orbit around the sun, 186 000 000 miles, and the observed delay in Io's appearance is 1000 seconds, then the velocity of light is 186 000 000 miles/1000 seconds = 186 000 miles/second, or roughly 300 000 km/sec. (The data used here are more accurate than those that were available to Römer.)

Galileo also had believed that light traveled with a finite velocity and attempted to measure it by timing the travel of a pulse of light over a measured distance (Fig. 9–5). He used a hand-operated shutter to transmit flashes from a lantern to an assistant stationed on a hill about a mile away. When the light beam reached the assistant, he was to uncover his lantern; Galileo, upon seeing this light, would note the elapsed time.

Galileo failed in his purpose, although he did demonstrate that the velocity of light must be very great. One weakness of Galileo's method was the reaction time of the assistant. The physicists Hippolyte-Louis Fizeau (1819–1896) and Léon Foucault (1819–1868)

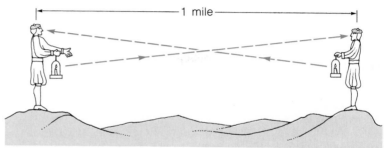

made refinements in the method that paved the way for modern precise measurements. They replaced the assistant with a mirror that returned the light to its point of origin (Fig. 9–6). They improved upon another drawback, the manually operated shutter, first by replacing it with a toothed wheel in front of a light beam that transmitted the light in short pulses, and later by using a rotating mirror, Knowing the distance of the round trip and the time of travel of light to the mirror and back, Foucault determined the velocity of light to be 298 000 km/sec.

Albert A. Michelson (1852–1931), a physicist at the University of Chicago, applied the Galileo-Fizeau-Foucault method with an unprecedented degree of precision. A rotating mirror was positioned on Mount Wilson in California; 22 miles away a fixed mirror was set up on Mount San Antonio (Fig. 9–7). The distance

Figure 9–5 With an assistant stationed on a hill about a mile away, Galileo attempted to measure the speed of light by flashing his lantern and recording the time elapsed to receive a flash of light from the assistant. The time was too short to detect.

Figure 9–6 In Fizeau's method, light from an arc source was reflected from a mirror (M₁) toward a rotating wheel. Bursts of light passed through the wheel and were reflected to the eye of the observer from a mirror (M₂) 5.36 miles away.

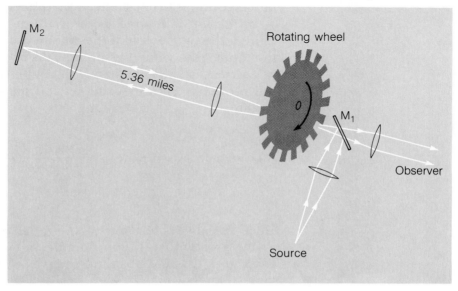

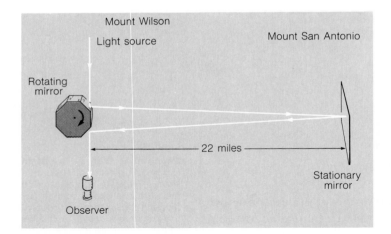

Figure 9–7 Albert A. Michelson used an octagonal mirror in rapid rotation to reflect light from an intense source to a stationary mirror 22 miles away. The speed of rotation of the mirror was adjusted so that the mirror moved one-eighth of a revolution while the light made the 44-mile round trip. The time for the mirror to make one-eighth of a revolution was then equal to the time for the light to travel 44 miles.

between the two mirrors, measured by the U.S. Coast and Geodetic Survey, was based upon a primary base line that may have been the most accurately measured line ever marked on the surface of the earth. Michelson determined the velocity of light to be 299 796 km/sec. The currently accepted value, based upon an average of several determinations, is 299 792 km/sec, or 186 282 mi/sec.

The velocity of light depends upon the nature of the medium through which it travels. Foucault established that the velocity of light is less in water than in air, and a group led by Michelson determined that it is slightly greater in a vacuum than in air.

The change in the velocity when light passes from one medium into another causes refraction. The *index of refraction* is related to the ratio of the velocity of light in the two media. If the first medium is a vacuum, the ratio of the velocities defines the *absolute index of refraction,* usually called just the index of refraction or the refractive index of the second medium, such as water.

Equation 9–1 $\text{Absolute index of refraction (water)} = \dfrac{\text{Velocity of light in a vacuum}}{\text{Velocity of light in water}}$

$$\mu = \frac{c}{v}$$

Since v is always less than c, the index of refraction is always greater than 1. For example, in water μ is 1.333. Table 9–1 lists the absolute indexes of refraction for some selected materials.

Table 9–1 ABSOLUTE INDEXES OF
REFRACTION OF SELECTED MATERIALS

MATERIALS	μ
Air	1.0003
Water (20°C)	1.333
Alcohol (ethyl)	1.362
Glycerine	1.473
Benzene	1.501
Glass (crown)	1.517
Quartz (crystalline)	1.544
Mica	1.561
Carbon disulfide	1.63
Glass (flint)	1.65
Diamond	2.417

EXAMPLE 9–1

Calculate the velocity of light in benzene.

SOLUTION

Knowing the value of c, and obtaining the refractive index of benzene from Table 9–1, we determine the velocity of light in benzene from Equation 9–1.

μ(benzene) = 1.501

c = 3 × 10^8 m/sec

velocity of light in benzene = ?

$$\mu \text{ (benzene)} = \frac{\text{velocity of light in a vacuum}}{\text{velocity of light in benzene}}$$

$$\mu = \frac{c}{v}$$

$$v = \frac{c}{\mu}$$

$$v = \frac{3 \times 10^8 \frac{m}{sec}}{1.501}$$

$$v = 2 \times 10^8 \frac{m}{sec}$$

Figure 9–8 Photograph of printed material viewed through an Iceland spar crystal (calcite). The letters appear double.

9–3 POLARIZED LIGHT

While examining a crystal of calcium carbonate brought from Iceland to Copenhagen and known as calcite or Iceland spar, Erasmus Bartholinus (1625–1692) discovered a puzzling phenomenon. Looking through the crystal at an object, he observed two images rather than one (Fig. 9–8). He also found that if the crystal was rotated, one of the images, which he called the extraordinary one, rotated with the crystal, but the other, which he called the ordinary one, remained fixed. He called this phenomenon *double refraction*, because an incident beam of light seemed to be split into two beams (Fig. 9–9).

A satisfactory explanation of the phenomenon had to wait until other properties of double-refracting crystals had been discovered. The index of refraction of the ordinary ray in calcite is always 1.658; for the extraor-

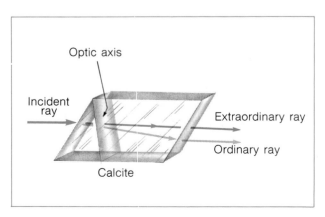

Figure 9–9 Double refraction by a calcite crystal. The crystal separates the E and O rays from incident light and transmits them in different directions and at different speeds.

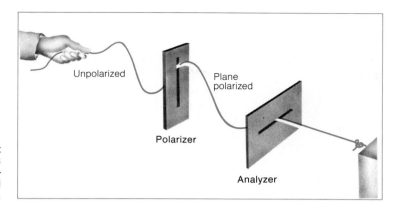

Figure 9–10 The vertical slit transmits only transverse waves vibrating in the vertical direction, thus acting as a vertical polarizer for waves along a rope.

dinary ray it varies with the angle of incidence. In some double-refracting crystals, one of the beams is nearly completely absorbed on its way through the crystal, and only the other is transmitted undiminished. Such a crystal is tourmaline, in which sections only 1 mm thick completely absorb one of the rays. Thomas Young suggested that these effects could best be explained if light is regarded as having a wave nature and, more specifically, as having a transverse rather than a longitudinal wave motion.

Ordinarily, light waves oscillate in every possible direction at right angles to the beam's path. If the vibrations are confined to just one of these directions, the light is plane polarized just as transverse waves along a rope are plane polarized when the vibrations are restricted to a single plane (Fig. 9–10). If light is a transverse wave motion, it should be possible to polarize it. Longitudinal waves cannot be polarized. Double-refracting crystals are best understood, according to Young, by assuming that the ordinary and extraordinary rays are polarized at right angles to each other.

Figure 9–11 In a Nicol prism, a calcite crystal has been cut at a particular angle and re-cemented with Canada balsam. The refractive indices of the polarized E and O rays are such that only the E ray passes through the cement and then through the crystal. A beam of plane-polarized light is thus obtained.

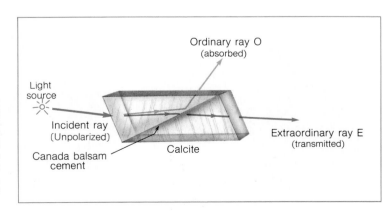

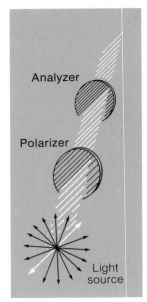

Figure 9–12 Principle of a polariscope. The amount of light transmitted depends upon the alignment of the polarizer and analyzer and on the medium between them.

Figure 9–13 Transmission of light through Polaroid filters. A maximum amount of light is obtained when the transmitted Polaroids are parallel, and a minimum amount when they are crossed.

Polarized light became a useful scientific tool when William Nicol (1768–1851), a physicist, prepared crystals in such a way that one of the rays, the ordinary ray, was absorbed by the crystal, and only the extraordinary, polarized ray emerged (Fig. 9–11). A Nicol prism alone constitutes a polarizer; a second crystal, an analyzer, transmits more or less of the polarized light, depending upon how it is aligned with reference to the polarizer. An arrangement consisting of a polarizer and an analyzer is called a *polariscope* (Fig. 9–12).

Glass, lucite, and many other transparent materials become double-refracting under stress. When placed between the polarizer and analyzer of a polariscope, a colored pattern resulting from the interference effects of polarized light is observed that can be related to strains in the material. By building models of structures with transparent materials, engineers can study stresses with polarized light.

Edwin H. Land (b. 1909), an inventor and physicist, succeeded in orienting large numbers of the minute, needle-shaped crystals of the organic compound quinine iodosulfate. He did this by suspending them in a plastic film, then stretching the film in one direction. The polarizing action of *Polaroid*, as the material is called, is based on the uniform alignment of the double-refracting crystals, which absorb all of one ray and transmit the other.

A sheet of Polaroid acts very much like an optical picket fence, transmitting plane-polarized light in which the vibrations of light are parallel to the crystals. For optical use, films are mounted between glass plates (Fig. 9–13). Used in sunglasses, Polaroid reduces glare from roads and bodies of water. Light reflected from

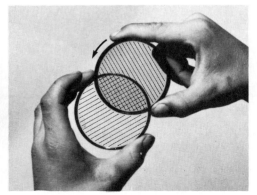

A   B

such surfaces is strongly polarized in a horizontal plane and is absorbed by the vertically oriented crystals in the Polaroid.

9-4 ELECTROMAGNETIC WAVES

The optical phenomena of diffraction and interference support the hypothesis that light has a wave nature, and polarization that light waves are transverse. But what is it that actually waves and what is the medium that carries this wave? Huygens and Young agreed that light was transmitted by mechanical waves in an elastic medium called the *luminiferous* ("light-bearing") *ether*, which was assumed to pervade all space.

James Clerk Maxwell gave a different answer to the first question. His masterpiece, *A Treatise on Electricity and Magnetism* (1873), contains his famous equations of the electromagnetic field. From his equations, Maxwell deduced the existence of electromagnetic waves in which changing fields produce other fields in their vicinity and the disturbance travels through space at the speed of 3×10^8 m/sec, the speed of light. Electromagnetic waves would have all the known properties of light waves. Light is thus just a form of electromagnetic radiation of a certain wavelength. The propagation of the waves involves a changing magnetic field that generates a changing electric field; the electric field in turn generates a changing magnetic field. In an electromagnetic wave, it is the electric and magnetic fields that propagate through space. Both fields are perpendicular to the direction of propagation, so the wave is transverse (Fig. 9–14).

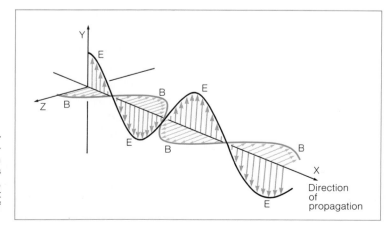

Figure 9–14 In electromagnetic waves, electric and magnetic fields propagate through space. The electric field, E, is vertical here; the magnetic field, B, is horizontal; and both are at right angles to the direction of propagation.

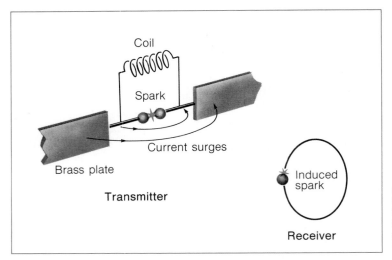

Figure 9–15 Hertz's experiment in which electromagnetic waves were generated and detected. Sparks in the transmitter cause waves that travel through space and induce sparks at the receiver.

For years there appeared to be no way to test experimentally Maxwell's theory of the electromagnetic nature of light. The problem was to prove the existence of a relation between light and electricity. Then Heinrich Rudolf Hertz (1857–1894) succeeded in producing and detecting other electromagnetic waves. Already convinced that every kind of light—sunlight, candlelight, the light from a glowworm—was basically electrical, he set out to prove it on the basis of Maxwell's equations.

Hertz's design was to generate waves electrically and to demonstrate that they traveled through the "ether" in the same way and with the same speed as light. He reasoned that such oscillations should be detected by a resonant receiver at some distance from the transmitting source. The receiver would pick up the oscillations, and the induced current would produce a spark across a gap.

Hertz's transmitter consisted of two polished metal spheres separated by a small air gap and connected to an induction coil which built up a large voltage across the gap and caused sparks (Fig. 9–15). The sparks produced current oscillations that, according to Maxwell's theory, generate electromagnetic waves that then propagate through space. The receiver—a piece of circular wire with a tiny spark gap formed by a little knob and a sharp point—was set up at the other end of a darkened physics lecture hall. The instant the transmitter was turned on and sparks jumped across it, tiny sparks —only a few thousandths of an inch long and lasting not more than a microsecond—jumped across the receiving "antenna."

To prove that the electromagnetic waves were like light waves, Hertz measured the wavelength and frequency of the radiation and determined that the speed was approximately 3×10^8 m/sec—the speed of light, as Maxwell had predicted. He showed further that the electromagnetic waves behaved like light waves in other respects. He reflected them from polished metal sheets and mirrors, refracted them with large prisms, made them interfere, and polarized them. In every case the new radiation behaved just as light did, except that it was not visible. Hertz proved that light and electromagnetic radiation were identical, and that they both are explained by the electromagnetic theory of light.

9–5 THE ELECTROMAGNETIC SPECTRUM: RADIO, TV, MICROWAVES

The waves in the various regions of the electromagnetic spectrum share similar properties, but differ in wavelength, frequency, and method of production. Radio and TV waves and microwaves are generated by oscillating electrical circuits and range in wavelength from 20 000 meters down to a fraction of a centimeter. Infrared waves originate in molecular systems; their wavelengths extend from a fraction of a centimeter to 8000 Å; visible light, from 8000 Å to about 4000 Å; ultraviolet radiation, as short as 50 Å; and x-rays, to 0.1 Å, arise from disturbances in the electronic structure of atoms. Gamma rays are as short as 10^{-5} Å, and originate in atomic nuclei (Fig. 9–16).

The transmission of a radio wave begins with a

Figure 9–16 The electromagnetic spectrum. (From Pasachoff, J. M., *Contemporary Astronomy*, W. B. Saunders Company, 1977.)

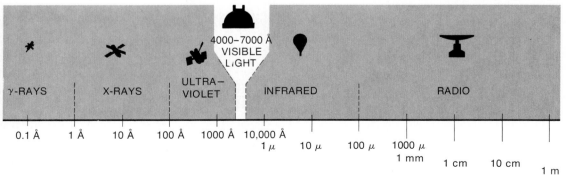

high-frequency (possibly 1 MHz) oscillating current in a long metal rod, an antenna. Each electron is accelerating since its position and velocity are changing. Because of the continually changing electric and magnetic fields, due to the oscillating electrons, electromagnetic waves are radiated. These changing fields induce other fields in space, the disturbance traveling outward with the speed of light.

When the wave reaches a receiving antenna, the oscillating electric field exerts an oscillating force on the electrons in the antenna, causing them to oscillate in the same way as the electrons in the transmitting antenna. Communication thus becomes possible through the intervening space as energy is carried by the electromagnetic waves. In amplitude modulation (AM) radio, an audible signal frequency of several hundred or thousand hertz is impressed upon a carrier

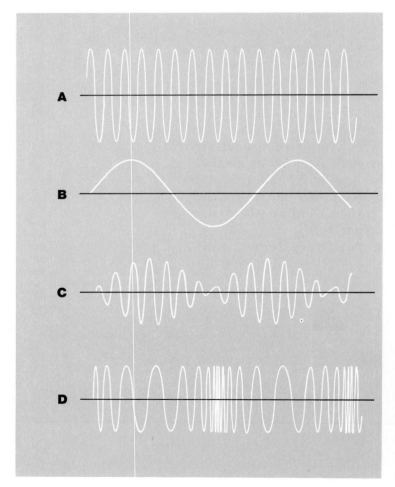

Figure 9–17 The carrier frequency in AM radio broadcasting ranges from 550 KHz to 1700 KHz. FM stations broadcast in the range from 88 to 108 MHz. Represented in this figure are an unmodulated "carrier" radio wave *(A)*; an audio-frequency wave *(B)*; an AM (amplitude modulation) wave *(C)*; and an FM (frequency modulation) wave *(D)*.

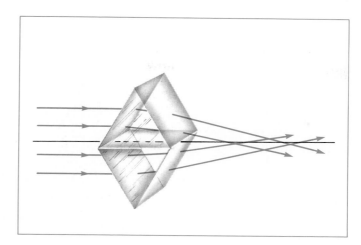

Figure 9–18 Parallel light rays are bent toward the thicker section of two prisms placed base to base. The action resembles that of a converging or convex lens.

wave of megahertz frequency, modulating its amplitude but having no effect on its frequency (Fig. 9–17). The signal is recovered at the receiver end, amplified, and converted to a sound wave. In frequency modulation (FM) only the frequency of the carrier wave is modulated, increasing and decreasing at the same rate as the impressed audible signal frequency. Since atmospheric static affects only the amplitude and not the frequency of the modulated wave, FM transmission is essentially noise free. In other regions of the spectrum, energy transfer is manifested in various ways, for example, in the visible range, the stimulation of appropriate nerve endings in the eye and the initiation of chemical reactions through suntanning or photosynthesis.

9–6 SIMPLE LENSES

The action of most optical instruments depends upon the refraction of light by lenses made of transparent glass or plastic. A convex or converging lens is thicker at the center than at the edges. Its structure is like that of two prisms placed base to base (Fig. 9–18). Parallel rays of light incident upon such a prism arrangement obey the law of refraction. Each ray is bent toward the normal as it enters a prism, and away from the normal as it emerges into the air. The net effect is that the rays of light are bent toward the thicker part of the prism, that is, they converge.

Similarly, parallel rays are refracted toward the thick part of a double convex lens and converge at a

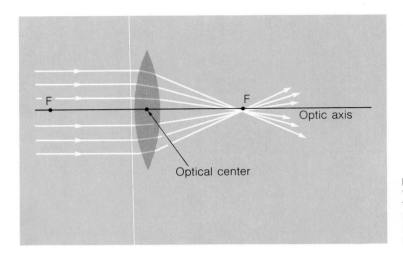

Figure 9–19 Action of a converging lens on light rays parallel to the optic axis. The rays are refracted and pass through the principal focus (F) on the optic axis.

point called the principal focus, F (Fig. 9–19). (There is a principal focus on each side of such a lens.) The distance between F and the optical center C of the lens is called the focal length, f. This is the primary characteristic of the lens. The optical center and the two principal foci are situated on an imaginary line through C, perpendicular to the plane of the lens. This imaginary line is known as the optic axis. Incident light rays that are parallel to the optic axis are all refracted through the point F.

The position of the image formed by a double convex lens of a small object standing on the optic axis on one side of the lens at a distance beyond 2f may be determined by selecting three rays from each point on the object. This is shown for a single point in Figure 9–20. A ray parallel to the optic axis passes through the focal point on the opposite side of the lens after refraction by the lens. A second ray from the same point on the object is selected to pass through the optical center of the lens; it passes through essentially undeviated.

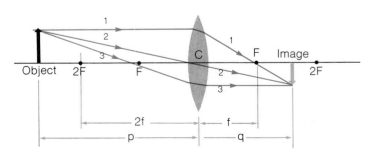

Figure 9–20 Ray diagram of image formation by a converging lens. With the object beyond 2F, the image is real, inverted, and smaller than the object.

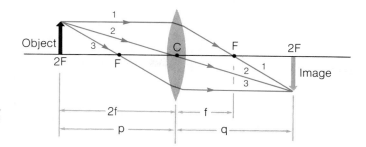

Figure 9–21 When the object is at 2F, the image is at 2F and is the same size as the object.

A third ray passing through the principal focus emerges from the lens parallel to the optic axis. The point at which the three rays converge locates the image of the point. Three rays are similarly selected for a point at the opposite end of the object, and the image of that point is located where the three rays intersect. The image formed by the lens is made up of millions of such point images. Since light rays actually go to this image, it is called a real image. In this case the image is real, inverted, and smaller than the object.

When the object is at a distance from the converging lens greater than the focal length, the image is always real and inverted. For distances greater than 2f, the image is smaller than the object. When the object distance is 2f, the image distance is also 2f and the image is the same size as the object (Fig. 9–21). If the object distance is between 2f and f, the image is enlarged. Try to draw the principal rays for this case yourself.

In a single lens motion picture projector the film is placed just beyond the focal length of the lens, that is, between f and 2f. By tracing the ray optics you will see that the image is real, inverted, and much larger than the object. The image on the screen is certainly much larger than the film. To have a picture that is right side up, though, the film must be inverted when placed in the projector.

A converging lens is unable to converge rays to a real focus when the object is nearer the lens than the principal focus. Since the rays from the object are so highly diverging when they are incident on the lens, they are still diverging after they pass through the lens, so that no real image is formed (Fig. 9–22). The rays appear to diverge from a point on an image that is virtual, erect, and enlarged. The image is virtual because there is no light at the image position and no image would form on a screen placed there. This is how a

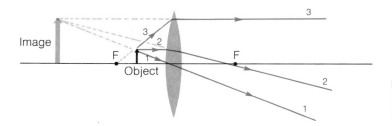

Figure 9–22 With a magnifying glass, the object is between F and the lens. A virtual image is formed that is right side up and enlarged.

magnifying glass works. You hold the magnifying glass close to an object and see an erect, magnified image.

A diverging lens is thinner at the center than at the edge. A ray from a point on the object that is parallel to the optic axis is refracted in such a way that it appears to come from the principal focus. A second ray passes through the center of the lens with essentially no deviation. After passing through the lens, these rays appear to come from a single point on an image on the same side as the object (Fig. 9–23). The image is virtual, upright as long as the object distance is greater than f, and inside the focal length regardless of the object distance. A diverging lens can never form a real image, because the rays passing through the lens can never be brought to a real focus.

9–7   THE OPTICS OF THE EYE

Light enters the eye at the curved front surface of the cornea, then passes through the aqueous humor, crystalline lens, and vitreous humor, before reaching the light-sensitive retina, which contains cone cells

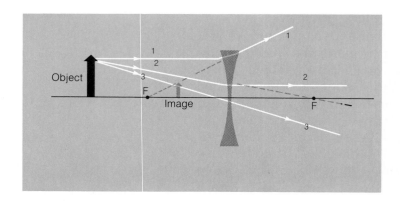

Figure 9–23 A diverging lens forms a virtual, upright image that is reduced in size.

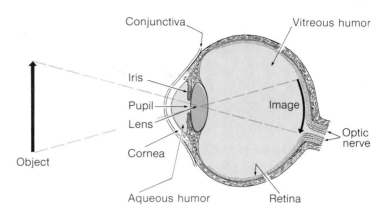

Figure 9–24 Structure and optics of the human eye.

and rod cells (Fig. 9–24). A sharp image must focus on the retina for clear vision to occur. Although the image is inverted, the brain interprets it as right side up. The distance between the lens and retina is fixed, but focusing can be accomplished by changing the curvature of the lens through the action of the muscles of the ciliary body on the ligaments around the rim of the lens. The lens is flexible, and by such accommodation can adapt itself to objects at various distances. The amount of light entering the eye is controlled by the iris diaphragm; its opening, the pupil, automatically contracts in bright light and enlarges in dim light.

Most of the ability of the eye to form an image is concentrated in the cornea and the aqueous humor, their combined power being approximately twice as great as that of the lens itself. Even if the lens must be removed, vision is still possible, although an auxiliary spectacle lens is necessary to compensate for its loss, and no accommodation can occur. Myopia, or nearsightedness, is a condition in which the eyeball is so lengthened that a sharp image of a distant object cannot be focused on the retina without the aid of diverging lenses (Fig. 9–25). The rays that should converge on the retina actually converge in front of the retina, so that the image on the retina is blurred.

The weakening of the eye muscles with age reduces the power of the lens, and farsightedness becomes common. Rays converge behind the retina in this case. This condition is corrected with auxiliary converging lenses (Fig. 9–26). In middle age, bifocal lenses are often fitted. The lower part is convex, and assists in converging light for near vision; the upper part is used in distant vision.

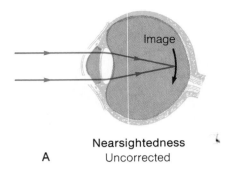

Nearsightedness
Uncorrected

A

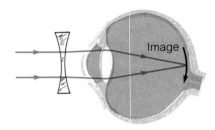

Nearsightedness
Corrected

B

Figure 9–25   Nearsightedness (myopia) is corrected with a diverging lens.

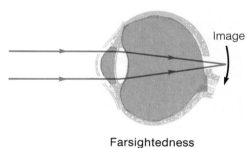

Farsightedness
Uncorrected

A

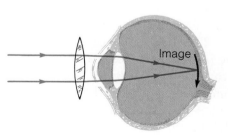

Farsightedness
Corrected

B

Farsightedness (hypermetropia) is corrected with a converging lens.

9-8 THE MICROSCOPE AND TELESCOPE

The principles of image formation by lenses are also applied in the microscope and telescope. The microscope is essentially a combination of two converging lenses (Fig. 9–27). The objective, the lens at the lower end of the microscope tube, acts as a projector, forming a magnified real image of the specimen (object) on the slide. The ocular, or eyepiece, the lens at the upper end of the tube, acts as a magnifying lens; it forms a magnified virtual image of the image formed by the objective. In more complex microscopes, the objective and the ocular are themselves composed of several lenses in order to eliminate distortions, but their function is the same as in the simpler case.

The refracting astronomical telescope is similar in principle to the microscope. It is a two-lens instrument, consisting of an objective and an ocular, designed to view objects at a distance. The objective in the telescope forms a real inverted image of a distant object in front of the ocular, the image being far smaller than the object. The ocular forms an enlarged virtual image that remains inverted. Telescopes are so long because large magnification requires an objective with a very long focal length. Since for viewing things on earth we prefer to have the image upright, an additional convex lens is inserted between the objective and the ocular to invert the image. These are often called terrestrial telescopes.

In very large telescopes, a concave mirror is used in place of the objective lens (Fig. 9–28). The purpose of the mirror, like that of the objective, is to gather as

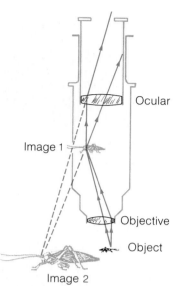

Figure 9–27 Principle of the compound microscope. The objective forms a real, enlarged image of the specimen in front of the ocular, and the ocular magnifies this image.

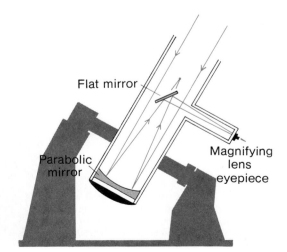

Figure 9–28 Reflecting telescope. A parabolic mirror reflects light to a focal point near the top of the tube. There a small flat mirror deflects the light out the side of the tube toward the magnifying lens eyepiece.

Figure 9–29 The 40-inch refracting telescope at Yerkes Observatory, Wisconsin. (Yerkes Observatory photograph.)

Figure 9–30 View of the 200-inch reflecting Hale telescope at Mount Palomar, California. The mirror, with a diameter of 200 in. and 27 in. thick, weighs 15 tons. (Courtesy of Hale Observatories.)

much light as possible and form an image that can be
magnified by the ocular. These are called reflecting
telescopes. The objective lens in the refracting telescope at the Yerkes Observatory, one of the largest in
use, is 40 in. in diameter (Fig. 9–29); the mirror in the
reflecting Hale telescope of Mount Palomar Observatory is 200 in. in diameter (Fig. 9–30).

9–9　COLOR

Whether a rainbow is in the sky, a garden spray, or
a fountain, droplets of water disperse white light into
its colors, each droplet acting like a prism. A beam of
light is refracted as it enters a droplet, dispersed into its
component colors, reflected internally, then refracted a
second time as it reemerges (Fig. 9–31). With the sun,
observer, and droplets positioned at an appropriate
angle, a bow is formed with red at the outer edge (42°2′),
since red is refracted least, and violet at the inner edge
(40°17′).

The colorful display on the surface of soap bubbles,
oil slicks, and similar thin films, as well as iridescence
of the wings and bodies of many insects, is due to interference effects. Light reflected from the two surfaces of
a thin film travels different path lengths (Fig. 9–32). As
they recombine, the light waves undergo constructive

Figure 9–31　A primary rainbow
formed from sunlight refracted
and internally reflected by water
droplets.

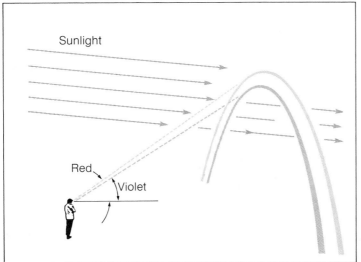

A

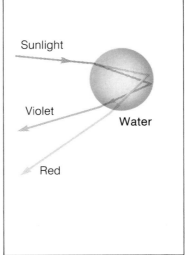

B

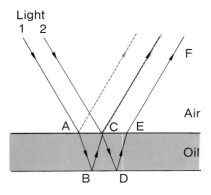

Figure 9–32 Iridescent colors of an oil slick arise from the constructive and destructive interference of light waves traveling along different path lengths. The difference is twice the thickness of the oil film. If rays 1 and 2 meet in phase at C, a bright spot appears. If there is destructive interference for a particular wavelength, the color associated with that wavelength will be absent from the reflected light.

and destructive interference depending upon whether they are in or out of phase.

The process of light scattering—the reflection of light in many directions—accounts for the blueness of the sky (Fig. 9–33). Particles that are small compared with the wavelength of light, such as the molecules and other small particles in the air, scatter light inversely as the fourth power of the wavelength; The shorter, blue waves are scattered far more than the longer, red waves. Large particles scatter light of all wavelengths about equally, and therefore appear white. That is why snow, salt, sugar, powders, soapsuds, clouds, and seafoam are white.

Most of the colors we see—in leaves, flowers, birds, fabrics, paints, dyes, minerals—are due to selective absorption. This is a subtractive process in which cer-

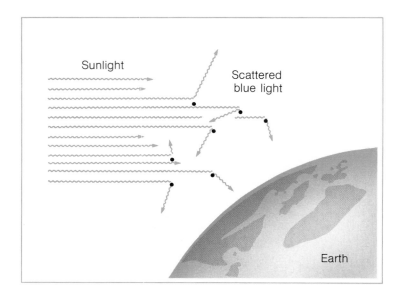

Figure 9–33 The blue color of the sky is produced by the greater amount of scattering of the shorter wavelengths of sunlight by air molecules. Thus, the light from the sky is predominantly scattered blue light.

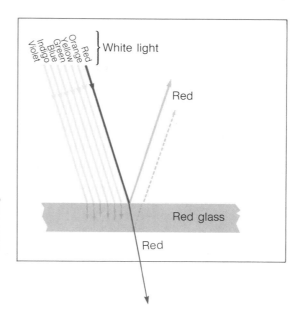

Figure 9–34 Colors of most objects are due to selective reflection and absorption of light. Red glass appears red because it transmits the red component present in white light while absorbing all other colors, and reflects some of the incident light from the back surface.

tain wavelengths are removed from the incident light (Fig. 9–34). The chlorophyll in green leaves and grass absorbs much of the light that enters the cells, allowing only green light to escape. A green leaf illuminated by light lacking green wavelengths appears black, since all of the incident light is absorbed and none is reflected. The petals of red flowers absorb most of the wavelengths except red. The color of cloth therefore depends upon whether it is illuminated by natural or various forms of artificial illumination.

Three spectral hues or pure colored lights—red, green, and blue—combine to form white. They are called the primary light colors (Fig. 9–35). All the other colors in the visible spectrum and many that are not can be produced by combinations of these three. Complementary colors are any two colored lights that blend to give white. On a color triangle they are opposite each other: blue and yellow, green and magenta, red and cyan (Fig. 9–36).

Mixing pigments is different from mixing spectral hues. The primary pigments of artists are yellow, cyan (blue-green), and magenta (blue-red), the complementaries of the primary light colors. Yellow reflects the red and green components of incident white light, while absorbing the blue. Cyan reflects blue and green, and absorbs red. If equal amounts of yellow and cyan are mixed, the result is green. The reason is that both yellow and cyan reflect green. Although cyan also reflects blue,

Figure 9–35 Additive mixing of colored lights from more than one source.

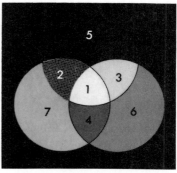

1=White 2=Magenta 3=Yellow 4=Cyan 5=RED 6=GREEN 7=BLUE

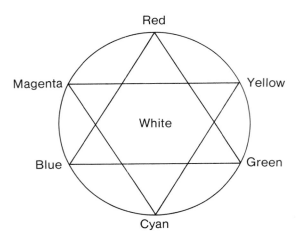

Figure 9–36   Complementary hues are any two hues that produce white when blended in some proportion, for example, blue and yellow.

the blue is absorbed by the red in the yellow pigment; similarly, the red in the yellow is absorbed by the blue in cyan. When all three primary pigments are combined, the result is black if the pigments are pure.

According to the Young-Helmholtz hypothesis of color vision, three light-sensitive pigments are distributed among three different kinds of receptor cells in the retina, each sensitive in a different region of the spectrum. One is primarily responsible for sensing red light, one for sensing green, and one for blue: the R-cones, G-cones, and B-cones. The 3-receptor hypothesis is supported by recent spectrophotometric measurements of individual cone cells in the retinas of the goldfish, the rhesus monkey, and human being. But how this information is coded in the retina, transmitted to the brain, and decoded there are questions that remain to be answered.

When pure spectral yellow enters the eye, the R- and G-cones respond equally, and the sensation is yellow. A similar sensation can be produced when no spectral yellow is presented. Pure red and green cause both the R- and G-cones to respond equally, and the sensation is again yellow. When blue and green enter the eye, the stimulation of B- and G-cones produces the sensation of cyan. The absence of R-cones in the retina leads to a type of color-blindness called protanopia. Only two primary colors, blue and green, are at the disposal of these individuals who can see only to 6800 Å instead of the normal 7600 Å. The number of hues that the color-blind person can see is therefore reduced.

CHECKLIST

EXERCISES

9-1.   Explain the formation of a rainbow.

9-2.   How do Polaroid sunglasses reduce glare from a road or a body of water?

9-3.   What experimental evidence strengthened Maxwell's hypothesis that light and electromagnetic waves are similar?

9-4.   List the forms of behavior in which light manifests wave properties.

9-5.   How long does it take light to travel from the sun to the earth, 93 million miles away?

9-6.   Which color has the higher frequency, red or blue?

9-7.   Which radiation has the shorter wavelength, radio waves or X rays?

9-8.   A light-year is a unit of length equal to the distance light travels in a year. What is the length of a light-year in (a) kilometers? (b) miles?

9-9.   The star Arcturus is 40 light-years from the earth. What is this distance in (a) kilometers? (b) miles?

9-10.   How far away is an object if the reflection from a radar pulse is received in 93 microseconds?

9–11.   Determine the frequency of light that has a wavelength of 5890 ångstrom units.

9–12.   The speed of light in a clear plastic material is 1.5×10^8 m/sec. What is the index of refraction?

9–13.   What spectral colors are complementary to (a) blue; (b) yellow; (c) magenta?

9–14.   Discuss the evidence for the transverse-wave nature of light.

9–15.   (a) What color would the sky be if there were no atmosphere? (b) What proof is there?

9–16.   If they are visible, what colors would you assign the following wavelengths: (a) 400 Å; (b) 4000 Å; (c) 6000 Å; (d) 40 000 Å?

9–17.   Compare the number of "octaves" to which the eye responds with the number to which the ear responds.

9–18.   Explain why the mixing of red and blue lights gives a different color than the mixing of equal amounts of red and blue pigments.

9–19.   *Multiple Choice*

A. Compared to the velocity of light, the velocity of radio waves is
(a) less   (c) the same
(b) greater   (d) infinite

B. Polarized light is explained by the hypothesis that light consists of
(a) transverse waves   (c) particles
(b) longitudinal waves (d) an ether

C. Color is related to
(a) amplitude   (c) velocity
(b) frequency   (d) quality

D. The bending of a ray of light as it passes at an angle from air into water is explained by
(a) reflection   (c) refraction
(b) diffraction   (d) interference

E. The pupil of the eye adjusts for
(a) long-distance vision
(b) color
(c) size of object
(d) amount of light

Connes, Pierre, "How Light is Analyzed." *Scientific American,* September, 1968.
   Much of our information about the world is carried by light. Describes instrumental spectroscopy, the technique of separating light into its constituents.
Feinberg, Gerald, "Light." *Scientific American,* September, 1968.
   Surveys various answers to the question, "What is light?"
Land, Edwin H., "The Retinex Theory of Color Vision." *Scientific American,* December, 1977.
   The eye has evolved to see the world in unchanging colors. How does it achieve this feat?
MacAdam, David L., "Color Science and Color Photography." *Physics Today,* January, 1967.
   James Clerk Maxwell invented three-color photography and explained the principle on which modern color photography, color printing, and color television are based. Modern workers in these fields still have much to learn from him.
Shankland, R. S., "Michelson: America's First Nobel Prize Winner in Science." *The Physics Teacher,* January, 1977.
   Discusses the famous experiment that disproved the hypothesis of a "luminiferous ether."

SUGGESTIONS FOR
FURTHER READING

ELECTRICITY AND MAGNETISM

10-1 INTRODUCTION

On the muggy evening of July 13, 1977, two bolts of lightning struck two transmission towers, causing New York City to experience its second major blackout in twelve years. Eight million persons lost their electricity, and subways and elevators went dead. The great blackout that descended upon the populous northeastern sector of the United States on the evening of November 9, 1965, had given convincing proof of the electrical basis of modern civilization (Fig. 10–1). Normal living patterns ground to a sudden and unexpected halt, indicating how vulnerable that basis is and how dependent we have become upon something that was developed for the benefit of humanity only about a century ago. There is no assurance that such an event will not happen again, as the increasing demand for electrical energy taxes our ability to provide it, a condition described by the term "energy crisis."

It is interesting that we are not dealing with a new phenomenon. Four natural manifestations of electric-

Figure 10–1 The fragility of our electrical age was dramatically demonstrated on the evening of November 9, 1965, when a faulty electrical relay caused a power failure and plunged the northeastern part of the United States into total darkness. (*Left,* courtesy of Con Edison of New York; *right,* photograph by Bob Gomel, LIFE MAGAZINE, © Time Inc.)

ity have been known for centuries: lightning; the ability of amber to attract objects when rubbed; "animal electricity" as it occurs in the discharge of the electric eel and other species; and St. Elmo's fire, a discharge of light seen near pointed objects in storms.

Our electrical technology, which has superseded the steam age, is based on magnetism. Electric power generators, motors, radio and television, the telephone—all depend upon magnetism and its relation to electricity.

10–2   THE AMBER PHENOMENON

You are probably acquainted with some of the simplest facts of electrical force. A balloon rubbed with wool, flannel, or dry hair will stick to a wall or ceiling. Moreover, a balloon "electrified" in this way attracts bits of paper and other small objects, and is attracted to the material that was used to electrify it (Fig. 10–2). You can easily demonstrate that an electrified balloon repels another balloon treated in the same way, and that these effects are more pronounced on a dry day than when it is humid. How would you explain these phonomena?

The basic discoveries of frictional electricity were made over a period of more than 2000 years. The philosopher and scientist Thales of Miletus is credited with

Figure 10–2 The basic experimental facts of electricity are illustrated. *A*, A balloon rubbed with wool attracts bits of paper and (*B*) is attracted to the wool. *C*, Two electrified balloons repel each other.

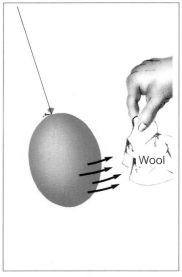

Wool

Bits of paper

A   B   C

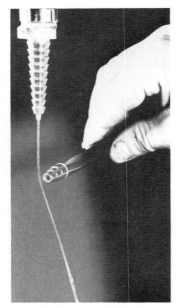

Figure 10–3   An electrified ball-point pen bends a stream of water.

Figure 10–4   If the charge on a glass rod is shared by several pith balls, the similar charges cause mutual repulsion.

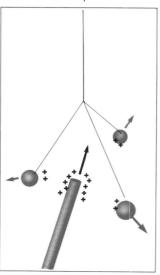

the discovery (about 600 B.C.) that when the resin amber is rubbed with wool or fur, it temporarily attracts small bits of matter, such as straw, leaves, and hair. The Greek word for amber is "elektron." But neither the Greeks nor others did much more with electricity than to make this and similar observations. A basic difficulty is that frictional electricity is not usually produced in great amounts and does not lend itself to many kinds of experiments. Moreover, it is quickly lost in damp surroundings.

The modern development of *electrostatics*—electricity at rest—dates from the publication in 1600 of the book *De Magnete* by Sir William Gilbert (1544–1603), chief physician to Queen Elizabeth I. In this classic of scientific literature, Gilbert brought together everything that was known about magnetism and electricity, described many of his own experiments, and set down his conclusions and speculations. He demonstrated that the amber phenomenon was not limited to amber and a few other materials, but that many types of bodies could be electrified, among them diamonds, sapphires, glass, sulfur, sealing wax, and mica. Moreover, electrified bodies attracted not only chaff and twigs, but metals, wood, stones, even water and oil. You can demonstrate one of these effects by rubbing a ball-point pen on wool and placing it near a gentle stream of water from a faucet (Fig. 10–3). The pen diverts the direction of the stream.

Electrified bodies can be divided into two classes. Materials such as glass and porcelain are included in one class. The second class includes amber, silk, and other materials. In the terminology proposed by Benjamin Franklin (1706–1790), statesman, scientist, and philosopher, the electrical condition glass acquires when it has been rubbed with silk is called *positive* electric charge. The kind of charge that hard rubber acquires when it has been rubbed with cat's fur is called *negative* electric charge. The *law of electric charges* (Fig. 10–4) is based upon the experimental finding: *like charges repel each other; unlike charges attract.*

Our present view of electricity is based on the *electron theory of matter*. Matter is composed of atoms with a positive nucleus surrounded by a cloud of orbiting negative electrons. Charges are associated with particles and are properties of those particles just as their masses are. Ordinarily, an atom has equal num-

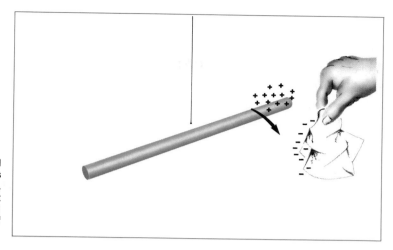

Figure 10-5 When a glass rod is rubbed with silk, electrons pass from the glass to the silk. The glass rod is left with a net positive charge, the silk an equal negative charge. Charges of opposite sign attract.

bers of positively charged protons and negatively charged electrons, and is electrically neutral. However, when two different materials, such as rubber and wool, are rubbed together, the one with a greater affinity for electrons (rubber) attracts some from the other (wool). The body that gains electrons acquires an excess negative charge, and the body that loses electrons acquires a net positive charge (Fig. 10–5). Whenever one body acquires a negative charge, a second body will acquire an equal positive charge. This observation is embodied in the principle of the *conservation of charge:*

> *Charge cannot be created or destroyed; the*
> *total charge in the universe is constant.*

Although rubbing causes a charge transfer, the mere contact between two unlike materials causes them to become electrically charged to some extent. Rubbing, by bringing more areas into contact, increases the effect by making many points of contact instead of the relatively few formed when two rough surfaces are placed in contact. Your shoes scuffing over the pile of a rug, the tires of a moving car, a comb passing through the hair—all acquire electrical charges in this way.

In Table 10–1, various materials are listed on the basis of the electrostatic charges they acquire upon contact with other materials. Each material becomes more positive than the one below it. If glass is in contact with silk, glass acquires a positive charge and silk a negative charge, showing that it is easier to remove electrons from glass than from silk. But if sulfur is in

Table 10–1 Electrostatic Series

Glass
Mica
Wool
Quartz
Cat's fur
Silk
Felt
Cotton
Wood
Cork
Amber
Graphite
Rubber
Sulfur

contact with silk, silk becomes positive with respect to sulfur; that is, it is easier to remove electrons from silk than from sulfur.

10–3 CONDUCTORS, SEMI-CONDUCTORS, AND INSULATORS

Since many ordinary metals can transmit electricity readily, they are classified as good electrical conductors. On the other hand, many other materials— glass, rubber, amber—do not conduct electricity well and are called *insulators*. When placed on an insulator, a charge does not migrate but remains approximately where it is placed. Because they are good conductors, metals are used in electric wires to take electricity where we want it to go, while glass, fabrics, and plas-

Table 10–2 Electrical Conductors
and Insulators

CONDUCTORS	INSULATORS
Aluminum	Amber
Copper	Fur
Gold	Glass
Iron	Mica
Mercury	Rubber
Sea water	Sealing wax
Silver	Silk
	Wool

tics are used to insulate wires, that is, to prevent electricity from going where it is not wanted.

Materials do not always fall neatly into these two classes. On a dry winter day, for example, air behaves as an insulator. As we walk across a rug and accumulate charge on our bodies, the charge remains fixed and we may experience a shock as we reach for an electric wall switch or metal doorknob. On a damp day this is less likely to happen; then air is more conducting, and the moisture in the air drains off charge from our bodies. Some typical conductors and insulators are listed in Table 10–2.

A third class of materials, called semi-conductors, include materials such as carbon, silicon and germanium, which ordinarily are slightly conducting. Semi-conductors are not good insulators, but at the same time they do not conduct electricity well enough to be considered good conductors. However, when light falls upon them or when their temperature is raised, they conduct well. This property allows them to play an important technological role.

10–4 THE ELECTROSCOPE

In the early days of experiments with electrical charges, the amount of the charge was estimated roughly by the size of the sparks produced, or by the size of the shock the experimenter received upon touching the electrified body. Today a gold-leaf electroscope may be used to tell whether there is a charge, whether it is positive or negative, and, more or less, the amount.

An *electroscope* has two thin gold or aluminum leaves mounted at the end of a metal rod with a metal knob at the top (Fig. 10–6). The rod and metal leaves are enclosed in a container made either of glass or of metal with glass windows, to avoid the effects of air currents and to prevent any electrical effects on the leaves other than those produced on the knob. The rod is mounted in a plug of rubber or some other good insulator to separate it from the box.

If a negative charge from a rubber rod that has been rubbed with fur is placed directly on the knob of the electroscope—a process called charging by conduction or contact—it spreads over the knob, rod, and leaves. This is because the negative charge consists of electrons that are free to move through metallic parts.

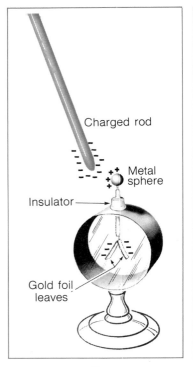

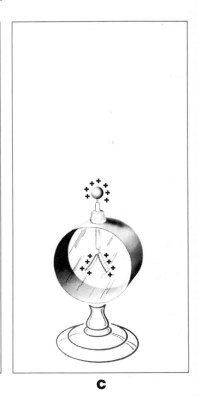

A   **B**   **C**

Figure 10–6 Charging an electroscope by induction. *A,* A negative rod brought near the electroscope repels electrons to the leaves, which diverge. *B,* Electrons are removed by grounding when the knob is touched; the leaves collapse. *C,* With the ground connection broken first and then the inducing charge removed, positive charges on the leaves cause them to diverge. The electroscope is now charged positively.

The electrons would like to get as far away from the rod as possible. In this case the leaves are the farthest point. Each leaf acquires a net negative charge. Since like charges repel, the leaves repel each other and diverge. The amount of divergence depends upon the quantity of charge on each leaf and the mass of each leaf. The electroscope is now charged negatively. It may be charged positively in a similar way with a charging body such as a glass rod rubbed with silk. In this case, electrons migrate from the leaves, rod, and knob to the glass rod, and each leaf acquires a positive charge. The leaves again diverge.

The electroscope is a sensitive instrument. The amount of charge required to produce a deflection of the leaves is small, owing to the small mass of the thin leaves. It is therefore often preferable to charge it by another method—*induction*—to prevent damage to the leaves. In this process there is no direct contact between the charging body and the electroscope (Fig. 10–6). If a negatively charged rubber rod is brought near the knob of the electroscope, some of the electrons in the knob are repelled to the leaves, which then diverge. You will note that no electrons are added to the electroscope in this process, and that the electroscope is still neutral.

With the rubber rod still in position, the knob is momentarily touched with a finger. Since there is now a connection between the electroscope and the ground through another conductor (the body), this step is called *grounding*. When grounded, electrons can get even farther away. Some of the electrons in the electroscope are repelled from the leaves to the rod, knob, finger, and ground. Having lost their excess electrons, the leaves collapse. When the finger is removed, followed by the removal of the rubber rod, the leaves diverge again. The reason is that the electroscope has lost some electrons to the ground and, because it is deficient in electrons, has acquired a positive charge. Although the charging body, the rubber rod, is negative, the electroscope has acquired the opposite charge by induction; it becomes positive. To charge an electroscope negatively by induction, the charging body must be positive.

With a charged electroscope, we can determine the presence of a charge on any body brought near it and also the sign of the electric charge. Suppose the electroscope is charged negatively (Fig. 10–7). If a body is brought near the knob and the leaves diverge still further, we can be sure that the body has a negative charge; more electrons are repelled from the knob onto the leaves, giving the leaves a greater negative charge and hence greater repulsion. On the other hand,

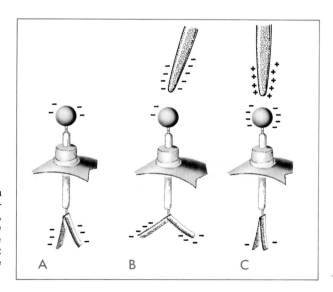

Figure 10–7 Determining the sign of a charged body with a negatively charged electroscope (A). If the leaves diverge farther, more electrons are driven from the ball to the leaves, so the rod must be negative (B). If the leaves draw together, we conclude that electrons are attached from the leaves to the rod, which must be positive (C).

A   B   C

if the leaves collapse to some extent, we conclude that the body has a positive charge, since some of the electrons are withdrawn from the leaves and the force of repulsion between them is reduced. Once charged, an electroscope may retain its charge for some hours, depending upon the moisture content of the air.

Electroscopes are used with more modern devices to study radiation from X rays, radioactivity, and cosmic rays. Such radiations charge particles in the air, and the charged particles, in turn, cause the discharge of an electroscope. The rate of discharge is related to the amount of radiation.

10–5   FORCES BETWEEN ELECTRIC CHARGES

Electrical repulsions and attractions are pushes and pulls, and are therefore forces. We would expect F to decrease as the distance increases. Perhaps it would be reasonable to assume that as the distance doubled, the force would become half as strong. Using a torsion balance, the physicist Charles-Augustin de Coulomb (1736–1806) discovered how electrical forces varied with the distance between the charges.

The torsion balance was set up as follows. By means of a fine thread, Coulomb suspended a light rod that had a metal sphere at each end (Fig. 10–8). At the top, he put a support that he could turn and a means to measure the amount of turn. When he charged two spheres in the same way, the repulsive forces between them caused the one on the rod to move. Since the rod was attached to the thread, the thread twisted; but it could be twisted back by turning it at the top until the two charged spheres returned to their original distance apart. The amount of twist became a measure of the force of repulsion. Later, using another arrangement, Coulomb measured the force of attraction between two charges.

Coulomb found that the electrical force of attraction or repulsion between two bodies varies inversely with the square of the distance between them. He also found that the electrical force is proportional to the amount of electric charge on each of the bodies. *Coulomb's law of electrostatic force* is expressed as follows:

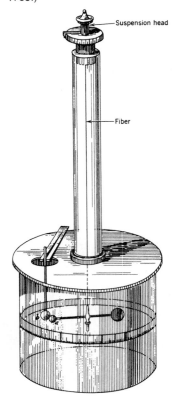

Figure 10–8 Coulomb's torsion balance. The repulsion of two like charges causes the suspending wire to twist. The angle of twist is related to the force of repulsion. (Drawing from Coulomb's mémoire to the French Academy of Sciences, 1785.)

Suspension head

Fiber

The force between two electric charges, q_1 and q_2, is proportional to the magnitude of each of the charges and inversely proportional to the square of the distance between them.

In equation form:

$$F = k\frac{q_1 q_2}{r^2}$$

Equation 10-1

where, in SI units,

F = the force of attraction or repulsion in newtons (N)
q_1 and q_2 = the electric charges in coulombs (C)
r = the distance separating the charges in meters (m)
$$k = 9 \times 10^9 \frac{N - m^2}{C^2}$$

When q_1 and q_2 have like signs, the force F is positive and is interpreted as a repulsion; when q_1 and q_2 have unlike signs, F is negative and is interpreted as an attraction.

The coulomb is a very large amount of charge. A body with one coulomb of charge one meter away from a similar body would exert a force on it of 9×10^9 newtons (about 2 billion pounds). A lightning discharge may transfer 200 coulombs of charge. Therefore, charges are often expressed in terms of the submultiples, the micro-or nanocoulomb. The charge of a comb rubbed on the sleeve is less than one microcoulomb.

Electrical charges cannot be subdivided indefinitely. Robert A. Millikan (1868–1953), a physicist at the University of Chicago and the California Institute of Technology, and others proved that the minimum possible electric charge is equal to 1.6×10^{-19} coulomb. This is the charge associated with the electron (negative) and proton (positive). All charges possessed by the material particles are integral multiples of this minimum charge.

EXAMPLE 10-1

What is the force between two electrons separated by 10^{-10} m (1 ångstrom unit)?

SOLUTION

Knowing the values of the charges and the distance separating them, apply Equation 10–1 to calculate the force.

$q_1 = -1.6 \times 10^{-19}$ coul

$q_2 = -1.6 \times 10^{-19}$ coul

$r = 1 \times 10^{-10}$ m

$F = ?$

$$F = \frac{9 \times 10^9 \text{ N-m}^2}{\text{coul}^2} \frac{q_1 q_2}{r^2}$$

$$= \frac{9 \times 10^9 \text{ N-m}^2}{\text{coul}^2} \frac{(-1.6 \times 10^{-19} \text{ coul})(-1.6 \times 10^{-19} \text{ coul})}{(1 \times 10^{-10} \text{ m})^2}$$

$$= \frac{9 \times 10^9 \text{ N-m}^2}{\text{coul}^2} \frac{(2.56 \times 10^{-38}) \text{ coul}^2}{1 \times 10^{-20} \text{ m}^2}$$

$$= \frac{23.04 \times 10^9 \times 10^{-38}}{10^{-20}} \text{ N}$$

$$= 2.3 \times 10^{-8} \text{ newton}$$
(repulsion)

10–6   LIGHTNING

The nature of lightning was in considerable doubt until Benjamin Franklin demonstrated that a lightning stroke was an electric discharge and, therefore, a natural phenomenon that could be described by scientific laws. Franklin carried out his famous kite experiment in the summer of 1752. He constructed a kite from a large silk handkerchief and cedarwood, attached to it a length of twine, held it by a silk ribbon tied to the lower end of the twine, and attached a door key where the ribbon and twine were joined. Standing under a shed to keep himself dry, he raised the kite, and after a while saw the loose fibers of the twine stand erect; this meant that they were electrified and that mutual repulsion existed between them. When he placed a knuckle to the key, he received a spark and a shock. Franklin concluded that lightning too is an electric spark. (That his experiment was a dangerous one to perform was proved a year later when Georg Richmann (1711–1753) was struck dead in St. Petersburg while he experimented with lightning.)

It occurred to Franklin that buildings and ships

could be protected from lightning by placing on the highest parts of those structures upright sharp-pointed iron rods with a wire running from the base of the rod down the side of the structure into the subsoil or water. He thought that such rods would safely draw electricity from a cloud and thus prevent a damaging lightning stroke. Or, if a stroke did occur, it would be conducted away harmlessly. Lightning rods soon appeared on many buildings here and abroad. Seldom has an application of a scientific discovery moved so swiftly from the realm of speculation to the arena of practice.

The way in which lightning discharges are produced has been difficult to explain. The basic mechanism—how water droplets in a thundercloud become electrified—is still not clear. However, it is known that a weather system that produces updrafts of warm, moist air reaching a velocity of 160 miles/hour or more and rising above the freezing levels in the atmosphere becomes a huge generator of electricity. The atomizing action of the wind currents on water droplets probably produces a fine, negatively charged spray, leaving the larger droplets positively charged. Within a cloud, there is a separation of large masses of positively and negatively charged water droplets.

As a charged thunderhead approaches, trees, buildings, and other objects on the ground become oppositely charged by induction. A corona discharge, popularly known as St. Elmo's fire, sometimes issues from a church steeple or an airplane wing during such times. The electric charge within a cloud is much greater than a charge on the ground, and lightning discharges occur between cloud and earth. Many more discharges occur within a cloud, however, than between a cloud and the earth. Such discharges are often concealed by the cloud and account for the so-called sheet or heat lightning visible in the distance on dark nights. Tall structures such as skyscrapers, chimneys, and bridges are struck scores of times, though usually without damage. (Fig. 10–9)

10–7 THE ELECTRIC BATTERY

Aside from the amber phenomenon, lightning, and St. Elmo's fire, natural electricity has long been known to be associated with some animals. Certain fishes of the Nile and other waters can deliver paralyzing elec-

Figure 10–9 A bolt of lightning striking the Empire State Building. (Courtesy of The New York Times/Steven Blumrosen.)

tric shocks. In South American waters there is an eel that shocks handlers when the top of the head and the underside of its body are touched with both hands. In Senegal there is an electric catfish, and in the Mediterranean the electric ray, or torpedo.

Luigi Galvani (1737–1798), a physician and professor of anatomy whose interests lay in the actions of nerves and muscles, studied the effects of electric discharges on the nervous system of the frog. He found that the amputated hind legs of a frog would kick convulsively if they were made part of an electric circuit. Galvani established that pairs of different metals caused the legs to twitch. He found, for example, that a fork with one iron and one copper tong caused the leg to contract each time contact was made with the nerve and muscle. Galvani explained the effect by assuming that electricity was a special animal property; that animal tissue contained a vital force which he named "animal electricity"; and that to produce the currents, living tissue must touch the metals.

Galvani sent a copy of his results to his physicist friend Alessandro Volta (1745–1827), a pioneer in electricity. Volta repeated Galvani's experiments and constructed the most sensitive electroscope of his time. He became convinced that the contraction of the frog's legs was an inorganic phenomenon, and that he could produce an electric current with two wires of different metals and a salt-water solution. In honor of his friend, Volta called the phenomenon *galvanism*. He proposed that the electricity came from the two dissimilar metals, not from the frog.

Volta proved his point by inventing the first electric battery, or "voltaic pile," consisting of a series of alternating copper and zinc disks separated by pads moistened with salt water (Fig. 10–10). A stack with 20 or so pairs produced a vigorous deflection of the leaves of his electroscope, as well as other effects. A wire, for example, connected across the opposite plates became hot, as did the plates themselves. With the development of the battery, electricity became available for study in the laboratory. On of the many discoveries that followed was the electrolysis of water.

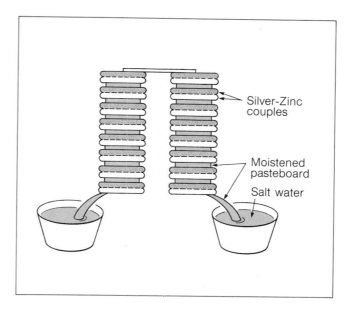

Silver-Zinc couples

Moistened pasteboard

Salt water

Figure 10–10 Voltaic pile, the first electric battery, consisted of disks of two dissimilar metals and absorbent material immersed in salt water.

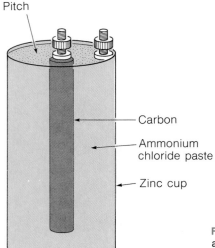

Pitch

Carbon

Ammonium
chloride paste

Zinc cup

Figure 10-11   A "dry cell" generates electricity from the chemical
action of an electrolyte and two dissimilar electrodes. A battery
consists of a group of cells.

Galvani and Volta were both partly right in their
views of animal and metallic electricity. Electrical
forces do arise at the junctions of dissimilar metals and
a solution, and living cells do produce electricity.
Michael Faraday (1791–1867), a chemist and physicist,
proved that the electricity from electric fishes and voltaic piles is the same. Many examples of animal electricity have been measured, such as that produced by
the heart and brain. The electrocardiograph and the
electroencephalograph have become powerful tools in
medical diagnosis and research.

The ordinary "dry" cell used in flashlights is the
best-known type of voltaic cell. It is not really dry (Fig.
10–11). The cell consists of two dissimilar materials
called *electrodes*, and a conducting solution, the *electrolyte*. The electrodes can be zinc in the form of a cylindrical can, and a solid carbon rod in the center of
the cell but not touching the cylinder. In place of the
salt water, a moist paste of ammonium chloride and
zinc chloride often serves as the electrolyte. It is embedded in a mixture of powdered carbon, sawdust, and
manganese dioxide. The cell is sealed at the top with
pitch or sealing wax, to prevent evaporation of the
moisture. When the cell is operating, a chemical reaction between the zinc and ammonium chloride converts chemical energy into electrical energy. Some
metallic zinc dissolves while the ammonium chloride
solution decomposes, and an excess of electrons accumulates on the zinc. If a wire connects the zinc and
carbon terminals, an electric current flows through this

external circuit, and operates a lamp or other devices in the circuit. Many combinations of chemical substances are used in the design of cells for different purposes. A group of connected cells is called a *battery*.

10–8 ELECTROSTATIC GENERATORS

Benjamin Franklin and his predecessors had no other way to produce electrical charges than by rubbing dissimilar materials. For the continuous generation of charge, a circular glass plate with a pad of silk pressing against it was mounted so that it could be rotated. Later, some electrostatic generators were constructed which depended on the induction principle. The *electrophorus*, invented by Volta, generates large electrostatic charges in this manner.

The electrophorus consists of a flat disk made of sealing wax, hard rubber, or some other insulator, and a metal plate with an insulating handle (Fig. 10–12). When the wax is rubbed with fur or wool, it acquires a negative charge. Placing the metal plate in contact with the charged disk causes electrons in the plate to be repelled to the upper surface by electrons in the disk, leaving the lower surface of the plate positively charged. The plate is then grounded by touching with a finger or grounded wire; electrons escape from the plate to the ground, leaving the plate with a net positive charge. When the plate is lifted from the disk, it becomes a source of strong positive charge that may be used for a variety of purposes, such as producing sparks or discharging a neon tube. The process may be repeated many times. During this process, the mechanical energy of rubbing the disk is converted into electrical energy.

One type of electrostatic machine is based on the principle that when an electric charge is introduced into the interior of a hollow metal sphere, the charge becomes distributed along the outer surface of the sphere. The reason is that the mutual repulsion between electrons drives them as far away as possible. Called the Van de Graaff generator, after the inventor-physicist Robert J. Van de Graaff (1902–1967), it consists of a hollow metal sphere mounted on an insulated hollow cylinder, the cylinder containing a running belt that is continuously charged at the bottom and discharged after entering the sphere (Fig. 10–13). The transferred

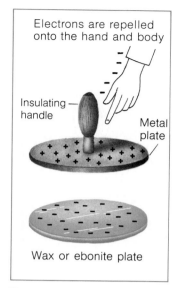

Electrons are repelled onto the hand and body

Insulating handle

Metal plate

Wax or ebonite plate

Figure 10–12 The electrophorus is a means of generating electrostatic charge.

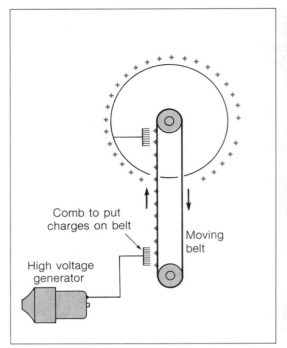

A

B

Figure 10–13 Principle of a Van de Graaff generator. Charges from a small generator are continuously carried upward by the moving belt and transferred to the metal sphere at the top of the insulated column.

positive charge accumulates on the outside of the sphere. Van de Graaff electrostatic generators are adapted to such varied uses as cancer therapy, industrial radiography, and research with electron and X-ray beams.

10–9   ELECTROSTATICS TODAY

The scope of modern electrostatics goes far beyond the literal definition of electric charges at rest. A main area of application has been the prevention of air pollution by the electrostatic precipitation of particulate industrial wastes, including fly ash, dust, and fumes. Although many communities continue to produce a great deal of water pollution by dumping sewage into our streams and rivers, electrostatic precipitators keep the air clean and thus make life livable near cement mills and plants that process mineral ores (Fig. 10–14). Evidence of their effectiveness in keeping our canopy of air breathable is the fact that, in the United States alone, electrostatic precipitators remove fly ash from coal-burning power plants at a rate of approximately 20 million tons per year.

Frederick G. Cottrell (1877–1948), a physical chemist, developed the first effective electrostatic precipitator in 1905. It is based on the principle that unlike charges attract each other. A grounded duct, the passive electrode, carries the flue or other gas loaded with waste particles. Gas ions (charged particles such as oxygen molecules that have captured an electron) are produced around a central wire, the active electrode (Fig. 10–15). The ions in turn charge the waste par-

Figure 10–14 Electrostatic precipitators prevent air pollution by removing particulates from a smokestack. These pictures indicate how effective they can be in operation. (Courtesy of Eastman Kodak Company.)

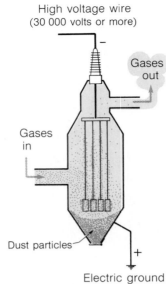

High voltage wire
(30 000 volts or more)

Gases out

Gases in

Dust particles

Electric ground

Figure 10-15 Simplified diagram of an electrostatic precipitator. An industrial precipitator consists of a large assembly of units.

ticles to which they may become attached, and the composites move across the gas stream and collect on the walls of the duct. Periodically, the duct is rapped to shake the residue into a hopper. In practice, an industrial precipitator consists of an assembly of such units. Efficiencies can exceed 99 per cent by weight. If the particles are liquid, as they are in most fumes, the residue runs down the duct walls.

Perhaps the best-known application of modern electrostatics is *xerography*, the dry-copy imaging process invented by Chester Carlson (1906–1968) in the 1930's. In a typical copying machine, a rotating drum coated with a semiconductor such as selenium is charged in the dark (Fig. 10–16). Then an optical system focuses an image of the page to be copied on the drum. The semiconductor has a photoelectric property—it releases electrons when light strikes it; the light removes all the charge except where the images of black areas appear. These images then attract a black dust called the toner. Precharged paper is fed in, makes contact with the drum, attracts the toner from the drum, and moves through a heating stage that fuses the toner to the paper to make a permanent copy.

Electrostatic effects can be annoying and dangerous as well as useful. It is impossible to move sugar, flour, or any similar dry powder through ducts without charging the powder. Belts running over pulleys often generate sparks. If this occurs with flammable materials, such as paper or plastic films, fires and explosions

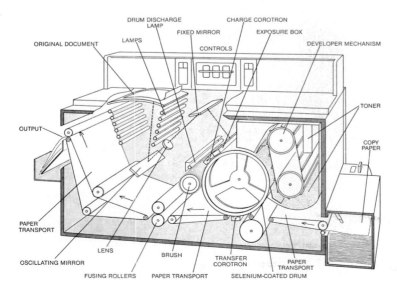

Figure 10–16 Typical Xerox dry-copying machine. Copyright © 1965 by Ziff-Davis Publishing Co.

can be set off. Excess charge can sometimes be re-moved by simply raising the humidity, thereby in-creasing the conductivity of the air. Airplanes can be-come highly charged flying through dust, sleet, or high-ly charged clouds, and their communications and con-trol systems can be damaged. Solutions to this problem have not yet been found.

10–10 MAGNETS

Just as amber was involved in so many of the early experiments in electricity, so has the *lodestone* (*lode*, to lead or attract) been associated with magnetism. One story tells of a shepherd named Magnes who on the slopes of Mount Ida on the island of Crete found his iron-tipped crook attracted to the ground. Digging to find why, he discovered the magnetic stones called lodestones.

Lodestones, now recognized as a fairly common magnetic iron ore called *magnetite*, are found in many parts of the world (Fig. 10–17). Their property of at-tracting iron was noted by various civilizations early in their Iron Age. Until 1820 the lodestone was the only source of magnetism available.

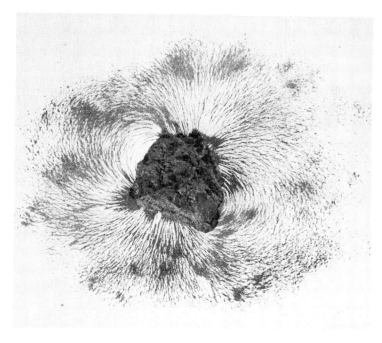

Figure 10–17 A lodestone is a natural magnet.

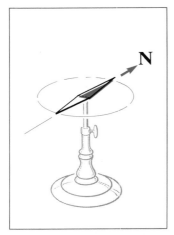

Figure 10–18 A magnetized needle, supported so that it can swing freely horizontally, turns so that it is aligned with the earth's magnetic field. One end then points north.

A stone possessing such an almost magical property became prime material for spinners of yarns, and it was sometimes difficult to separate fact from fancy. The belief arose that there were mountains of lodestone which attracted or repelled iron. The mountains were then placed on a magnetic island that attracted ships put together with iron nails and held them there. Magnets were also said to pull nails out of ships, and Archimedes (287–212 B.C.), a mathematician and inventor, is supposed to have sunk enemy ships in this way.

Iron becomes magnetic when rubbed with lodestone. Both lodestone and iron rubbed in this way turn to the north-south orientation when suspended and allowed to swing freely horizontally (Fig. 10–18). The tendency of a magnet to orient itself in this way is also exhibited when it is fastened to a piece of wood and floated on water. Put to navigational use in the mariner's compass, this property enabled seafarers to venture confidently far out to sea.

A magnet dipped into a pile of iron filings will collect the filings around regions called *magnetic poles* (Fig. 10–19). Poles come in pairs, north and south. Like poles repel and unlike poles attract (Fig. 10–20). Whenever a magnet is cut, opposite poles appear on either side of the cut, and each segment has two dissimilar poles (Fig. 10–21). Sprinkling iron filings on a piece of paper or glass placed over a magnet causes the filings to arrange themselves in curved paths (Fig. 10–22).

Magnetism seemed at first to be limited to lodestone and iron. But the elements cobalt and nickel also exhibit magnetic attraction, and certain mixtures of elements (alloys) are magnetic. Even non-magnetic elements, such as copper, tin, and manganese, are present in a group of magnetic alloys known as Heusler alloys.

Permanent magnets are made in many different shapes: bar, circular, U, V, and the familiar horseshoe. A rod of magnetic material can be magnetized with a permanent magnet by stroking the rod from one end to the other with one pole of the magnet, always starting at the same end. Another method is to hammer the rod while it is aligned in a north-south direction. Water pipes and steel girders in buildings become magnetized merely by standing in the earth's magnetic field. You verify this fact with a magnetic compass. A magnetic substance can also become magnetized by heat-

Figure 10–19　Iron filings cling mostly to the poles of the magnet. (From Huggins, Elisha R., *Physics One.* Menlo Park, Calif.: W. A. Benjamin, Inc., 1968, p. 328.)

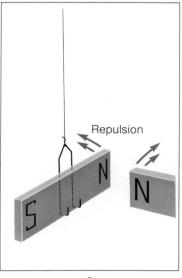

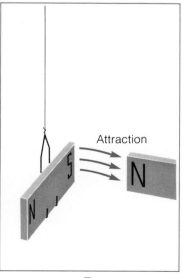

Figure 10–20　Like magnetic poles (*A*) repel; unlike poles (*B*) attract.

A

B

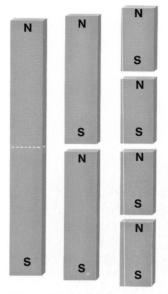

Figure 10–21 Cutting a magnet produces new magnets, with a pair of N and S poles at each break. The poles do not become isolated.

ing, then cooling, in the presence of a magnetic field, or by being placed in a coil or wire through which a current is passed.

10–11   THE EARTH AS A MAGNET

The end of a magnetic needle that points north is called a north pole, and the opposite end is called a south pole. Since a north pole is attracted by a south pole, there must be a south (magnetic) pole in the northern hemisphere of the earth. This south magnetic pole of the earth and the north magnetic pole (in the southern hemisphere) form a magnetic axis that does not coincide with the geographical axis. Thus, except in very restricted parts of the earth, compasses do not point to true geographical north (Fig. 10–23). At present, the magnetic pole in the northern hemisphere lies considerably below the surface of the earth at a point above the latitude of 70 degrees in the islands to the north of Canada; the magnetic pole in the southern hemisphere is near the Ross Sea in Antarctica.

Figure 10–22 Iron-filing pattern produced by a bar magnet. (Courtesy of the Education Development Center, Inc., 39 Chapel St., Newton, Mass.)

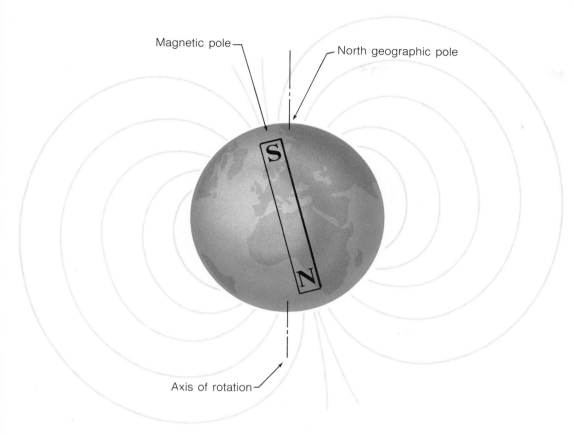

Magnetic pole

North geographic pole

S

N

Axis of rotation

Figure 10–23 The earth's magnetism is like that of a bar magnet at the center of the earth, with the S pole near the north geographic pole. The magnetic poles are more than 1000 miles from the geographic poles.

Since the magnetic poles and the geographical poles do not coincide (they are separated by about 1000 miles), a compass needle in most places points to the east or west or true north. (There is a story that Columbus, by shifting the compass card at night, averted a mutiny of his sailors when they saw the compass needle deviate 10 degrees from true north. Another tale is that he persuaded the sailors that the North Star had moved, not the compass.) The angle between magnetic north and geographic north for any locality on the earth's surface is called the *angle of declination* for that locality (Fig. 10–24).

Through surveys of the magnetic conditions of the earth, carried out over the past several hundred years, it has been learned that the angle of declination at a given locality is not constant, but changes with time, is some places several degrees in a century. Since 1600 the magnetic declination in London has varied from 11°E to 24°W. On the extreme east and west coasts of the United States, the declination may be 15 degrees or

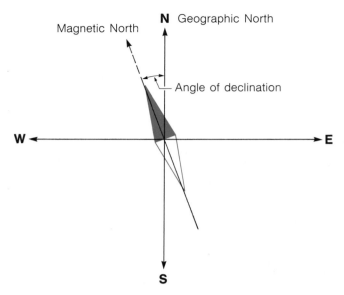

N Geographic North

Magnetic North

Angle of declination

W ◄─────────────────────► E

S

Figure 10–24   Magnetic declination. The angle of declination is measured in degrees east or west of geographic north. It is 11°50' W for New York City; 2°56' E for Chicago.

more. For Boston, the declination currently is about 15°W; for Los Angeles, 15°E.

A magnetized needle not only points approximately north and south but, if suspended at its center so as to be free to swing in a vertical north-south plane, it sets itself at an angle (Fig. 10–25). At the earth's magnetic poles the needle dips straight down; the angle of dip or inclination is 90° at those two locations. At points along the magnetic equator the angle of dip is 0°, and elsewhere it is between 0° and 90° measured from the horizontal. Magnetic surveys have shown that the angle of dip for a given locality is not constant but may change as much as several degrees in a century. The angle of dip is currently about 72° for New York City and 55° for Los Angeles.

Sir William Gilbert shaped a lodestone in the form of a sphere, called a "terrella" ("little earth") to simulate the earth, and found that a magnetized needle placed on the surface acts in the same way as a compass needle does at different places on the earth's surface. He demonstrated inclination with his terrellas and tiny magnets, and predicted that the angle of dip would increase as one traveled north, a prediction soon verified by explorers (Fig. 10–26). From this he deduced that the earth itself is a giant lodestone with a covering of rocks, soil, and water. Terrestrial magnetism, according to this view, arises within the earth and is not caused by the sun, the North Star, or any other source outside the earth.

Figure 10–25   A magnetic dipping needle determines the inclination or angle of dip.

This view of terrestrial magnetism was novel at the time, but it has since been amply substantiated. Karl Friedrich Gauss (1777–1855), a mathematician and astronomer, derived a mathematical expression for the strength and shape of the earth's magnetic field that fitted the observed values rather closely. If the earth's magnetism were caused by some external agency, the field would have a shape very different from the one it has.

10–12 MAGNETIC POLES AND FORCES

Magnetic poles behave differently from electric charges. Magnetic poles always occur in pairs, while electric charges can be separated. In theory there seems to be no reason why monopoles should not exist, but a search has thus far failed to reveal them. In a long thin magnet, the north and south poles may be well separated from each other at opposite ends of the magnet, although they can never be completely separated and isolated.

Using the same torsion balance that he used to study electrostatic forces, Coulomb suspended such a magnet from the thread and placed a similar magnet near it. He proved that the law that applies to electrostatic forces also holds for magnetic forces and is also analogous to Newton's law of gravitational force—all three are inverse-square laws. According to Coulomb's law of force between magnetic poles, like poles repel, unlike poles attract, and the force between magnetic poles varies directly with the pole strengths and inversely with the square of the distance between them.

Despite the similarity of the laws of magnetic and electric force, the magnetic and electrical properties of matter are quite different. Magnetism is a property of iron and only a relatively few other substances, but *any* substance may be electrified. Magnetic poles come in pairs that cannot be separated; positive and negative charges can be separated and isolated. A substance may retain its magnetism indefinitely, but it rapidly loses its electric charge. Magnetic pole strength is confined to certain regions in a magnet, while electric charge can flow from one place to another more or less freely. Despite these differences, there is evidence that magnetism and electricity are related phenomena.

Magnetized needle

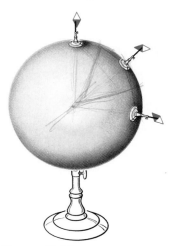

Figure 10–26 A magnetized needle placed on the surface of Gilbert's "terella" behaves like a compass needle on the earth's surface.

10–13   MAGNETIC EFFECT OF A CURRENT

Having heard of Volta's work with electricity, Hans Christian Oersted (1777–1851), a professor at the University of Copenhagen, constructed an electric pile of his own. At the time some felt that there was a connection between electricity and magnetism. In a classic case of serendipity, Oersted discovered a missing link between the two.

During a lecture in his course on electricity and magnetism, Oersted placed his voltaic pile on the lecture table. There was a compass needle nearby. Upon connecting a platinum wire to the battery, thus causing a current to flow, and holding the wire above the compass needle, Oersted observed the compass needle swing around. Instead of orienting itself in the north-south direction, the needle, to his surprise, moved and came to rest in a direction perpendicular to the wire (Fig. 10–27).

Oersted followed up the experiment after class to make certain that he had really found something. When he placed the compass needle above the wire, the needle moved in the opposite direction. Reversing the direction of the current 180 degrees caused the compass needle to turn 180 degrees. Oersted was then certain that he had discovered a connection between electricity and magnetism: electric current flowing through a wire exerted a force on a magnet—the compass needle—and the direction in which the compass needle was oriented depended upon the direction of current in the wire (Fig. 10–28). By Newton's third law, the magnet should exert an equal and opposite force on the

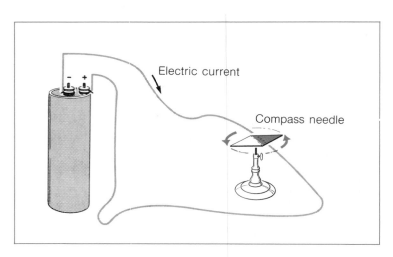

Electric current

Compass needle

Figure 10–27 Oersted's discovery. Electric current in the wire caused the compass needle to swing around at a right angle to the wire.

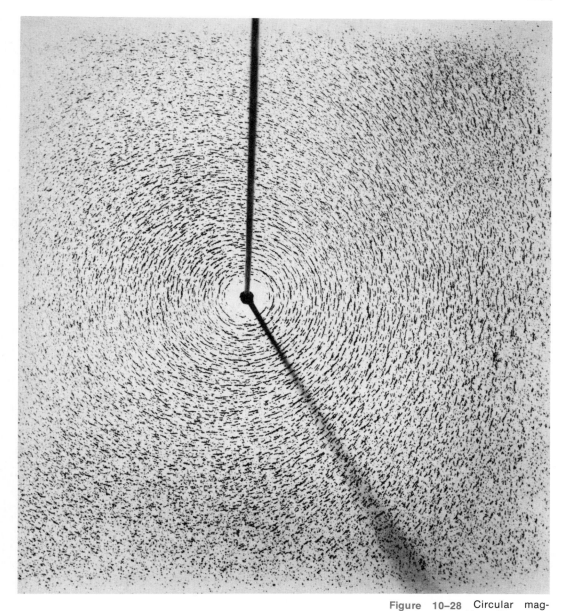

Figure 10–28 Circular magnetic pattern of iron filings surrounding a wire conducting an electric current. (From Physical Science Study Committee, *Physics*. Lexington, Mass.: D. C. Heath & Co., 1965.)

wire carrying the current, and Oersted discovered this effect as well.

Oersted was surprised by the fact that in this interaction the needle was neither attracted nor repelled directly, but set itself at right angles to the direction of the current. He repeated the experiment with many variations, then wrote a report based on his discovery.

André Marie Ampère (1775–1836), a physicist, was among those who were inspired by Oersted's findings to investigate the interactions of electricity

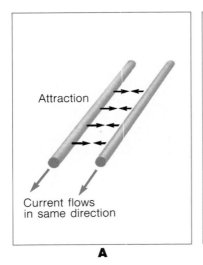

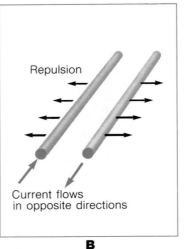

Figure 10–29 Ampère's discovery of the magnetic force of one electric current on another. *A,* Parallel currents in a common direction attract each other; *B,* currents having opposite directions repel.

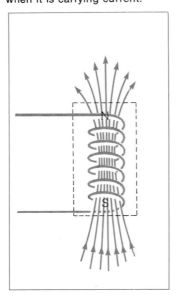

Figure 10–30 A solenoid has the properties of a bar magnet when it is carrying current.

and magnetism. Within a few weeks, Ampère established that not only does an electric current act on a magnetic needle, but two electric currents interact with each other. If the currents in two parallel wires are in the same direction, the wires attract each other, and if the currents travel in opposite directions, the wires repel each other (Fig. 10–29). Ampère attributed these interactions to magnetic force.

Ampère also established that two copper coils conducting an electric current interact in the same way as two bar magnets. He coined the word "solenoid" ("canal") for a wire bent into the form of a helix that showed magnetic properties, since it acted to channel and thus to intensify the magnetism (Fig. 10–30). Ampère's work led him to the hypothesis that natural magnetism arises from electric currents in the atoms and molecules of magnetized bodies. In modern terms, atoms contain electrons which move rapidly in orbits around the nuclei of the atoms and spin on their own axes as well. Since the electrons are charged, each of these motions constitutes an electric current. Each atom becomes, in effect, a tiny magnet with a north and south pole. In most materials, however, the magnetic effects of these atomic currents cancel out, because the electron orbits are oriented at random.

From experiments we learn that the individual atomic magnets of iron do not act separately but in groups called *magnetic domains,* in which whole blocks of atoms have a common orientation. The domains are small in size, with about a million domains per cubic millimeter, but each domain may have 10^{12} to 10^{15} atoms. In an unmagnetized sample of such material, the domains are randomly oriented (Fig. 10–31). To make the material magnetic, an external force must overcome thermal agitation and line up the domains in the same direction.

There is visual evidence for the existence of domains in microphotographs of magnetite powder in colloidal suspension applied to the polished surface of a magnetic crystal. When observed through a powerful microscope, the powder is found to have collected along well-defined lines, the domain boundaries. With this technique, the process of magnetization can be monitored. When an external magnetic field is applied, there is a shift in the wall or boundary between one domain and the next. Domains with favorable orientation grow at the expense of those less favorably oriented, and the latter shrink.

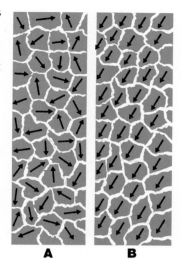

A   **B**

Figure 10–31 Magnetic domains. *A,* In unmagnetized iron, the domains point in different directions. *B,* In magnetized iron, they form a parallel array.

10–14 FIELDS

Many scientists have felt uncomfortable with the concept of "action-at-a-distance" between pairs of practically isolated bodies (for example, between the sun and the earth) between which various kinds of forces are exerted and yet there is no apparent contact. It has seemed better to say that each body sets up a "field" of some sort and that each body is disturbed by the field of the other.

A *field* is a region of space in which certain physical effects exist and can be detected. We speak of gravitational fields, electrostatic fields, and magnetic fields. For example, a gravitational field is a region in which gravitational forces act on bodies. The gravitational field at any point is defined as the force per unit mass acting on a body at that point. If the acceleration due to gravity, g, is 9.8 m/sec^2, then a force of 9.8 newtons acts on a 1-kilogram body at that point. The strength of the earth's gravitioanl field is, therefore, 9.8 newtons/kilogram at that point. Here we have the characteristic property of a field. A field is described by sets of numbers, each set denoting the field strength and direction at a point in space.

An electric field exists in any region of space in which an electric charge would experience an electric force. The strength of the electric field at a point is defined as the electric force **F** that would be exerted on a very small positive test charge +q placed at that point. The force per unit test charge is known as the electric field intensity **E**, a vector quantity. The direction of the electric field at a point is defined as the direction of the force on a positive charge placed at that point, and always points away from the positive charge.

Equation 10–2 Electric field intensity $= \dfrac{\text{Force}}{\text{Quantity of charge}}$

$$\mathbf{E} = \frac{\mathbf{F}}{q}$$

where
  F = force in newtons
  q = electric charge in coulombs
  E = electric field intensity in newtons/coulomb

The magnetic field at any point in space is the force that a magnet would exert on a unit north pole placed at that point if one existed. The direction of the magnetic field is given by the direction of a compass needle placed at that point.

Figure 10–32 Electric field patterns. The lines of force for (A) a single point positive charge; (B) a single point negative charge; (C) a positive charge and negative charge of equal magnitude.

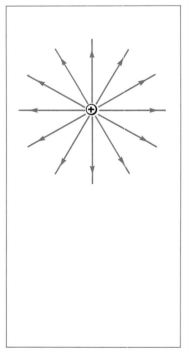
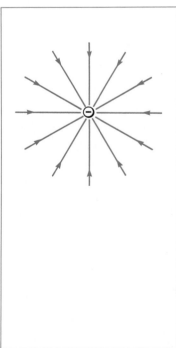
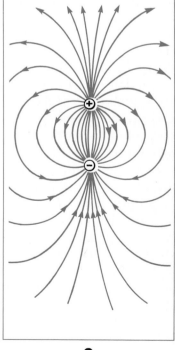

A   **B**   **C**

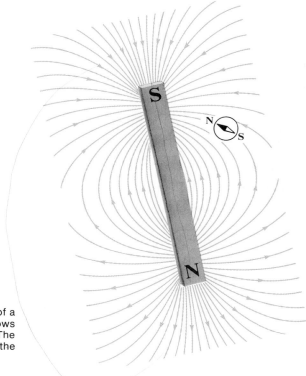

Figure 10-33 The field in the neighborhood of a magnet represented by lines of force. Arrows indicate the direction an N-pole would move. The N-pole of a compass points in the direction of the field.

Michael Faraday proposed a useful approach for describing an electric field. It considers an electric charge as having associated with it a number of lines of force that extend outward from the charge in every direction (Fig. 10–32). The lines of force start at a positive charge and end on a negative one. At every point these lines give the direction of the force that would be exerted on a small positive test charge if placed at that point. The arrowheads always point away from a positive charge and toward a negative charge. When the lines are closely packed, the electric force is large and the field is strong. In the same way, magnetic fields exist near magnets (Fig. 10–33). The lines of force of a magnet originate at a north pole and enter at a south pole. The field strength decreases as the square of the distance from a charge or pole.

The classical gravitational and electromagnetic field theories satisfactorily explain all large-scale physical phenomena. They are correct as long as they are applied to objects much bigger and heavier than a single atom, but they fail to describe the behavior of individual atoms. For the small-scale arena of physics, other theories give more accurate explanations.

10-15   ELECTRICITY FROM MAGNETISM

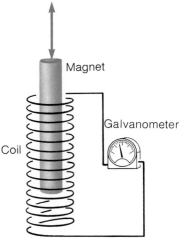

Figure 10-34 Faraday's discovery of electromagnetic induction proved that magnetism could produce electricity. By either thrusting a magnet into a coil of wire or withdrawing it, an electric current was generated.

Although he is regarded as the father of the modern electrical age, Michael Faraday was interested in science for its own sake and not in its practical applications. He no doubt realized the practical importance of his discoveries, but he left the development of the generator and motor, for example, to others. The Prime Minister of Great Britain, perhaps impatient with pure science, upon seeing the wires, coils, and magnets and learning about electromagnetic induction from Faraday, remarked: "That's interesting, Mr. Faraday, but of what possible use is it?" Faraday replied: "Well, one day, Mr. Prime Minister, you will be able to tax it."

Oersted's discovery that an electric current produced a magnetic field motivated Faraday to search for the reverse effect: the production of electricity by magnetism. After several years of trying, the breakthrough came in 1831 with his discovery of electromagnetic induction. Faraday connected a coil, consisting of many turns of wire, to a galvanometer, an instrument for measuring electric current. When he inserted a magnet into the coil, the needle of the galvanometer moved over and back momentarily, indicating that a current flowed. Then, as he withdrew the magnet, the needle moved in the opposite direction (Fig. 10–34). Faraday repeated this process many times and established that the key to the production of an electric current by magnetism was the relative motion between the coil and the magnet. The result was unexpected, for in Oersted's discovery, a steady current produces a steady magnetic field.

Faraday's discovery of electromagnetic induction showed that it was feasible to produce electricity by mechanical means. Several days after making his dis-

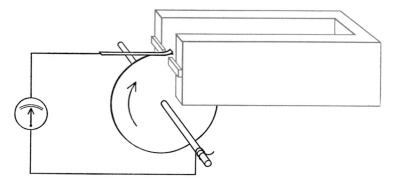

Figure 10–35 The first dynamo or electric generator. Faraday rotated a copper disk between the poles of a magnet. A steady electric current was induced across the disk.

covery, Faraday built the first *dynamo*, or electric generator, which produced a continuous flow of electricity. A 12-inch copper disk mounted on an axle was made to rotate between the poles of a powerful magnet, and an electric current was induced in the disk (Fig. 10–35). The energy of mechanical motion is converted into electric current in the disk. Fifty years were to elapse, however, before Faraday's discovery was applied to commercial generators. The production of electricity on a large scale then became possible from such sources as the potential energy of water at high elevations and the mechanical motion produced by a steam engine.

It does not detract from Faraday's greatness in any way to point out that he was not quite the first to produce an electric current by magnetism. The same discovery was made by Joseph Henry (1799–1878) at almost the same time. Then a teacher at Albany Academy, in New York, Henry made his discovery during a one-month summer vacation, but he did not publish his work until a year later. By then, Faraday had published his results following an exhaustive investigation into many aspects of the subject.

10–16   SOURCE OF THE EARTH'S MAGNETISM

After centuries of research, the earth's magnetic field is one of the best described but least understood of all planetary phenomena. Its origin and existence are still unexplained, although Gilbert and Gauss did establish that it arises within the earth and not outside it. However, Gilbert's idea that the field might be produced by a large body of permanently magnetic material inside the earth had to be dropped when it was realized that the temperature of the earth's interior is too high to allow any material to retain its magnetism. Although no theory of the earth's magnetism has yet been generally accepted as correct, the *dynamo theory* proposed by Walter M. Elsasser and Sir Edward C. Bullard seems to account for many of the observed facts.

The Dynamo Theory. An electric current produces a magnetic field not only in a bar of iron but in a gas or other fluid as well. The earth's iron-and-nickel core is analogous to the copper disk of Faraday's dynamo and acts like a huge natural dynamo.

According to the dynamo theory, the radioactive content of the earth's core supplies heat to produce motion, or convection currents, in the liquid of the core. The convection currents cause the inner part of the core to rotate at a faster rate than the outer part. This difference in rotation can produce electric currents if a magnetic field like the earth's is already present; the electric currents then cause and maintain the earth's field. The process may have begun with a stray magnetic field early in the earth's formative period, or with small currents produced by some kind of battery-like action. Support for the dynamo theory comes from the strong magnetic fields of various stars. The stars rotate; their gaseous matter conducts electricity well; their magnetic fields change rapidly—all of these factors point to a dynamic rather than a static explanation of their fields.

Any theory of the earth's magnetism must explain the curious reversal of its magnetic field that has occurred from time to time in the past. When molten volcanic rocks cool and solidify, the magnetic materials in them are magnetized in the direction of the earth's magnetic field. Bernard Brunhes found some volcanic rocks that were magnetized not in the direction of the earth's present field but in exactly the opposite direction, and concluded that the field must have reversed. It has been established that the earth's magnetic field has two stable states: it can point either toward the North Pole as it does today, or toward the South Pole, and it has alternated between the two orientations. It is estimated that there have been at least 9 reversals of the earth's field in the past 3.6 million years.

10–17   ELECTRIC CURRENT

Any motion of an electric charge constitutes an electric current. In this sense electric currents include charged water droplets rising or falling in a cloud, the charged moving belt of a Van de Graaff generator, the migration of ions (charged particles) in an electric cell, or a lightning stroke. Usually, however, we think of an electric current as a flow of electrons in a metal wire. This is the kind of current that operates our radios, toasters, and light bulbs.

The flow of electrons in a conductor is comparable to the flow of water in a pipe. We can express the rate of flow of electric charge quantitatively, just as we do the rate of flow of water. The rate of flow of electrons is a measure of the electric current. When the rate of flow past a given point is 1 coulomb per second (1 coulomb = 6.24×10^{18} electrons), we say that 1 ampere of current is flowing.

$$\text{Current} = \frac{\begin{array}{c}\text{Quantity of charge}\\ \text{flowing by a point in a conductor}\end{array}}{\text{Time required for the flow}}$$ Equation   10–4

$$I = \frac{Q}{t}$$

$$1 \text{ ampere} = \frac{1 \text{ coulomb}}{1 \text{ second}}$$

EXAMPLE   10–2

What quantity of charge passes through the wire filament in a light bulb in 30 minutes if the current is 0.5 ampere?

SOLUTION

Since the time unit in the definition of the ampere is the second, convert 30 minutes to seconds. The current is given, leaving one unknown, the quantity of charge.

$I = 0.5 \text{ ampere}$ $I = \dfrac{Q}{t}$

$t = 30 \text{ min} \times 60\dfrac{\text{sec}}{\text{min}} = 1800 \text{ sec}$ $Q = I \cdot t$

$Q = ?$ $= 0.5 \text{ ampere} \cdot 1800 \text{ sec}$

        $= 900.0 \text{ coulombs}$

Just as a force must be exerted on water to move it through a pipe, so a force must be exerted on the electrons to cause them to flow through a conductor. This means that work is done on the electrons. The driving force behind the flow of electrons is the difference in electrical condition between one point and another in a conductor. Electrons flow from a region of high potential, where they have more energy per unit charge, to a region of low potential, where they have less. (Compare this with water, which flows from a region of high pressure to one of low pressure.) The electrical potential or *voltage* is defined as the work done in moving a charge between two points, divided by the quantity of charge. An automobile battery supplies either 6 or 12 volts, and an electric power company typically provides a potential difference of 110 volts in an electrical outlet.

$$\text{Electric Potential} = \frac{\text{Work}}{\text{Charge}}$$

Equation 10–5

$$V = \frac{W}{Q}$$

$$1 \text{ volt} = \frac{1 \text{ joule}}{1 \text{ coulomb}}$$

EXAMPLE 10–3

An electron is moved from one terminal of an accelerator to another. The two terminals are separated by a potential difference of 3 million volts. How much work is expended?

SOLUTION

The charge on an electron is 1.60×10^{-19} coulomb. Solve for the work done on the electron.

$V = 3 \times 10^6$ volts

$Q = 1.60 \times 10^{-19}$ coulomb

$W = ?$

$$V = \frac{W}{Q}$$

$$\therefore W = V \cdot Q$$

$$= 3 \times 10^6 \text{ volts} \cdot 1.60 \times 10^{-19} \text{ coulomb}$$

$$= 4.80 \times 10^{-13} \text{ joule}$$

Georg Simon Ohm (1787–1854), a physicist and mathematician, studied electric currents in circuits. With voltaic piles and a glavanometer to measure the strength of the electric current, he conducted experiments with wires made of different metals and of different lengths and thicknesses. He found that the current was greater for short wires than for long wires, and greater for thick wires than for thin wires. The current also depended on the nature of the material in the wire and the potential difference between the ends of the wire. He found a relationship that is called *Ohm's law:*

At a given temperature the current through a conductor is directly proportional to the potential difference between the ends of the conductor and inversely proportional to a property of the wire that is called its electrical resistance.

Ohm's law is expressed in equation form as follows:

$$\text{Current} = \frac{\text{Potential difference}}{\text{Resistance}}$$

$$I = \frac{V}{R}$$

Equation   10–6

$$1 \text{ ampere} = \frac{1 \text{ volt}}{1 \text{ ohm}}$$

The unit of resistance, the *ohm,* is defined as the resistance of a conductor such that when the current through it is exactly 1 ampere the potential difference between its ends is exactly 1 volt.

EXAMPLE   10–4

A current of 7.5 amperes flows through an air conditioner. If the voltage in the wall outlet is 110 volts, what is the resistance of the air conditioner?

SOLUTION

Knowing the current I and the voltage V, use Ohm's law to determine the resistance R.

I = 7.5 amperes $I = \dfrac{V}{R}$

V = 110 volts

$\therefore R = \dfrac{V}{I}$

R = ?

$$R = \frac{110 \text{ volts}}{7.5 \text{ amperes}}$$

$$= 14.7 \text{ ohms}$$

Every conductor opposes the flow of current through it to an extent which depends upon the length, thickness, material, and temperature of the wire. The moving electrons undergo collisions with the vibrating atoms in the conductor. Doubling the length of the wire doubles the number of collisions and halves the current for a given potential difference. When the temperature is raised, the atoms vibrate with a greater amplitude and make a greater number of collisions with the "current-carrying" electrons, thereby impeding the forward motion of the electrons. Therefore, the current is reduced in most substances. There are many situations in which high-resistance wires are desired since the heat generated from collisions between the electrons and atoms can be put to practical use. Such is the case with electric heaters, stoves, toasters, and flatirons; for these purposes, a wire made of the alloy Nichrome is commonly used.

10–18   ELECTRIC CIRCUITS

When the same current flows through several conductors, the conductors are said to be connected in *series*. The resistance of the combination is the sum of the individual resistances, since each device in the circuit resists the flow of current (Fig. 10–36).

Equation   10–7 $$R_{total} = R_1 + R_2 + R_3 + \cdots$$

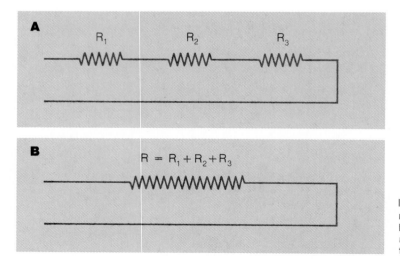

A

R_1 R_2 R_3

B

$R = R_1 + R_2 + R_3$

Figure 10–36   *A,* Resistors connected in series. *B,* An equivalent circuit consists of one resistor equal to the sum of the three.

EXAMPLE 10–5

Three coils with resistances of 5, 10, and 15 ohms are connected in series. If a voltage of 120 volts is maintained across the group, how much current will flow?

SOLUTION

Find the total resistance first, then apply Ohm's law to determine the current.

$R_1 = 5$ ohms

$R_2 = 10$ ohms

$R_3 = 15$ ohms

$V = 120$ volts

$R_{total} = ?$

$I = ?$

$$R_{total} = R_1 + R_2 + R_3$$
$$= 5 \text{ ohms} + 10 \text{ ohms} + 15 \text{ ohms}$$
$$= 30 \text{ ohms}$$

$$I = \frac{V}{R}$$
$$= \frac{120 \text{ volts}}{30 \text{ ohms}}$$
$$= 4 \text{ amperes}$$

In a divided or parallel circuit, more than one path is provided for the current, which divides and passes through each resistance independently (Fig. 10–37). Each additional path makes it easier for current to flow, just as a multiple-lane highway makes it easier for traffic to flow. The total resistance in a parallel circuit is less than any single resistance, and is given by the expression

$$\frac{1}{R_{total}} = \frac{1}{R_1} + \frac{1}{R_2} + \frac{1}{R_3} + \cdots \qquad \text{Equation 10–8}$$

Consider what would happen if light bulbs, toasters, electric stoves, and the many other appliances in a home were connected in series. The combined resist-

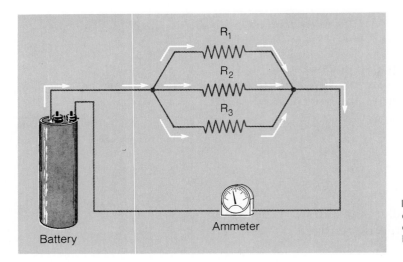

Figure 10-37 Three resistors connected in parallel with each other and then in series with a battery and ammeter.

Figure 10-38 In a parallel circuit, each appliance provides a separate path for the flow of electric current. The combined resistance is less than any individual resistance in the circuit. (From Highsmith, P., *Physics, Energy and Our World,* W. B. Saunders Company, 1975, p. 255.)

ance would be so great that there would probably not be current enough to operate the appliance that had the lowest resistance. Moreover, in a series circuit, to operate any one appliance, all would have to be used; otherwise, the circuit would be opened and current could not flow. In a parallel circuit, on the other hand, each unit is connected separately from the others to the two main supply lines by two wires, and can be operated without using any of the other appliances (Fig. 10-38). The connections in wiring a home, therefore, are parallel circuits, each circuit being connected separately between the main supply lines.

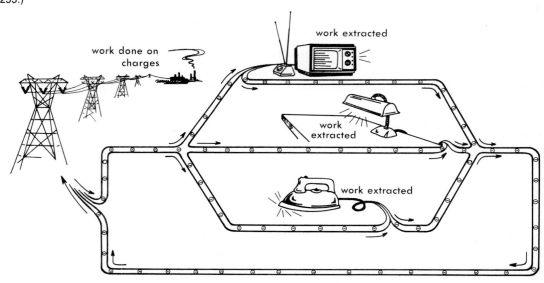

work done on charges

work extracted

work extracted

work extracted

EXAMPLE 10–6

Assume that the three coils in Example 10–5 are connected in parallel, and that there is a potential difference of 120 volts across their common terminals. (a) What is the combined resistance of the three coils? (b) What is the current in the overall circuit? (c) What is the current through each coil?

SOLUTION

Find total resistance. Then apply Ohm's law to the overall circuit and to each coil separately.

(a) $\dfrac{1}{R_{total}} = \dfrac{1}{R_1} + \dfrac{1}{R_2} + \dfrac{1}{R_3}$

$\dfrac{1}{R_{total}} = \dfrac{1}{5 \text{ ohms}} + \dfrac{1}{10 \text{ ohms}} + \dfrac{1}{15 \text{ ohms}}$

$\dfrac{1}{R_{total}} = \dfrac{6 + 3 + 2}{30 \text{ ohms}} = \dfrac{11}{30 \text{ ohms}}$

$11\,R = 30 \text{ ohms}$

$R = \dfrac{30 \text{ ohms}}{11} = 2.72 \text{ ohms}$

(b) $I = \dfrac{V}{R}$

$= \dfrac{120 \text{ volts}}{2.72 \text{ ohms}}$

$= 44 \text{ amperes}$

(c) $I_1 = \dfrac{V}{R_1}$ $\qquad I_2 = \dfrac{V}{R_2}$ $\qquad I_3 = \dfrac{V}{R_3}$

$= \dfrac{120 \text{ volts}}{5 \text{ ohms}}$ $\quad = \dfrac{120 \text{ volts}}{10 \text{ ohms}}$ $\quad = \dfrac{120 \text{ volts}}{15 \text{ ohms}}$

$= 24 \text{ amperes}$ $\quad = 12 \text{ amperes}$ $\quad = 8 \text{ amperes}$

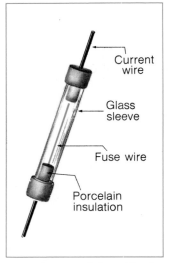

Current
wire

Glass
sleeve

Fuse wire

Porcelain
insulation

Figure 10–39   A fuse. The low
melting point of the metallic
element causes it to melt and
break the circuit if the current
should become excessive.

If the current passing through home wiring reaches a level too high for these wires to carry, there is the danger of the wires overheating to the point of causing a fire. To avoid this, fuses are connected in series with the supply line. A fuse consists of a strip of a metal that has a low melting point, housed in a porcelain or glass receptacle (Fig. 10–39). When the current reaches a predetermined value, say 20 amperes, the wire melts; the circuit is thus opened and current stops flowing. The fuse in this way protects the circuit from an electrical fire. Most houses have several different circuits, each individually fused. In new homes, a circuit breaker is generally used instead of a fuse to accomplish the same thing. When the overload is removed, the circuit breaker can be reset by pressing a lever, whereas a fuse must be replaced.

10–19   ELECTRIC POWER AND ENERGY

The production of heat is one of the best-known effects of electric current. It is the basis of such appliances as the toaster and electric blanket. In a power transmission line, however, heat represents a waste of electrical energy, and in a faulty section of wire in the home it is a potential source of fire.

In Joule's experiment involving the conversion of electrical energy into heat (discussed in Chapter 7), the heating element was a coil of wire immersed in a known mass of water in a calorimeter. A current I was maintained by a known potential difference V for a time t; the electrical energy VIt was measured and compared to the heat energy absorbed by the water and calorimeter. Joule found in experiments of this kind that the rate at which heat was developed, or power delivered to a conductor, is proportional to the square of the current flowing through the conductor. Known as *Joule's law*, this relationship is expressed as follows:

Power = (Current)^2 × Resistance = Voltage × Current

Equation 10–9 $P = I^2R = VI$

1 watt = 1 volt × 1 ampere

In SI units, when V is in volts and I is in amperes, P is in joules/sec or watts.

Table 10–3   AVERAGE HOUSEHOLD APPLIANCE USAGE AND COSTS

	ESTIMATED KILOWATT-HOUR USAGE IN 2 MONTHS	ESTIMATED COST FOR 2-MONTHS' USE
Food Preparation		
Blender	2.5	$ .09
Broiler	16.7	.60
Carving knife	1.3	.05
Coffee maker	17.7	.60
Deep fryer	13.8	.50
Dishwasher	60.5	2.18
Frying pan	31.0	1.12
Hot plate	15.0	.54
Mixer	2.2	.08
Oven, self-cleaning	191.0	6.88
Range	195.8	7.05
Roaster	34.2	1.23
Toaster	5.5	.20
Waste disposer	5.0	.18
Food Preservation		
15 cu ft Freezer	199.2	7.17
frostless	293.5	10.57
12 cu ft Refrigerator	121.3	4.37
frostless	202.8	7.30
14 cu ft Refrigerator/Freezer	189.5	6.82
frostless	304.8	10.97
Laundry		
Clothes dryer	165.5	5.96
Iron (hand)	24.0	.86
Washing machine (automatic)	17.1	.62
Water heater	870.0	13.22
Comfort & Health		
Air conditioner (room)	231.5	8.33
Dehumidifier	62.8	2.26
Fan (attic)	48.5	1.75
Fan (circulating)	7.2	.26
Fan (window)	28.3	1.02
Hair dryer	2.3	.08
Humidifier	27.2	.98
Shaver	.3	.01
Lights (equivalent of five 150-watt bulbs burning 5 hours a day)	225.0	8.10
Home Entertainment		
Radio	14.3	.51
Radio/record player	18.2	.66
Television (b & w)	60.3	2.17
Television (color)	83.7	3.01
Housewares		
Clock	2.8	.10
Sewing machine	1.8	.06
Vacuum cleaner	7.7	.28

A watt-meter is a device that measures V and I simultaneously, and automatically performs the multiplication to get P. Watt-hour-meters in house electrical connections go one step beyond and accumulate readings of power over time. These values correspond to the total electrical energy consumed. This is the quantity—energy—that is reflected in the electric bill. Electricity is not consumed in a circuit; we pay the utility company for the energy of the electric current we use that is forced through appliances yielding heat, light, etc. (Table 10–3). The filament of an ordinary light bulb is designed to be heated to incandescence by the conversion of electrical energy into heat and light.

EXAMPLE   10–7

An electric light bulb is marked 60 watts, 120 volts. (a) What current flows through the bulb? (b) What is the resistance of the filament?

SOLUTION

Calculate the current from Equation 10–9, then apply Ohm's law to obtain the resistance.

$P = 60$ watts   (a) $P = VI$ (b) $I = \dfrac{V}{R}$

$V = 120$ volts

$\therefore I = \dfrac{P}{V}$

$I = ?$ $R = \dfrac{V}{I}$

$R = ?$

$= \dfrac{60 \text{ watts}}{120 \text{ volts}}$

$= \dfrac{120 \text{ volts}}{0.5 \text{ ampere}}$

$= 0.5$ ampere

$= 240$ ohms

EXAMPLE   10–8

An electric clothes dryer is connected to a 110-volt outlet. How much current does it use if it requires 2000 watts of electric power?

SOLUTION

The power and voltage are given. Calculate the current from Equation 10–9.

$$P = 2000 \text{ watts} \qquad P = VI$$

$$V = 110 \text{ volts} \qquad \therefore I = \frac{P}{V}$$

$$I = ?$$

$$= \frac{2000 \text{ watts}}{110 \text{ volts}}$$

$$= 18.2 \text{ amperes}$$

10–20   SUPERCONDUCTIVITY

In the normal conduction of an electric current by a metal, electrons are pumped through the crystal lattices of the metal under the influence of an electromotive force supplied by a cell, battery, generator, or the like. The motion of the electrons is impeded by collisions with the atoms in the lattice; as we have seen, this impedance is responsible for the conductor's electrical resistance. The resistance increases as the temperature increases, because the more energetic vibrations of the atoms interfere with the electrons' motion more strongly.

It seems reasonable to assume that if the atoms' vibrations are completely stopped by reducing the temperature to absolute zero, electrical resistance might vanish. To test the idea, Heike Kamerlingh-Onnes (1853–1926), a physicist at the University of Leiden, passed an electric current through some frozen mercury and discovered that all detectible resistance to the flow of current disappeared at 4.15 K. This phenomenon is called superconductivity.

Superconductivity is not a rare phenomenon; 26 elements and several hundred compounds and alloys are already known to be superconductors. In each case the electrical resistance disappears at a particular temperature, for example, 20.7 K for an alloy of aluminum, germanium, and niobium. It is not necessary to go to absolute zero.

In the superconducting state, a stream of electrons can flow without encountering any resistance. Such

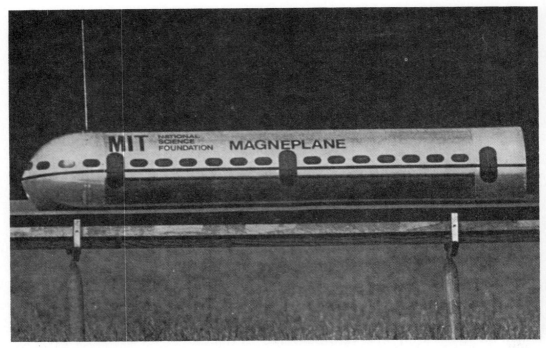

Figure 10–40 Model of the
magneplane lifted and propelled
by electromagnetic forces.
Possible high-speed ground
transportation of the future.
(Courtesy of Milford Small,
Photography, New York.)

currents, once started, persist for long periods with
little or no energy. In one experiment, an electric current induced in a small ring of metal was still circulating a year later, and had not noticeably diminished
in strength.

Practical applications of superconductivity are
being made. Superconducting magnets are the largest
class of devices in use, and large electric motors using
superconductors have been built. They have been proposed for trains, which could travel 300 miles per hour
by eliminating the problem of friction; a train equipped
with such a motor would, in effect, be lifted off the
tracks by using the force between current elements
(Fig. 10–40). Superconductors have also been proposed for amplifiers, particle accelerators, and computers.

CHECKLIST		
Action-at-a-distance	Coulomb
Amber	Coulomb's law
Ampere	Current
Angle of declination	Dry cell
Battery	Dynamo
Conductor	Electric intensity
Conservation of charge	Electroscope

Electrostatic generator
Electrostatic precipitator
Field
Induction
Insulator
Lines of force
Lodestone
Magnetic domain
Magnetic field reversal
Negative
North pole
Ohm's law

Parallel circuit
Positive
Potential difference
Resistance
Semiconductor
Series circuit
South pole
Superconductivity
Volt
Voltaic pile
Xerography

EXERCISES

10–1.   Explain why you may receive a shock from a metal door handle when you slide across a seat in an automobile.

10–2.   Although rubbing a ball-point pen with a piece of flannel is not essential for electrifying the pen, it acquires a stronger charge than when it is merely in contact with the flannel. Explain.

10–3.   If a positively charged glass rod attracts a pith ball suspended by a silk thread, must the pith ball be negatively charged? Explain.

10–4.   At toll booths on thruways, a thin metal wire sticks up from the road and makes contact with cars before they reach the toll collectors. Explain how this works.

10–5.   Why is repulsion rather than attraction a conclusive test of electrification?

10–6.   An unknown charge is brought near the knob of a positively charged electroscope and the leaves converge. What is the nature of the unknown charge?

10–7.   Assuming that an object is charged, explain how you would decide whether the charge is positive or negative.

10–8.   Why is it more difficult to maintain an electric charge on an electroscope on a damp day than on a dry day?

10-9.   Explain how you would charge an electroscope negatively by induction.

10-10.   Explain how a charged body attracts an uncharged one.

10-11.   The atoms of conductors and insulators are composed of the same constituents: electrons, protons, and neutrons. In what way, then, do conductors and insluators differ?

10-12.   (a) In what ways are electrical forces and gravitational forces similar? (b) In what ways do they differ?

10-13.   In hydrogen, the simplest atom, the electron and proton (nucleus) are separated by 0.53 ångstrom units. If the charge on each is 1.6×10^{-19} coulomb ($-$ for the electron, $+$ for the proton), what is the force between them?

10-14.   Calculate the force between two electrons that are separated by 10^{-9} meter. (Charge on the electron is -1.6×10^{-19} coulomb.)

10-15.   Volta's invention of the electric cell revolutionized the study of electricity. Why?

10-16.   Explain the discharge of a lightning stroke between a negatively charged cloud and the earth.

10-17.   How does an automobile that has a metal top and body shield its occupants from lightning?

10-18.   Why is standing in an open field during a thunderstorm a dangerous practice?

10-19.   Why is the electrical energy generated by a dry cell relatively expensive?

10-20.   (a) In what ways are electric charges and magnetic poles similar? (b) Different?

10–21.   Why may heating reduce the strength of a magnet?

10–22.   (a) Two bar magnets are placed with their south poles facing each other. Represent the magnetic field between them in a diagram. (b) Repeat with the north pole of one magnet opposite the south pole of the second magnet.

10–23.   Discuss the meaning of the term electric field.

10–24.   Calculate the magnitude and direction of the electric field 3 meters from a positive charge of 10^{-5} coulomb.

10–25.   A transistor radio draws 0.5 ampere of current. If it uses a 9-volt battery, what is the resistance of the radio?

10–26.   Discuss the significance of Faraday's and Henry's discovery of the production of an electric current by means of magnetism.

10–27.   Three incandescent lamps with resistances of 220, 440, and 660 ohms are connected in parallel across a 110-volt line. (a) What is their combined resistance? (b) What total current is drawn by the three lamps? (c) How much current does each lamp draw?

10–28.   Explain why operating too many electrical devices at one time may cause a fuse to blow.

10–29.   A 100-watt lamp draws 0.8 ampere in operation. What is its resistance?

10–30.   What is the current through a 60-watt lamp that operates at 120 volts?

10–31.   Which uses more energy: a 250-watt TV set in 1 hour, or a 1200-watt toaster in 10 minutes?

10–32.   A 100-watt lamp was accidentally left burning in a storeroom for one week. How much did it cost at 10 cents per kilowatt-hour?

10–33.   An electric range draws 20 amperes at 240 volts. What does it cost per hour to operate at 10 cents per kilowatt-hour?

10–34.   How many electrons flow through the filament of a 120-volt, 60-watt light bulb per second?

10–35.   Multiple Choice

1. The force of attraction between two charged spheres can be calculated from
   (a) Newton's second law
   (b) Coulomb's law
   (c) Joule's law
   (d) Ohm's law

2. An instrument for determining the magnitude of a charge is the
   (a) Van de Graaff generator
   (b) electrostatic precipitator
   (c) electroscope
   (d) electrophorus

3. An electric current in a wire is a flow of
   (a) atoms
   (b) protons
   (c) ions
   (d) electrons

4. Regions where the magnetic fields of atoms are lined up in the same direction are known as
   (a) superconductors
   (b) magnetic domains
   (c) currents of spinning electrons
   (d) dynamos

5. A magnetic field is produced by
   (a) static negative charges
   (b) an electron current
   (c) static positive charges
   (d) falling water

SUGGESTIONS FOR FURTHER READING

Burrill, E. Alfred, "Van de Graaff, the Man and His Accelerators." *Physics Today,* February, 1967.
   A sketch of Robert J. Van de Graaff and a class of electrostatic generators that have become associated with his name.
Devons, Samuel, "Benjamin Franklin as Experimental Philosopher." *American Journal of Physics,* December, 1977.
   Presents Franklin's contributions to science, and to electricity in particular.
MacDonald, D. K. C., *Faraday, Maxwell, and Kelvin.* Garden City, N.Y.: Doubleday & Company, 1964.

Illustrated personal and scientific biographies of three illustrious scientists.

Moore, A. D., "Electrostatics." *Scientific American,* March, 1972. Modern electrostatics goes far beyond the limited domain of non-moving electric charges implied by the literal definition of the word.

Schwartz, Brian B. and Simon Foner, "Large-Scale Applications of Superconductivity." *Physics Today,* July, 1977. Superconductivity plays a role in generating and transmitting power, propelling ships, and levitating trains.

Van Vleck, J. H., "Quantum Mechanics: The Key to Understanding Magnetism." *Science,* Vol. 201, July 14, 1978. Quantum mechanics is the master key that unlocks each door to understanding magnetism.

THE QUANTUM
THEORY OF
RADIATION AND
MATTER

11-1 INTRODUCTION

Classical physics, the physics of Newton and his successors to the year 1900, seemed so magnificent that it left many scientists with the feeling that the essential task of physical science had been completed. There were no new worlds to conquer, only "mopping up" operations, such as extending the accuracy of measurement to the fourth, fifth, or sixth decimal place.

At the same time, however, cracks appeared in the structure of physics. Although the transmission of light, with its associated phenomena such as interference diffraction, polarization, and its finite velocity, was satisfactorily accounted for by the wave theory, the emission and absorption of radiation were not. The wave theory also provided no satisfactory explanation for either atomic spectra or the photoelectric effect. Then, a series of discoveries in the 5-year period preceding the year 1900—X rays, radioactivity, the electron—and theoretical breakthroughs produced an upheaval in thought that is not resolved to this day. Chief among them was the quantum theory.

11-2 SPECTROSCOPY

It is through the analysis of light from the sun, planets, stars, and galaxies that we know their chemical composition. This has to be considered an achievement of the first rank. There were many who believed that such knowledge would be forever denied to us. No elements have been found to occur in these bodies that do not also occur on the earth. An apparent exception—the discovery on the sun of a hitherto unknown element, helium (*helios,* sun)—was eventually discovered to exist on earth as well.

Such useful applications of spectroscopy flowed from two developments: (1) the introduction of refinements that led to an increase in the resolving power of spectroscopes, instruments that spread light out by wavelength and made finer analysis possible, and (2) the interpretation of the spectra. Joseph von Fraunhofer (1787–1826), an optician and physicist whose skillful use of lenses and slits indicated the presence in the solar spectrum of many dark lines, called Fraunhofer lines, mapped hundreds of them. The interpretation of these lines by Robert Bunsen (1811–1899), a chemist (also known for the Bunsen burner), and Gustav Kirchhoff (1824–1887), a physicist, pointed up the potential value of spectroscopy. They found that each chemical element gives rise to a characteristic pattern of lines—a fingerprint, as it were—of the element. As the sun's rays pass through its outer atmosphere, certain elements absorb rays at sharp and well-defined wavelengths. For example, a pair of closely spaced dark lines in the yellow region of the spectrum, the Fraunhofer D lines, are due to the absorption of sunlight by sodium atoms in the sun's atmosphere.

Fraunhofer's invention of the diffraction grating gave scientists a tool that far exceeded the prism in its ability to disperse light (Fig. 11–1). In transmission gratings, a large number of parallel rulings is made on glass, plastic, or other transparent material, and light goes through the unscratched areas. A reflection grating employs polished metal, with reflection coming off the unscratched areas. Typical gratings have from 10 000 to 30 000 lines or slits per inch and may be as much as 6 inches in width. In either case, the constituent waves of a mixture constructively interfere (see Chapter 8) in different directions depending on their wavelength and are thereby separated. In place of the subjective concept of color, we get wavelength, a precisely measurable quantity (Table 11–1).

Table 11–2 gives the wavelengths of some selected lines.

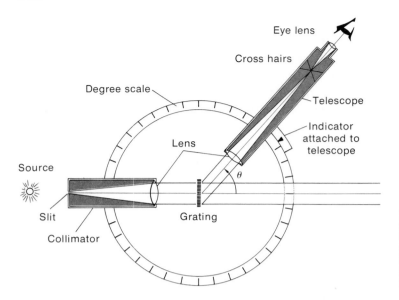

Figure 11–1 The essential parts of a type of spectrometer that uses a diffraction grating. (From Greenberg, L. H., *Physics With Modern Applications.* Philadelphia: W. B. Saunders Company, 1978, p. 351.)

11–3   BALMER'S FORMULA AND THE HYDROGEN SPECTRUM

The light given off by an element when it is made to burn is found to contain only certain colors in the form of bright lines. The apparent simplicity of the four prominent lines in the visible region of the hydrogen spectrum aroused the curiosity of Johann Jakob Balmer (1825–1898), a teacher of mathematics. Following years of attempting to discover a numerical relationship among the lines, Balmer in 1885 published a paper with the title "Notice Concerning the Spectral Lines of Hydrogen." It contained a formula that yielded the observed lines to the most remarkable degree of accuracy. In modern notation, Balmer's formula is written:

Equation   11–1

$$\frac{1}{\lambda} = R_H \left(\frac{1}{2^2} - \frac{1}{n^2} \right)$$

Table 11–1   WAVELENGTH RANGES OF COLORS

COLOR	WAVELENGTH (Å)
Violet	3800–4500
Blue	4500–4900
Green	4900–5600
Yellow	5600–5900
Orange	5900–6300
Red	6300–7600

Table 11–2 WAVELENGTHS OF CERTAIN
SPECTRAL LINES

ELEMENT	REGION OF SPECTRUM	WAVELENGTH (Å)
Hydrogen	Red	6562.8
	Blue	4861.3
	Violet	4340.5
	Violet	4101.7
Sodium	Yellow	5895.9
	Yellow	5890.0
Mercury	Yellow	5790.7
	Green	5460.7
	Blue	4916.0
	Violet	4358.3

in which λ is the wavelength in centimeters; R_H is a constant called the Rydberg constant for hydrogen, equal to 109678 cm^{-1}, and n is a running integer that takes the values 3, 4, 5, 6, ... A comparison of the wavelengths as given by Balmer's formula and as measured by the astonomer and physicist Anders Ångström (1814–1874) is given in Table 11–3.

Balmer predicted the existence at the extreme violet end of the hydrogen spectrum of a fifth line corresponding to n = 7 and a wavelength of 2969.65 Å. This line had not been seen before because the human eye is not very sensitive in that range. Its prompt discovery by the astronomer Sir William Huggins (1824–1910) and the chemist Hermann Vogel (1834–1898), and the discovery of other lines for values of n greater than 7, gave strong support to his formula. Balmer further speculated that the term $1/2^2$ in his formula might be replaced by $1/1^2$ or $1/3^2$, and predicted the existence

Table 11–3 WAVELENGTHS OF LINES IN VISIBLE
REGION OF HYDROGEN SPECTRUM

LINE	η	ACCORDING TO BALMER'S FORMULA (Å)	MEASURED BY ÅNG-STROM (Å)	DIFFERENCE
Hα	3	6562.08	6562.10	+0.02
Hβ	4	4860.8	4860.74	−0.06
Hγ	5	4340	4340.1	+0.1
Hδ	6	4101.3	4101.2	−0.1

Table 11–4 Complete Hydrogen Spectrum $\frac{1}{\lambda} = R_H\left(\frac{1}{m^2} - \frac{1}{n^2}\right)$

SERIES	YEAR DISCOVERED	REGION	FORMULA	m	n
Lyman	1906–14	ultraviolet	$\frac{1}{\lambda} = R_H\left(\frac{1}{1^2} - \frac{1}{n^2}\right)$	1	2,3,4, . . .
Balmer	1885	visible	$\left(\frac{1}{2^2} - \frac{1}{n^2}\right)$	2	3,4,5, . . .
Paschen	1908	infrared	$\left(\frac{1}{3^2} - \frac{1}{n^2}\right)$	3	4,5,6, . . .
Brackett	1922	infrared	$\left(\frac{1}{4^2} - \frac{1}{n^2}\right)$	4	5,6,7, . . .
Pfund	1924	infrared	$\left(\frac{1}{5^2} - \frac{1}{n^2}\right)$	5	6,7,8, . . .

of lines in the ultraviolet and infrared regions of the spectrum. In time, these regions were explored with apparatus sensitive in these regions and other series of lines were discovered as predicted. They are shown in Table 11–4 and correspond to the general formula

Equation 11–2

$$\frac{1}{\lambda} = R_H\left(\frac{1}{m^2} - \frac{1}{n^2}\right)$$

in which m is 1 for the Lyman series, 2 for the Balmer series, 3 for the Paschen series, etc., and n is a running integer for each respective series.

Although the Balmer formula applies only to the hydrogen spectrum, Johannes Robert Rydberg (1854–1919) developed a general formula that applies to a wide variety of line spectra. The fact that each element has its own distinct spectrum is striking enough. Since the line spectra of elements fit into a general pattern, speculation grew that they were produced by some common mechanism. The fundamental and harmonic frequencies of a musical note suggested a similar regularity in the spacings of the line spectra, but no such relationship could be found.

EXAMPLE 11–1

Calculate the wavelength of the first spectrum line of the Lyman series of hydrogen.

SOLUTION

Apply Balmer's formula. From Table 11–4 we see that m = 1 for the Lyman series, and n = 2 for the first line. Knowing that for hydrogen, R_H = 109 678 cm^{-1} = 1.1×10^5 cm^{-1} (rounded off for convenience), we solve for the wavelength λ.

m = 1

n = 2

$R_H = 1.1 \times 10^5$ cm^{-1}

λ = ?

$$\frac{1}{\lambda} = R_H \left(\frac{1}{m^2} - \frac{1}{n^2} \right)$$

$$\frac{1}{\lambda} = \frac{109\,678}{\text{cm}} \left(\frac{1}{1^2} - \frac{1}{2^2} \right)$$

$$\frac{1}{\lambda} = \frac{1.1 \times 10^5}{\text{cm}} \left(1 - \frac{1}{4} \right)$$

$$\frac{1}{\lambda} = \frac{1.1 \times 10^5}{\text{cm}} \left(\frac{3}{4} \right)$$

$$\frac{1}{\lambda} = \frac{3.3 \times 10^5}{4 \text{ cm}}$$

$$\lambda = \frac{4 \text{ cm}}{3.3 \times 10^5}$$

$$\lambda = 1.2 \times 10^{-5} \text{ cm} \times \frac{10^8 \text{ Å}}{\text{cm}}$$

$$\lambda = 1200 \text{ Å (ultraviolet)}$$

11–4 THE ELECTRON

When Heinrich Geissler (1814–1879), a glass-blower and mechanic, invented the mercury diffusion pump—the first major improvement in the vacuum pump in 200 years—he opened up a new field of study. Geissler was able to remove enough gas from a strong glass tube to reduce the pressure to 1/10 000 of standard atmospheric pressure. He prepared the tubes with such great skill that evacuated tubes became known as Geissler tubes. By sealing a wire into each end of a tube, the wire ending in a metal plate called an electrode, and running the wires to a battery, the mathematician and physicist Julius Plücker (1801–1868) found to his surprise that electricity flowed through the tube

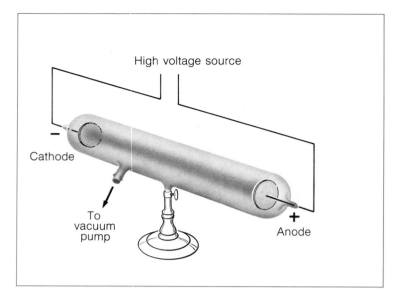

High voltage source

Cathode

To
vacuum
pump

Anode

Figure 11–2 Discharge of electricity through a gas at low pressure is accompanied by various effects as the vacuum is increased.

(Fig. 11–2). It was the forerunner of present-day fluorescent and neon tubes and television tubes.

When a Geissler tube is connected to an induction coil or electrostatic machine and an electric discharge passes between the negative terminal (cathode) and the positive terminal (anode), a dark region appears around the cathode that gradually spreads through the tube. The glass itself at the opposite end glows with a color that depends upon the nature of the glass making up the tube.

Sir William Crookes (1832–1919), a physicist and

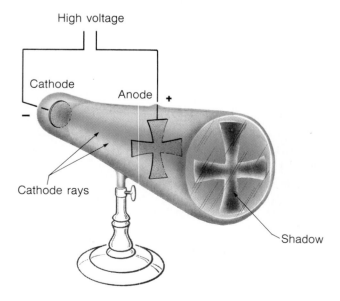

High voltage

Cathode

Anode +

Cathode rays

Shadow

Figure 11–3 Crookes' tube showed that cathode rays travel in straight lines and cast sharp shadows.

chemist, studied these effects with tubes of his own design (Fig. 11–3). He introduced into one a metal barrier in the form of a Maltese cross and observed that instead of an intense glow, a shadow appeared at the end of the tube. Emanating from the cathode were what Eugen Goldstein (1850–1930), a physicist, called *cathode rays,* which caused the glow to appear when they reached the glass. From the shadows formed, Crookes concluded that cathode rays traveled in straight lines, like light rays. Unlike light rays, however, he found that a magnetic field deflects the path of cathode rays. Further, he found that cathode rays produced from many different materials are similar.

There was intense speculation concerning the nature of cathode rays in the last quarter of the 19th century. Were they waves or particles? In 1895, Jean Perrin collected some on an insulated conductor, and with an electroscope proved that they carry a negative charge. Then in 1897 Sir Joseph John Thomson (1856–1940), who developed the Cavendish research laboratory at Cambridge University, made a quantitative study of cathode rays.

Thomson found that by applying an electric field across two plates, or a magnetic field around a Crookes tube, he could deflect the cathode rays. The deflection showed that they act like charge particles rather than light (Fig. 11–4). Further, the deflection is in the direc-

Figure 11–4 Diagram of the apparatus with which J. J. Thomson measured the deflection of cathode rays by electric and magnetic fields, their velocity, and ratio of charge to mass (e/m). (From Jones, M. M., et al., *Chemistry, Man and Society,* 2nd ed. Philadelphia: W. B. Saunders Company, 1975, p. 113.)

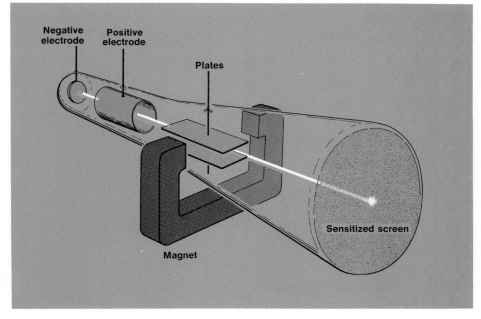

tion to be expected for particles carrying a negative charge. He was able to calculate the ratio of the charge of a particle to its mass, e/m, and found that the rays from cathodes made of different materials all had the same value: 1.76×10^{11} coulombs/kilogram. This showed conclusively that particles could be split off from atoms and that the atom was, therefore, not the ultimate subdivision of matter as had been believed. Thomson identified the cathode rays as a building-block of atoms and called them electrons.

Robert A. Millikan (1868–1953), then at the University of Chicago, between 1909 and 1916 measured the quantity of charge on the electron in his famous oil-drop experiment. When oil is sprayed in a chamber, the atomizer effect on the droplets gives them an electrical charge (Fig. 11–5), some positive and others negative. The force of air resistance on the minute droplets causes them soon to acquire a very low terminal speed of a fraction of a centimeter per second. In a properly illuminated chamber, the droplets appear as bright stars of light, and their motion can be followed with a telescope for hours. By applying an electric field to the drops, Millikan could alter their motion, balancing it against the force of gravity and even causing the droplets to rise. When the electric field is adjusted to balance the downward force of gravity,

$$F_{\text{electric}} = F_{\text{gravitational}}$$
$$eE = mg \text{ and}$$
$$e = \frac{mg}{E}$$

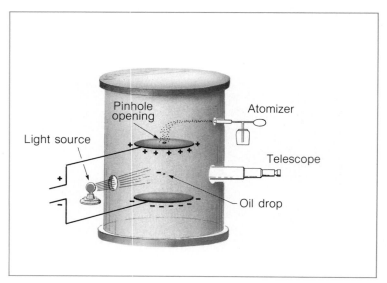

Figure 11–5 Millikan's oildrop experiment for determining the charge on an electron. The pull of gravity on the drop is balanced by the upward electrical force.

Pinhole opening

Atomizer

Light source

Telescope

Oil drop

where e is the electric charge, E is the electric field strength, m is the mass of the droplet, and g is the acceleration due to gravity.

Millikan found that the minimum electric charge that an oil drop can acquire is 1.6×10^{-19} coulomb, which he assumed to be the charge on the electron. Electric charge comes in multiples of this unit: 3.2×10^{-19} coulomb, 4.8×10^{-19} coulomb, and so on. By combining Thomson's charge/mass value for the electron, e/m, with Millikan's value for the charge, e, the mass of the electron is 9.1×10^{-31} kilogram, about 1840 times smaller than the mass of hydrogen, the lightest atom.

11–5   X RAYS

Wilhelm Konrad Roentgen (1845–1923), a physicist, was also curious about the cathode rays produced in a Crookes tube. Testing the opacity of black paper for experiments he was planning, he happened on November 8, 1895, to cover a Crookes tube with pieces of it, then darkened the laboratory and turned on the current. There was darkness except for a glimmer originating on something on a little bench that was about a meter away. Roentgen lit a match and discovered that the light was coming from a screen coated with barium platinocyanide, a fluorescent material, lying on the bench.

What could have caused the fluorescence? Ordinarily, barium platinocyanide fluoresces when illuminated with ultraviolet light, but there was no source of ultraviolet light in Roentgen's apparatus. Might cathode rays have caused it? Cathode rays had not been known to travel more than a few centimeters through air, and the screen was well beyond their range. Roentgen was forced to conclude that the fluorescence was due to a new kind of ray, which he named X ray, X standing for unknown. During the following weeks and months he studied the properties of X rays intensively, reporting his first results to a medical society seven weeks later in a paper entitled, "On a New Kind of Ray," and publishing two additional papers describing the properties of this new radiation the following year.

Roentgen established that X rays originate in the glass walls of a Crookes tube where the cathode rays strike. Since they are not deflected by a magnetic field or an electric field, they cannot be charged particles.

The rays travel in straight lines, darken a photographic plate, and have a remarkable penetrating ability, passing easily through a 1000-page book, a double pack of cards, tin foil, blocks of wood, and many other materials. X rays differ from visible light only in that their wavelength is considerably shorter, like ultra-ultra-violet rays, and their frequency is very great. Like light waves, X rays can be reflected, refracted, polarized, and diffracted. Roentgen found that if he held his hand between a Crookes tube and the screen, the bones appeared as a dark shadow within the slightly dark shadow image of the hand itself (Fig. 11–6).

Few scientific discoveries have had such an immediate impact on the public. Early in January, 1896, some of Roentgen's X-ray photographs were exhibited at a meeting of the Berlin Physical Society. The first press reports appeared on the following day, and the news was carried around the world in newspapers and periodicals. Roentgen's experiments were repeated and extended in many laboratories, since cathode-ray tubes were readily available and X rays are produced easily. Within weeks, X rays were being used to photograph broken bones or to locate bullets in wounds.

In a modern X-ray tube, X rays are produced when a beam of electrons is accelerated across a voltage of

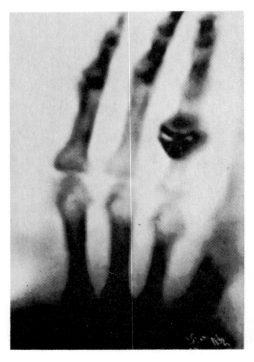

Figure 11–6   X-ray of the human hand. Roentgen took this radiograph of his wife's hand on December 22, 1895. (Courtesy of H. Klickstein, Philadelphia, Pa.)

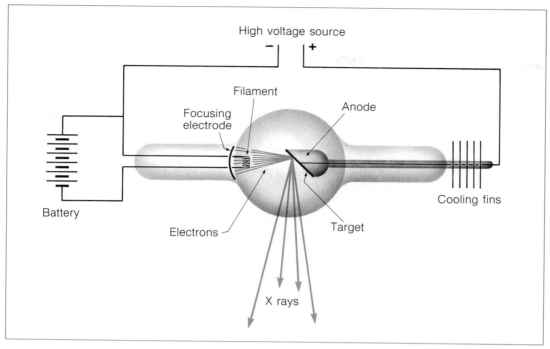

High voltage source

Filament

Focusing electrode

Anode

Battery

Electrons

Cooling fins

Target

X rays

Figure 11–7 X-ray tube. Electrons emitted by the heated cathode are accelerated and strike the metal target at high speeds. X rays are emitted from the target.

30 000–40 000 volts (Fig. 11–7). Stopped suddenly by a metal target—the anode—the electrons undergo a rapid deceleration and radiate X rays, a form of electromagnetic radiation.

X rays are used in medical diagnosis and treatment; in industrial diagnosis for possible defects; in the detection of artificial gems; in the analysis of crystal structure; and in many other ways. One of the more unusual applications is in the field of art, to confirm the age and integrity of paintings and to detect forgeries and fakes (Fig. 11–8). With X rays, information may be obtained about the structure of the painting that lies below the surface. The old masters used white leaded paint to shape the underlying structure of their paintings. Since lead absorbs X rays very well, this underlying structure can be explored. Modern paints differ in chemical content from older paints, leading to differences in X-ray absorption. Zinc-oxide white was not used before 1870, and titanium-oxide white not until the 20th century; therefore, they should not appear in a painting that is supposedly 300 years old. On the basis of X rays, Hans Van Meegeren, an art dealer, was discovered in 1945 to have forged 300-year-old paintings by the artist Vermeer that were worth more than $2 million.

Figure 11–8   X rays in art. At the left is a photograph of the head of a man in a 15th-century Flemish painting. At the right is an x-ray photograph of the same area which discloses the features of a younger man. The present head was painted on top of another. (From *Science in the Art Museum,* by Rutherford J. Gettens. Copyright © 1952 by Scientific American, Inc. All rights reserved.)

11–6   RADIOACTIVITY

Having heard a report of Roentgen's discovery of X rays, the physicist Antoine Henri Becquerel (1852–1909) wondered whether they could be associated with fluorescence and *phosphorescence*—the ability of certain crystals to glow in the dark. In particular, he was aware of the pronounced phosphorescence of potassium uranyl sulfate, a uranium salt, when excited by ultraviolet light.

Becquerel knew that uranium salts when exposed to sunlight emit radiations that pass through aluminum, cardboard, and other materials, and blacken a photographic plate, just as X rays do. During a spell of cloudy weather in February, 1896, he wrapped his photographic plates in black paper and stored them along with some uranium salts in a cabinet drawer. Believing that sunlight or ultraviolet light was necessary to excite the radiation, Becquerel did not expect to find the photographic plates exposed when he developed them a few days later. To his great surprise he found that they were intensely exposed.

Becquerel continued to experiment for several months and obtained other remarkable results. He found that the salts continue to emit radiations when kept in complete darkness for months. They radiate whether they are in crystal form or in solution, and with an intensity proportional to the uranium content. He established that the element uranium radiates even more than its salts, and attributed this new radiation to the presence of uranium.

Coming as it did on the heels of the more spectacular discovery of X rays, Becquerel's discovery evoked little interest, with one important exception. The husband-and-wife team of physical chemists Pierre Curie (1859–1906) and Marie Sklodowska Curie (1867–1934) pursued the matter further and found that the element thorium displayed a radiation similar to that of uranium. The Curies introduced the name *radioactivity* to describe it. Finding that pitchblende, an ore rich in uranium, was far more radioactive than its uranium content suggested, the Curies in 1898 isolated minute amounts of two new elements from a ton of pitchblende, which they named polonium and radium. They found polonium to be 400 times as radioactive as uranium itself, and radium a million times. They also showed that radioactivity is a spontaneous process in certain elements and is unaffected by pressure, heat, or chemical combination.

It was at first believed that the Becquerel rays were probably weak X rays. But Ernest Rutherford (1871–1937), then at McGill University, using magnetic and electric fields, found that radioactivity consisted of two components. Rutherford designated them alpha particles and beta particles, from the first two letters of the Greek alphabet. Alpha particles are heavy, carry a positive charge, can be stopped by a thick sheet of paper, and bear a resemblance to helium. Beta particles are light, negatively charged, and can be stopped by a thin piece of aluminum; they were later shown to be high-energy electrons. A third component, gamma rays, was discovered by Paul Villard (1860–1934); they have no observable charge or mass, and are highly penetrating, resembling high-energy X rays (Fig. 11–9).

11-7 RUTHERFORD'S MODEL OF THE ATOM

In a scattering experiment, a narrow beam of particles or rays is aimed at a target usually consisting of a thin foil or film of some material. As the beam strikes the target, some of the projectiles are deflected or scattered from their original direction. The scattering is the result of interactions between the projectile and the

Figure 11–9 Separation of alpha, beta, and gamma rays from a radioactive source by an electric field. Alpha and beta rays are deflected in opposite directions; gamma rays are undeflected.

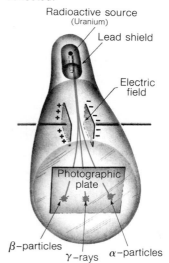

atoms of the material. From a knowledge of the mass, velocity, and direction of the projectiles, and the scattering angles, the properties of the atoms that scattered them can be deduced.

The alpha particle joined the growing arsenal of projectiles—electrons and X rays—that could be used to probe the structure of matter. No one put alpha particles to more effective use than Ernest Rutherford. Under his direction, Hans Geiger (1882–1945) and Ernest Marsden (1889–1970) carried out studies of the scattering of alpha particles by thin metal foils (Fig. 11–10). Most of the particles went right through with little or no deflection, as observed with a scintillation screen and microscope. The screen contained a phosphor, zinc sulfide, that glowed momentarily when struck by an alpha particle. Surprisingly, however, a small fraction of the alpha particles were scattered through very large angles, a few even bounding straight back. To Rutherford, "It was quite the most incredible event that has ever happened to me in my life. It was almost as incredible as if you fired a 15-inch shell at a piece of tissue paper and it came back and hit you." Strong forces are needed to turn back or deflect an alpha particle moving with a speed greater than 10^9 cm/sec and having a mass thousands of times as great as an electron.

The results of the alpha-scattering experiments could not be explained by the widely accepted atomic theory of J. J. Thomson. In this model, the positive charge is distributed uniformly over the atom, and neg-

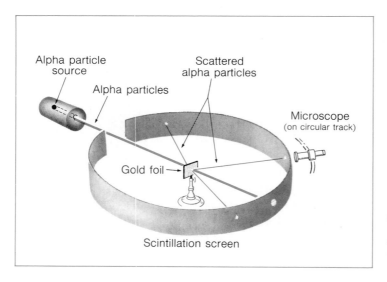

Figure 11–10 Experiment of Rutherford, Geiger, and Marsden, showing that alpha particles are scattered by a metal foil in all directions.

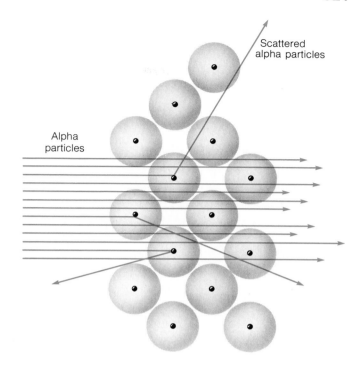

Figure 11-11 Alpha-particle scattering according to Rutherford's nuclear model of the atom. Most of the particles pass through the foil undeflected. Those particles that approach a positive nucleus closely experience Coulomb repulsive forces and are scattered through various angles.

ative electrons are imbedded in it like raisins in a cake. The positive charge is so diffuse, however, that an alpha particle could never be close to more than a small part of it, and any deflection of a positive alpha particle would be minimal. The experimental results called for a concentration of positive charge small enough to permit a close approach of an alpha particle.

Rutherford was convinced by the experimental evidence that a new model of the atom was called for. He proposed one in which the positive charge is confined to a small sphere, the nucleus, which has a radius no larger than 10^{-14} m, and which also contains most of the mass of the atom. The radius of an atom is 10^{-10} m, approximately 10 000 times as great as that of the nucleus. An atom, therefore, is almost entirely empty space. Light negative electrons in a number that will balance the positive charge on the nucleus are distributed over a region 10^{-10} m from the center. Thus, hydrogen has a positive charge of 1 unit on its nucleus, and one extranuclear electron; helium has a charge of $2+$ on the nucleus, and two electrons; gold, a nuclear charge of $79+$ and 79 electrons. Most of the alpha particles pass through the largely empty atom undeflected; those few that encounter a tiny nucleus account for the observed large-angle scattering (Fig. 11-11).

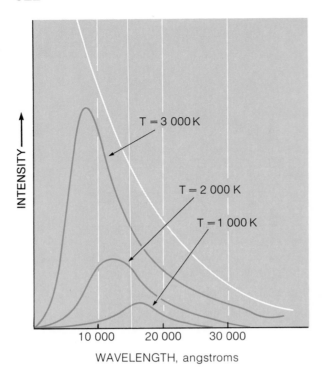

Figure 11–12　Radiation curves of a "black body." At a particular temperature, the radiation has a characteristic distribution of wavelengths (solid curves). The hotter the body, the more it radiates at shorter wavelengths. However, the classical theory predicted infinite radiation at short wavelengths (white line).

Every atomic model must account for the stability of the atom. How is the hydrogen atom held together? Gravitational forces are far too small. If we assume only electrical forces, a possibility is that the electron is stationary outside the nucleus. But this arrangement is unstable: the electrostatic attraction between the unlike charges of the nucleus and the electron would cause them to merge, destroying the atom.

What if the electron could travel in orbit around the nucleus, like a charged solar system with electrical rather than gravitational forces? The planetary model, alas, had a fatal flaw, and it comes from Maxwell's theory that predicts that an electron orbiting an atom would radiate energy in the form of electromagnetic radiation with a frequency equal to its frequency of revolution, because it is undergoing acceleration. As the electron loses energy by radiation, it should spiral closer and closer to the nucleus, radiating more and more energy until it reached the nucleus within billionths of a second. During this process, radiation should be given off in a continuous range of frequencies, a band spectrum contrary to the observed sharp and discrete line spectrum. The Rutherford nuclear atom was, therefore, inherently unstable whether the electron was stationary or in orbital motion around the nucleus.

11–8 PLANCK'S QUANTUM HYPOTHESIS

The failure of classical radiation theory to explain the stability of the atom was but one of the shortcomings of that theory. Another that troubled physicists in the last quarter of the 19th century concerned a prediction that it made about radiation.

A hot piece of iron emits electromagnetic radiation. If hot enough, much of this radiation is in the visible part of the spectrum. Thus, as the iron is heated it glows first a dull red, then a brighter red, and if heated to a high enough temperature, white like the tungsten filament in a light bulb. The actual spectrum of colors emitted from a very hot object differs completely from the prediction of classical physics. According to the theory, the radiation should always be in the form of very short wavelength or high-frequency electromagnetic waves, such as ultraviolet light waves and X rays. Lord Rayleigh (John William Strutt, 1842–1919) and Sir James Jeans (1877–1946) gave the name "ultraviolet catastrophe" to this prediction. If borne out by experience, it would mean that if you opened the door of a hot oven you would be bombarded with deadly, high-energy X rays. The oven actually emits mostly red light.

The solution to the radiation dilemma came in 1900. Max Planck (1858–1947), a physicist at the Univeristy of Berlin, discovered a formula that correctly describes the distribution of intensity with respect to wavelength of a "black body"—a perfect absorber or emitter of radiation (Fig. 11–12). To explain why the formula worked, Planck had to assert that radiant energy comes in discrete amounts, or "quanta," and that the energy content of each quantum was directly proportional to the frequency.

Energy of a quantum = a constant × frequency of quantum   Equation 11–3

$$E = hf$$

or, since $f = \dfrac{c}{\lambda}$

$$E = \frac{hc}{\lambda}$$

The constant, h, known as *Planck's constant,* has since proved to be a fundamental constant of nature. By matching the theory with observations, it was determined to be 6.62×10^{-34} joule-second. Planck's theory avoided the ultraviolet catastrophe by limiting the energy of an incandescent body to a finite number of sources. The high-frequency quanta require more energy; hence, fewer of them would be radiated.

Planck's hypothesis had an important virtue: it worked; that is, it agreed with experiment. The theoretical energy-distribution curve for radiation matched the experimental curve. The price that was paid for this success was a revolution in the way people thought about electromagnetic radiation: not as waves, but as discrete bundles or quanta of energy.

To see how great a departure Planck's quantum hypothesis is from classical theory, recall that in classical theory, the energy of a wave is related to the amplitude. Large waves, such as ocean waves, have a large energy. There is no necessary relation between the energy and the frequency. One can have waves of low energy and high frequency, or of high energy and low frequency. According to Planck, however, each electromagnetic wave carries a minimum energy that depends on its frequency. It took a quarter-century for this strange idea to be fully assimilated into the mainstream of scientific thought. Planck himself tried for years to reconcile the quantum hypothesis with classical theory, but did not succeed. He had created a revolution, almost to his dismay.

EXAMPLE　11-2

Calculate the energy of an X ray of 1.0 Å.

SOLUTION

Convert the wavelength, given in angstrom units, to meters. Knowing Planck's constant, h, and the velocity of light, c, apply Equation 11-3.

$$E = hf = \frac{hc}{\lambda}$$

$h = 6.62 \times 10^{-34}$
　　joule-sec

$\lambda = 1.0 \text{ Å} = 10^{-10} \text{ m}$

$$= \frac{\left(6.62 \times 10^{-34} \text{ joule-sec}\right)\left(3 \times 10^8 \frac{\text{m}}{\text{sec}}\right)}{10^{-10} \text{ m}}$$

c = 3 × 10⁸ m/sec
(Velocity of light)

$$= 19.86 \times 10^{-34+8+10} \text{ joule}$$

c = f · λ
(wave equation)

$$= 19.86 \times 10^{-16} \text{ joule}$$

$$= 1.98 \times 10^{-15} \text{ joule}$$

$$\therefore f = \frac{c}{\lambda}$$

E = ?

11-9   EINSTEIN'S PHOTOELECTRIC EQUATION

Planck's quantum hypothesis was soon applied successfully to something that had long been mystifying, the *photoelectric effect* (*photo*, light) (Fig. 11–13). Discovered by Heinrich Hertz and Wilhelm Hallwachs (1859–1922), the photoelectric effect has these features:

1. Electrons are emitted from the surfaces of certain metals when exposed to visible or ultraviolet light.

2. Increasing the intensity of the light increases the number of photoelectrons, but not the velocity with which they leave the surface of the metal.

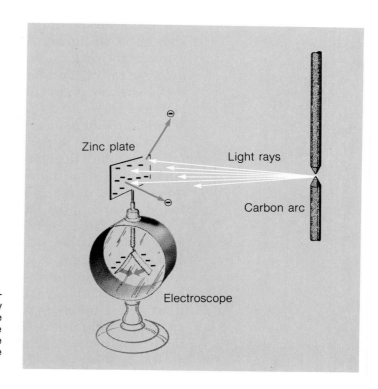

Figure 11–13 The photoelectric effect. Light striking the negatively charged zinc plate causes it to lose charge by emitting electrons. The negatively charged electroscope connected to the zinc is therefore discharged.

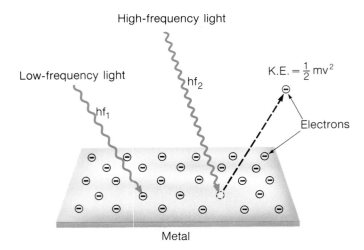

Figure 11–14 Light below the threshold frequency for a given substance does not eject photoelectrons; high-frequency radiation does.

3. There is a threshold frequency for each substance below which the effect does not occur.

4. The higher the frequency of the light, the greater the kinetic energy of the photoelectrons.

The mystifying aspects of the photoelectric effect, in terms of classical theory, come from the lack of a relation between the intensity of the incident light and the velocity of the photoelectrons. No matter how weak the illumination, provided that it exceeds the threshold frequency, the emission of photoelectrons takes place instantly. This fact defied explanation.

From the point of view of classical theory, there should be no threshold frequency. Given time, electrons might soak up enough energy from the electric field vector of the incident light to escape from the metal surface at any applied frequency. According to classical wave theory, the velocity of photoelectrons should depend upon the amplitude of the electric field vector in the incident light, and therefore upon the intensity rather than the frequency, just as a cork on the surface of a pond bobs weakly if the height—amplitude—of the wave is small. Classical theory utterly failed to explain the facts.

Albert Einstein (1879–1955), the theoretical physicist, resolved those difficulties when he extended Planck's quantum hypothesis to light. He proposed that light is not only emitted in discrete packets of energy, hf, but is also absorbed as such. (The quanta of light were later given the name *photons* by Gilbert N. Lewis.) A bright red light is regarded as a stream of many relatively weak photons; a faint blue light as a relatively few energetic photons.

The energy of a light wave, according to Einstein, is not spread out all along the wave front but is concentrated in photons. The energy transferred from a photon to an electron in the photoelectric effect is therefore related only to the frequency, not the intensity. No matter how intense the illumination, if the frequency is too low—below the threshold—a photon cannot transfer enough energy to an electron to eject it from the surface (Fig. 11–14). At a given frequency of incident light, the more intense the light, the greater the number of photoelectrons emitted with the same energy. These ideas are expressed in Einstein's photoelectric equation:

$$\begin{pmatrix} \text{Energy of} \\ \text{incident} \\ \text{photon} \end{pmatrix} = \begin{pmatrix} \text{Energy required} \\ \text{to free electron} \\ \text{from metal} \end{pmatrix} + \begin{pmatrix} \text{Kinetic energy of} \\ \text{electron as it leaves the} \\ \text{metal surface} \end{pmatrix} \quad \text{Equation} \quad 11\text{–}4$$

$$E \quad = \quad \underset{\text{work function}}{W} \quad + \quad KE$$

Robert A. Millikan tested Einstein's theory in a series of experiments between 1912 and 1917. Using sodium and other metals as targets, he illuminated a target with light of various frequencies, and studied the maximum kinetic energy of photoelectrons as a function of the frequency of the incident light (Fig. 11–15). His results were in excellent agreement with

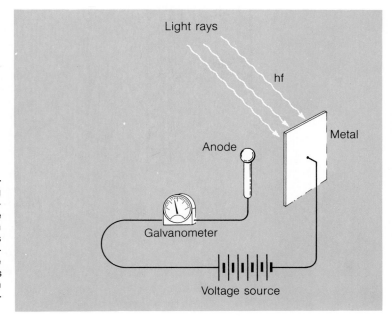

Figure 11–15 Diagram showing arrangement for measuring the kinetic energy of photoelectrons. The retarding voltage is adjusted until the current in the galvanometer just becomes zero, indicating that photoelectrons are not reaching the anode. This retarding voltage is a measure of the maximum kinetic energy of the photoelectrons.

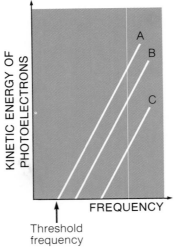

KINETIC ENERGY OF PHOTOELECTRONS

A
B
C

FREQUENCY

Threshold frequency

Figure 11–16 Plots of maximum kinetic energy of photoelectrons against frequency of incident light for different metals are parallel, although threshold frequencies vary.

Figure 11–17 Use of a photocell in a sound motion picture projector. The varying current from the sound track is converted to sound by an amplifier and speaker.

Einstein's theory. Millikan also determined the value of Planck's constant from his measurements and found it to be in close agreement with values derived by other methods. By showing that the straight lines obtained for different metals all had the same slope (Fig. 11–16), Millikan proved that Planck's constant is the same in all cases. Einstein's theory thus passed the quantitative test with flying colors, as did the quantum hypothesis upon which it rested.

The photoelectric effect is applied in photocells, which are used to open store doors when patrons cut off a light beam, to sound burgular alarms, to turn on street lights, and to produce pictures in television cameras and the soundtracks of motion pictures. In one type of photocell, the cathode is coated with cesium, a light-sensitive material that emits electrons when illuminated. The electrons collect at a second electrode, the anode, the current being proportional to the illumination. The sound information is stored on the film in the form of spots of varying widths. When a film is run between a light source and the photoelectric cell, variations in the light intensity reaching the cell cause pulsations in the current that, after being amplified,

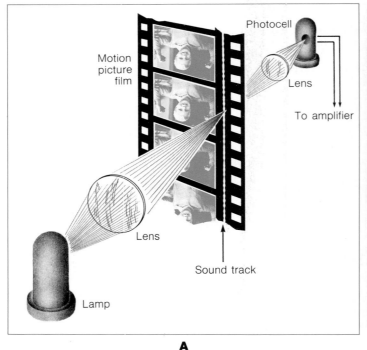

Motion picture film

Photocell

Lens

To amplifier

Lens

Sound track

Lamp

A

Sound track

B

activates a loudspeaker and reproduces the sound (Fig. 11–17).

11–10   BOHR'S ENERGY-LEVEL ATOM

The success of Planck and Einstein with the quantum theory was extended by the physicist Niels Bohr (1885–1962) in 1913 to the dilemma of the stability of the hydrogen atom. Bohr's model of the atom combined classical physics, the quantum theory of light, and some new ideas.

Bohr assumed Rutherford's nuclear model of the hydrogen atom of an electron orbiting around a tiny nucleus and held there by Coulomb's law. Reasoning that there must be definite energy levels to account for the discrete spectrum observed, Bohr proposed that:

1. Of all the possible circular orbits, only orbits with certain radii are allowed. (Bohr could not say why; the idea just "worked.")

2. When the electron is in one of the allowed orbits, it does not radiate energy despite the fact that it is undergoing acceleration.

3. When an electron makes a transition from a large orbit to a smaller one, the energy lost by the electron in falling to a lower level appears as a photon of light.

To obtain photons with frequencies that were in agreement with those observed, Bohr discovered that the allowed radii corresponded to specific values of angular momentum:

$$mvr = \frac{nh}{2\pi}$$

where n is any integer: 1,2,3, ... The Balmer formula now had meaning. For the Lyman series,

$$\frac{1}{\lambda} = R_H \left(\frac{1}{1^2} - \frac{1}{n^2} \right)$$

"1" represents the "ground state" for the series, and the values of n = 2,3,4, . . . represent higher or "excited" energy levels (Fig. 11–18). Ordinarily, the electron occupies the lowest or ground state, but by absorbing a quantum of energy it is promoted to a higher level where it remains briefly (of the order of 10^{-8} seconds) before falling back to a lower level. When an electron

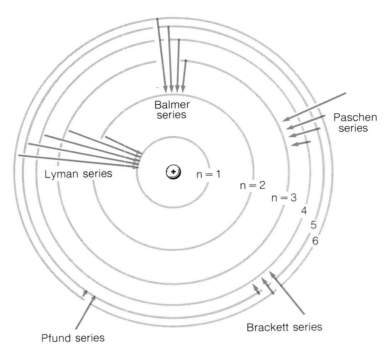

Figure 11-18 Bohr model of the hydrogen atom. The circular electron orbits have radii in the ratios of 1:4:9:16. An electron emits light of a characteristic frequency (indicated by arrows) when it falls from an orbit of higher energy to one of lower energy. A group of energy transitions gives rise to a spectral series. (Not drawn to scale.)

returns from the second orbit to the first, the difference in energy between the two orbits is emitted as a photon with a wavelength of 1200 Å (in the ultraviolet). Bohr's theory accurately explained the hydrogen spectrum, and provided a physical model for a stable atom.

11-11   THE COMPTON EFFECT

In the 1920's, Arthur Holly Compton (1892–1962), then at Washington University, carried out scattering experiments using a beam of X rays having a single well-defined frequency. Since earlier experiments had established the wave nature of X rays, it could be assumed that the scattered X rays would have the same wavelength as the rays of the incident beam. Ordinary light is scattered by air particles in all directions with no change in wavelength.

Compton did find such scattering, the "primary" scattering, but he also detected a different kind in which the scattered X rays had a longer wavelength than the rays of the original beam. Compton could not explain the presence of "secondary" scattering by classical theory. He could interpret the results by treating X rays as photons having energy, and momentum, and the scattering as the result of collisions between a photon and an electron in the target. In effect, in scattering experiments an X-ray beam behaves like a stream of particles.

Before a collision, an electron is assumed to be stationary in a target atom; the energy that an X-ray photon brings to the collision sets the electron in motion. The collision causes the X ray to be deflected with reduced energy, the energy difference appearing as the recoil energy of the electron (Fig. 11–19). The laws of the conservation of momentum and energy apply in every photon-electron collision. In a typical result, with incident X rays having a wavelength of 0.708 Å, Compton found an increase of 0.022 Å in wavelength for secondary X rays scattered at 90°, corresponding to a decrease in photon energy. The difference in energy was imparted to an electron.

The Compton effect indicated that radiation has a wave-particle duality. Light behaves in some cases—interference, polarization—like a wave, and in other cases, as in the photoelectric effect and the Compton effect, like a particle. It is wavelike in interactions with radiation, and particle-like in its interactions with matter.

11–12 MATTER WAVES

Compton's proof that X rays have a particle as well as a wave character suggested to Louis de Broglie (b.

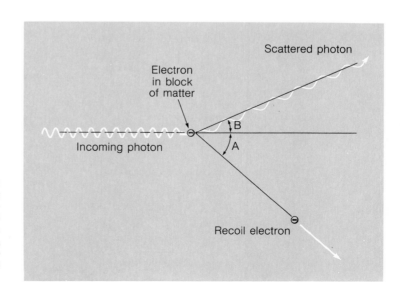

Figure 11–19 The Compton effect explained how X rays passing through matter may increase in wavelength, producing "secondary" scattering. An X-ray photon, like a particle, collides with an electron, is deflected, and loses energy to the electron.

1892) that duality might extend to electrons and other atomic particles, since nature is in general symmetrical. Now that it had been shown that radiation had wave and particle properties, perhaps matter did likewise. De Broglie introduced the hypothesis that there is a wavelength associated with a moving particle and that the wavelength is governed by the following relation:

Equation 11–5

$$\text{Wavelength} = \frac{\text{Planck's constant}}{\text{Momentum}}$$

$$\lambda = \frac{h}{mv}$$

EXAMPLE 11–3

What is the de Broglie wavelength associated with an electron traveling at 3×10^7 m/sec (1/10 the speed of light)?

SOLUTION

Since Planck's constant and the mass and velocity of the electron are known, apply Equation 11–5 to determine the wavelength of the electron. Reduce joules to fundamental units in order to express the result in a proper unit of length.

$h = 6.62 \times 10^{-34}$ joule · sec $\lambda = \dfrac{h}{mv}$

$m = 9.11 \times 10^{-31}$ kg

$v = 3 \times 10^7$ m/sec

$\lambda = ?$

joule $= \mathbf{F} \times d$

$\mathbf{F} = ma$

$\therefore$ joule $= m \times \mathbf{a} \times d$

$= \dfrac{\text{kg} \cdot \text{m} \cdot \text{m}}{\text{sec}^2}$

$= \dfrac{\text{kg} \cdot \text{m}^2}{\text{sec}^2}$

$= \dfrac{6.62 \times 10^{-34}\text{joule} \cdot \text{sec}}{\left(9.11 \times 10^{-31}\text{kg}\right)\left(3 \times 10^7\dfrac{\text{m}}{\text{sec}}\right)}$

$= \dfrac{0.224 \times 10^{-10}\dfrac{\text{kg} \cdot \text{m}^2}{\text{sec}^2}\text{sec}}{\text{kg}\dfrac{\text{m}}{\text{sec}}}$

$= 2.24 \times 10^{-11}\text{m} \times 10^{10}\dfrac{\text{Å}}{\text{m}}$

$= 0.224$ Å

(which is in the X-ray range)

George P. Thomson (son of J. J., b. 1892), set out to test de Broglie's hypothesis of the wave nature of particles. He designed an experiment to observe interference effects for an electron beam. His approach was to direct a beam of high-energy electrons through a thin gold foil. Since electrons would have about the same wavelength as X rays, he expected to get an electron diffraction picture similar to X-ray diffraction pictures (Fig. 11–20).

Thomson demonstrated the wave nature of the electron in 1927, just 30 years after his father had established its particle nature and several months after Clinton J. Davisson (1881–1958) and Lester H. Germer (1896–1971) at the Bell Telephone Laboratory obtained an electron-diffraction pattern from a beam of electrons incident on a nickel crystal. The experiments of Davisson, Germer, and Thomson showed that electrons do have wave properties, and that their wavelengths are correctly given by de Broglie's formula. Matter, like radiation, was thus shown to have a dual nature: particle and wave.

Wave properties have since been discovered for other particles—protons, neutrons, alpha particles, even whole atoms and molecules. They are all accurately described by de Broglie's formula. Interference effects have not been observed with such things as ping-pong balls or baseballs, simply because their

Figure 11–20 *A,* Diffraction pattern made by a beam of x-rays passing through thin aluminum foil. *B,* Diffraction pattern made by a beam of electrons passing through the same foil. (From Holton, G., et al., *The Project Physics Course Text.* New York: Holt, Rinehart & Winston, Inc., 1970, Unit 5, Chapter 20, p. 94. Photographs from the P.S.S.C. film *Matter Waves.*)

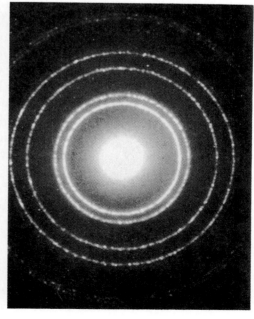

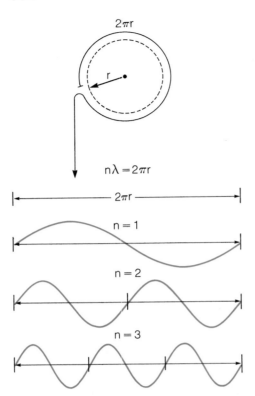

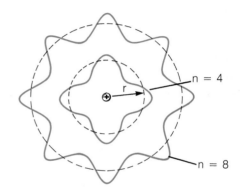

Figure 11–21 De Broglie electron waves in an atom. Each Bohr orbit, circumference = $2\pi r$, accommodates an electron standing wave pattern that exactly fits the equation $n\lambda = 2\pi r$, for which n = 1,2,3, etc. If n = 2r, the wave interferes destructively with itself and cannot exist.

wavelengths are so short as to be unobservable. The de Broglie wavelength of a 0.25-kilogram baseball traveling at 20 m/sec is 1.32×10^{-34} meters, or only about 10^{-27} as large as a light wave.

De Broglie's hypothesis changed Bohr's picture of the atom (in which the electron is a locatable particle orbiting the nucleus at a radius r with a specified velocity v) to one in which the electron becomes a wave along that orbit still located at distance r from the nucleus. De Broglie drew a comparison between the set of discrete energy levels of an atom and the set of discrete standing waves observed in vibrating strings and air columns. Wave motion that is confined in any way, as on a string fixed at both ends, results in standing waves, for which only certain wavelengths are possible. If an electron is confined around a nucleus, then it too must be present as a standing wave that closes on itself; otherwise destructive interference will cause the amplitude of the wave to drop to zero and the wave will die out.

In the atom, the possible modes correspond to Bohr's orbits, and only those orbits for which standing electron waves are possible. The condition is that the circumference of an orbit, 2π r, must be equal to a whole number of wavelengths λ: λ, 2λ, 3λ, 4λ, ... The length of the whole circumference of an orbit at radius r must be a whole wavelength, or two shorter wavelengths, or three still shorter wavelengths, and so on (Fig. 11–21).

The wave properties of electrons are applied in the electron microscope (Fig. 11–22). To see an object, an agent, such as light, that is smaller in wavelength than the object is required; otherwise the light might pass by the object as though it did not exist. Since electron wavelengths are shorter than light wavelengths, electrons can reveal a wealth of detail about the structure of such specimens as bacteria and viruses, which are beyond the capability of even the best optical microscopes. For a 50 000-volt accelerating voltage, common in electron microscopy, the electron wavelength is about 5×10^{-11} meters, roughly 1/10 000 the wavelength of visible light. An electron microscope has a

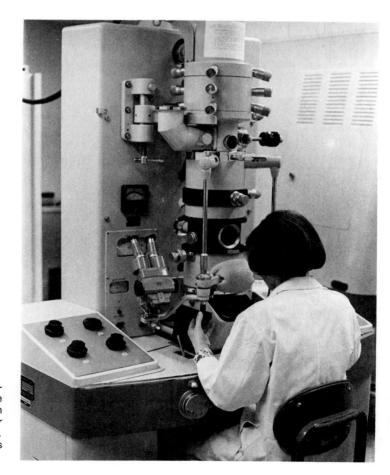

Figure 11–22 Electron microscope depends upon the wave properties of electrons. (From Greenberg, L. H., *Physics for Biology and Pre-Med Students.* Philadelphia: W. B. Saunders Company, 1975, p. 554.)

resolving power about a thousand times better than an ordinary microscope, and can resolve objects down to about a hundred-millionth of an inch. An ordinary microscope can reveal details down to only about a hundredth-thousandth of an inch.

In an electron microscope, a beam of high-speed electrons is passed through a sample thin enough to transmit the beam. Instead of optical lenses, electric fields or magnetic fields are used to refract the beam. A greatly magnified image can be photographed or made visible on a fluorescent screen.

11–13   QUANTUM THEORY AND LUMINESCENCE

A fluorescent lamp and a television picture tube have this property in common: light is produced by exciting a phosphor, a substance that absorbs electromagnetic energy and then re-emits it as visible light. Wurtzite (zinc sulfide) and fluorite (calcium fluoride) are two of the many natural phosphors. The difference between the devices lies in the agent used for excitation. In a fluorescent lamp, the phosphor is excited by ultraviolet rays and in a television picture tube by a beam of high-speed electrons. The key to the emission of light is the ability of electrons, upon absorbing a quantum of energy, to jump to an "excited" state. On returning to a lower state, they re-emit the absorbed energy. In a phosphor, part of this energy is emitted as light.

The fluorescent lamp was first exhibited in the United States at the New York World's Fair in 1939, and now challenges the incandescent lamp in popularity. It is essentially a mercury-vapor lamp in a long glass tube with a phosphor coated on the inside sur-

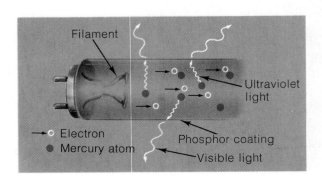

Filament

Ultraviolet light

Electron
Mercury atom

Phosphor coating

Visible light

Figure 11–23   Fluorescent lamp.

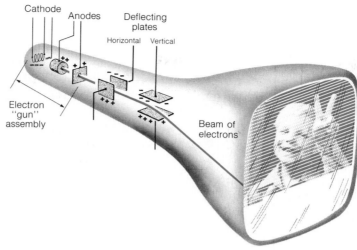

Cathode
Anodes
Deflecting
plates
Horizontal   Vertical

Electron
"gun"
assembly

Beam of
electrons

Figure 11–24 Diagram of a picture tube in a television receiver. High-speed electrons produce fluorescence.

Fluorescent screen

face of the tube (Fig. 11–23).When the two filaments built into the ends of the tube are heated, they emit electrons that move through the tube, colliding with mercury atoms. The primary source of light is from the excitation of the mercury atoms. The collisions also yield ultraviolet light, part of which is absorbed by the phosphor on the wall of the tube and reradiated as visible radiation. By a suitable selection of phosphors, light of various wavelengths can be added to the primary source from the mercury to produce any desired color.

Fluorescence and phosphorescence differ in the persistence of the emission. In fluorescence, the emission of light persists for but few hundred-millionths of a second after the source is removed. This is the time required for an electron to make a transition in an atom and emit a photon. The picture tube of a black and white television receiver is covered with a fluorescent phosphor that glows with a slightly bluish light when sprayed with a jet of electrons (Fig. 11–24). The more intense the jet, the brighter the glow. Phosphors that fluoresce with different colors are used in color television. If the emission persists for minutes or even hours, it is described as phosphorescence.

11–14   LASER LIGHT

"Laser" is an acronym for "*l*ight *a*mplification by *s*timulated *e*mission of *r*adiation." Laser radiation has three main characteristics: the waves are coherent—all in step; highly monochromatic—all with the same wavelength; and can be propagated over long distances in the form of well-collimated beams. The light emitted by an incandescent lamp, in contrast, consists of uncoordinated waves of many different wavelengths. The first operating laser was constructed in 1960 by Theodore H. Maiman (b. 1927) of the Hughes Research Laboratory.

Stimulated emission lies at the heart of the laser principle. By absorbing a photon, an atom is raised to an excited quantum state. As Albert Einstein suggested in 1917, the atom while still in the excited state can be stimulated to emit a photon if it is struck by an outside photon that has precisely the energy of the photon that would otherwise be emitted spontaneously (Fig. 11–25). The incoming photon is thereby augmented by the one given up by the excited atom. What is remarkable is that the wavelengths of the two photons are precisely in phase.

With an active medium, most of the atoms can be placed in an excited state so that an electromagnetic wave of the right frequency passing through them will stimulate a cascade of photons, thereby forming an in-

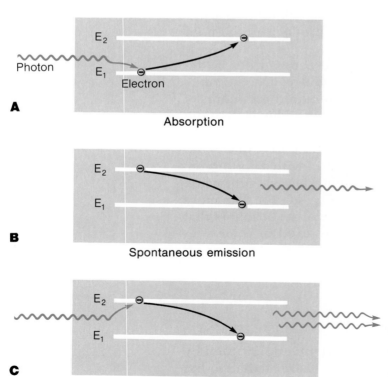

Figure 11–25   Comparison of (A) absorption, (B) spontaneous emission, and (C) stimulated emission of photons. Stimulated emission is the basis of the laser principle.

tense light wave. Through the activating process called pumping, an excess of excited atoms—a population inversion—is created. In a gas laser or a solid-state laser, this is done by passing an electric current through it or by illuminating it. In a typical laser, a ruby crystal (aluminum oxide) to which 0.05 per cent of the active ingredient, chromium atoms, is added, optical pumping is achieved from a xenon flash lamp that emits white radiation. Reflection of the coherent light back and forth from a pair of mirrors through the excited atoms also augments the cascade process. After the laser light has built up, it emerges at one end.

Lasers have many applications, although their total potential remains to be tapped in the years ahead. Surveying, communications, holography, surgery, photography, spectroscopy, and Doppler-shift measurements are some of the areas in which they have been applied. The lunar laser reflector placed on the moon in 1969 by the astronauts of Apollo 11, Neil A. Armstrong and Edwin E. Aldrin, Jr., had made it possible to measure the distance between the earth and the moon within about 6 inches. The distance is determined by aiming the intense light from a laser at the moon and measuring the time required for a brief pulse to travel to the lunar laser reflector on the moon and back. Studies over a period of years may reveal important information about changes in the location of the North Pole, and whether the gravitational constant G is constant or is weakening.

CHECKLIST

Balmer formula
Black-body radiation
Compton effect
Crookes tube
De Broglie wavelength
Fluorescence
Laser
Monochromatic
Nuclear atom
Oil-drop experiment

Phosphorescence
Photoelectric effect
Photon
Quantum
Radioactivity
Stimulated emission
Ultraviolet catastrophe
Wave-particle duality
X ray

EXERCISES

11–1. Discuss the basis for a claim that there is gold on the sun.

11–2. Compare radioactivity and X rays.

11–3. What was the experimental basis of Rutherford's concept of a nuclear atom?

11–4. Discuss two instances in which classical electromagnetic theory failed to account for the experimental observations.

11–5. In what way are quanta of radiation a return to the corpuscular theory of light?

11–6. In the photoelectric effect, the energy of a photoelectron is less than that of the incident photon. Explain.

11–7. Yellow light can induce the photoelectric effect in a certain metal, but orange light cannot. Would blue light? Explain.

11–8. The human eye is most sensitive to light of approximately 5500 Å in wavelength. What is the energy in joules of such a photon?

11–9. Which photon has the greater energy—an infrared or ultraviolet?

11–10. How did Bohr's model of the atom account for the Balmer and other spectral series of the hydrogen atom?

11–11. Although hydrogen has but a single electron, the complete hydrogen spectrum consists of numerous lines. How do you account for this?

11–12. Calculate the wavelength of the second line of the Balmer series for hydrogen.

11–13. Calculate the wavelength of the third line of the Paschen series for hydrogen.

11–14. How did de Broglie's hypothesis help to explain the stability of the atom?

11–15.   What is the de Broglie wavelength of a 0.5-kilogram football traveling at 15 m/sec?

11–16.   Calculate the de Broglie wavelength associated with the earth, if its mass is 6×10^{24} kilograms and it revolves around the sun with a velocity of 3×10^4 m/sec.

11–17.   How is the wave-particle duality related to the concept of symmetry in nature?

11–18.   Discuss the meaning of the dual nature of radiation and matter.

11–19.   Account for the considerably greater detail seen in a bacterial specimen revealed by an electron microscope as compared to the same specimen revealed by a light microscope.

11–20.   Why does the "fingerprint" of an atom have more than one characteristic frequency of light?

11–21.   A laser beam is aimed at a reflector on the moon 240 000 miles away. How long does it take for the reflection to return?

11–22.   *Multiple Choice*

1. When light shines on a metal surface, the energy of the photoelectrons depends on the
   (a) intensity of the light
   (b) frequency of the light
   (c) velocity of the light
   (d) all of these

2. A bright-line spectrum of an element
   (a) is concentrated at the red end
   (b) is evenly distributed
   (c) is concentrated at the blue end
   (d) is characteristic of the element

3. Radioactivity is a property of
   (a) excited electrons
   (b) ultraviolet light
   (c) atomic nuclei
   (d) X rays

4. The quantum theory does not explain
   (a) interference
   (b) the photoelectric effect
   (c) the Compton effect
   (d) line spectra

5. Compared to a light microscope, an electron microscope
   (a) uses less energy
   (b) works faster
   (c) can magnify in greater detail
   (d) uses longer wavelengths

SUGGESTIONS FOR
FURTHER READING

Andrade, E. N. da C., *Rutherford and the Nature of the Atom.* Garden City, N.Y.: Doubleday & Co., 1964.
> An account of the life and work of the foremost experimental nuclear scientist of the first third of the 20th century.

Banet, Leo, "Balmer's Manuscripts and the Construction of His Series." *American Journal of Physics,* July, 1970.
> Analyzes the development of the Balmer formula and shows its role in Bohr's atomic theory.

Bragg, Sir Lawrence, "The Start of X-Ray Analysis." *Chemistry,* December, 1967.
> Sir William Bragg and Sir Lawrence Bragg, father and son, determined the structure of many substances and started the science of X-ray crystallography.

Compton, A. H., "The Scattering of X-Rays as Particles." *American Journal of Physics,* December, 1961.
> Reviews the experiments and theory that led to the discovery and interpretation of the Compton effect.

Gehrenbeck, Richard K., "Electron Diffraction: Fifty Years Ago." *Physics Today,* January, 1978.
> A look at the experiment that established the wave nature of the electron, and at the investigators, Clinton Davisson and Lester Germer.

Medicus, Heinrich A., "Fifty Years of Matter Waves." *Physics Today,* February, 1974.
> Louis de Broglie's theory of matter waves inaugurated the era of modern quantum mechanics.

Schawlow, Arthur L., "Laser Light." *Scientific American,* September, 1968.
> An authoritative account of lasers by one who was intimately involved in their development.

Thomson, George, *J. J. Thomson, Discoverer of the Electron.* Garden City, N.Y.: Doubleday & Co., 1966.
> An eminent physicist in his own right describes his father's work and life at the Cavendish Laboratory that "J. J." headed for many years.

ATOMIC STRUCTURE AND THE PERIODIC TABLE OF ELEMENTS

12–1 INTRODUCTION

It had long been suspected that electricity and chemistry were closely connected. Long before the discovery of the electron, Michael Faraday had experimentally shown interactions between electricity and chemical activity. The nuclear model of the atom with its electrical basis turned out to be very successful. Then when Niels Bohr applied the quantum theory to the periodic table of the elements and showed that atomic structure is the basis of the periodic table, he cleared a path toward understanding how atoms combine to form molecules and other aggregates of matter. This is the concern of chemistry.

12–2 THE CHEMICAL ELEMENTS— ACTINIUM TO ZIRCONIUM

The idea that a few elementary substances combine to form the many thousands of substances known has always been attractive in attempting to answer the question, "What is the world made of?" Of course, there was not universal agreement as to which substances were "elements," although Empedocles' (about 500–430 B.C.) theory of the four elements— earth, air, fire, and water—was widely adopted.

The modern concept of chemical elements stems

Table 12–1   THE 20 MOST ABUNDANT ELEMENTS ON EARTH

ELEMENT	PER CENT BY WEIGHT
Oxygen	49.5
Silicon	25.7
Aluminum	7.5
Iron	4.7
Calcium	3.4
Sodium	2.6
Potassium	2.4
Magnesium	1.9
Hydrogen	0.9
Titanium	0.6
Chlorine	0.2
Phosphorus	0.1
Manganese	0.09
Carbon	0.08
Sulfur	0.06
Barium	0.04
Nitrogen	0.03
Fluorine	0.03
Nickel	0.02
Strontium	0.02

from the work of Robert Boyle (1627–1691). In his book, *The Sceptical Chemist* (1661), Boyle proposed instead of the theory of the four elements the operational definition that an element is a pure substance that cannot be decomposed into simpler substances by

Table 12–2   SOME COMMON ELEMENTS AND THEIR USES

ELEMENT	USE
Aluminum	House siding; boats; foil; cans
Chlorine	Swimming pools; household bleach
Chromium	Automobile trim; furniture
Copper	Electric wire; plumbing
Fluorine	Water supplies; toothpaste
Gold	Jewelry; coins; foil
Helium	Balloons; diving
Iodine	Antiseptic; iodized salt
Iron	Structural steel; automobiles; farm machinery
Lead	Engine antiknock additive; batteries
Magnesium	Ladders; flares; airplanes
Mercury	Thermometers
Neon	Advertising signs
Oxygen	Diving; flying; rockets; welding
Silver	Silverware; coins; medallions
Sulfur	Matches; sulfuric acid
Tungsten	Incandescent light bulb filaments; x-ray tubes
Zinc	Brass; batteries; plating

Table 12–3 RELATIVE ABUNDANCE OF ELEMENTS
IN THE HUMAN BODY

ELEMENT	PER CENT (ATOMS)
Hydrogen	60.3
Oxygen	25.5
Carbon	10.5
Nitrogen	2.42
Sodium	0.730
Calcium	0.226
Phosphorus	0.134
Sulfur	0.132
Potassium	0.036
Chlorine	0.032
Magnesium	0.010
Manganese	trace
Iron	,,
Copper	,,
Zinc	,,
Cobalt	,,

ordinary chemical or physical means. Since neither hydrogen nor oxygen is decomposable into simpler substances by ordinary methods they are classified as elements. (Water can be decomposed into hydrogen and oxygen, so water is not considered to be an element.) On this basis, there are today 100-odd chemical elements, each with its set of characteristics and out of which everything else in the universe is believed to be composed. Eighty-eight of the elements occur naturally; the rest can be produced in a laboratory. (The list of elements will be increased if experiments leading to the synthesis of a number of "superheavy" elements in nuclear accelerators are successful.) The 20 most abundant elements on earth are listed in Table 12–1, some familiar elements in Table 12–2, and the elements required for life in Table 12–3.

12–3 KNOW THEM BY THEIR SYMBOLS

The chemical elements are represented by a set of symbols accepted throughout the world. Compared to a time when as many as 14 different symbols were used for lead and 20 for mercury, this is fortunate. You only need learn, for example, that the symbol for lead is Pb and the symbol for mercury is Hg. We owe the existence of the present system to Jöns Jakob Berzelius

(1779–1848), a chemist at the University of Stockholm. Alchemical signs such as the ones shown in Table 12–4—circles, squares, dots, arrows, and various combinations of these—had all been tried before the modern symbols were adopted.

In place of the signs that were then in vogue, Berzelius proposed that letters be used for chemical symbols, since they are simple and easy to write. A symbol may be the capitalized first letter of the name of the element, such as H for hydrogen. In those cases in which two or more elements share the same first letter, a second letter in lower case is added, so that He is the symbol for helium. Elements for which there were Latin names often have symbols derived from them. For example, the Latin name for mercury was "hydrargyrum" ("quicksilver" or "liquid silver"), and the chemical symbol for mercury is Hg. The symbols of some elements that are unrelated to their common English names are given in Table 12–5. Nearly all the symbols suggested by Berzelius are in use today. Most of the elements, however, have been discovered since Berzelius' time, but their symbols are based upon his suggestions.

Those who discover an element have the right to name it. Elements have been named for the sun, moon, and planets: Helium (He), sun; Selenium (Se), moon; Tellurium (Te), earth; Neptunium (Np); Plutonium (Pu); Uranium (U). Various qualities or properties have also served as the basis of the names of elements: Argon (Ar), inactive; Beryllium (Be), sweet; Bromine (Br), stench; Chlorine (Cl), light green; Hydrogen (H), water-former; Neon (Ne), new; Oxygen (O), acid-former; Phosphorus (P), light-bearing: Technetium

---

Table 12–4   SOME SYMBOLS OF ELEMENTS USED BY ALCHEMISTS

Antimony	Lead (Saturn)
Arsenic	Quicksilver (Mercury)
Copper (Venus)	Silver (Moon)
Gold (Sun)	Sulfur
Iron (Mars)	Tin (Jupiter)

Table 12–5   ORIGINS OF SOME CHEMICAL SYMBOLS

ELEMENT	SYMBOL	ORIGIN OF SYMBOL
Antimony	Sb	Stibium ("mark")
Copper	Cu	Cuprum (Cyprus, source of copper)
Gold	Au	Aurum ("shining dawn")
Iron	Fe	Ferrum
Lead	Pb	Plumbum ("heavy")
Mercury	Hg	Hydrargyrum ("liquid silver")
Potassium	K	Kalium ("potash")
Silver	Ag	Argentum
Sodium	Na	Natrium
Tin	Sn	Stannum
Tungsten	W	Wolfram (wolframite, a mineral)

(Tc), artificial; and Xenon (Xe), stranger. A recent trend is to name the elements after people or places, as are those shown in Table 12–6.

12–4   THE ATOMIC WEIGHT SCALE; ISOTOPES; ATOMIC NUMBER

The lightest atom has a mass of about 1.6×10^{-24} gram; the heaviest, only about 250 times as much. Since it is impossible to measure such masses directly, a relative scale is used to compare the masses of individual atoms and molecules. The atomic masses (or weights) of the elements are their relative masses (weights) compared to an arbitrary standard. Since 1961, by international agreement, the atomic mass unit has been defined as 1/12 the mass of a certain variety of

Table 12–6   RECENTLY DISCOVERED ELEMENTS

ELEMENT	SYMBOL	YEAR DISCOVERED	DERIVATION
Americium	Am	1944	The Americas
Curium	Cm	1944	Pierre and Marie Curie
Berkelium	Bk	1949	Berkeley (California)
Californium	Cf	1950	California
Einsteinium	Es	1952	Albert Einstein
Fermium	Fm	1952	Enrico Fermi
Mendelevium	Md	1955	Dmitri Mendeléev
Nobelium	No	1958	Alfred Nobel
Lawrencium	Lr	1961	Ernest Lawrence
Kurchatovium (proposed)	Ku	1964	Igor Kurchatov
Hahnium (proposed)	Ha	1971	Otto Hahn

carbon atom, carbon-12, taken as exactly 12.00000. On this scale, the lightest atom, hydrogen, is about 1 atomic mass unit (amu).

The relative masses of atoms are precisely determined with the *mass spectrometer* (Fig. 12–1). Francis William Aston (1877–1945) used it to make an exhaustive study of the elements. The mass spectrometer is based on the principle that a moving charged particle is deflected into a curved trajectory by a magnetic field perpendicular to the direction of motion. A magnetic field exerts a force on any moving charged particle (Fig. 12–2). The force is perpendicular to the direction of motion of the electric charge, and to the magnetic field (not attractive or repulsive to either pole of the magnet). The magnitude of the force **F** on the charged particle is directly proportional to the charge q on the particle, to the velocity v of the particle, and to the magnetic field strength **B**, and is given by:

Equation 12–1

$$\mathbf{F} = q \, v \, \mathbf{B}$$

The heated filament emits electrons that bombard the gas sample, creating charged particles called *positive ions*. When accelerated by a voltage and subjected to a magnetic field, the positive ions are bent in a path governed by their charge-to-mass (q/m) ratio. Ions of different q/m values are curved into different paths and separated at the collector. The more massive the particle, the less it is deflected. The largest q/m value for positive ions is obtained when hydrogen is the gas, and is approximately 1/1837 of the q/m value for electrons.

Figure 12–1 *Mass Spectrometer.* A gas sample bombarded by electrons becomes a source of positively charged ions that are curved into different paths by a magnetic field and separated depending upon their charge-to-mass (q/m) ratio.

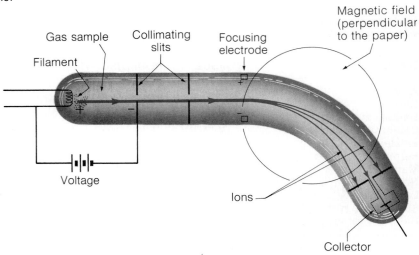

Gas sample Collimating slits Focusing electrode Magnetic field (perpendicular to the paper)

Filament

Voltage

Ions

Collector

With neon in a mass spectrometer, the particles have three different masses relative to carbon-12: 20, 21, and 22 (Fig. 12–3). We have to conclude that three varieties of neon atoms are present. The chemist Frederick Soddy (1877–1956) proposed the name *isotopes* (*iso*, same; *topos*, place) for the different varieties of the atoms of an element. Knowing the relative abundances of the isotopes of an element and their masses compared to carbon-12, we can calculate the *atomic weight* of the element—a weighted average of the isotopes as they occur naturally (see Table 12–7).

Atoms are composed of a nucleus containing protons and neutrons, and electrons outside the nucleus. The charge and mass of these particles are listed in Table 12–8.

The number of protons in the nucleus is called the atomic number of the element, designated Z. All atoms of a given element have the same atomic number, ranging from 1 for hydrogen to 105 for hahnium; for neon, Z = 10. The nuclear charge determines which element an atom represents. In a neutral atom electrons are distributed outside the nucleus in a number to balance the positive charge on the nucleus. Each neon isotope thus has 10 protons in the nucleus and 10 extranuclear electrons. We consider an element to be a substance, therefore, in which all the atoms have the same atomic number.

The neon isotopes differ in mass, owing to the presence in the nucleus of a varying number of uncharged particles called *neutrons* (Fig. 12–4). These particles were discovered by James Chadwick (1891–1974) in 1932, and have practically the same mass as protons. The sum of the protons and neutrons in a nucleus is called the mass number, A. Isotopes are denoted by their symbols with a subscript to the left to represent the atomic number and a superscript to the left to denote the mass number, as in:

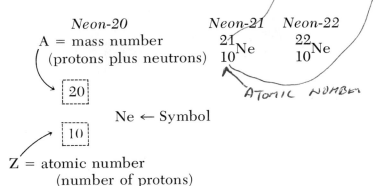

Neon-20
A = mass number
(protons plus neutrons)

20

Neon-21   Neon-22
$^{21}_{10}Ne$ $^{22}_{10}Ne$

ATOMIC NUMBER

Ne ← Symbol

10

Z = atomic number
(number of protons)

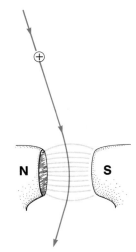

Figure 12–2 A magnetic field exerts a force upon a moving electrically charged particle, changing its direction of motion.

Figure 12–3 The mass spectrum of neon. The higher the peak, the more abundant the particular isotope.

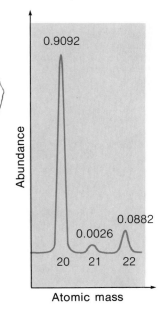

0.9092

0.0882

0.0026

20   21   22

Atomic mass

Table 12–7   CALCULATING THE ATOMIC WEIGHT OF NEON

ISOTOPE	MASS COMPARED TO CARBON-12 AS 12.00 amu		PROPORTION OF ALL NEON ATOMS		TOTAL MASS CONTRIBUTED (amu)
Neon-20	19.99	×	0.9092	=	18.17
Neon-21	20.99	×	0.0026	=	0.05
Neon-22	21.99	×	0.0882	=	1.94
			Atomic Weight of Neon:		20.16 amu

Table 12–8   SUBATOMIC PARTICLES

PARTICLE	CHARGE	MASS IN ATOMIC MASS UNITS (amu)	RELATIVE MASS
Electron	−1	0.00055	1/1837
Proton	+1	1.00728	1
Neutron	0	1.00867	1

Figure 12–4 Structures of the neon isotopes. Protons and neutrons are located in the nucleus; electrons are extranuclear. (Not drawn to scale.)

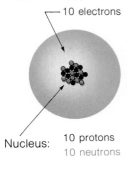

10 electrons

Nucleus: 10 protons
10 neutrons

Neon-20

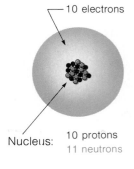

10 electrons

Nucleus: 10 protons
11 neutrons

Neon-21

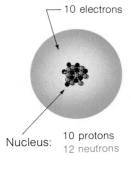

10 electrons

Nucleus: 10 protons
12 neutrons

Neon-22

Table 12–9   ATOMIC WEIGHTS AND ISOTOPES
OF SEVERAL ELEMENTS

ELEMENT	ATOMIC NUMBER	ATOMIC WEIGHT	ISOTOPES, MASS NUMBER
Helium	2	4.0026	^{3}He ^{4}He ^{5}He ^{6}He
Carbon	6	12.01115	^{10}C ^{11}C ^{12}C ^{13}C ^{14}C ^{15}C
Nitrogen	7	14.0067	^{12}N ^{13}N ^{14}N ^{15}N ^{16}N ^{17}N
Oxygen	8	15.9994	^{14}O ^{15}O ^{16}O ^{17}O ^{18}O ^{19}O
Sodium	11	22.9898	^{20}Na ^{21}Na ^{22}Na ^{23}Na ^{24}Na ^{25}Na

Although the number of elements at present is somewhat over 100, about 1400 different isotopes are known.
Even hydrogen occurs in three isotopic forms: $^1_1$H, $^2_1$H,
and $^3_1$H. Since the atomic number and the chemical
symbol both denote the element, the subscript can be
omitted. The atomic number, atomic weight, and isotopes of several elements are given in Table 12–9.

EXAMPLE   12–1

Give the values of the atomic number Z and
the mass number A, and the numbers of electrons,
protons, and neutrons, for $^{39}_{19}$K.

SOLUTION

The subscript 19 denotes the atomic number
Z, which gives the number of electrons and the
number of protons; therefore, there are 19 electrons and 19 protons. The superscript 39 denotes
the mass number A, the sum of protons and neutrons. Having determined that 19 protons are present, the number of neutrons must be A minus Z,
or $39 - 19 = 20$ neutrons.

The atomic weights of some environmentally important elements are listed in Table 12–10.

Table 12–10 ATOMIC WEIGHTS OF SOME ELEMENTS
IMPORTANT IN THE ENVIRONMENT

ELEMENT	SYMBOL	ATOMIC NUMBER	ATOMIC WEIGHT
Hydrogen	H	1	1.008
Helium	He	2	4.0026
Carbon	C	6	12.011
Nitrogen	N	7	14.0067
Oxygen	O	8	15.9994
Fluorine	F	9	18.9984
Sodium	Na	11	22.9898
Magnesium	Mg	12	24.305
Aluminum	Al	13	26.9815
Silicon	Si	14	28.086
Phosphorus	P	15	30.9738
Sulfur	S	16	32.06
Chlorine	Cl	17	35.453
Potassium	K	19	39.102
Calcium	Ca	20	40.08
Iron	Fe	26	55.847
Copper	Cu	29	63.54
Arsenic	As	33	74.9216
Strontium	Sr	38	87.62
Cadmium	Cd	48	112.40
Iodine	I	53	126.9045
Mercury	Hg	80	200.59
Lead	Pb	82	207.2
Uranium	U	92	238.03

12–5 THE PERIODIC TABLE

Julius Lothar Meyer (1830–1895), a chemist at the
University of Tübingen, and Dmitri Ivanovitch
Mendeléev (1834–1907), a chemist at the University
of St. Petersburg, discovered independently that when
the elements are arranged in a table ordered from light
to heavy, the chemical and physical properties recur at
definite intervals. The periodicity involves such properties as density, solubility, melting point, boiling
point, ionization energy, and hardness. Meyer plotted
curves that graphically illustrate the periodicity of each
property, such as in Figure 12–5. When the physicist
H. G. J. Moseley (1887–1915) found that the atomic
number of an element is a fundamental property more
important than the atomic weight, he formulated the
new periodic law, or *Moseley's law,* which states that
*the properties of the elements are a periodic function
of their atomic numbers.*

A periodic table is useful to the extent that it correlates the physical and chemical properties of groups

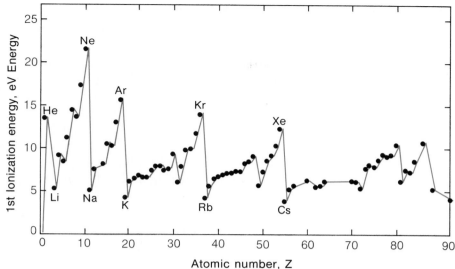

Figure 12–5 Variation of first ionization energy of the elements with atomic number. Symbols of elements with very high and very low ionization energy are shown. (Modified from Rock, Peter A., and George A. Gerhold, *Chemistry: Principles and Applications.* Philadelphia: W. B. Saunders Company, 1974, p. 63.)

of elements. In his 1871 table, the forerunner of modern classification, Mendeléev left some of the spaces vacant rather than insert elements that did not fit properly into a family. He made the bold prediction that new elements would be discovered that would fit. He also predicted some of the properties of three such elements, which he designated "ekaboron," "ekaaluminum," and "ekasilicon." Within 15 years, all three were discovered and were named scandium, gallium, and germanium. A comparison of the properties Mendeléev predicted for ekasilicon and those observed for the new element germanium is given in Table 12–11. The periodic table can still be used to

Table 12–11 Some Properties of Germanium

PROPERTY	PREDICTED FOR EKASILICON (MENDELÉEV, 1871)	OBSERVED FOR GERMANIUM (WINKLER, 1886)
Atomic weight	72	72.3
Atomic volume	13 cm³	13.22 cm³
Boiling point of chloride	100°C	86°
Boiling point of ethyl derivative	160°C	160°C
Color	Dark Gray	Grayish white
Specific gravity	5.5	5.469
Specific gravity of chloride	1.9	1.887
Specific gravity of oxide	4.7	4.703

show where undiscovered elements are to be expected and provides a means for predicting their chemical and physical properties.

The usual version of the modern periodic table is divided into *periods*—horizontal rows running from left to right across the table (Fig. 12–6). There are seven periods, the first consisting of only two elements, hydrogen and helium. The second and third periods contain 8 elements each. The fourth and fifth periods have 18 elements each. The sixth period has 32 elements, 14 of which are usually shown in a separate block at the bottom of the table. The seventh and last period appears incomplete.

Periodicity in chemical properties is illustrated by the fact that, except for the first period, each period begins with a very reactive metal. Successive elements within a period are less and less reactive and increasingly non-metallic, until a very reactive non-metal is encountered. Each period finally closes with a noble gas.

The vertical columns are occupied by elements

Figure 12–6 The periodic table of the elements. (From Brescia, F., et al., *Chemistry: A Modern Introduction*, 2nd ed. Philadelphia: W. B. Saunders Company, 1978, p. 226.)

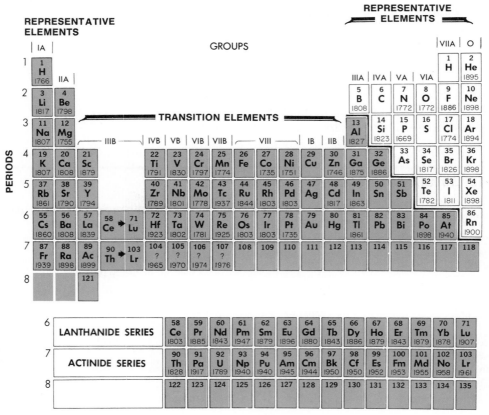

Table 12–12 PHYSICAL PROPERTIES OF ALKALI METALS

PROPERTY	Li	Na	K	Rb	Cs
Melting point, °C	180	98	63.4	38.8	28.7
Boiling point, °C	1336	889	757	679	690
Density, g/ml at 20°C	0.535	0.971	0.862	1.53	1.90
Electronegativity	1.0	0.9	0.8	0.8	0.7
Atomic radius, Å	1.52	1.86	2.31	2.44	2.62
Ionic radius, Å	0.60	0.95	1.33	1.48	1.69

that have related chemical and physical properties and make up a chemical family or group. The members of a group designated IA to VIIA are called main-group elements. Those in a column headed by a Roman numeral and the letter B, IB to VIIB, are known as transition metals. The elements within a given subgroup—for example, Li, Na, K, Rb, and Cs in group IA—resemble one another more closely than they resemble the members of the other family subgroup—Cu, Ag, and Au in group IB. The elements in group IA (except H) are very reactive metals that never occur uncombined in nature. On the other hand, Cu, Ag, and Au, the elements in group IB, are relatively inert and can be recovered from their compounds with such ease that they are among the oldest metals known. The gradation in physical properties in passing successively from one element to the next in a group is typified by the alkali metals in Table 12–12.

12–6 WAVE MECHANICS

Why do the elements in a given subgroup in the periodic table exhibit similar properties? The answer lies in the electronic structures of their atoms. The chemical and physical properties of an atom depend on the arrangement of its electrons. Elements whose atoms have similar electronic structures in their outer shells have many properties in common.

Many of the lines in the emission spectrum of an element are not individual lines but closely spaced *groups* of lines. The spectral lines are split into additional lines when the light-emitting source is placed in a strong magnetic field. Although Bohr's theory of the atom correctly predicts the wavelengths of the principal lines of the hydrogen spectrum, it does not account for the "fine structure" and "hyperfine structure" of many of the lines. The existence of this structure implies that the principal energy levels consist of groups of sublevels that differ slightly in energy.

Erwin Schrödinger's (1887–1961) wave equation resolves some of these difficulties. Building upon de Broglie's discovery that matter has a wavelike character, Schrödinger assumed the wave nature of electrons and described electrons as standing waves in the atom. This is the basis of *wave mechanics,* or quantum mechanics. The Schrödinger equation can be solved exactly for the electron in the hydrogen atom, and approximately for the electrons in other atoms.

It is not necessary here to know the Schrödinger equation or how to solve it; only that the solutions are used to describe the arrangement of electrons in atoms. A solution is expressed as a set of permitted values of four "quantum numbers"—$n, l, m,$ and s—that express the energy of the electrons and their most probable location in the atom. The first three are mathematically related to one another. Electrons are located in energy levels around the nucleus and their energies are *quantized;* that is, they can take on only certain values. Each prinicipal energy level or shell consists of one or more sublevels or subshells. These subshells, in turn, have one or more orbitals. The arrangements of electrons in atoms can be predicted from the quantum numbers.

The *principal quantum number, n,* determines the main energy level in which the electron is located. Corresponding to Bohr's numbers for orbits, n can take on any positive integral value; for the elements known at present the values range from $n = 1$ to $n = 7$. The shell for which $n = 1$ is closest to the nucleus and has the lowest energy.

The *angular momentum quantum number, l,* designates the subshell within a main energy level. It may have any integral value from $l = 0$ to $l = (n - 1)$. The subshells for which $l = 0, 1, 2,$ and 3 are denoted by the letters s, p, d, and f (from spectral lines that were originally designated *s*harp, *p*rincipal, *d*iffuse, and *f*undamental). Thus, an energy level for which $n = 1$ and $l = 0$ is designated ls; for $n = 3$ and $l = 1$, 3p; and so on.

Each subshell consists of one or more sub-subshells, or orbitals, defined by a set of permitted values of $n, l,$ and the magnetic quantum number, m. According to the *Heisenberg uncertainty principle,* if we can measure the energy of an electron precisely, we cannot

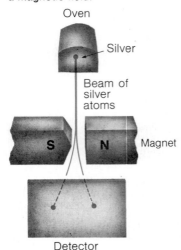

Figure 12–7   The Stern-Gerlach experiment. A beam of silver atoms is split into two beams by a magnetic field.

Oven

Silver

Beam of silver atoms

S   N   Magnet

Detector

simultaneously determine the exact position of the electron. Since the energy of an electron in a given orbital is specified precisely by its quantum numbers, the position of the electron is uncertain. We can therefore talk only about the probability of the electron being in a certain region. An *orbital* is defined as a *three-dimensional region in space around the nucleus where there is the greatest probability of locating a particular electron.* The probability may be expressed diagrammatically as an "electron cloud" that has a shape and density. The *magnetic quantum number (m)* describes the possible orientations of an electron cloud in space in the presence of an external magnetic field. The permitted values of m range from $-l$ to $+l$, including 0. For p orbitals, therefore, where $l = 1$, m can have three values: -1, 0, and 1; thus there are three p orbitals, designated p_x, p_y, and p_z.

The *spin quantum number, s,* was formulated not from the wave equation but from a hypothesis put forth to explain certain experiments. Otto Stern (1887–1969) and Walther Gerlach (b. 1889) had shown that if a beam of silver atoms is allowed to pass by the poles of a magnet, the beam is split into two beams (Fig. 12–7). Similar results are obtained with H, Na, and other elements. Something in the atom is apparently affected by the magnet. George E. Uhlenbeck (b. 1900) and Samuel A. Goudsmit (b. 1902) proposed that an electron spins about its own axis like a top as it moves about the nucleus, generating a magnetic field and behaving like a

Figure 12–8 (a) The two assumed spins of the electron, "the micromagnet," relative to a magnetic field: one spinning west to east, with the north pole of the electron facing the north pole of the external magnet; the second spinning in the opposite direction, with the north pole of the electron facing the south pole of the external magnet. Any other position, such as those illustrated in (b), is forbidden. Therefore, we can say that the spin of an electron is quantized. (From Brescia, F., et al., *Chemistry: A Modern Introduction,* 2nd ed. Philadelphia: W. B. Saunders Company, 1978, p. 205.)

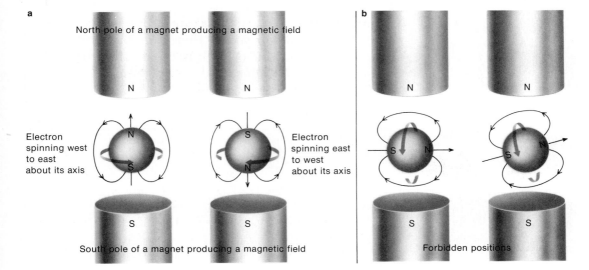

a

North pole of a magnet producing a magnetic field

N

N

Electron spinning west to east about its axis

Electron spinning east to west about its axis

S

S

South pole of a magnet producing a magnetic field

b

N

N

S

S

Forbidden positions

Table 12–13   CHARACTERISTICS OF THE ELECTRON QUANTUM NUMBERS

SYMBOL	NAME	DESCRIPTION	PERMITTED VALUES
n	Principal quantum number	Main shell	1,2,3, . . .
l	Angular momentum quantum number	Subshell	0,1,2,3, . . . $(n-1)$ s p d f
m	Magnetic quantum number	Orbital	$-l$, 0, $+l$
s	Spin quantum number	Electron spin	$+\frac{1}{2}$, $-\frac{1}{2}$

tiny magnet. The spin is quantized in one of two equivalent directions, clockwise and counterclockwise, accounting for the two beams produced (Fig. 12–8). An external magnet therefore interacts with a spinning electron. For each value of m, there are two values of s: $+\frac{1}{2}$ and $-\frac{1}{2}$. The two spin orientations will be designated by arrows pointing up (↑) and down (↓). Table 12–13 summarizes the properties of the four electron quantum numbers.

EXAMPLE   12–2

For the principal quantum number $n = 2$, how many orbitals are possible?

SOLUTION

In the main shell ($n = 2$), the number of possible subshells depends upon the permitted values of l: 0(s), 1(p). In the s subshell, $l = 0$; the only possible value for m is 0; therefore, the s subshell consists of just one orbital. In the p subshell, $l = 1$, and m may take the values -1, 0, and $+1$; therefore, three p orbitals are possible. In all, $1 + 3 = 4$ orbitals are possible when $n = 2$.

The relations among the quantum numbers n, l, and m are shown in Table 12–14 for the first three energy levels.

Table 12–14 DERIVATION OF ORBITALS FROM QUANTUM NUMBERS

QUANTUM NUMBER			
n	l	m	ORBITAL
1	0	0	1s
2	0	0	2s
	1	−1	$2p_x$
		0	$2p_y$
		+1	$2p_z$
3	0	0	3s
	1	−1	$3p_x$
		0	$3p_y$
		+1	$3p_z$
	2	−2	$3d_{xy}$
		−1	$3d_{xz}$
		0	$3d_{yz}$
		+1	$3d_{x^2y^2}$
		+2	$3d_{z^2}$

12–7 BUILDING UP THE PERIODIC TABLE

In the ground state of an atom the electrons occupy the lowest energy levels available and the atom is stable. Any other arrangement corresponds to an excited state of the atom. Starting with a nucleus, which has a positive charge equal to the atomic number of the element, we can get the electronic configuration of the ground state of an atom by adding electrons equal in number to the atomic number. Niels Bohr referred to this process as the "Aufbau" principle, that is, the building-up principle of the periodic system of the elements.

Why are not all of the electrons in an atom found in the lowest energy orbital? It is a law of nature that *no two electrons in the same atom can have the same set of four quantum numbers*. The statement of this law is known as the *Pauli exclusion principle*, after Wolfgang Pauli (1900–1958). Every electron must have a separate set of quantum numbers. Since the quantum numbers n, l, and m specify an orbital, an orbital can accommodate only two electrons, and they must have opposite spins, clockwise and counterclockwise. It is impossible for any two electrons in an atom to exist in the same state, that is, to have identical sets of quantum numbers.

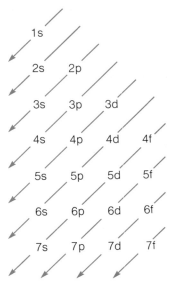

Figure 12–9 Electronic configuration sequence.

The order in which electrons enter the energy levels of an atom is determined by their energies. The electrons of lowest energy enter shells closest to the nucleus. The order of increasing energy for atomic orbitals is given by following the arrows in the scheme in Figure 12–9. Following this sequence, we see that when the 3p energy level has been filled, the next electron enters the 4s level. Only when this level has been filled do electrons enter the 3d level. The sequence, therefore, is: 1s 2s 2p 3s 3p 4s 3d 4p 5s 4d. . . . On this basis we can predict the location of electrons in atoms provided that we also know the electron capacity of each subshell. The capacity depends on the number of orbitals, as given in Table 12–15. The electron capacities of shells 1 to 4 are shown in Table 12–16.

To write the electronic configuration for an element, show the number of electrons that occupy each subshell, listed in order of increasing energy. A superscript denotes the number of electrons occupying a subshell. The sum of the superscripts is the number of electrons in the atom. The electronic configuration of hydrogen, which has one electron, is expressed as follows:

symbol and atomic number of hydrogen $= {}_1H$

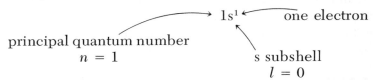

principal quantum number
$$n = 1$$
$\quad$ s subshell
$\quad l = 0$

This notation means that the hydrogen electron is in the s subshell of the first shell. For sodium, the electronic configuration is:

$${}_{11}Na \qquad 1s^2 2s^2 2p^6 3s^1$$

Table 12–15 ELECTRON CAPACITIES OF SUBSHELLS

SUBSHELL	ANGULAR MOMENTUM QUANTUM NUMBER, l	MAGNETIC QUANTUM NUMBER, m	ORBITALS	ELECTRON CAPACITY
s	0	0	1	2
p	1	$-1,0,+1$	3	6
d	2	$-2,-1,0,+1,+2$	5	10
f	3	$-3,-2,-1,0,+1,+2,+3$	7	14

Table 12–16   ELECTRON CAPACITIES OF SHELLS 1 TO 4

MAIN SHELL n	SUBSHELL l	NUMBER OF ELECTRONS	TOTAL NUMBER OF ELECTRONS IN SHELL
$n = 1$	s	2	2
$n = 2$	s	2	
	p	6	8
$n = 3$	s	2	
	p	6	
	d	10	18
$n = 4$	s	2	
	p	6	
	d	10	
	f	14	32

It is often convenient to group the electrons by main shell, so that we may express the configuration of sodium as 2,8,1, a total of 11 electrons.

Another way of representing the electronic configuration of sodium is [Ne] $3s^1$. Neon, atomic number 10, is a noble gas that precedes sodium in the periodic table and has the configuration 2,8. With the exception of helium, all of the noble gases have an octet of electrons in their outermost shell, a particularly stable configuration. According to the Pauli exclusion principle,

Table 12–17   ELECTRONIC CONFIGURATIONS OF THE FIRST 36 ELEMENTS

$_1$H	$1s^1$		$_{19}$K	[Ar] $4s^1$
$_2$He	$1s^2$		$_{20}$Ca	[Ar] $4s^2$
$_3$Li	$1s^2 2s^1$		$_{21}$Sc	[Ar] $3d^1 4s^2$
$_4$Be	$1s^2 2s^2$		$_{22}$Ti	[Ar] $3d^2 4s^2$
$_5$B	$1s^2 2s^2 2p^1$		$_{23}$V	[Ar] $3d^3 4s^2$
$_6$C	$1s^2 2s^2 2p^2$		$_{24}$Cr	[Ar] $3d^5 4s^1$
$_7$N	$1s^2 2s^2 2p^3$		$_{25}$Mn	[Ar] $3d^5 4s^2$
$_8$O	$1s^2 2s^2 2p^4$		$_{26}$Fe	[Ar] $3d^6 4s^2$
$_9$F	$1s^2 2s^2 2p^5$		$_{27}$Co	[Ar] $3d^7 4s^2$
$_{10}$Ne	$1s^2 2s^2 2p^6$		$_{28}$Ni	[Ar] $3d^8 4s^2$
$_{11}$Na	[Ne] $3s^1$		$_{29}$Cu	[Ar] $3d^{10} 4s^1$
$_{12}$Mg	[Ne] $3s^2$		$_{30}$Zn	[Ar] $3d^{10} 4s^2$
$_{13}$Al	[Ne] $3s^2 3p^1$		$_{31}$Ga	[Ar] $3d^{10} 4s^2 4p^1$
$_{14}$Si	[Ne] $3s^2 3p^2$		$_{32}$Ge	[Ar] $3d^{10} 4s^2 4p^2$
$_{15}$P	[Ne] $3s^2 3p^3$		$_{33}$Ar	[Ar] $3d^{10} 4s^2 4p^3$
$_{16}$S	[Ne] $3s^2 3p^4$		$_{34}$Se	[Ar] $3d^{10} 4s^2 4p^4$
$_{17}$Cl	[Ne] $3s^2 3p^5$		$_{35}$Br	[Ar] $3d^{10} 4s^2 4p^5$
$_{18}$Ar	[Ne] $3s^2 3p^6$		$_{36}$Kr	[Ar] $3d^{10} 4s^2 4p^6$

the first 10 electrons of sodium have the same quantum numbers and configuration as the electrons of neon, that is, $1s^22s^22p^6$. The 11th electron is characteristic of sodium. Therefore, the symbol [Ne], the "core," designates the 10 electrons of sodium that have the same quantum numbers as the 10 electrons of neon. To represent the core of a nearby noble gas in the electronic configuration of an element is particularly useful with elements of higher atomic number, since this saves both space and time. Table 12–17 gives the electronic configurations of the first 36 elements.

EXAMPLE   12–3

Identify the element that has the following electronic structure: $1s^22s^22p^63s^23p^64s^23d^{10}4p^65s^2$.

SOLUTION

From the sum of the superscripts we get the atomic number of the element, 38. The element is strontium.

EXAMPLE   12–4

Show the electronic configuration of $_{53}$I.

SOLUTION

We have to assign the 53 electrons of iodine to their respective subshells. Following the sequence shown in Figure 12–9, the configuration is:

$$1s^22s^22p^63s^23p^64s^23d^{10}4p^65s^24d^{10}5p^5$$

Figure 12–10 Electronic configurations of group IA elements.

Li
3
$1s^2\ 2s^1$
Na
11
[Ne]$3s^1$
K
19
[Ar]$4s^1$
Rb
37
[Kr]$5s^1$
Cs
55
[Xe]$6s^1$
Fr
87
[Rn]$7s^1$

The Aufbau principle explains the periodicity of chemical and physical properties. Families of elements have similar electronic configurations in their highest energy levels. For example, note the similarities in the electronic arrangments of the alkali metals in Group IA (Fig. 12–10). Each member of the family has an unfilled shell of one electron surrounding the nucleus and one or more filled shells of electrons.

EXAMPLE 12–5

The elements neon, argon, and krypton are typical members of the noble gas family. Do they share any similarity in their electronic configurations?

SOLUTION

Write the spdf structures of these elements directly from their atomic numbers (or from Table 12–17).

$_{10}$Ne $1s^2 2s^2 2p^6$ $= 2,8$

$_{18}$Ar $1s^2 2s^2 2p^6 3s^2 3p^6$ $= 2,8,8$

$_{36}$Kr $1s^2 2s^2 2p^6 3s^2 3p^6 4s^2 3d^{10} 4p^6 = 2,8,18,8$

Grouping the electrons by main shells reveals the structural similarity of eight electrons in the outermost shell surrounding inner, filled shells of electrons.

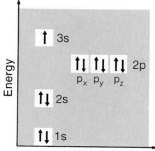

Figure 12–11 Energy-level diagram of the ground state of sodium.

The electronic configuration of an atom can also be expressed in an energy-level diagram (Fig. 12–11). A circle or box is used to represent an orbital; one for an s orbital, three for p orbitals, five for d orbitals, and seven for f orbitals. An arrow represents an electron, pointing up for one spin orientation and down for the other spin orientation. Any orbital has a capacity for two oppositely spinning electrons. For sodium, $_{11}$Na, the 11 electrons are arranged as follows: $1s^2 2s^2 2p^6 3s^1$. Two electrons of opposite spin are accommodated in the 1s orbital; two in 2s; six in 3p—two each in $2p_x$, $2p_y$, and $2p_z$; and the 11th electron in the 3s orbital.

An energy-level diagram provides a pictorial method of applying another important principle of electron configurations. Consider the arrangement of the electrons in carbon, $_6$C: $1s^2 2s^2 2p^2$. Although it seems reasonable to expect the fifth and sixth electrons to fill the $2p_x$ orbital, the evidence is that this does not happen and that the sixth electron enters a different $2p$ orbital. The rule, known as *Hund's rule of maximum multiplicity*, is that *electrons do not enter into joint occupancy of an orbital until all of the available or-*

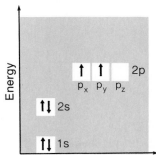

Figure 12–12 Hund's rule as applied to carbon. The $2p_x$ and $2p_y$ orbitals are half occupied.

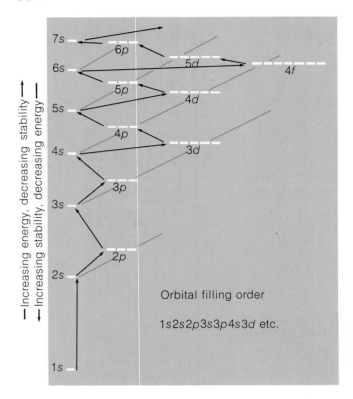

Orbital filling order

1s2s2p3s3p4s3d etc.

Figure 12–13 Order of energy of atomic orbitals. (From Brescia, F., et al., *Chemistry: A Modern Introduction*, 2nd ed. Philadelphia: W. B. Saunders Company, 1978, p. 198.)

bitals of the same type in that shell are half occupied (Fig. 12–12). When several orbitals of the same type are available, a single electron is assigned to each orbital before any electrons are allowed to pair. Figure 12–13 gives the order of the energy of atomic orbitals.

CHECKLIST

Angular momentum quantum number	Neutron
Atomic number	Noble gas
Atomic weight	Orbital
Aufbau principle	Pauli exclusion principle
Element	Period
Ground state	Periodicity
Group	Principal quantum number
Heisenberg uncertainty principle	Proton
Hund's rule of maximum multiplicity	Shell
	spdf configuration
Isotopes	Spin quantum number
Magnetic quantum number	Subshell
Mass spectrometer	Symbol
	Wave mechanics

12–1.   Give the number of electrons, protons, and neutrons   EXERCISES for these environmentally important elements: (a) fluorine; (b) sulfur; (c) arsenic; (d) strontium; (e) mercury.

12–2.   Determine the nuclear structure of the following isotopes: (a) $^{35}_{17}Cl$; (b) $^{36}_{17}Cl$; (c) $^{37}_{17}Cl$.

12–3.   The abundances of the isotopes of magnesium (based upon the mass of ^{12}C as 12.00000) are as follows:

Isotope	Mass	%Abundance
^{24}Mg	23.9847	78.6
^{25}Mg	24.9858	10.1
^{26}Mg	25.9814	11.3

Determine the atomic weight of magnesium.

12–4.   Name the elements associated with the following electronic configurations:
  (a) $1s^2 2s^2$
  (b) [Ne] $3s^1$
  (c) [Ar] $4s^2 3d^7$
  (d) $1s^2 2s^2 2p^6 3s^2 3p^6 4s^2 3d^{10} 4p^5$
  (e) [Kr] $5s^2 4d^{10} 5p^2$

12–5.   What experimental evidence is there for the idea that electronic spin is "quantized" in two possible directions?

12–6.   On the basis of one property that you select, discuss the meaning of the term "periodic" in the expression "periodic table."

12–7.   Experiments are under way to synthesize element 114. In which group of the periodic table would you expect it to be located?

12–8.   Sodium initiates the third period, and argon closes it. What feature of atomic structure do the elements in this period share?

12–9.   In the energy level $n = 3$, (a) How many orbitals are possible? (b) What is the electron capacity of this level?

12–10.   What similarity in electronic arrangement is shared by the members of the nitrogen family?

12–11.   Write the electronic configurations for (a) $_5$B; (b) $_9$F; (c) $_{21}$Sc; (d) $_{33}$As; (e) $_{56}$Ba.

12–12.   The highest energy electron of a chlorine atom is designated $3p^5$. (a) Explain why. (b) Deduce from this notation the values of the four quantum numbers (n, ℓ, m, s).

12–13.   (a) Construct an "electron-in-the-box" energy level diagram for phosphorus, $_{15}$P. (b) What principle dictates the arrangement of the outermost electrons?

12–14.   Relate the Heisenberg uncertainty principle to the idea of an electron orbital.

12–15.   *Multiple Choice*

    A. When the electron of a hydrogen atom is in the energy level $n=2$, we say that
      (a) it emits light
      (b) it is in the ground state
      (c) it is in an excited state
      (d) It ionizes the atom

    B. The atomic number of gold is 79 and the mass number of one of its isotopes is 197. The number of neutrons is
      (a) 79   (c) 118
      (b) 197   (d) 158

    C. Elements in the same group in the periodic table
      (a) have similar chemical properties
      (b) are called isotopes
      (c) have consecutive atomic numbers
      (d) make up a period of elements

    D. The principle that a maximum of two electrons is allowed in any orbital is known as
      (a) Bohr's aufbau principle
      (b) Heisenberg's uncertainty principle
      (c) Hund's rule
      (d) Pauli's exclusion principle

    E. The total electron capacity of the energy level $n=4$ is
      (a) 8   (c) 18
      (b) 16   (d) 32

Block, Felix, "Heisenberg and the Early Days of Quantum Mechanics." *Physics Today,* December, 1976.

> Heisenberg's first doctorate student, later a Nobel laureate, reminisces about the formative years of the new mechanics.

Goudsmit, Samuel A., "It Might as Well Be Spin." *Physics Today,* June, 1976.

> Co-discoverer with George Uhlenbeck of electron spin, the author recalls the atmosphere in the "springtime of modern atomic physics."

Jaffe, Bernard, *Moseley and the Numbering of the Elements.* Garden City, N. Y.: Doubleday & Co., 1971.

> An absorbing account of the discoverer of the new periodic law.

Kragh, Helge, "Chemical Aspects of Bohr's 1913 Theory." *Journal of Chemical Education,* April, 1977.

> Discusses the chemical content of the 1913 theory that revolutionized our concept of matter.

Longuet-Higgins, H. C., "A Periodic Table: The Aufbauprinzip as a Basis for Classification of the Elements." *Journal of Chemical Education,* January, 1957.

> Presents a periodic table that shows how the position of an element in the table is related to the electronic structure of the atom.

Spronsen, Jan W. von, "The Priority Conflict between Mendeléev and Meyer." *Journal of Chemical Education,* March, 1969.

> Discusses one of the classic priority disputes in science.

Szokefalvi-Nagy, Zoltan, "How and Why of Chemical Symbols." *Chemistry,* February, 1967.

> Although chemical symbols have been in use for thousands of years, their full potential has been realized only recently.

CHEMICAL BONDING AND MOLECULAR GEOMETRY

13-1 INTRODUCTION

We usually encounter the 100-odd elements not as pure substances but as constituents of compounds. Bulk matter occurs in combination rather than in atomic form and is held together by forces called chemical bonds. Simple or complex, all compounds—water, sugar, salt—are formed of aggregates of atoms in a definite ratio represented by a chemical formula, as shown in Table 13–1. Chemical bonds are the "glue" that holds the atoms together and with molecular geometry determines the properties of the compound.

13-2 THE NOBLE GASES

Although most of the noble gases were discovered in the 1890's, until 1962 it was believed that they were

Table 13–1 SOME FAMILIAR COMPOUNDS

NAME	FORMULA	NAME	FORMULA
Alcohol (ethyl)	C_2H_5OH	Milk of Magnesia	$Mg(OH)_2$
Aspirin	$C_7H_6O_3$	Photographer's hypo	$Na_2S_2O_3$
Baking soda	$NaHCO_3$	Plaster of Paris	$CaSO_4 \cdot \frac{1}{2}H_2O$
Borax	$Na_2B_4O_7 \cdot 10H_2O$	Sugar (sucrose)	$C_{12}H_{22}O_{11}$
Dry ice	CO_2	Table salt	$NaCl$
Epsom salt	$MgSO_4 \cdot 7H_2O$	TNT	$C_7H_5N_3O_9$
Lime	CaO	Vinegar	$C_2H_4O_2$
Lye	$NaOH$	Water	H_2O

368

Table 13–2 ELECTRONIC CONFIGURATIONS OF THE NOBLE GASES

ELEMENT	SYMBOL AND ATOMIC NUMBER	ELECTRONIC CONFIGURATION	NUMBER OF ELECTRONS IN SHELL					
			1	*2*	*3*	*4*	*5*	*6*
Helium	$_2$He	$1s^2$	2					
Neon	$_{10}$Ne	[He] $2s^22p^6$	2	8				
Argon	$_{18}$Ar	[Ne] $3s^23p^6$	2	8	8			
Krypton	$_{36}$Kr	[Ar] $3d^{10}4s^24p^6$	2	8	18	8		
Xenon	$_{54}$Xe	[Kr] $4d^{10}5s^25p^6$	2	8	18	18	8	
Radon	$_{86}$Rn	[Xe] $4f^{14}5d^{10}6s^26p^6$	2	8	18	32	18	8

chemically inert. They were not known to take part in the formation of any compounds whatsoever and appeared to be models of chemical stability. Then Neil Bartlett of the University of British Columbia and others prepared compounds of the heavier noble gases, Kr, Xe, and Rn, with fluorine, and found certain of these compounds also to be quite stable.

Their almost complete lack of chemical reactivity suggests that the noble gases have little inclination to lose, gain, or share electrons in order to form chemical bonds. As their electronic configurations in Table 13–2 indicate, their orbitals are filled to capacity, a condition that apparently confers unusual stability upon the atoms involved. In all of the noble gases except helium there are eight electrons in the outermost or valence shell; in helium there are two electrons. Before an atom can accept one or more electrons, there must be orbital vacancies available to accommodate them. Since all of the orbitals in noble gas atoms are filled, there are no orbitals available for bond formation.

The electrons of the noble gas atoms are held tenaciously. A measure of the tightness with which an atom holds on to its electrons is the *ionization energy*—the energy necessary to remove an electron from an atom in the gaseous state. Ionization energy is usually expressed in units of electron volts/atom or kilocalories/mole (1 eV/atom = 23.1 kcal/mole).

The first electron of an atom is removed with less difficulty than a second or third, since the latter are removed from positively charged ions. The energies are referred to as the first ionization energy, the second ionization energy, and so on, as diagrammed in Table 13–3). Note the trend in the first ionization energy for

Table 13–3   IONIZATION ENERGIES*

ATOMIC NUMBER	ATOM	IONIZATION ENERGIES (eV)							
		1st	2nd	3rd	4th	5th	6th	7th	8th
1	H	13.6							
2	He	24.6	54.4						
3	Li	5.4	75.6	122.4					
4	Be	9.3	18.2	153.9	217.7				
5	B	8.3	25.1	37.9	259.3	340.1			
6	C	11.3	24.4	47.9	64.5	392.0	489.8		
7	N	14.5	29.6	47.4	77.5	97.9	551.9	666.8	
8	O	13.6	35.1	54.9	77.4	113.9	138.1	739.1	871.1
9	F	17.4	35.0	62.6	87.2	114.2	157.1	185.1	953.6
10	Ne	21.6	41.1	64	97.2	126.4	157.9	—	—
11	Na	5.1	47.3	71.7	98.9	138.6	172.4	208.4	264.2
12	Mg	7.6	15.0	80.1	109.3	141.2	186.9	225.3	266.0
13	Al	6.0	18.8	28.4	120.0	153.8	190.4	241.9	285.1
14	Si	8.1	16.3	33.5	45.1	166.7	205.1	246.4	303.9
15	P	10.6	19.7	30.2	51.4	65.0	220.4	263.3	309.3
16	S	10.4	23.4	35.0	47.3	72.5	88.0	281.0	328.8
17	Cl	13.0	23.8	39.9	53.5	67.8	96.7	114.3	348.3
18	Ar	15.8	27.6	40.9	59.8	75.0	91.3	124.0	421
19	K	4.3	31.8	46	60.9	—	99.7	118	155
20	Ca	6.1	11.9	51.2	67	84.4	—	128	147

*From Johnston et al., *Chemistry and the Environment.* Philadelphia, W. B. Saunders Company, 1973, p. 60.

the first 20 elements. In each period, a maximum is reached with the noble gas that completes that period.

13–3   THE IONIC BOND

Although the noble gases are quite inert, they are flanked on both sides in the periodic table by very active groups of elements. The halogens (a group of non-metals) and the alkali metals readily enter into reactions with one another, forming compounds called salts (*halogen,* salt-former) of which ordinary table salt, sodium chloride, is an example. These salts are crystalline solids; they are hard and brittle, have relatively high melting (600°C–1000°C) and boiling points, are generally soluble in water, and are good conductors of electricity when molten or in water solution (but not in the crystalline state). Table 13–4 shows the electronic structures of atoms of the halogen and alkali metal families.

Atoms of the halogens and the alkali metals can acquire a stable noble gas structure by gaining or losing

Table 13–4 ᴇʟᴇᴄᴛʀᴏɴɪᴄ Sᴛʀᴜᴄᴛᴜʀᴇs ᴏғ
ᴛʜᴇ Hᴀʟᴏɢᴇɴs ᴀɴᴅ Aʟᴋᴀʟɪ Mᴇᴛᴀʟs

HALOGENS		*NOBLE GASES*	*ALKALI METALS*	
		$_2$He	$_3$Li	[He] 2s^1
$_9$F	[He] 2s^{2}2p^5	$_{10}$Ne	$_{11}$Na	[Ne] 3s^1
$_{17}$Cl	[Ne] 3s^{2}3p^5	$_{18}$Ar	$_{19}$K	[Ar] 4s^1
$_{35}$Br	[Ar] 4s^{2}4p^5	$_{36}$Kr	$_{37}$Rb	[Kr] 5s^1
$_{53}$I	[Kr] 4d^{10}5s^{2}5p^5	$_{54}$Xe	$_{55}$Cs	[Xe] 6s^1
$_{85}$At	[Xe] 4f^{14}5d^{10}6s^{2}6p^5	$_{86}$Rn	$_{87}$Fr	[Rn] 7s^1

one electron, respectively. Thus, a sodium atom can
acquire the neon electronic structure by losing an electron and forming an ion with a single positive charge,
Na$^+$. By gaining an electron, a chlorine atom acquires
an octet of electrons in its valence shell and becomes
the negatively charged chloride ion, Cl$^-$, with the electronic configuration of argon. When the elements sodium and chlorine are brought together, electrons are
transferred to form Na$^+$ and Cl$^-$ ions (Fig. 13–1). The
stable ions in NaCl are held together by the Coulomb
force of attraction between oppositely charged ions,
and form an ionic or electrovalent bond. From such
observations the "rule of eight" or *octet rule* has been
developed: *Atoms enter into combination to obtain a
structure that has eight electrons in the outermost
shell.*

Bond formation usually involves only those electrons that occupy the highest-energy orbitals of the
atom, called the *valence electrons*. Electrons in fully
occupied lower-lying energy orbitals are ordinarily
held too tightly by the nucleus to participate in ionic
bonding. The highest-energy s and p orbitals are
valence orbitals, containing from one to eight valence

Figure 13–1 Ionic bonding.
Formation of sodium chloride by
the transfer of electrons between atoms.

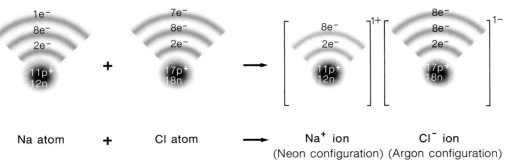

Na atom $+$ Cl atom $\longrightarrow$ Na$^+$ ion   Cl$^-$ ion
(Neon configuration) (Argon configuration)

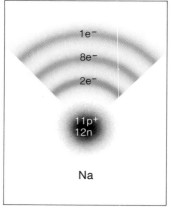

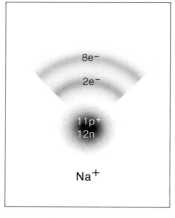

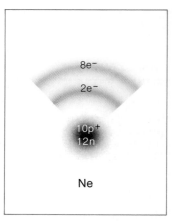

A

B

C

Figure 13–2 Comparison of the sodium atom *(A)*, sodium ion *(B)*, and neon atom *(C)*.

electrons. The number of charges on the ion defines its electrovalence or oxidation number and represents the combining power of the element. Thus, sodium has a positive valence of +1 while chlorine has a valence of −1.

In acquiring the electronic structures of neon and argon, sodium and chlorine are not changed into the noble gas atoms. The sodium atom, sodium ion, and neon atom are related as shown in Figure 13–2. Since the nuclear charge determines the nature of the element, the Na^+ ion with 11 positive charges on the nucleus is different from neon, which has only 10 positive charges.

Table 13–5 SELECTED SETS OF ISOELECTRONIC IONS AND NOBLE GASES

NOBLE GAS	NUMBER OF ELECTRONS	ION
Neon, Ne	10	Oxide, O^{2-} Fluoride, F^- Sodium, Na^+ Magnesium, Mg^{2+} Aluminum, Al^{3+}
Argon, Ar	18	Sulfide, S^{2-} Chloride, Cl^- Potassium, K^+ Calcium, Ca^{2+}
Krypton, Kr	36	Selenide, Se^{2-} Bromide, Br^- Rubidium, Rb^+ Strontium, Sr^{2+}

Table 13–6   Properties of Sodium and Chlorine in the Free and Combined States

SUBSTANCE	STATE AT ROOM TEMPERATURE	MELTING POINT, °C	BOILING POINT, °C
Sodium	Silvery, soft solid	98	892
Chlorine	Greenish, irritating gas	−101	−35
Sodium chloride	Colorless, crystalline solid	801	1413

The charge of +1 on the sodium ion comes from the imbalance between the number of positively charged protons in the nucleus and the number of negatively charged electrons:

Electric Charges in the Sodium Ion

11 + from 11 protons
10− from 10 electrons
1+ net charge on sodium ion

Since the sodium ion and the neon atom have the same number of electrons, they are said to be *isoelectronic* (*iso*, same) with each other.

There is a limit to the number of electrons that an atom can lose or gain. Positive ions (cations) commonly have charges of +1, +2, or +3 at most. On the other hand, a non-metallic atom may gain an extra one, two, or perhaps three electrons, but electrovalences lower than −3 are unknown. The noble gas structures of various ions are shown in Table 13–5.

The often remarkable differences in the properties of the uncombined elements and the compounds that they form are evident in sodium chloride, as Table 13–6 shows.

Table 13–7 gives the names and chemical formulas of some common ionic compounds.

Table 13–7   Representative Ionic Compounds

COMMON NAME	FORMULA	CHEMICAL NAME
Baking soda	$NaHCO_3$	Sodium bicarbonate
Bleaching powder	$Ca(ClO)_2$	Calcium hypochlorite
Limestone	$CaCO_3$	Calcium carbonate
Rock salt	NaCl	Sodium chloride
Silver bromide	AgBr	Silver bromide

13-4 THE COVALENT BOND

Although ionic bonding accounts for the properties of many important compounds, there are many thousands more for which there is no evidence to indicate that the elements exist as ions. These compounds may be gaseous, liquid, or solid at room temperature. They usually have rather low melting and boiling points. They do not conduct an electric current, and they are not very soluble in water as a rule. The experimental evidence indicates that they are made up of discrete molecules, a *molecule* being the smallest particle of a substance that possesses all of the properties of that substance.

Gilbert Newton Lewis (1875–1946) at the University of California studied the electronic structure of molecular compounds. He proposed that a stable noble gas structure could be achieved by the sharing of electrons between atoms. Irving Langmuir (1881–1957), a chemist at the General Electric Research Laboratory, elaborated upon Lewis's theory and introduced the term "covalent bond" to describe the Lewis electron-pair bond.

Lewis introduced a convenient notation to represent atoms, ions, and molecules. In electron-dot symbols, called *Lewis structures,* the valence electrons are shown as dots surrounding the symbol of the element. The symbol itself represents the "kernel," consisting of the atom's nucleus and inner electrons. Except for helium, the number of valence electrons of the elements in the first three periods is equal to the number of their periodic group. For example, aluminum in Group III has 3 valence electrons. The Lewis structures of the first 18 elements are shown in Table 13–8.

Table 13–8   LEWIS STRUCTURES OF HYDROGEN THROUGH ARGON

I	II	III	IV	V	VI	VII	VIII
H·							He:
Li·	·Be·	·Ḃ·	·Ċ·	·N̈·	:Ö·	:F̈·	:N̈e:
Na·	·Mg·	·Ȧl·	·S̈i·	·P̈·	:S̈·	:C̈l·	:Är:

Let us apply the Lewis-Langmuir approach to the hydrogen molecule. An individual H atom has one electron and is therefore one electron short of the noble gas configuration of helium. The stable diatomic hydrogen molecule (H_2) results from the sharing of two electrons of opposite spin by the two hydrogen atoms. In this way, each H atom is surrounded by two electrons and acquires the stable helium structure. The shared pair of electrons, represented by two dots or a dash, is called a covalent bond.

$$H : H \quad H-H$$

covalent bond

The Lewis-Langmuir model of the covalent bond accounts for the diatomic nature of fluorine (F_2) and other halogen molecules (Cl_2, Br_2, I_2). The valence shell of a fluorine atom has 7 electrons ($2s^2 2p^5$), so it is one electron short of the noble gas configuration of neon ($2s^2 2p^6$). A stable diatomic molecule (F_2) results from the sharing of a pair of electrons by two fluorine atoms. In this way, each fluorine atom acquires an octet of electrons in its valence shell. The formation of Cl_2, Br_2, and I_2 is accounted for in the same way.

$$:\ddot{F}:\ddot{F}: \quad F-F$$

covalent bond

Some atoms acquire a noble gas configuration by sharing more than one pair of electrons. In these cases, the atoms in the molecules are linked by multiple bonds. Two nitrogen atoms, for example, become bonded in a nitrogen molecule by the sharing of three pairs of electrons. Each N atom has the structure $1s^2 2s^2 2p_x^1 2p_y^1 2p_z^1$. The three p electrons of one N atom pair with three electrons of opposite spin from the second N atom.

$$:N::N: \quad N{\equiv}N$$

Nitrogen (N_2)

The sharing of two pairs of electrons leads to a double bond; three pairs, a triple bond. A molecule having

Table 13–9   LEWIS STRUCTURES OF SOME COVALENT MOLECULES

SUBSTANCE	LEWIS STRUCTURE	DASH FORMULA
Bromine	:B̈r : B̈r:	Br—Br
Chlorine	:C̈l : C̈l:	Cl—Cl
Fluorine	:F̈ : F̈:	F—F
Iodine	:Ï : Ï:	I—I
Hydrogen	H : H	H—H
Oxygen	·Ö :: Ö·	O=O
Nitrogen	:N ⫶ N:	N≡N
Carbon dioxide	·Ö :: C :: Ö·	O=C=O
Ammonia	H : N̈ : H H	H—N—H H
Methane	H : C̈ : H (H above and below)	H—C—H (H above and below)
Carbon tetrachloride	:C̈l : C̈ : C̈l: (with C̈l above and below)	Cl—C—Cl (with Cl above and below)

only single bonds is said to be saturated, while one with double or triple bonds is unsaturated.

Several covalent molecules are listed in Table 13–9 with their Lewis structures.

EXAMPLE   13–1

Draw an electron dot diagram (Lewis structure) for a molecule of hydrogen fluoride, HF.

SOLUTION

Since Lewis structures involve only valence electrons and "kernels," write the electronic configurations of H and F to determine the number of electrons in the outermost shell of each.

$_1$H   1s^1 H ·

$_9$F   1s^{2}2s^{2}2p^5 · $\ddot{\text{F}}$:

Hydrogen has one valence electron, while fluorine has (5 + 2) = 7 valence electrons. Fluorine acquires an octet by sharing the electron of hydrogen. The Lewis structure for hydrogen fluoride is, therefore,

H : $\ddot{\text{F}}$:

In the hydrogen molecule (H_2) and chlorine molecule (Cl_2), two identical atoms are joined by a covalent bond:

H : H   : $\ddot{\text{C}}$l : $\ddot{\text{C}}$l :

and both atoms share equally in the electrons of the bond. The covalent bond in each molecule is *nonpolar*, because the region of greatest electron density is exactly midway between the nuclei. Non-polar substances do not conduct electricity, since they do not have charged particles to respond to an electric field. Although the atoms in the molecules are held together by strong covalent bonds, the forces between the molecules are very weak, and the molecules are easily separated. Non-polar covalent compounds are therefore often gases or liquids, or solids that sublime easily, and tend to have rather low melting points (under 300°C) and boiling points (under 500°C).

13–5   POLAR MOLECULES

In a molecule of hydrogen fluoride (HF), the fluorine nucleus exerts a greater attraction for the shared pair of electrons than the hydrogen nucleus. The region of maximum electron density therefore lies closer to fluorine:

H$^{\delta+}$: $\ddot{\text{F}}$: $^{\delta-}$

Since one end of the bond is negative and the other end positive, owing to the partial separation of electronic charge, the bond is called a polar covalent bond. Bond polarity can be indicated by the symbols $\delta+$ and $\delta-$, meaning "partially positive" and "partially negative," respectively.

A measure of the ability of a bonded atom in a molecule to attract electrons is given by a positive number called the *electronegativity*. On the electronegativity scale proposed by Linus Pauling (b. 1901) at the California Institute of Technology, fluorine is assigned the highest value, 4.0. Fluorine is the element with the greatest electron attracting power, followed by oxygen, nitrogen, and chlorine. Cesium gives up an electron most readily and has the lowest value, 0.7, as shown in Table 13–10. Electronegativities increase from left to right across the period, and decrease from the top to the bottom of a group. Thus, the most electronegative elements are located in the upper right corner of the periodic table; the least electronegative in the lower left corner.

With the electronegativity scale we can predict the polarity of molecules. The greater the electronegativity difference between two bonding atoms, the greater the polarity of the bond. In the hydrogen molecule or chlorine molecule the electronegativity difference

Table 13–10   ELECTRONEGATIVITIES* (PAULING'S SCALE)

1 H 2.1						
3 Li 1.0	4 Be 1.5	5 B 2.0	6 C 2.5	7 N 3.0	8 O 3.5	9 F 4.0
11 Na 0.9	12 Mg 1.2	13 Al 1.5	14 Si 1.8	15 P 2.1	16 S 2.5	17 Cl 3.0
19 K 0.8	20 Ca 1.0	31 Ga 1.6	32 Ge 1.8	33 As 2.0	34 Se 2.4	35 Br 2.8
37 Rb 0.8	38 Sr 1.0		50 Sn 1.7	51 Sb 1.9	52 Te 2.1	53 I 2.4
55 Cs 0.7	56 Ba 0.9		82 Pb 1.7	83 Bi 1.9		

*Elements in the darker area are more electronegative.

Table 13-11 BOND TYPE PREDICTED FROM ELECTRONEGATIVITY DIFFERENCE

ELECTRONEGATIVITY DIFFERENCE	BOND TYPE
0	Non-polar covalent
0 to 1.7	Polar covalent
Greater than 1.7	Ionic

between the two atoms is zero; hence the bond is non-polar, with equal sharing of electrons.

	H—H	Cl—Cl
Electronegativity:	2.1 2.1	3.0 3.0
Difference:	0	0
	(non-polar covalent)	(non-polar covalent)

With an electronegativity difference of 0.9 between H and Cl, the hydrogen chloride molecule is polar covalent.

	H—Cl
Electronegativity:	2.1 3.0
Difference:	0.9 (polar covalent)

When the electronegativity difference is greater than 1.7, the shift in electron density is great enough to describe the bond as ionic, rather than covalent.

	Na Cl
Electronegativity:	0.9 3.0
Difference:	2.1 (ionic)

The relation between electronegativity difference and bond type is given in Table 13-11.

Electronegativity values can also be used to estimate the "ionic character" of a bond. The bond between most atoms is neither completely ionic nor completely covalent. A percentage describes the ionic or covalent character of a bond, as in Table 13-12. When the electronegativity difference is greater than 1.7, the bond is more than 50 per cent ionic.

Table 13–12 ELECTRONEGATIVITY DIFFERENCE AND PER CENT IONIC CHARACTER OF A CHEMICAL BOND

ELECTRONEGATIVITY DIFFERENCE	% IONIC CHARACTER	ELECTRONEGATIVITY DIFFERENCE	% IONIC CHARACTER
0.2	1	1.8	55
0.4	4	2.0	63
0.6	9	2.2	70
0.8	15	2.4	76
1.0	22	2.6	82
1.2	30	2.8	86
1.4	39	3.0	89
1.6	47	3.2	92
		3.4	96

EXAMPLE 13–2

Predict (a) whether a carbon-chlorine bond (C—Cl) is relatively ionic or covalent, and (b) which atom is relatively positive and which negative.

SOLUTION

First determine the electronegativity difference between the atoms from Table 13–10. Then refer to Table 13–12 for the per cent ionic character of the bond.

$$
\begin{array}{cc}
 & \text{C—Cl} \\
\textit{Electronegativity:} & \underbrace{2.5 \quad 3.0} \\
\textit{Difference:} & 0.5
\end{array}
$$

(a) An electronegativity difference of 0.5 corresponds to approximately 7 per cent ionic character (or 93 per cent covalent character). A C—Cl bond is, therefore, primarily covalent.

(b) The greater attraction that chlorine has for electrons creates an unbalanced electron distribution, and makes the chlorine side of the molecule more negative than the carbon side:

$$C^{\delta+} - Cl^{\delta-}$$
$$\longmapsto$$

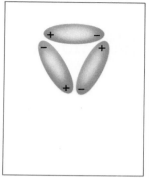 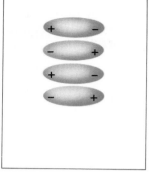 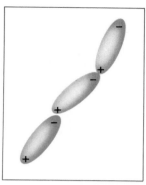

Figure 13–3 Possible associations of polar covalent molecules in the liquid state.

The presence of polar bonds increases the tendency of molecules to stick to or associate with one another in the liquid state. The positive end of one molecule attracts the negative end of another to form clumps of several molecules (Fig. 13–3). Polar covalent compounds do not conduct electricity however. The negative and positive charges of the dipole are contained in the same particle and the particle is attracted by the same amount in opposite directions by the electric field. Polar covalent compounds such as hydrogen chloride (HCl) have higher melting points and boiling points than comparable non-polar molecules such as silane (SiH_4). In the solid state the crystals of polar molecules are harder than non-polar molecular crystals.

13–6 DIPOLE MOMENT

In a polar molecule such as HCl, the centers of positive and negative charges do not coincide. Such a molecule is called a dipole, represented by the symbol $\mapsto$ with the arrow pointing from the positive toward the negative pole. Quantitatively, the polar character of a molecule is expressed by its dipole moment, μ, defined as the product of the effective charge (either positive or negative) and the distance between the centers of opposite charges.

Dipole moment = Charge × Distance   Equation 13–1

$$\mu = q \times d$$

Table 13–13 DIPOLE MOMENTS OF SOME MOLECULES (DEBYE UNITS) ($ID = 1 \times 10^{-18}$ esu-cm)							
H_2	0	HF	1.9	H_2O	1.84	NH_3	1.46
O_2	0	HCl	1.0	H_2S	0.92	PH_3	0.55
CH_4	0	HBr	0.8	CH_3Cl	1.86	PCl_3	0.78
CCl_4	0	HI	0.4	CH_2Cl_2	1.59	PF_3	1.025

The dipole moments of various molecules are given in Table 13–13 in Debye units (D), after Peter Debye (1884–1966), a pioneer in the field.

The dipole moments of H_2 and O_2 are zero, since the electronic charge is symmetrically distributed about two similar atoms in these molecules. The zero dipole moment of CH_4 is surprising, however. The electronegativities of C and H are 2.5 and 2.1. Each C—H bond, therefore, has an electronegativity difference of 0.4, corresponding to 4 per cent ionic character. Although each C—H bond is slightly polar, in the molecular architecture of CH_4 the centers of negative and positive charges of the molecule as a whole coincide; hence there is no net polarity. In other words, the vector sum of the four dipole moments is zero (Fig. 13–4).

For the same reason, the dipole moment of CCl_4 is zero. But if one, two, or three of the hydrogens in methane are replaced by chlorine, the polar nature of the molecules is rather pronounced. Thus, the dipole moment of methyl chloride (CH_3Cl) is 1.86D; dichloromethane (CH_2Cl_2), 1.59D; chloroform ($CHCl_3$), 1.15D. The dipole moment of water, 1.84D, leads us to conclude that the molecule is bent rather than linear, for if it were linear, the dipole moment would be zero (Fig. 13–5).

Figure 13–4 A nonpolar molecule, CH_4, $\mu = 0$. The vector sum of the dipole moments is zero, because the four C—H bonds are symmetrically distributed in space. All of the bond dipole moments are cancelled, and the center of positive charge coincides with the center of negative charge.

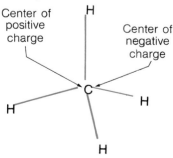

Center of positive charge

Center of negative charge

13–7   MOLECULAR GEOMETRY

Oxygen is bonded covalently to two hydrogen atoms in the water (H_2O) molecule. The fact that water has a permanent dipole moment (1.84D) as well as other evidence leads us to believe that the water mole-

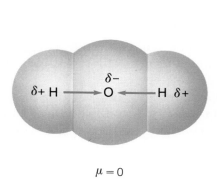

$\mu = 0$

Not true structure

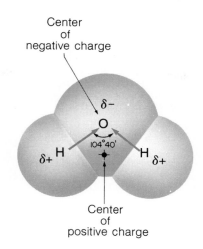

Center of negative charge

Center of positive charge

$\mu = 1.84D$

True structure

Figure 13-5 The water molecule. If the structure were linear, the dipole moment μ would be 0. Since $\mu = 1.84D$, the molecule must be bent; the centers of positive and negative charge do not coincide.

cule is bent and that the O—H bonds form a bond angle of 105° with respect to one another. The bond angles in molecules determine how the atoms are arranged in space—that is, the geometry of the compound. Since molecular structure has a decided influence upon the properties of the substance, it is important to know that structure. Only a few geometrical structures are necessary to account for most of molecular chemistry.

The *valence shell electron pair repulsion (VSEPR) theory*, proposed by R. J. Gillespie at McMaster University, allows us to predict the molecular geometry of a wide variety of compounds. The VSEPR theory says that the number of valence-shell electron pairs surrounding an atom determines the arrangement of the bonds around the atom. The electron pairs may be bound or unbound. The lowest energy, and therefore preferred, arrangement of the electron pairs is assumed to be the molecular geometry in which there is a minimum of electron-electron repulsion. This condition is satisfied when the electron pairs are kept as far apart as possible.

Consider the structure of beryllium hydride, BeH_2. The two valence electrons of beryllium form electron pairs with the electrons from two hydrogen atoms, in what may be considered two positions on a sphere representing the beryllium atom. The bond

Figure 13-6 Structure of beryllium hydride, BeH_2. All three atoms lie in the same straight line. The bond angle is 180°.

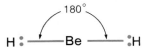

180°

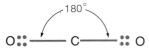

Figure 13–7 Structure of carbon dioxide, CO_2. The two electron locations are as far apart as possible: 180°.

angle is at a maximum 180° when the distance between the two electron pairs is greatest, since this is as far apart as two locations on a sphere can be—at opposite ends of an axis. The bonds, therefore, are in the same straight line, and the molecule is linear: H—Be—H (Fig. 13–6).

The carbon atom of carbon dioxide (CO_2) is linked to each oxygen atom by a double bond. All four valence electrons of carbon are used; two for each bond. For the two electron positions on the carbon atom in CO_2 to be as far apart as possible, the oxygen atoms must be at opposite sides of the carbon. The result, again, is a linear molecule, O—C—O, with a bond angle of 180° (Fig. 13–7).

Although two double bonds are also present in a molecule of sulfur dioxide (SO_2), these bonds are not linear but bent. In this case, a lone pair of electrons is also present, since sulfur has six valence electrons, but it is uninvolved in the bonding. There are, then, three electron locations. The repulsion between the bonding electrons and the lone-pair electrons forces the three locations as far apart as possible on the surface of a sphere, namely, to the corners of an equilateral triangle, with the center of the sphere in the plane of the triangle. The angles are each 120° (Fig. 13–8).

Boron forms three covalent bonds with chlorine in boron trichloride (BCl_3). Since all three valence shell electrons of boron are involved in the bonds, the electron pairs are located at only three positions on the atomic surface. The farthest apart these three positions can be is a trigonal planar arrangement with bond angles of 120°, as for SO_2.

In ammonia (NH_3), nitrogen forms three covalent bonds with hydrogen. However, the nitrogen atom also has a lone pair of electrons that is not involved in the three covalent bonds. The mutual repulsion among these four electron locations directs the electron pairs to the four corners of a tetrahedron. The observed bond angle in ammonia is 107°, close to that for a tetrahedral arrangement (109°28′). The hydrogen atoms are located at three corners and a lone pair of electrons at the fourth; in effect, this causes the ammonia structure to be a low pyramid, with the nitrogen atom over a triangle of hydrogen atoms (Fig. 13–9). Table 13–14 summarizes the geometries of some simple molecules, including some with five bonds (trigonal bipyramid) and six bonds (regular octahedron).

Figure 13–8 Structure of sulfur dioxide, SO_2. The three electron locations define the corners of an equilateral triangle. The atoms lie in the same plane, at bond angle of 120°.

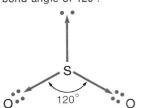

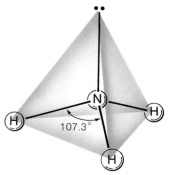

EXAMPLE 13–3

Predict the molecular structure of PCl_3.

SOLUTION

The electronic structure of phosphorus can be determined from its atomic number, 15: $1s^2 2s^2 2p^6 3s^2 3p^3$. The five valence electrons are represented as follows in the Lewis structure:

$$\cdot \overset{\cdot}{\underset{\cdot}{P}} \cdot$$

There are thus three electrons and a lone pair. Chlorine, a halogen, has the Lewis structure:

$$: \overset{\cdot \cdot}{\underset{\cdot \cdot}{Cl}} \cdot$$

Phosphorus forms three covalent bonds with chlorine. We have, therefore, three bonds and a lone pair, for a total of four electron-pair locations. The best separation in this case gives a pyramidal structure, with the phosphorus atom over a triangle of chlorine atoms.

Figure 13–9 Structure of the ammonia molecule, NH_3. Four electron locations are involved. The observed bond angle is 107.3°, close to that for a tetrahedron (109°28′).

13–8 COORDINATE COVALENT BOND

Soon after the nature of the covalent bond was explained, Langmuir and others, including Nevil V. Sidgwick (1873–1952) at Oxford University realized

Table 13–14 MOLECULAR STRUCTURES

NUMBER OF BONDS	GEOMETRY OF MOLECULE	BOND ANGLE	EXAMPLES
2	Linear	180°	BeF_2, $MgCl_2$
2	Angular	90°–120°	H_2O, SO_2
3	Trigonal planar (equilateral triangle)	120°	BF_3, SO_3
3	Pyramidal	90°–114°	NH_3, PBr_3
4	Tetrahedral	109°28′	CH_4, SiF_4
5	Trigonal bipyramidal	90° and 120°	PCl_5, $AsBr_5$
6	Regular octahedral	90°	SF_6, SeF_6

that there was nothing to prevent both electrons of an electron-pair bond to come from the same atom. For example, the nitrogen atom in an ammonia molecule (NH_3) has a lone pair of electrons that it can share with the boron atom of boron trifluoride (BF_3):

$$H \quad\quad :\ddot{F}: \quad\quad\quad H:\ddot{F}:$$
$$H:\ddot{N}: + \quad \ddot{B}:\ddot{F}: \rightarrow H:\ddot{N}:\ddot{B}:\ddot{F}:$$
$$H \quad\quad :\ddot{F}: \quad\quad\quad H:\ddot{F}:$$

A bond of this type is called a *coordinate covalent bond,* or *dative bond.* In some molecules, a polarity is created, the donor atom becoming slightly positive and the acceptor atom slightly negative. This fact is represented by an arrow pointing away from the atom that has furnished the bonding pair of electrons, as in $H_3N \rightarrow BF_3$. The nitrogen atom is the donor, the boron atom the acceptor; the arrow indicates the relationship of donor to acceptor.

Once formed, however, a coordinate covalent bond is similar to any other covalent bond. The "donation" is a special case of electron sharing and does not involve a complete transfer of electrons. Nevertheless, the donor atom, nitrogen, no longer has full possession of the pair of electrons and acquires a "formal" positive charge. The acceptor atom, boron, now has a share in more electrons than it did before the bond was formed, and acquires a "formal" negative charge.

13–9 STRUCTURE OF THE HYDROGEN MOLECULE

Quantum mechanics was first applied to the problem of the structure of the hydrogen molecule in the 1920's by Walter Heitler and Fritz W. London, who calculated the energy and thus the strength of the covalent bond. Imagine two H atoms approaching one another (Fig. 13–10). When they are far apart, the attractive force exerted by each upon the other is slight. At shorter distances, this force increases while the energy of the system decreases.

When they are close, each electron is attracted not only by its own nucleus but also by that of the other atom. At distances closer than 0.74 Å, however, the repulsive forces between the two electrons and between the two protons increase greatly, and the energy

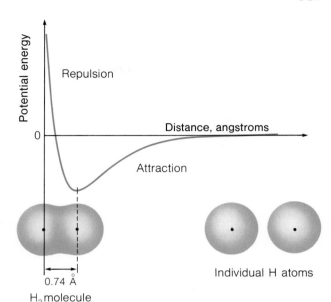

Figure 13-10 Potential energy diagram for two hydrogen atoms. At a distance of 0.74 Å, the potential energy is at a minimum and stability is at a maximum. The H_2 molecule forms at this position.

rises. At that distance there is a minimum in potential energy; the attractive force between the electron of one atom and the proton of the other far exceeds the repulsive forces and the system reaches a state of maximum stability, the hydrogen molecule, H_2. The orbitals merge or overlap, provided that the electron spins are opposed, and there is then a chance of finding each electron in the vicinity of either nucleus. The electrons now travel about both nuclei, the resulting mutual attractions of the two electrons and the two nuclei binding the atoms together. The pair of electrons is "shared" by the two atoms and the result is a covalent bond.

13-10   THE METALLIC BOND

The structure and properties of a great majority of the elements in the periodic table (about 75 per cent) are due to the metallic bond. These elements exist as metallic solids. A typical metal, such as copper, is opaque, except in very thin layers, and lustrous; that is, it reflects light of any wavelength. It is relatively dense and strong. It can be rolled or hammered into shape—it is malleable—and it can be drawn into a wire —it is ductile. It is a good conductor of heat and electricity, and has a high melting point. We use metals pri-

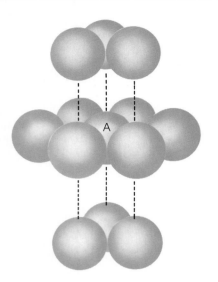

Figure 13–11 Hexagonal close-packed structure, as in the metals magnesium and zinc. A given atom is surrounded by 12 nearest neighbors: 6 in the same plane, 3 in the plane above, and 3 in the plane below.

marily for these mechanical, electrical, and thermal properties.

Because of close packing, the mass per unit volume of metals—the density—is high. Metals typically crystallize in either a face-centered cubic or hexagonal close-packed lattice, in which each metal atom is surrounded by 12 nearest neighbors, or in a slightly more open structure, the body-centered cubic lattice, in which each atom is surrounded by 14 others: eight nearest neighbors and six others slightly farther away (Fig. 13–11). Such arrangements are efficient for filling space, far more so than in non-metals. For example, in a non-metallic covalent crystal like the diamond, each

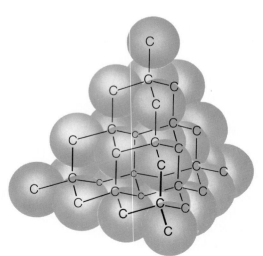

Figure 13–12 Diamond structure. Each carbon atom is covalently bonded to four other carbon atoms.

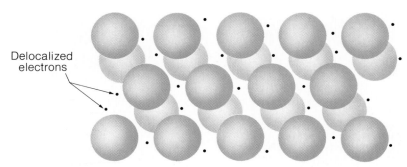

Figure 13-13 Simple diagram of a metallic crystal: an array of cations in a "sea of electrons."

Delocalized electrons

carbon atom is bonded to only four other carbon atoms (Fig. 13–12).

The atoms of metals must be bonded quite differently from those in covalent substances, since the one, two, or three valence electrons in metals are not nearly enough to form electron-pair bonds with 12 or 14 adjacent atoms. With only one valence electron per atom, the metal lithium crystallizes in a body-centered cubic lattice in which each atom is bound to 14 other atoms tightly enough to give it a melting point of 186°C.

A simple picture of a metal crystal visualizes it as an array of positive metal ions—cations—suspended in a "sea of electrons" (Fig. 13–13). The valence electrons are considered to be *delocalized;* that is, they no longer belong to individual atoms but to the crystal as a whole. The sea of negatively charged, mobile electrons overcomes the repulsive forces between the cations and acts as a kind of "glue" that holds the crystal together by electrostatic attraction.

The easy mobility of the electrons in metals accounts for their high thermal and electrical conductivities. Metals conduct electricity because the electrons are free to move in an electric field. Metals are opaque and lustrous because free electrons absorb and radiate back most of the light energy that falls on them. Metals conduct heat effectively because electrons can transfer thermal energy.

The electron-sea model also explains the mechanical properties of malleability and ductility. The non-directional bonding in metals enables the metallic ions to give up old loyalties and form new bonds easily. A group of ions in a metal may move past another group without changing the environment of the positive ions (Fig. 13–14). The shape of the crystal can therefore be changed without destroying its integrity. Metals can be rolled or hammered into shape, and welded and

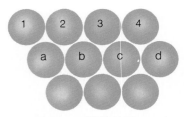 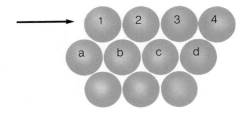

Figure 13-14 Malleability and ductility of metals are explained by the movement of a plane of ions with respect to another plane, resulting in no net change of environment.

alloyed because of the unselectivity of the metallic bond. A comparison of physical properties and the type of bonding force is given in Table 13–15.

13-11   BAND THEORY OF SOLIDS

The *band theory* provides a good working model for solids in general, including conductors (metals), semi-conductors, and insulators (non-metals). It assumes that isolated electron energy levels of atoms are broadened into energy bands that belong to the crystal as a whole and not to individual atoms, and that the Pauli exclusion principle applies to each energy level. The lower bands are completely filled by electrons, while the bands of higher energy—the valence

Table 13–15   RELATION OF BONDING FORCE AND
PHYSICAL PROPERTIES

	IONIC	*COVALENT*	*METALLIC*
Structural Unit	Ions	Atoms	Positive ions
Attractive Force	Strong; ionic or electrostatic	Strong; sharing of electron pairs	Moderate to strong; electrostatic attraction between positive ions and "sea of electrons"
Melting Point	High	Low to very high	Medium to high
Boiling Point	High	Low to high	Medium to high
Hardness	Hard and brittle	Soft to very hard	Soft to medium
Electrical Conductivity	Very low	Very low	Good
Examples	Calcium fluoride (CaF_2) Silver bromide (AgBr) Sodium chloride (NaCl)	Carborundum (SiC) Diamond (C) Quartz (SiO_2)	Copper (Cu) Iron (Fe) Mercury (Hg)

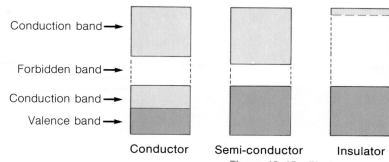

Conductor   Semi-conductor   Insulator

Figure 13-15 Electron energy bands in metals, semi-conductors, and insulators.

band and the conduction band—are separated by an energy gap (Fig. 13-15).

The electron bands in metals are incompletely filled. In $_3$Li, for example, (electronic configuration $1s^2 2s^1$), the band originating from the broadening of the 1s atomic orbital is fully occupied (Fig. 13-16). The 2s band, however, is only half filled, since each lithium atom has one electron in the 2s level, but this level has a capacity for two electrons; this is the conduction band of lithium. When an electric field is applied to the metal, electrons at the top of the partially filled 2s band gain small amounts of energy and migrate into the available states lying immediately above. Electrons in the unfilled valence band of a metal can also absorb wavelengths of light over a broad range and be promoted to the unfilled energy states immediately above the filled states, accounting for their opaqueness and luster. That is why most metals have no color of their own; they absorb and re-emit all colors of light.

In metals that have a filled s band, the close packing causes the p band to combine with it or overlap so that there is no "forbidden" energy zone between them. Electrons are then easily promoted from the s band to the empty p band. The large energy barrier between the valence and conduction bands of an insulator such as diamond (about 6 eV) effectively prevents the promotion of electrons into the conduction band. The forbidden zone separating the bands in silicon, a semi-conductor, is small enough (about 1.1 eV) to permit bridging by electrons under the influence of an electric field, at high temperatures, or even by light. The electrical conductivity of semi-conductors thus increases as the temperature increases. Germanium and silicon are two of the important semi-conductors used in transistors and other "solid-state" devices.

Figure 13-16 Band model of lithium, showing partly filled conduction band.

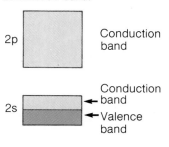

2p   Conduction band

2s   ← Conduction band
      ← Valence band

1s

CHECKLIST

Anion	Isoelectric
Band theory	"Kernel"
Bond angle	Lewis structure
Cation	Lone-pair electrons
Close packing	Metallic bond
Coordinate covalent bond	Molecular geometry
Covalent bond	Molecule
Dipole	Multiple bonds
Dipole moment	Noble gases
Double bond	Octet rule
Electronegativity	Per cent ionic character
Electron pair	Polar covalent
"Electron-sea" model	Semi-conductor
Electrovalent bond	Triple bond
Ionic bond	Valence shell
Ionization energy	VSEPR theory

EXERCISES

13–1. Account for the fact that the second ionization energy of lithium (75.6 eV) is greater than the first (5.4 eV).

13–2. Explain why copper wire is a good conductor of electricity.

13–3. Compare the atoms of metals and non-metals with respect to (a) ionization energy; (b) electronegativity.

13–4. How does a coordinate covalent bond differ from an ordinary covalent bond?

13–5. Write Lewis structures for the following compounds:
- (a) Sodium fluoride: NaF
- (b) Carbon tetrachloride: CCl_4
- (c) Boron trifluoride: BF_3
- (d) Molecular nitrogen: N_2
- (e) Carbon monoxide: CO

13–6. Predict (a) whether the bonds between the following pairs of elements would be ionic or covalent, and (b) if covalent, the per cent ionic character of the bond:
- (a) Ba, O
- (b) Zn, S
- (c) N, Cl
- (d) C, S
- (e) Si, C

13-7. Although fluorine is the most electronegative element, silicon tetrafluoride (SiF_4) does not have a dipole moment. Explain.

13-8. (a) List four ions that are isoelectronic with the noble gas xenon, and (b) give the charge of each.

13-9. What is the relation between the octet rule and a noble gas structure?

13-10. Why would it be misleading to refer to "molecules" of sodium chloride (NaCl) in the crystalline state?

13-11. The following substances are isoelectronic with one another. Write a Lewis structure for each.
    (a) CH_4 (b) NH_3 (c) H_2O (d) HF   (e) Ne

13-12. Explain why the halogens (fluorine, chlorine, bromine, and iodine) occur as diatomic molecules.

13-13. Based upon the electronegativities of the atoms, arrange the following bonds in order of increasing ionic character:

      B—Cl   P—F   H—S   Mg—O   C—N

13-14. What geometric structure would you predict for each of the following substances?
    (a) CF_4 (d) AsH_3
    (b) $BeCl_2$ (e) SF_6
    (c) BBr_3

13-15. (a) List three types of solids based on the nature of the bonding forces, and (b) give an example of each.

13-16. (a) What bond angle would you predict when all of the valence electrons of an atom are involved in three single bonds? (b) Would the presence of a lone pair of electrons in addition to three bonds change your prediction? If so, how?

13-17. Account for the characteristic luster of silver metal.

13–18. On the basis of molecular structure, explain why NH₃ has a dipole moment but BF₃ does not.

13–19. *Multiple Choice*

A. In the Lewis structure for fluorine, the number of dots surrounding the symbol for fluorine is
   (a) one   (c) five
   (b) four   (d) seven

B. The most electronegative element on the Pauling electronegativity scale is
   (a) oxygen   (c) fluorine
   (b) neon   (d) cesium

C. The kind of bonding in sodium chloride is
   (a) ionic   (c) coordinate covalent
   (b) covalent   (d) metallic

D. The compound that has the greatest covalent character is
   (a) H_2O (c) Br_2
   (b) NaCl   (d) Fe_2O_3

E. The bond between two nitrogen atoms in nitrogen gas is
   (a) single   (c) triple
   (b) double   (d) electrovalent

SUGGESTIONS FOR FURTHER READING

Benfey, Theodor, "Geometry and Chemical Bonding." *Chemistry,* May, 1967.
   Emphasizes the continuum of bond types from non-polar covalent to polar covalent to ionic, and from metallic to both ionic and covalent.
Cottrell, A. H., "The Nature of Metals." *Scientific American, September, 1967.*
   Correlates the close-packed crystal structure of metals with their mechanical properties.
Gillespie, R. J., "The Electron-Pair Repulsion Model for Molecular Geometry." *Journal of Chemical Education,* January, 1970.
   Applies the valence shell electron pair repulsion (VSEPR) theory to the problem of molecular shape.
Gillespie, R. J., "A Defense of the Valence Shell Electron Pair Repulsion (VSEPR) Model." *Journal of Chemical Education,* June, 1974.
   Asserts that the VSEPR theory is more reliable for predicting molecular shape than any other theory we have at present.
Hyman, Herbert H., "The Chemistry of the Noble Gases." *Journal of Chemical Education,* April, 1964.
   Discusses the experimental background of noble gas chemistry and theoretical approaches to explain it.
Mott, Sir Nevill, "The Solid State." *Scientific American,* September, 1967.
   Quantum mechanics as applied to the solid state is not restricted to pure science but has become a working tool of technology.

CHEMICAL FORMULAS AND EQUATIONS

14–1 INTRODUCTION

Chemical formulas make up the vocabulary of chemistry. To understand the structure of the materials in our environment, we have to become familiar with that vocabulary. The common names for many materials are carryovers from an earlier time when knowledge of their chemical composition was limited. In this chapter we take up some principles of systematic chemical nomenclature. Then we look at the quantitative significance of formulas, and finally we apply this vocabulary to the writing of statements about chemical reactions called *chemical equations*.

14–2 CHEMICAL COMPOUNDS AND FORMULAS

We have seen that every element is assigned a unique name that identifies it. In systematic chemical nomenclature our aim is to give each *compound* a unique name, too, that indicates its chemical composition. The names run well into the hundreds of thousands.

A *compound* is *a pure substance composed of two or more elements that are chemically combined in a definite proportion by weight*. Table salt, for example, is composed of 39.3 per cent by weight of sodium and 60.7 per cent by weight of chlorine in chemical combination. A compound is represented by a *formula*—a combination of chemical symbols that tells what elements are present and how many atoms of each are involved in the compound. The formula NaCl for table salt shows that sodium and chlorine are present and that they are present in a 1:1 ratio. The *systematic name*, sodium chloride, indicates the chemical composition of the substance and identifies it uniquely.

Sodium chloride is an example of a *binary compound*, one that consists of two different elements. Many binary compounds contain a metal or hydrogen united with a non-metal. Both in writing the formulas of binary compounds and in naming them, we place the more metallic element first. The suffix *-ide* is added to the stem of the name of the more non-metallic element, as in:

$$KBr \ = potassium\ bromide$$
$$ZnO \ = zinc\ oxide$$
$$HF \ = hydrogen\ fluoride$$
$$CaCl_2 = calcium\ chloride$$

EXAMPLE   14–1

Name the compound BaS.

SOLUTION

Note that BaS is a binary compound. Write the name of the first element, barium, and modify the name of the second, sulfur, in order to add the suffix *-ide*. The name is barium sulfide.

14–3 OXIDATION NUMBERS

The same two elements can often form two or more binary compounds. Iron and chlorine, for example, are combined in ferrous chloride ($FeCl_2$) and ferric chloride ($FeCl_3$). If we know some things about oxidation numbers we can write and name such chemical formulas.

The *oxidation number,* or *valence,* of an element is the positive or negative charge of its atoms in a compound. In ionic compounds the charge comes from the transfer of electrons from one atom to another, while in covalent compounds it comes from the unequal sharing of electrons between atoms. The oxidation number, therefore, is a measure of the combining capacity of an element, and indicates the number of electrons that one of its atoms has gained, lost, or shared in forming chemical bonds with other atoms.

Since an iron atom transfers an electron to each of two chlorine atoms in $FeCl_2$, iron is assigned an oxidation number of +2 in this compound, and chlorine an oxidation number of −1. In $FeCl_3$, the oxidation number of chlorine is again −1, but iron is +3. The oxidation number of an atom in an ionic compound is thus equal to the charge of the ion. For covalent compounds, a negative oxidation number is assigned to the more electronegative atom of a bonded pair, and a positive charge to the other atom. Thus, in HF, H is assigned an oxidation number of +1 and F an oxidation number of −1. Since a compound is electrically neutral, the sum of the oxidation numbers of all the atoms must be zero. These rules apply to oxidation numbers:

1. The oxidation number of a free, uncombined element is zero.
2. For a neutral substance, the sum of the oxidation numbers of all the atoms in it is zero.
3. The oxidation number of a simple ion (a single atom that has lost or gained electrons) is equal to the charge of the ion.
4. The sum of the oxidation numbers of all the atoms in a complex ion (that is, one consisting of two or more atoms) is equal to the charge on the ion.
5. In most compounds, hydrogen is assigned an oxidation number of +1.
6. In most compounds, oxygen is assigned an oxidation number of −2.

A charged chemical radical is a group of atoms having either a positive or negative charge that occurs as a unit in compounds but is not stable by itself. An example is the sulfate radical, SO_4^{2-}, that is found in compounds such as Na_2SO_4. The sulfur and oxygen

Table 14–1 Oxidation Numbers of Some Elements and Chemical Radicals

+1		−1	
H	Hydrogen	F	Fluoride
Au	Gold (I) or aurous	Cl	Chloride
Ag	Silver	Br	Bromide
Cu	Copper (I) or cuprous	I	Iodide
Hg	Mercury (I) or mercurous	ClO	Hypochlorite
K	Potassium	ClO_2	Chlorite
Li	Lithium	ClO_3	Chlorate
Na	Sodium	ClO_4	Perchlorate
NH_4	Ammonium	NO_2	Nitrite
		NO_3	Nitrate
		CN	Cyanide
		$C_2H_3O_2$	Acetate
		HCO_3	Bicarbonate
		MnO_4	Permanganate
		OH	Hydroxide

+2		−2	
Ba	Barium	O	Oxide
Ca	Calcium	S	Sulfide
Cd	Cadmium	CO_3	Carbonate
Mg	Magnesium	C_2O_4	Oxalate
Sr	Strontium	CrO_4	Chromate
Zn	Zinc	Cr_2O_7	Dichromate
Co	Cobalt (II) or cobaltous	SO_3	Sulfite
Cu	Copper (II) or cupric	SO_4	Sulfate
Fe	Iron (II) or ferrous	SiO_3	Silicate
Pb	Lead (II) or plumbous		
Mn	Manganese (II) or manganous		
Hg	Mercury (II) or mercuric		
Ni	Nickel (II) or nickelous		
Sn	Tin (II) or stannous		

+3		−3	
Al	Aluminum	N	Nitride
Bi	Bismuth	P	Phosphide
B	Boron	AsO_4	Arsenate
As	Arsenic (III) or arsenious	PO_3	Phosphite
Sb	Antimony (III) or antimonous	PO_4	Phosphate
Cr	Chromium (III) or chromic		
Co	Cobalt (III) or cobaltic		
Au	Gold (III) or auric		
Fe	Iron (III) or ferric		

+4	
Pb	Lead (IV) or plumbic
Mn	Manganese (IV) or manganic
Sn	Tin (IV) or stannic

atoms within the radical are bonded covalently and behave as a single unit in many chemical reactions:

$$Na^+ \left[\begin{array}{c} :\overset{..}{\overset{..}{O}}: \\ :\overset{..}{O}: S :\overset{..}{O}: \\ :\overset{..}{\underset{..}{O}}: \end{array} \right] Na^+$$

Although one sulfur and four oxygen atoms have (5×6)
or 30 outer shell or valence electrons, the sulfate radical has 32 valence electrons and carries a charge of -2.
The two excess electrons have come from the two sodium atoms that form ionic bonds with the SO_4^{2-} radical.

Certain metals have two different oxidation numbers. The classic method of distinguishing between
the two is to add the suffix *-ous* to the base of the name
of the metal for the lower oxidation number, and the
suffix *-ic* for the higher, as in the following examples:

Fe^{2+} Ferrous

Fe^{3+} Ferric

Sn^{2+} Stannous

Sn^{4+} Stannic

For example, as previously mentioned, $FeCl_2$ is ferrous
chloride, and $FeCl_3$ is ferric chloride. In another system, proposed by Alfred Stock (1876–1946), the oxidation number of the metal is indicated by Roman numerals enclosed in parentheses following the name of the
metal. Suffixes for the metals are not used. The Stock
name for $FeCl_2$ is iron (II) chloride; $FeCl_3$, iron (III)
chloride. Table 14–1 gives the names, formulas, and
oxidation numbers of some common ions.

14–4 WRITING CHEMICAL FORMULAS FROM OXIDATION NUMBERS

To write a formula for a compound of aluminum
and sulfur refer to Table 14–1. The oxidation number
of Al is $+3$, and the oxidation number of S is -2.

$$Al^{3+} S^{2-}$$

Applying the principle of electroneutrality—that a
chemical compound is electrically neutral—we see

that two Al^{3+} are needed to balance three S^{2-} ions for the sum of the charges to equal zero. Write the number of each ion as a subscript to the right of the symbol for the simplest formula of aluminum sulfide:

$$Al_2S_3$$

EXAMPLE 14–2

Write the chemical formula for iron (III) nitrate.

SOLUTION

We know from the Roman numeral III Stock name that the oxidation number of iron in this compound is +3. In Table 14–1 the nitrate radical has an oxidation number of -1 that applies to the entire radical and not to the oxygen or nitrogen alone: NO_3^{1-}. Enclose the nitrate radical within parentheses:

$$Fe^{3+}(NO_3)^{1-}$$

Next apply the electroneutrality principle: one Fe^{3+} ion will balance three NO_3^{1-} ions. If more than one ion is needed, write a subscript to the right of the ion. The formula is therefore:

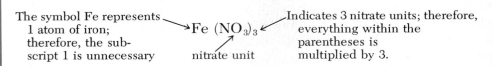

The symbol Fe represents 1 atom of iron; therefore, the subscript 1 is unnecessary   →Fe (NO₃)₃←   nitrate unit   Indicates 3 nitrate units; therefore, everything within the parentheses is multiplied by 3.

Since iron (III) nitrate contains three different kinds of atoms—Fe, N, and O—it is an example of a *ternary compound.* Many ternary compounds are composed of a metal in combination with a radical consisting of another element and oxygen. If the radical occurs in the formula of a compound more than once, as does the NO_3^{1-} in $Fe(NO_3)_3$, it is enclosed in parentheses. A subscript following the parentheses indicates the number of times the radical occurs in the compound. Since the subscript outside the parentheses acts as a multiplier of every atom within the parentheses, we might have written the formula FeN_3O_9. We prefer the form $Fe(NO_3)_3$, however, since it preserves the identity of the radical, which behaves as a unit within the compound.

EXAMPLE 14–3

Give an appropriate name for $SnSO_3$.

SOLUTION

The formula signifies a ternary compound: tin combined with the SO_3 radical. In Table 14–1 the oxidation number of SO_3 is -2. According to the principle of electroneutrality, the oxidation number of Sn in this compound must be $+2$:

$$Sn^{2+} SO_3^{2-}$$

We can call the Sn^{2-} ion either tin (II) or stannous. So, two appropriate names for the compound are:

Tin (II) sulfite

and

Stannous sulfite

14–5 MOLECULAR AND FORMULA WEIGHTS

The *molecular weight* is the sum of the atomic weights of all the atoms in a molecule. For a diatomic molecule such as H_2, N_2, or O_2, the molecular weight is twice the atomic weight. For phosphorus, which forms P_4 molecules, the molecular weight is four times the atomic weight of phosphorus.

The concept of molecular weight cannot be applied to ionic compounds, since they exist as ions, not molecules. For them, the formula weight may be computed instead. The *formula weight* of a compound is the sum of the atomic weights of all the atoms in its formula. Since molecular and formula weights are computed in the same way, the more inclusive term, formula weight, applies whether a substance is molecular or ionic.

EXAMPLE   14–4

What is the formula weight of aspirin, $C_9H_8O_4$?

SOLUTION

The formula weight of aspirin represents the mass of an aspirin molecule relative to Carbon-12. Simply add the atomic weights of each atom in the molecule.

Atomic weight of C = 12.0×9 = 108.0 amu (contributed by C)

Atomic weight of H = 1.0×8 = 8.0 amu (contributed by H)

Atomic weight of O = 16.0×4 = $\underline{64.0}$ amu (contributed by O)
Formula weight   = total   = 180.0 amu

14–6   PERCENTAGE COMPOSITION OF COMPOUNDS

The formula weight of a compound represents its total mass, or 100 per cent of the compound. The mass of each element, therefore, represents a percentage of the total mass, just as a piece of pie represents a percentage of the whole pie. The percentage composition of a compound is the weight-percent of each element present. It is determined by dividing the total mass of each element by the formula weight and multiplying by 100 per cent.

EXAMPLE   14–5

Calculate the percentage composition of sugar (sucrose), $C_{12}H_{22}O_{11}$.

SOLUTION

First determine the formula weight of $C_{12}H_{22}O_{11}$, then divide the total mass of each element by it.

$$12\ C = 12 \times 12.0 = 144.0\ \text{amu} \qquad \%C = \frac{144\ \text{amu}}{342\ \text{amu}} \times 100\% = 42.1\%C$$

$$22\ H = 22 \times 1.0 = 22.0\ \text{amu} \qquad \%H = \frac{22\ \text{amu}}{342\ \text{amu}} \times 100\% = 6.4\%H$$

$$11\ O = 11 \times 16.0 = \underline{176.0}\ \text{amu} \qquad \%O = \frac{176\ \text{amu}}{342\ \text{amu}} \times 100\% = \underline{51.5\%O}$$

Formula
weight = total   = 342.0 amu   100.0%

14–7 GAY-LUSSAC'S LAW OF COMBINING VOLUMES

Joseph Louis Gay-Lussac (1778–1850), a chemist and physicist, found that two volumes of hydrogen gas react with one volume of oxygen gas to produce two volumes of steam when the gases are measured under the same conditions of temperature and pressure: 100°C and 1 atmosphere pressure (Fig. 14–1).

He also found that two volumes of carbon monoxide gas combine with one volume of oxygen gas to form two volumes of carbon dioxide gas. From these and other experiments, he proposed a generalization known as *Gay-Lussac's law of combining volumes:*

When gases combine to form new compounds, the volumes of the reactants and products are in the ratio of small whole numbers.

14–8 AVOGADRO'S HYPOTHESIS

Amadeo Avogadro (1776–1856), a chemist and physicist, saw in Gay-Lussac's law of combining volumes a clue to the molecular formulas of reactants and products. He assumed that:

1. Equal volumes of all gases under the same con-

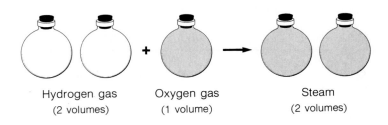

Hydrogen gas   Oxygen gas   Steam
(2 volumes)   (1 volume)   (2 volumes)

Figure 14–1   Gay-Lussac's law.

ditions of temperature and pressure contain the same number of molecules.

2. Molecules of certain elements consist of two identical atoms that separate when the element reacts to form a compound.

If two hydrogen molecules each contain two identical hydrogen atoms and produce two water molecules, then each water molecule must contain two hydrogen atoms. If each oxygen molecule contains two oxygen atoms and produces two water molecules, then each water molecule must contain one oxygen atom. The molecular formula of water must therefore be H_2O.

14–9 THE MOLE: AVOGADRO'S NUMBER

Atomic, molecular, and formula weights are all expressed in atomic mass units (amu). By converting these units to grams, chemists have created a useful unit of both weight and number.

One *gram-atomic weight* (gram-atom) is the weight of the element in grams that is numerically equal to its atomic weight. One gram-atom of any element contains the same number of *atoms* as one gram-atom of any other element (Table 14–2).

One *gram-molecular weight* of a substance is the weight of the substance in grams that is numerically equal to the molecular weight of the substance. One gram-molecular weight of any substance contains the same number of *molecules* as one gram-molecular weight of any other substance (Table 14–3).

One *gram-formula weight* of a substance is the weight in grams numerically equal to the sum of the atomic weights appearing in the formula. One gram-formula weight of any substance contains the same number of *formula units* as one gram-formula weight of any other substance (Table 14–4).

The unit, *mole,* is used interchangeably with

Table 14–2 GRAM-ATOMIC WEIGHTS

ELEMENT	ATOMIC WEIGHT	VALUE OF ONE GRAM-ATOM
Carbon	12.01	12.01 g
Gold	196.97	196.97 g
Silver	107.87	107.87 g
Titanium	47.9	47.9 g

Table 14–3 GRAM-MOLECULAR WEIGHTS

SUBSTANCE	MOLECULAR WEIGHT	VALUE OF ONE GRAM-MOLECULAR WEIGHT
H_2O	18.0	18.0 g
O_2	32.0	32.0 g
CCl_4	154.0	154.0 g
S_8	256.5	256.5 g

gram-atomic weight, gram-molecular weight, and gram-formula weight. Thus, one mole, one gram-atom of any element, one gram-molecular weight of any substance, and one gram-formula weight of any compound, whether in the solid, liquid, or gaseous state, all contain the same number of atoms, molecules, or formula units. This number, called *Avogadro's number* (symbol N), has been evaluated experimentally in various ways, and they all yield the same value: 6.02×10^{23}. *PARTICLES*

For example, one gram-atom of oxygen, 16.0 g, is the weight of 6.02×10^{23} oxygen atoms. One mole of HCl, molecular weight 36.5, weighs 36.5 g and contains 6.02×10^{23} molecules of HCl. One mole of KCl, an ionic compound, contains N formula units, or 6.02×10^{23} K^+ ions and 6.02×10^{23} Cl^- ions. A chemical formula, therefore, may refer to one atom, molecule, or formula unit, or to a mole of atoms, molecules, or formula units. Avogadro's number is also the number of molecules of a gas confined in 22.4 liters at 0°C and 1 atmosphere pressure, a volume called the *gram-molecular volume*. Finally, since they represent relative numbers of atoms, molecules, or other units, quantities expressed in moles mean more than quantities expressed in grams. Thus, chemical symbols and formulas have both qualitative and quantitative significance, as Table 14–5 shows.

Table 14–4 GRAM-FORMULA WEIGHTS

SUBSTANCE	FORMULA WEIGHT	VALUE OF ONE GRAM-FORMULA WEIGHT
$CaCl_2$	111.1	111.1 g
Na_2CO_3	106.0	106.0 g
KOH	56.1	56.1 g
$BaSO_4$	233.4	233.4 g

Table 14–5 INFORMATION CARRIED BY CHEMICAL SYMBOLS AND FORMULAS

THE SYMBOL	*SIGNIFIES*
Ni	1. The element nickel 2. One atom of nickel 3. One atomic weight of nickel: 58.7 amu 4. One gram-atom of nickel: 58.7 g 5. One mole of nickel atoms: 6.02×10^{23} atoms (Avogadro's number)

THE FORMULA	*SIGNIFIES*
CO_2	1. Carbon dioxide 2. One molecule of carbon dioxide 3. One molecular weight of carbon dioxide: 44.0 amu 4. One gram-molecular weight of carbon dioxide: 44.0 g 5. One mole of carbon dioxide molecules: 6.02×10^{23} molecules (Avogadro's number)

EXAMPLE 14–6

How many molecules are there in 40.0 g of I_2?

SOLUTION

Knowing that a mole of iodine (I_2) contains N iodine molecules, we may use dimensional analysis to determine the number of molecules in 40.0 g. The atomic weight of iodine is 127, and we remember that iodine is diatomic.

$$40.0 \text{ g } I_2 \times \frac{6.02 \times 10^{23} \text{ molecules}}{(127 \times 2)\text{g } I_2}$$

$$= 40.0 \text{ g } I_2 \times \frac{6.02 \times 10^{23} \text{ molecules}}{254 \text{ g } I_2}$$

$$= 0.158 \, (6.02 \times 10^{23}) \text{ molecules}$$

$$= 0.950 \times 10^{23} \text{ molecules}$$

$$= 9.5 \times 10^{22} \text{ molecules}$$

EXAMPLE 14–7

How many moles are there in 798 g of octane, C_8H_{18}?

SOLUTION

Evaluate the molecular weight (M.W.) of C_8H_{18}, and apply dimensional analysis to get the number of moles of octane.

C_8H_{18} ＄＄1 mole C_8H_{18} = 114 g

$8C = 8 \times 12 = 96$ $798 \text{ g } C_8H_{18} \times \dfrac{1 \text{ mole}}{114 \text{ g } C_8H_{18}} = 7 \text{ moles}$

$18H = 18 \times 1 = \underline{18}$
$\text{M.W.} = \overline{114}$

14–10 EMPIRICAL AND MOLECULAR FORMULAS

The *empirical formula* is the simplest formula of a compound and gives the relative numbers of each kind of atom in the compound. The *molecular* or *true formula* represents the total number of atoms of each element present in one molecule of the compound, and may or may not be the same as the empirical formula. The empirical and molecular formulas of H_2O are exactly the same. If the molecular formula is not the same, it will be an integral multiple of the empirical formula.

For example, CH (M.W. 13.0) is an empirical formula derived from a chemical analysis of a compound that showed 92.3 per cent C and 7.7 per cent H. The compounds acetylene and benzene both have this empirical formula. Other information will be necessary to decide which compound we have. Thus, the molecular weight of acetylene is known to be 26.0. This is equal to 2×13.0; the molecular formula is therefore $2 \times CH$, or C_2H_2. From the molecular weight of benzene, 78.0 (6×13.0), we conclude that the molecular formula is $6 \times CH$, or C_6H_6.

$$CH = \text{Empirical formula}$$

$$2 \times CH = C_2H_2 = \text{Acetylene (molecular formula)}$$

$$6 \times CH = C_6H_6 = \text{Benzene (molecular formula)}$$

The empirical formula of a compound is calculated from its percentage composition. By assuming that we have 100 g of material, we can express the percentage of each element in grams and convert to gram-atoms. The relative number of gram-atoms is then divided by the smallest relative number to give whole numbers. These small whole numbers become the subscripts in the formula.

EXAMPLE 14–8

Calculate the empirical (simplest) formula of the refrigerant Freon, which analysis shows contains 9.92 per cent C, 58.68 per cent Cl, and 31.40 per cent F.

SOLUTION

Step 1: Assume that we have 100 g of Freon and express the percentage of each element in grams.

$$C = 9.92\% = 9.92 \text{ g}$$

$$Cl = 58.68\% = 58.68 \text{ g}$$

$$F = 31.40\% = 31.40 \text{ g}$$

Step 2: Convert each weight to gram-atoms through dimensional analysis.

$$9.92 \text{ g C} \times \frac{1 \text{ gram-atom C}}{12.0 \text{ g C}} = 0.83 \text{ g-atoms C}$$

$$58.68 \text{ g Cl} \times \frac{1 \text{ gram-atom Cl}}{35.5 \text{ g Cl}} = 1.65 \text{ g-atoms Cl}$$

$$31.40 \text{ g F} \times \frac{1 \text{ gram-atom F}}{19.0 \text{ g F}} = 1.65 \text{ g-atoms F}$$

Step 3: Express the relative numbers of gram-atoms as whole numbers by dividing each by the smallest, 0.83.

$$C = \frac{0.83 \text{ g-atoms}}{0.83} 1.00 \text{ g-atom C} = 1$$

$$Cl = \frac{1.65 \text{ g-atoms}}{0.83} 1.99 \text{ g-atoms Cl} = 2$$

$$F = \frac{1.65 \text{ g-atoms}}{0.83} 1.99 \text{ g-atoms F} = 2$$

The empirical formula is CCl_2F_2.

EXAMPLE 14–9

The molecular weight of Freon is 121.0. What is its molecular formula?

SOLUTION

In Example 14–8, the empirical formula of Freon was found to be CCl_2F_2. Calculate the molecular weight corresponding to the empirical formula as follows:

$$1 \text{ C} = 1 \times 12.0 = 12.0$$
$$2 \text{ Cl} = 2 \times 35.5 = 71.0$$
$$2 \text{ F} = 2 \times 19.0 = \underline{38.0}$$
$$121.0$$

Therefore, the empirical and molecular formulas of Freon are the same: CCl_2F_2.

14–11 CHEMICAL EQUATIONS

A *chemical equation* is a symbolic expression of a chemical reaction that shows what has taken place qualitatively and quantitatively. It tells what substances react, what substances are produced, and in what relative amounts. The reactants and products are represented by symbols and formulas. The reactants are on the left and the products on the right; the two sides of the equation are separated by a single arrow, →, a double arrow, ⇄, or an equal sign, =. The physical states of substances taking part in the reaction may be designated by subscripts following the formula or symbol: (s), (l), and (g), representing solid, liquid, and gas. The escape of a gaseous product may be shown by an upward arrow (↑), and a solid product precipitating out of a solution during the reaction by an arrow pointing downward (↓) or by underscoring. Symbols commonly used in equations are summarized in Table 14–6.

In chemical reactions, bonds are made and broken between atoms. Atoms are not created or destroyed by the reaction process; they are rearranged. This is implied by the law of conservation of mass, first stated by Antoine Lavoisier (1743–1794). Each atom that enters into a chemical reaction must therefore be accounted for in the end.

An equation is balanced by equalizing the number of atoms of each element participating in the reaction with those appearing in the products. The atom inventory of the reactants and products must be identical when the process is completed. The required number of each kind of atom is obtained by writing numbers called coefficients in front of formulas or symbols. The subscripts of a formula are never changed, since this would change the nature of the substance. The *coefficient* is a multiplier for the formula; for example, 3

Table 14–6 SYMBOLS USED IN CHEMICAL EQUATIONS

SYMBOL	INTERPRETATION
+	Plus
→	Yields; produces
⇄	The reaction is reversible
=	Equilibrium between reactants and products
△	Heat
↑	Gaseous product
↓	Precipitate; solid product
(aq)	Aqueous solution
(s)	Solid
(l)	Liquid
(g)	Gas

$CaSO_4$ indicates 3 Ca atoms, 3 S atoms and 12 O atoms. The coefficients of an equation are reduced to their lowest possible integral values. Consider the following reaction:

Aluminum + Iron (III) oxide $\rightarrow$ Aluminum oxide + Iron

$$Al + Fe_2O_3 \rightarrow Al_2O_3 + Fe \text{ (not balanced)}$$

Our atom inventory shows the following:

Atom Inventory:	Reactants	Products
	1 Al	2 Al
	2 Fe	1 Fe
	3 O	3 O

The number of oxygen atoms is 3 on the left and 3 on the right; therefore, oxygen is balanced. There is one Al atom on the left, but two on the right; by writing the coefficient 2 before the symbol Al on the left, we balance aluminum. Finally, there are two Fe atoms on the left but only one on the right; we write the coefficient 2 in front of Fe, and this balances iron. The balanced equation is:

$$2Al + Fe_2O_3 \rightarrow Al_2O_3 + 2Fe \text{ (balanced)}$$

Atom Inventory:	Reactants	Products
	2 Al	2 Al
	2 Fe	2 Fe
	3 O	3 O

Most chemical reactions can be classified as belonging to one of four types:

1. *Direct Combination or Synthesis Reactions.* Atoms or molecules combine to form larger molecules.

$$2Al + N_2 \rightarrow 2AlN$$

$$SO_3 + H_2O \rightarrow H_2SO_4$$

2. *Simple Decomposition or Analysis Reactions.* A compound is broken down into simpler substances.

$$2HgO \rightarrow 2Hg + O_2 \uparrow$$

$$2KNO_3 \rightarrow 2KNO_2 + O_2 \uparrow$$

3. *Simple Replacement or Substitution Reactions.* An atom or radical replaces another atom or radical in a compound.

$$2Al + 6HCl \rightarrow 2AlCl_3 + 3H_2 \uparrow$$

$$Zn + 2AgNO_3 \rightarrow Zn(NO_3)_2 + 2Ag$$

4. *Double Replacement or Metathesis Reactions.* A double exchange is involved.

$$2NaCl + H_2SO_4 \rightarrow Na_2SO_4 + 2HCl$$

$$Fe_2(SO_4)_3 + 3Ca(OH)_2 \rightarrow 2Fe(OH)_3 + 3CaSO_4$$

CHECKLIST

Analysis	Mole
Avogadro's number	Molecular formula
Binary compound	Molecular weight
Empirical formula	Oxidation number
Formula	Percentage composition
Formula weight	Product
Gram-atomic weight	Radical
Gram-formula weight	Reactant
Gram-molecular weight	Simple replacement
Law of combining volumes	Stock system
Law of conservation of mass	Synthesis
Metathesis	Ternary compound

EXERCISES

14-1. Write formulas for the following compounds:
(a) Calcium phosphide
(b) Barium nitrate
(c) Magnesium bromide
(d) Tin (II) nitrite
(e) Ferric dichromate
(f) Cupric sulfate
(g) Lead (II) perchlorate
(h) Potassium phosphate
(i) Mercury (II) nitride
(j) Stannous chloride

14-2. Write formulas for the following compounds:
(a) Ferrous nitrate
(b) Calcium hypochlorite
(c) Zinc acetate
(d) Copper (I) oxide
(e) Lead (II) chromate
(f) Barium nitrite
(g) Cupric phosphate
(h) Tin (II) fluoride
(i) Potassium permanganate
(j) Sodium bicarbonate

14–3.   Name the following compounds:
(a) $CaCO_3$ (f) $Fe(C_2H_3O_2)_2$
(b) $BaSO_4$ (g) $Zn(NO_2)_2$
(c) Na_2SO_3 (h) $Fe(ClO_4)_3$
(d) $Sr_3(PO_4)_2$ (i) BF_3 ·
(e) Mg_3N_2 (j) AlP

14–4.   Name the following compounds:
(a) $FeSO_4$ (f) $Ca(C_2H_3O_2)_2$
(b) $Zn_3(PO_4)_2$ (g) NH_4Cl
(c) $Pb(CrO_4)_2$ (h) SnF_4
(d) $Sn(CO_3)_2$ (i) CS_2
(e) $Co(NO_3)_3$ (j) $Na_2C_2O_4$

14–5.   Determine the molecular weight of $C_7H_5N_3O_6$ (TNT).

14–6.   What is the weight of 1 atom of platinum?

14–7.   How many atoms are present in 1 kilogram of silver?

14–8.   What is the weight of 1 molecule of sucrose, $C_{12}H_{22}O_{11}$?

14–9.   An insecticide, parathion, has the formula $C_2H_{14}PNSO_5$. What is its percentage composition?

14–10.   What is the percentage composition of cholesterol, $C_{22}H_{45}OH$?

14–11.   Determine the percentage of iodine in thyroxine, $C_{14}H_{11}O_4NI_4$, a compound produced in the thyroid gland.

14–12.   Tetraethyl lead, $Pb(C_2H_5)_4$, is used as an additive in gasoline to increase the octane rating. Determine the percentage of lead present.

14–13.   Which of the following compounds contains the highest percentage of nitrogen?
(a) KNO_3 (c) $CO(NH_2)_2$
(b) NCl_3 (d) $Fe(CN)_2$

14–14.   The empirical formula of dextrose is CH_2O, and the molecular weight is 180. What is the molecular formula of dextrose?

14–15.   The empirical formula of butane is C_2H_5. If the molecular weight of butane is 58, what is its molecular formula?

14–16.   Ethyl alcohol shows the following analysis: 52.2 per cent C, 13.0 per cent H, 34.8 per cent O. What is its molecular formula if it has a molecular weight of 46?

14–17.   Urea shows the analysis 20.0 per cent C, 6.7 per cent H, 46.7 per cent N, 26.6 per cent O. What is its empirical formula?

14–18.   Sodium forms an ionic compound with sulfur. (a) Give the chemical formula of the compound. (b) Name the compound. (c) Write a balanced equation for the reaction.

14–19.   Aluminum cans react with oxygen to form aluminum oxide. Write the balanced equation for this reaction.

14–20.   The following reaction represents the combustion of a hydrocarbon (octane):

$$C_8H_{18} + O_2 \rightarrow CO_2 + H_2O$$

Balance the equation.

14–21.   Iron rust forms when iron reacts with oxygen to form iron (III) oxide. Write the balanced equation for this reaction.

14–22.   Excess carbon dioxide is removed from the atmosphere of a spacecraft by lithium hydroxide. The products are lithium carbonate and water. Write a balanced equation for this reaction.

14–23.   Aluminum sulfate reacts with calcium hydroxide to form a gelatinous compound, aluminum hydroxide, and calcium sulfate. Suspended solids in water supply systems are removed by the aluminum hydroxide and settle to the bottom. Write a balanced equation for the reaction.

14-24.   Balance the following equations:
(a) $P + O_2 \rightarrow P_4O_{10}$
(b) $Al + O_2 \rightarrow Al_2O_3$
(c) $SO_2 + O_2 \rightarrow SO_3$
(d) $NH_3 \rightarrow N_2 + H_2$
(e) $H_2O_2 \rightarrow H_2O + O_2$
(f) $NH_4NO_3 \rightarrow N_2O + H_2O$
(g) $Zn + H_3PO_4 \rightarrow Zn_3(PO_4)_2 + H_2$
(h) $Al + H_2SO_4 \rightarrow Al_2(SO_4)_3 + H_2$
(i) $NaNO_3 + H_2SO_4 \rightarrow$
(j) $FeCl_3 + H_3PO_4 \rightarrow$

14-25.   *Multiple Choice*

   A. When silicon reacts with chlorine, the compound
       is
       (a) $SiCl$ (c) $SiCl_3$
       (b) $SiCl_2$ (d) $SiCl_4$

   B. The formula for lead (II) sulfide is
       (a) PbS (c) Pb_2S
       (b) PbS_2 (d) Pb_4S

   C. The name of the compound $CuSO_4$ is
       (a) copper sulfate
       (b) cuprous sulfate
       (c) copper (I) sulfate
       (d) copper (II) sulfate

   D. A mole of sodium chloride contains
       (a) 6.02×10^{23} ions
       (b) 6.02×10^{23} molecules
       (c) 1.20×10^{24} ions
       (d) 1.20×10^{24} chloride ions

   E. The numbers needed to balance the equation
       $Al + F_2 \rightarrow AlF_3$ are
       (a) 1, 3, 2   (c) 4, 3, 2
       (b) 2, 1, 2   (d) 2, 3, 2

Brasted, Robert C., "Revised Inorganic (Stock) Nomenclature for the General Chemistry Student." *Journal of Chemical Education,* March, 1958.
    Chemists must be "bilingual" in chemical nomenclature. This article reviews the general principles of the Stock system.
Feifer, Nathan, "The Relationship Between Avogadro's Principle

SUGGESTIONS FOR
FURTHER READING

and the Law of Gay-Lussac." *Journal of Chemical Education,* August, 1966.

Relates Avogadro's hypothesis to concepts introduced by Lavoisier, Dalton, and Gay-Lussac.

Hawthorne, Robert M., Jr., "Avogadro's Number: Early Values by Loschmidt and Others." *Journal of Chemical Education,* November, 1970.

Discusses various methods used to arrive at a value for Avogadro's number.

Kolb, Doris, "The Mole." *Journal of Chemical Education,* November, 1978.

An excellent discussion of the mole concept.

Peters, Edward I., *Problem Solving for Chemistry.* Philadelphia: W. B. Saunders Company, 1971.

Chapter 3 in this self-tutorial text is concerned with problems involving chemical formulas.

CHEMICAL REACTION RATES AND EQUILIBRIUM

15–1 INTRODUCTION

Although a balanced chemical equation tells us what the reactants and products are, it tells us nothing about *how* the reaction occurs, that is, the mechanism of the reaction. *Chemical kinetics* is the name given to the study of the rate and mechanism of chemical reactions. By *rate* we mean the number of moles of a reactant converted to products per unit time. The rate of a reaction depends on a series of individual steps by which the reactants change to products. A single step may be involved or, more frequently, a sequence of steps. The single step or the sequence of steps is called the *mechanism* of the reaction. As more is learned of the mechanism of chemical reactions, there is a greater ability to control them. Examples are the design of improved internal combustion engines and solutions to the problems of air and water pollution.

15–2 RATE OF A REACTION

Chemical reaction rates range from microseconds to centuries. Some, such as a dynamite explosion and the mixing of silver nitrate and sodium chloride, occur almost instantaneously (Fig. 15–1). Others, such as the souring of milk or the rusting of iron, proceed slowly. Even similar reactions often proceed at different rates. Zinc releases hydrogen faster from hydrochloric acid

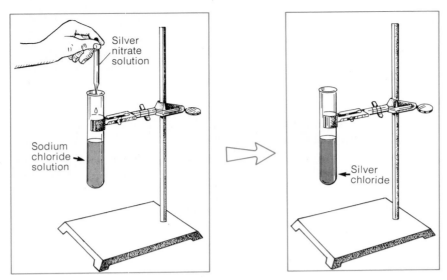

Figure 15–1 Silver chloride, AgCl, precipitates when silver nitrate, AgNO₃, reacts with sodium chloride, NaCl.

(HCl) than from acetic acid (HC₂H₃O₂). Magnesium reacts faster with dilute sulfuric acid (H₂SO₄) than with concentrated sulfuric acid.

The first studies of reaction rates dealt with systems that reacted in minutes or hours. The best method is to observe the change in some physical property of the system that varies with the concentration of reactants—color changes, pressure changes, variations in electrical conductivity, and so on (Fig. 15–2). The first

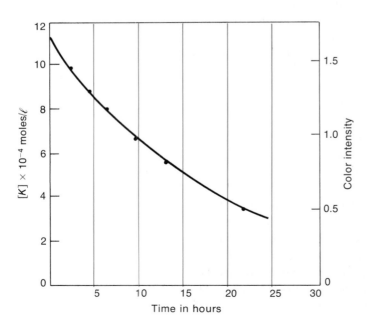

Figure 15–2 Change in color intensity of solution is proportional to change in connection of potassium in ammonia. A plot of concentration and color intensity against time yields the same graph. (From Brescia, F., et al., *Chemistry: A Modern Introduction,* 2nd ed. Philadelphia: W. B. Saunders Company, 1978, p. 507.)

published account of a rate measurement for a chemical reaction was by Ludwig Wilhelmy (1812–1864), a physicist at Heidelberg, who studied the rate at which sucrose (cane sugar) is converted in aqueous solution to glucose (dextrose) and fructose (fruit sugar) according to the equation:

$$C_{12}H_{22}O_{11} + H_2O \rightarrow C_6H_{12}O_6 + C_6H_{12}O_6$$
$$\text{sucrose} \qquad\qquad \text{glucose} \quad \text{fructose}$$

Equation 15–1

To follow the course of the reaction, Wilhelmy took advantage of differences in the optical properties of the substances. Using a polarimeter, he followed the change in rotation of polarized light from *dextro* to *levo* (right to left) as the sucrose was changed to invert sugar, a mixture of glucose and fructose. He observed that the rate was proportional to the concentration of sucrose, which deceases with time. The nature of the reacting substances, temperature, catalysts, and pressure are other factors that affect the rates of chemical reactions.

15–3 NATURE OF THE REACTANTS

Reaction rates depend first of all upon the nature of the reacting substances. A given chemical change, such as oxidation in a flame, takes place very slowly with copper or silver, while magnesium under the same conditions burns very rapidly. White phosphorus ignites spontaneously in air; red phosphorus does not. Some wood burns quickly and easily, but other wood is difficult to ignite and burns very slowly.

The state of subdivision of the reactants is another factor. For example, granulated sugar dissolves in water more rapidly than does lump sugar. Powdered zinc reacts more rapidly with an acid than do lumps of zinc. For solid substances, only that portion which is exposed can react, since reactions occur only at the surface boundary between the reacting substances. Doubling the surface area doubles the rate of reaction.

When a given mass is subdivided into smaller particles, the surface area is increased and the rate of reaction increases. A finely powdered solid presents vastly more surface than a few large chunks of the same solid (Fig. 15–3). For this reason, powders react much more rapidly than larger aggregates. For example, piles of

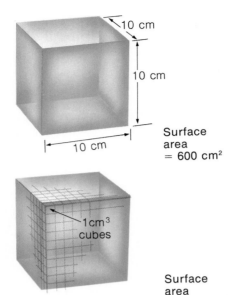

Surface
area
= 600 cm²

Surface
area
= 6000 cm²

Figure 15-3   Particle size and surface area. Dividing a solid into smaller particles increases the surface area. For a cube 10 cm on each side, the surface area is 600 cm². Dividing it into smaller cubes measuring 1 cm on each side yields 1000 cubes, each having a surface area of 6 cm², for a total of 6000 cm².

wheat flour or coal dust burn slowly after being ignited. However, if either flour or coal dust is dispersed in air, a mere spark can cause a disastrous explosion. In the dust cloud, the surface exposed to oxygen is much larger than in the piles, and the combustion reaction occurs much more rapidly. In this reaction, the sudden production of expanding hot gases (CO_2, H_2O) quickly increases the pressure as much as 40 atmospheres/second and produces a violent explosion. Some coal mine disasters and the explosive destruction of grain elevators are a result of such rapid reactions.

15-4   EFFECT OF TEMPERATURE

Temperature has a striking effect on the rate of chemical reactions. Reaction rates that are negligible at ordinary temperatures may become appreciable and even explosive at high temperatures. For example, a mixture of hydrogen and oxygen at room temperature may remain mixed for years without reacting much. At 400°C there is some reaction, at 600°C the reaction is fast, and at 700°C the mixture explodes. This is typical of many reactions: at low temperatures the chemical reaction is so slow that for all practical purposes it does not occur; in a range of intermediate temperatures the reaction is moderately rapid; and at high temperatures it becomes instantaneous.

A rough but useful approximation is that the rate of many chemical reactions is doubled for an increase of 10 Celsius degrees in temperature. Not all reaction rates behave in the same way with regard to temperature change, however. Most rates increase, but some actually decrease. In the latter class, for example, are enzyme reaction rates, which fall off at high temperatures as the enzyme decomposes. The reactions in which the enzyme is involved then proceed at lower rates. In some cases, a fall-off in rate occurs because of a "back reaction," the rate of which increases faster than the rate of the "forward reaction" as the temperature rises. Therefore, for the same temperature change there may be different changes in rate for different reactions.

15–5 COLLISION THEORY

The effect of temperature on reaction rates is explained by molecular motions and *collision theory.* The molecules have velocities of translation in the range of 4000 to 40 000 cm/sec, and as they move they collide with other molecules. The collisions last about 10^{-11} seconds at room temperature.

The number of colliding molecules calculated from the kinetic theory of gases is enormous, on the order of 10^{32} molecules per liter per second at standard conditions. Each molecule experiences, on the average, 4 billion collisions each second. If collisions were the only requirement for reaction, all gaseous reactions would proceed at practically the same explosive rate. But experiments with gases at the same concentration at the same temperature indicate that they react at very different rates. Gaseous HI decomposes at the rate of 4.4×10^{-3} mole/liter/hour at 300°C, while gaseous N_2O_5 decomposes at the rate of 9.4×10^5 moles/liter/hour at this temperature. Therefore, collisions between molecules cannot be the only factor involved in determining the rate of a reaction.

A chemical reaction would be over within a small fraction of a second if all the collisions in which molecules are involved were effective. Instead, a reaction may require hours at this temperature, or it may not proceed discernibly at all. As the temperature increases, the kinetic energy of translation of the molecules also increases. More energy is transferred between molecules as they collide, causing them to vibrate more vigorously. Some chemical bonds that hold the atoms together in the original molecules break. At still higher temperatures, electrons are excited and molecules become ionized. However, an increase of 10°C at room temperature produces only a small increase—about 3 per cent—in the average kinetic energy of the molecules. Although the collision rate is increased only slightly, the reaction rate of many reactions is increased about 100 per cent.

To account for these facts, the chemist Svante Arrhenius (1859–1927) proposed a theory that says that as molecules encounter each other, the collisions between them will result in a chemical reaction only if the energy of the molecules exceeds a certain minimum threshold. This quantity is known as the *energy of activation*, E_A (Fig. 15–4). Every chemical reaction has a characteristic energy of activation. If the molecules collide with less energy than this critical amount, they recoil without undergoing chemical change. When the energy of activation is low (only a few kilocalories per mole), reactions proceed quickly and smoothly at room temperature. Reactions with activation energies in the range of 20 to 50 kcal/mole proceed at convenient rates for experimental study at some temperature in the range of 25°C to 500°C. If the

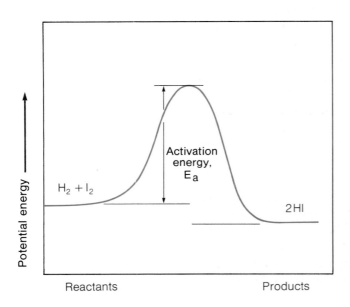

Figure 15-4 Activation energy profile. Reactants H_2 and I_2 must overcome the activation energy barrier, E_a, before they can react to form the product, HI.

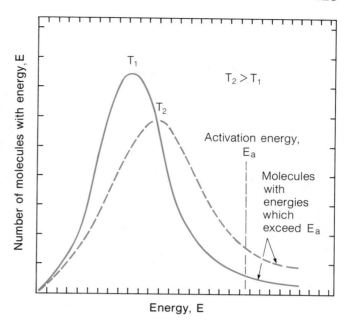

Figure 15–5 Energy distribution for gaseous molecules at two different temperatures. A small increase in temperature from T_1 to T_2 produces a large increase in the number of molecules with energies greater than the activation energy, E_a. The reaction, therefore, proceeds at a much faster rate.

energy of activation is high, the reaction might proceed at an infinitesimal rate, if at all, at room temperature.

According to Arrhenius' theory, only a small fraction of the molecules may possess enough energy of activation at a given temperature to react when they collide. Remember that although the average kinetic energy of the molecules is what determines the temperature, some molecules will be moving with much greater speeds than the average and others with much lower speeds than the average. Although reaction rates are frequently doubled by increasing the temperature 10 Celsius degrees, the number of collisions does not double, and in fact there are already many more collisions than reactions. An increase in temperature of this amount doubles the number of collisions that are effective in leading to chemical reaction, by endowing approximately double the number of molecules with the required energy of activation (Fig. 15–5).

In many reactions, however, even collisions between molecules possessing the required energy of activation do not all lead to reaction. The manner in which the molecules collide is also important (Fig. 15–6). Some molecules must be oriented in a very specific way for reaction to occur, while other molecules may react when colliding in any of a number of random orientations.

This idea led to a modification of classical collision

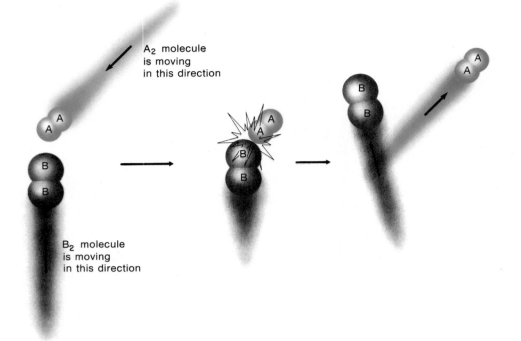

A₂ molecule
is moving
in this direction

B₂ molecule
is moving
in this direction

Figure 15–6   Collision between molecules results in no reaction if the orientation is poor. (From Brescia, F., et al., *Chemistry: A Modern Introduction,* 2nd ed. Philadelphia: W. B. Saunders Company, 1978, p. 512.)

theory by Henry Eyring (b. 1901) and others. In the *transition state theory,* as it is known, it is assumed that colliding molecules must first achieve a specific configuration with a definite energy before they can form the products of the reaction (Fig. 15–7). This "transition state" persists for only an instant (10^{-13} sec) and cannot be isolated. Once it is formed, it either returns to the original reactants or it proceeds to form products. If the two molecules approach each other with too little energy to reach the transition state, they do not react. Increasing the temperature of a reaction, then, has the effect of increasing the number of "activated complexes," a term that describes the species present in the transition state.

15–6   EFFECT OF CONCENTRATION ON REACTION RATE

Substances that burn in air burn much more quickly in pure oxygen, which is the actual burning agent. Since oxygen constitutes only one fifth of the air, the pure gas has a greater concentration of oxygen—five times as many oxygen molecules per cubic centimeter as air has (Fig. 15–8).

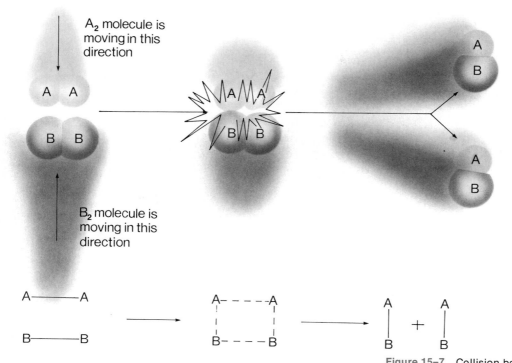

Figure 15–7 Collision between A_2 and B_2 molecules resulting in a reaction. Molecules possess sufficient energy and proper orientation. (From Brescia, F., et al., *Chemistry: A Modern Introduction,* 2nd ed. Philadelphia: W. B. Saunders Company, 1978, p. 511.)

The chemists Cato Maximilian Guldberg (1836–1902) and Peter Waage (1833–1900) were the first to see that the rate of a chemical reaction is usually proportional to the concentrations of the reactants expressed in moles per liter. Their generalization is known as the *law of mass action.* For the general overall reaction

$$A + B \rightarrow C$$

the rate of formation of C, or the rate of disappearance of A or B, is proportional to the concentration of A multiplied by the concentration of B:

$$\text{rate} \propto [A][B]$$

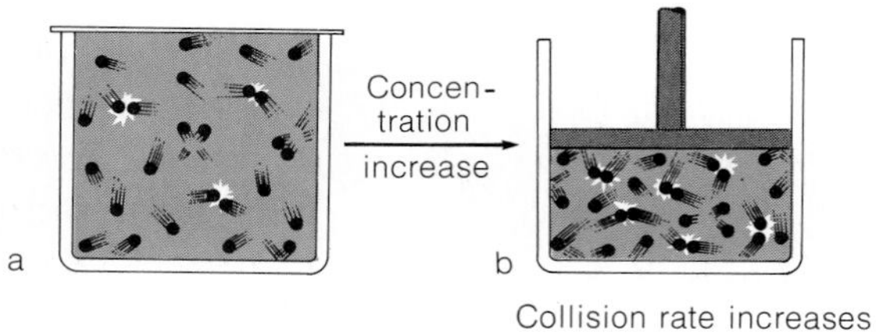

Figure 15–8 Effect of concentration on the rate of a chemical reaction. At the higher concentration, more collisions occur.

Concentration increase

Collision rate increases

or, the rate is equal to a constant multiplied by these concentrations:

Equation   15–2

$$rate = k[A][B]$$

in which k is a constant determined by experiment, and the brackets [   ] denote the concentration of A or B in moles per liter. For gaseous substances, concentration is directly proportional to partial pressures, and partial pressures of gases may be substituted for concentrations, as in

$$rate = kp_A p_B$$

15–7   EFFECT OF A CATALYST ON THE RATE OF A REACTION

The presence of small amounts of a foreign substance often changes the rate of a chemical reaction. The foreign substance is not changed by the reaction and may be recovered intact when the reaction is ended. Thus, hydrogen and oxygen combine rapidly to form water in the presence of finely divided platinum at low temperatures compared to those temperatures in a flame used to ignite a similar mixture, yet the platinum does not seem to change in the process. Michael Faraday studied this effect in great detail. He explained that the metal caused the gases (hydrogen and oxygen) to condense or become adsorbed, thereby bringing them into a condition such that they could readily react.

Berzelius reviewed all the examples of such effects that he could find in the literature and proposed a unifying theory. He suggested that some new force was acting in all these cases. Although he was not sure what this force was, he suggested that it be called "catalytic force," and the process "catalysis," from the Greek word for decomposition. A *catalyst* increases the rate of a chemical reaction by providing a path for the reaction that involves a lower activation energy than the one previously available (Fig. 15–9). More molecules, therefore, have the necessary energy for the new path, and the reaction speeds up. A negative catalyst, or *inhibitor*, slows down a reaction by blocking a reaction path, thus requiring a higher activation energy. The

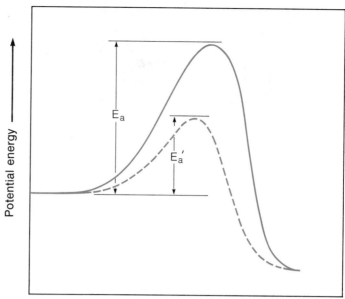

Figure 15-9 Effect of a positive catalyst on reaction rate. By lowering the required activation energy to E'_a, a positive catalyst speeds up a reaction. More molecules then have the required energy for reaction.

Figure 15-10 A surface catalyst. Gas molecules are adsorbed on the active sites, chemical bonds are weakened, and dissociation occurs. (From Masterton, W. L., and Slowinski, E. J., Chemical Principles, 4th ed., Philadelphia: W. B. Saunders Company, 1977, p. 399.)

RATES OF REACTION

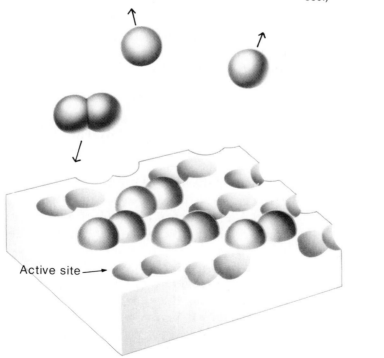

Active site →

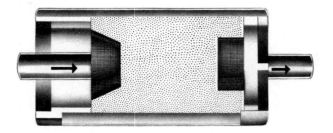

Figure 15–11   Catalytic exhaust converter.

use of catalysts has become important in industry. Many industrial processes, such as the manufacture of plastics, lubricating oils, motor fuel, fibers, and detergents, would be impossible without them. Industrial catalysts are, therefore, closely guarded secrets.

Solid catalysts used with reactions of gases are called contact catalysts, or heterogeneous catalysts, since they exist in a different phase from the gaseous reactants. They provide a large surface area at which reaction occurs much more rapidly than it does in the gas phase (Fig. 15–10). The catalytic muffler, an antipollution device installed in automobiles, is an example (Fig. 15–11). Its function is to accelerate the reaction between oxygen and carbon monoxide, and between oxygen and unburned hydrocarbons, escaping in the exhaust. The solid catalyst adsorbs molecules onto its surface, disrupts their bond structures to a certain extent, and brings together these activated molecules for reaction on the surface. The products of the reaction then "desorb" from the surface, and the process is repeated.

Equation 15–3
$$2CO + O_2 \xrightarrow{\text{catalyst}} 2CO_2$$

Equation 15–4
$$2C_8H_{18} + 25O_2 \xrightarrow{\text{catalyst}} 16CO_2 + 18H_2O$$

Finely divided platinum and nickel are among the many catalysts used in the "catalytic hydrogenation" of vegetable oils (corn oil, peanut oil, cottonseed oil) that converts them into the solid fat called "oleomargarine." In some cases, the homogeneous catalyst (a gas if the reactants are gaseous) forms an unstable intermediate compound with one of the reacting substances, which decomposes as the reaction proceeds.

Life as we know it would be impossible without the biological catalysts called *enzymes*. These are high–molecular-weight protein molecules that contain active sites within their structure that are endowed with highly specific catalytic activity toward some chemical reaction. They are more effective as catalysts than the best catalysts produced by chemists. Thus, starch reacts with water to form sugars, a reaction that takes days. However, a trace of the enzyme ptyalin in saliva catalyzes the conversion of starch to sugar in a fraction of this time.

The effectiveness of enzymes can be destroyed in various ways, including heating and poisoning. Heating at 80°C disrupts the protein structure and therefore the activity of enzymes. Nerve gases and insecticides can poison the active sites of enzymes and block nerve impulses, which may lead to unconsciousness and death. Many drugs used medically are believed to function by blocking or modifying the activity of enzymes in the body. Physiological poisons like mercuric chloride or rattlesnake venom react with enzymes, making them useless for essential biochemical reactions. They are examples of negative catalysts, or inhibitors.

Enzymes are highly specific in their action. There may be as many as 30 000 different enzymes in the human body. Each enzyme is capable of catalyzing only a particular reaction or a particular substance, called the *substrate*. This implies that spatial effects between the enzymes and substrate are important. When an enzyme serves as a catalyst in the decomposition of a substrate, the enzyme shape exactly accommodates the substrate molecule (Fig. 15–12). A complex is formed in which chemical bonds within the substrate are weakened and decomposition occurs more rapidly than it would without the enzyme.

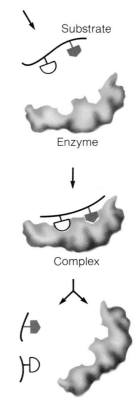

Figure 15–12 Similar enzyme and substrate shape may explain enzyme activity.

15–8 REACTION MECHANISM

The process by which a reaction occurs is called the *reaction mechanism.* Knowing the mechanism of a reaction, the chemist can select conditions that will produce a good yield of the product. The study of reaction mechanisms, however, is a tremendously complex subject. Compared to the enormous number of chemical reactions that are known, only a very few reaction mechanisms have ever been identified. For this reason, the study of mechanisms is a very active current research area.

Many reactions proceed as a *chain mechanism.* This concept appears to have originated in connection with the reaction between hydrogen and chlorine, a system that was investigated by early chemists such as John Dalton (1766–1844). This reaction, which appears to be simple enough to occur in one step, may actually take place in a series of steps.

Equation   15–5

$$H_2 + Cl_2 \longrightarrow 2HCl$$

At ordinary temperatures, hydrogen and chlorine form hydrogen chloride very slowly in the dark. But if the mixture of gases is exposed to bright light, the reaction proceeds explosively. This is an example of *photochemistry*—light inducing a chemical reaction.

Progress was made in understanding the hydrogen-chlorine reaction when Einstein introduced the quantum theory into photochemistry. According to Einstein's law of photochemical equivalence, one light quantum (or photon), $h\nu$, is absorbed for every molecule that is transformed. When Max Bodenstein (1871–1942), a physical chemist, applied this principle to experimental results, he discovered that one quantum brought about the union of 10^4 or 10^5 molecules of the reacting gases. These findings were confirmed by other experimenters and necessitated a revision of theory.

Bodenstein suggested a *chain reaction mechanism* to explain the high quantum efficiency of the reaction between hydrogen and chlorine. Walther Nernst (1864–1941) refined this idea and suggested that the light served to dissociate some of the chlorine molecules into chlorine atoms; the chlorine atoms then reacted with hydrogen molecules, forming hydrogen chloride and free hydrogen atoms. The chain finally ends with the direct union of the free chlorine and hydrogen atoms. Subsequent work has shown that this view is fundamentally correct as shown:

$$Cl_2 + h\nu \longrightarrow Cl_2{}^*$$
$$Cl_2{}^* \longrightarrow 2Cl$$

Equation   15–6

$$Cl + H_2 \longrightarrow HCl + H$$
$$H + Cl_2 \longrightarrow HCl + Cl$$
$$H + Cl \longrightarrow HCl$$

(*excited molecule)

Each photon (of wavelength about 4800 Å) absorbed by a chlorine molecule results in the formation of 10^5 hydrogen chloride molecules via a chain reaction. Although fast, chain reactions are not necessarily explosive. They become explosive when the heat evolved is great, thus increasing the rate of the reaction and the further liberation of heat until the reaction becomes explosive.

15–9 CHEMICAL EQUILIBRIUM

Some chemical reactions go "to completion" (that is, the reactants are entirely converted to products) and others do not. In the latter case, the products apparently react with one another to re-form the reactants. *Reversible reactions,* as these are called, were first carefully studied by William Williamson (1824–1904) at University College, London.

Williamson found situations in which, beginning with a mixture of A and B, the substances C and D were formed. If he began instead with a mixture of C and D, substances A and B were formed. In either case, there would be a mixture of A, B, C, and D in the end, with the proportions apparently fixed. The mixture would be at an equilibrium. Williamson felt that reaction did not cease at equilibrium, but that A and B were reacting to form C and D, and C and D, in turn, were reacting to form A and B at the same rate. This condition is called *dynamic equilibrium,* and is represented as follows, with a double arrow to indicate reversibility:

$$A + B \rightleftarrows C + D$$

Equation 15–7

The idea of dynamic equilibrium did not become immediately popular with chemists. Its significance was realized by Guldberg and Waage, who studied equilibrium reactions with systems containing solids in contact with solutions. They showed that an equilibrium is reached in incomplete reactions, and treated such reactions mathematically, expressing equilibrium conditions in terms of molecular concentration.

Let us assume that A, B, C, and D represent gases, and apply the law of mass action to the reversible reaction:

$$A + B \rightleftarrows C + D$$

If the system contains only A and B molecules at the outset, then the rate of the forward reaction is expressed in terms of the concentrations of A and B as follows:

Equation 15–8

$$R_f = k_f[A][B]$$

However, as molecules of the products C and D are formed, they begin to react, and the rate of this "back" reaction, R_b, is given by the expression:

Equation 15–9

$$R_b = k_b[C][D]$$

As the reaction between A and B proceeds, the concentrations of C and D increase, and the rate of the back reaction, R_b, increases, since it depends upon the concentrations of C and D. Meanwhile, the concentrations of A and B are becoming less and less, so that the rate of the "forward" reaction, R_f, decreases. The two reaction rates, forward and back, approach each other and finally become equal (Fig. 15–13). A condition of dynamic equilibrium is established in which the opposing reactions proceed at equal rates at a certain temperature. At equilibrium, with the rate of the forward reaction equal to the rate of the back reaction,

$$R_f = R_b$$

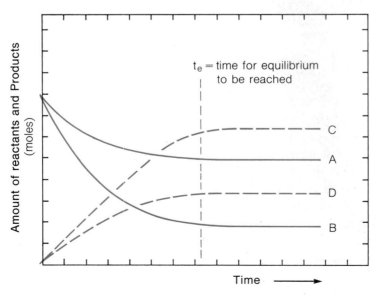

t_e = time for equilibrium to be reached

Figure 15–13 A reversible reaction: A + B ⇌ C + D. Neither of the initial reactants (A and B) is completely consumed. At time t_e, the forward and reverse reactions proceed with equal rates.

and we may write

$$k_f[A][B] = k_b[C][D]$$

or, by rearranging terms,

$$\frac{k_f}{k_b} = \frac{[C][D]}{[A][B]}$$

Because k_f and k_b are constants, k_f/k_b is also constant, and the expression may be written as

$$K = \frac{[C][D]}{[A][B]} \qquad \text{Equation } 15\text{–}10$$

K is called the equilibrium constant for the reaction and has a specific numerical value for each reaction at a specific temperature. For the general reaction,

$$aA + bB \rightleftarrows cC + dD \qquad \text{Equation } 15\text{–}11$$

the expression for the law of chemical equilibrium becomes

$$K = \frac{[C]^c[D]^d}{[A]^a[B]^b} \qquad \text{Equation } 15\text{–}12$$

The equilibrium expressions for several typical reactions are shown below:

$$H_2 + I_2 \rightleftarrows 2HI \qquad K = \frac{[HI]^2}{[H_2][I_2]} = 50.2 \ (448°C)$$

$$N_2 + 3H_2 \rightleftarrows 2NH_3 \qquad K = \frac{[NH_3]^2}{[N_2][H_2]^3} = 9.0 \ (350°C)$$

$$SO_2 + O_2 \rightleftarrows 2SO_3 \qquad K = \frac{[SO_3]^2}{[SO_2][O_2]} = 0.99 \ (1177°C)$$

A reaction that "goes to completion," then, is a reaction that has a very large value of K. The concentration of the reactants is small, having been largely converted to products. In a reaction that has a very small value of K at equilibrium, the concentration of the products is very small. Isolated chemical systems eventually come to equilibrium, in which the forward and reverse reactions occur at the same rate and neither goes to completion.

When chemical reactions reach a state of equilibrium instead of going to completion, equilibrium may be established when a reaction is very nearly complete or at some other point. Chemical manufacturers are in business to produce high yields of products. High yields depend upon reactions that go far toward completion. If a system reaches equilibrium before much of the compound has been formed, it may be possible to modify the equilibrium conditions so that the yield of the product will be increased.

15-10   LE CHÂTELIER'S PRINCIPLE

Le Châtelier's principle, formulated by Henri Le Châtelier (1850–1936), a chemist at a school of mines, is the most general principle that applies to systems at equilibrium. It states: *When a stress is imposed on a system at equilibrium, the equilibrium shifts in such a way as to minimize the effect of the stress.* In chemical systems, stress may be placed upon the equilibrium by altering concentrations of substances, changing conditions such as temperature and pressure, and adding a catalyst.

A. *Effect of Concentration Upon Equilibrium*
In the reaction

Equation 15-13 $$H_2 + I_2 \rightleftarrows 2HI \qquad\qquad K = \frac{[HI]^2}{[H_2][I_2]}$$

the addition of hydrogen to the equilibrium system is considered to constitute a stress. The stress is relieved by the speeding up of the forward reaction, which consumes the added hydrogen through reaction with iodine and results in the production of a greater amount of hydrogen iodide, HI. The concentration of HI increases until the rate of the backward reaction catches up with that of the forward reaction and equilibrium is restored. When equilibrium is again achieved, there is less I_2 and more HI than initially, and some of the H_2 that had been added and constituted a "stress" will have been used up. The position of equilibrium is said to have shifted to the right.

Table 15–1   CONCENTRATIONS OF H_2, I_2, AND HI INITIALLY AND AT
EQUILIBRIUM (440°C)

EQUI-LIBRIUM	INITIAL			EQUILIBRIUM			
	$[H_2]$	$[I_2]$	$[HI]$	$[H_2]$	$[I_2]$	$[HI]$	$\dfrac{[HI]^2}{[H_2][I_2]}$
1	1.000	1.000	0	0.218	0.218	1.564	51.5
2	0	0	2.000	0.218	0.218	1.564	51.5
3	0	0	1.000	0.109	0.109	0.782	51.5
4	1.000	0.500	0	0.532	0.032	0.936	51.5

Increasing the concentration of one substance in an equilibrium mixture causes the reaction to take place in the direction that consumes some of the material added. On the other hand, decreasing the concentration of a substance causes the production of more of it by the appropriate reaction. Thus, removing a part of the HI from the system would cause equilibrium again to shift to the right, forming more HI from the reaction between H_2 and I_2. Although the concentrations of H_2, I_2, and HI are changed in both cases, the proportions are such that the law of chemical equilibrium applies as before (Table 15–1). The original value of the equilibrium constant, K, is restored for that temperature. Unless the original value of K is restored, the system is not in equilibrium, and the concentrations will undergo further change until it is. Then the rates of the forward and reverse reactions become equal again.

B. Effect of Temperature Upon Equilibrium

Chemical changes involve the evolution or the absorption of heat. In a system at equilibrium, an endothermic and an exothermic reaction are taking place simultaneously, as in:

$$H_2 + I_2 \rightleftarrows 2HI + heat \qquad\qquad Equation \quad 15\text{--}14$$

In this case, the forward reaction results in the evolution of heat and it is said to be *exothermic*. The reverse reaction absorbs heat; it is *endothermic*.

If we add heat to this system and increase the temperature, Le Châtelier's principle predicts that the equilibrium will shift in such a way as to absorb the added heat. An increase in the temperature speeds up both the decomposition of and the formation of HI, but

speeds up the decomposition of HI more than the formation. Thus, when the two rates become equal at the
higher temperature, less HI is present than at the lower
temperature, since more of it has decomposed. In such
cases, a change in temperature results in a change,
usually large, in the value of the equilibrium constant.
The value of K is constant as long as the temperature
does not change, but it is different for different temperatures. Thus, for the reaction

$$H_2 + I_2 \rightleftarrows 2HI$$

the value for K is 54.8 at 425°C, but it is 871 at 25°C.
Therefore, the temperature at which a given value of K
is valid is reported with it.

C. Effect of Pressure Upon Equilibrium

In a system at equilibrium that involves a change
in the total number of gaseous molecules between the
reactants and the products, as in

Equation 15–15

$$2SO_2 \quad + \quad O_2 \quad \rightleftarrows \quad 2SO_3$$
$$\underbrace{\text{2 volumes} \quad \text{1 volume}}_{\text{3 volumes}} \quad \text{2 volumes}$$

a change in pressure causes a shift in the position of
equilibrium. In this system there are three volumes of
gaseous reactants and two volumes of product. If the
pressure is increased, then according to Le Châtelier's
principle the system will adjust in such a way as to absorb this stress. A gaseous system may absorb the stress
of increased pressure by reacting so as to decrease the
number of molecules, since the pressure of a gas depends upon the number of gaseous molecules present.
At high pressures, the equilibrium in this system is
shifted to the right, since the reduction from three to
two volumes of gas lowers the pressure and thereby
absorbs the stress. In a system in which there is no
change in the number of molecules, as in

Equation 15–16

$$2IBr \quad \rightleftarrows \quad I_2 \quad + \quad Br_2$$
$$\text{2 volumes} \quad \underbrace{\text{1 volume} \quad \text{1 volume}}_{\text{2 volumes}}$$

the equilibrium is not affected by a change in pressure.

D. Effect of a Catalyst Upon Equilibrium

The effect of a catalyst upon equilibrium is to
change the rate at which equilibrium is reached, with-

out affecting the value of the equilibrium constant. A catalyst is equally effective in increasing the rates of both the forward and reverse directions of the reaction. Thus, a catalyst that promotes the synthesis of HI from its elements is equally effective in increasing the rate of the reverse reaction, the decomposition of HI. The catalyst provides a reaction path with a lower energy barrier between the initial and final states. Its presence in a chemical system serves to accelerate the approach to equilibrium from either direction, thereby reducing the time required for the system to reach equilibrium. But the catalyst has no effect on the *composition* of the equilibrium mixture; equilibrium is merely reached more quickly.

15–11 THE HABER PROCESS: A CASE STUDY

The Haber process for the synthesis of ammonia, named after Fritz Haber (1868–1934), director of the Kaiser Wilhelm Institute for Physical Chemistry, is an outstanding example of the practical application of knowledge about the effects of temperature, pressure, concentration, and catalysis on a system in dynamic equilibrium. This process also illustrates the dual character of a major chemical discovery that may be applied to either constructive or destructive social purposes.

Nitrogen constitutes about 80 per cent of the atmosphere, being present there in its elementary form, N_2. Although it is abundant and essential for life, nitrogen is chemically inert, reacting to form compounds only with difficulty. To most organisms, however, it is useful only in compound form. Nitrogen occurs as nitrates in soil, but it is generally in short supply and must be supplied in the form of animal wastes or chemical fertilizers.

The earth's supply of nitrates is replenished partly through the activity of thunderstorms. The nitrogen and oxygen of the air in the vicinity of lightning bolts combine to form compounds that dissolve in raindrops and are brought to earth. Atmospheric nitrogen can also be "fixed" by certain soil bacteria in nitrogen compounds. The best natural source of nitrates is Chilean saltpeter, $NaNO_3$, from the Chilean desert, which is inadequate, however, to maintain the food supply of a rapidly growing world population.

At the same time, nitrogen compounds are used in very large quantities for the production of explosives:

nitroglycerine for dynamite, nitrocellulose ("guncotton"), trinitrotoluene (TNT), and trinitrophenol (picric acid).

Late in the 19th century, Sir William Crookes (1832–1919) expressed concern that a food shortage was imminent, and urged intensive cultivation to increase the yield of wheat per acre. In the absence of abundant, natural supplies of fixed nitrogen, he warned of possible depletion unless a commercial process for the fixation of atmospheric nitrogen could be developed. Otherwise, starvation appeared to be a real possibility.

Attempts were made to combine nitrogen and oxygen directly into nitric oxide, NO, as occurs in thunderstorms. But high temperatures requiring an adequate power source are required to bring about even a moderately favorable equilibrium concentration of nitric oxide. In Norway, Kristian Birkeland (1867–1917), a physicist, and Samuel Eyde (1866–1940), an engineer, did succeed in bringing such a process into commercial operation, because of the availability of cheap electricity there. The compound was eventually synthesized in calcium nitrate and sold for fertilizer as "Norwegian saltpeter." Since so much electricity was required in the process, however, it did not spread to other parts of the world.

The direct combination of nitrogen and hydrogen had been studied as a possibility, then dropped. Fritz Haber took up the problem when he began his studies of the synthesis of ammonia in 1904. Haber's theoretical work had proved to him that the synthesis was possible, but never before had anyone observed the reaction:

Equation   15–17

$$N_2 + 3H_2 \rightleftarrows 2NH_3$$

By 1908, Haber had overcome many difficulties and took out the key patent for the process that bears his name (Fig. 15–14).

The reaction of the Haber process for the synthesis of ammonia from atmospheric nitrogen is moderately exothermic:

Equation   15–18

$$N_2 + 3H_2 \rightleftarrows 2NH_3 + 22\,100 \text{ cal}$$

It would appear that ammonia could be produced in unlimited quantity at room temperature simply by

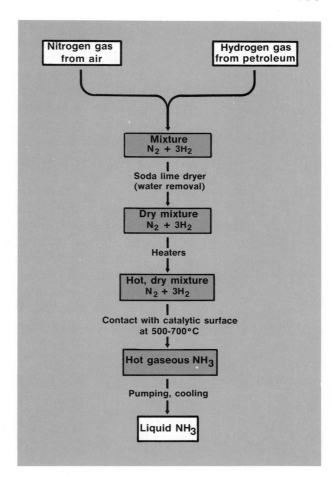

Figure 15-14   The Haber process for synthesizing ammonia. (From Jones, M. M., et al., *Chemistry, Man and Society,* 2nd ed. Philadelphia: W. B. Saunders Company, 1976, p. 269.)

bringing together nitrogen and hydrogen. But the reaction rate at room temperature is so slow that it is not of practical value; a mixture of nitrogen and hydrogen can be kept unchanged for years. If the reaction is carried out at high temperature, then according to Le Châtelier's principle the endothermic part of the reaction would be favored and an equilibrium with lower values for the equilibrium constant would be established in favor of greater dissociation of NH_3. At 1000°C, the reaction proceeds almost exclusively in the opposite, endothermic direction, favoring the decomposition of ammonia into nitrogen and hydrogen, and ammonia practically ceases to exist as a chemical compound.

Haber found that at 500°C, with an iron oxide catalyst, ammonia could be produced fast enough for practical purposes, even though at equilibrium only 8 per cent of the material was converted to ammonia. At higher temperatures, the reaction was faster, but the equilibrium too unfavorable. At lower temperatures the reaction was too slow. Thousands of experiments were necessary to discover a suitable catalyst and appropriate conditions for a good yield of ammonia. In practice, pressures of about 200 to 600 atmospheres are necessary at a temperature of about 500°C to achieve reasonable reaction rates and yields reaching 90 per cent.

At equilibrium, the mixture contains ammonia, nitrogen, and hydrogen. To separate the ammonia, the gas mixture is cooled, and liquid ammonia is produced and drawn off continuously. The residual mixture of nitrogen, hydrogen, and ammonia is again compressed, heated, and passed over a catalyst for recycling. Nitrogen and hydrogen are added as they are depleted.

Much of this synthetic ammonia was oxidized to nitric acid in a process developed by Wilhelm Ostwald (1853–1932):

Equation 15–19

$$4NH_3 + 5O_2 \xrightarrow{\text{Pt}} 4NO + 6H_2O$$

$$2NO + O_2 \longrightarrow 2NO_2$$

$$3NO_2 + H_2O \longrightarrow 2HNO_3 + NO$$

and the nitric acid went into the making of explosives. The rest was used for fertilizer.

By 1918, the production of synthetic ammonia reached 200 000 tons per year. Had it not been for the Haber process, Germany would probably not have been capable of continuing the war for more than six months. A blockade of Germany by the British fleet cut it off from its supply of Chilean saltpeter, but the Haber process—coming into commercial production just as the war was breaking out (1914)—saved Germany from premature defeat. It also saved millions of Germans from starvation during World War I. It is estimated that 100 million tons of fixed nitrogen produced by this process are consumed by the world each year.

The worldwide development of the Haber process had an adverse effect upon the economy of Chile, which for 75 years had been the world's principal source of fixed nitrogen. The specter of starvation due to a shortage of fixed nitrogen was alleviated by the develpment of the Haber process. Interest in developing new methods for an ammonia synthesis continues, nevertheless.

Much of the ammonia produced is shipped in liquid form in steel cylinders or tank cars and used directly as fertilizer. Some is converted to nitric acid by the Ostwald process, and part of that is reacted with ammonia to form ammonium nitrate (NH_4NO_3), which is used as a fertilizer or, when mixed with other materials, as an explosive. Explosives, of course, have peaceful civilian uses in mining and road construction, as well as military uses. The case history of the synthesis of ammonia illustrates how chemical science and technology can serve society for both military and peaceful purposes. The Haber process is an example of the neutrality of science, the fruits of which may be applied to forge weapons of war or instruments of peace. Which it shall be is a decision that society rather than science must make.

Catalyst	Heterogeneous catalyst	**CHECKLIST**
Catalytic hydrogenation	Inhibitor	
Chain reaction	Kinetics	
Collision theory	Law of mass action	
Dynamic equilibrium	Le Châtelier's principle	
Endothermic	Photochemistry	
Energy of activation	Reaction mechanism	
Enzyme	Reversibility	
Equilibrium constant	Transition state	
Exothermic	Transition state theory	

15–1. Write expressions for the equilibrium constants of **EXERCISES** the following reactions:

(a) $N_2 + O_2 \rightleftarrows 2NO$

(b) $CO_2 + H_2 \rightleftarrows CO + H_2O$

(c) $4NH_3 + 5O_2 \rightleftarrows 4NO + 6H_2O$

15–2. Apply Le Châtelier's principle to predict the effect of an increase in temperature on the degree of dissociation of ammonia gas to nitrogen and hydrogen if dissociation is endothermic:

$$2NH_3 + heat \rightleftarrows N_2 + 3H_2$$

15–3.　Explain why collisions between gaseous molecules, although great in frequency, rarely result in reaction at room temperature.

15–4.　For which of the following systems does the reaction go least to completion in the forward direction? (a) $K = 10^{-10}$; (b) $K = 1$; (c) $K = 10^4$.

15–5.　Discuss the principle of dynamic equilibrium as it applies to the system

$$3H_2 + SO_2 \rightleftarrows H_2S + 2H_2O$$

15–6.　For the equilibrium system

$$2SO_2 + O_2 \rightleftarrows 2SO_3 + \text{heat}$$

list four ways in which the concentration of SO_3 at equilibrium could be increased.

15–7.　For the reaction

$$N_2 + 3H_2 \rightleftarrows 2NH_3 + 22.0 \text{ kcal}$$

what is the effect of (a) an addition of N_2? (b) an increase in temperature? (c) an increase in pressure?

15–8.　Although a lump of coal burns slowly, a mine full of coal dust may explode. Explain.

15–9.　Will an increase in temperature cause the following equilibrium system to shift to the left, shift to the right, or have no effect?

$$N_2O_4 + \text{heat} \rightleftarrows 2NO_2$$

15–10.　Why does a 10°C rise in temperature have a marked effect on the rate of many chemical reactions?

15–11.　Predict the effect of (a) an increase in the concentration of one reactant; (b) an increase in pressure; (c) an increase in temperature; and (d) the introduction of a catalyst upon the following reactions at equilibrium:

1. $2CO_2 + O_2 \rightleftarrows 2CO_2 + heat$
2. $CO_2 + H_2 \rightleftarrows CO + H_2O + heat$

15–12. Explain why reactants are not converted completely to products in many chemical reactions.

15–13. Describe how a catalyst may accelerate a chemical reaction.

15–14. The rocket propellant hydrazine reacts with oxygen according to the following equation:

$$N_2H_4 + O_2 \rightleftarrows N_2 + 2H_2O$$

Write the equilibrium condition.

15–15. Why do iron filings rust much faster than an iron nail?

15–16. Discuss a method that might be used to measure the rate of a reaction.

15–17. Sulfur dioxide contributes to atmospheric pollution through its oxidation to sulfur trioxide:

$$2SO_2 + O_2 \rightleftarrows 2SO_3$$

Determine the value of the equilibrium constant, K, for this reaction if the equilibrium concentrations of a mixture of the gases are:

$$SO_2 = 4.2 \times 10^{-7} \text{ mole/liter}$$
$$SO_3 = 0.5 \text{ mole/liter}$$
$$O_2 = 2.1 \times 10^{-7} \text{ mole/liter}$$

15–18. Biochemical reactions occur in our bodies at atmospheric pressure and at about 37°C. Explain why higher

pressures and temperatures are usually required to carry
out similar reactions in the laboratory.

15–19. *Multiple Choice*

A. Which of the following would probably slow
   down the rate of a reaction?
   (a) increasing the temperature
   (b) increasing the concentration
   (c) using a larger beaker
   (d) lowering the temperature

B. The principal reason for the increase in reaction
   rate with temperature is
   (a) molecules collide more frequently
   (b) the pressure increases
   (c) the activation energy increases
   (d) the fraction of high-energy molecules in-
       creases

C. Which of the following changes will increase the
   yield of products at equilibrium?
   (a) an increase in temperature
   (b) an increase in pressure
   (c) addition of a catalyst
   (d) increasing reactant concentrations

D. What is the effect of increasing the concentration of B in the reaction $A + B \rightleftarrows C + D$?
   (a) the value of K decreases
   (b) the value of K increases
   (c) the equilibrium shifts to the right
   (d) the equilibrium shifts to the left

E. A catalyst speeds up a reaction by
   (a) increasing the energy released
   (b) decreasing the energy released
   (c) increasing the activation energy
   (d) decreasing the activation energy

SUGGESTIONS FOR
FURTHER READING

Asimov, Isaac, "Enzymes and Metaphor." *Journal of Chemical
Education,* November, 1959.
> Industrial chemists are constantly involved with catalysts;
> biochemists are concerned with the protein catalysts, en-
> zymes.
Bigeleisen, Jacob, "Chemistry in a Jiffy." *Chemical and Engineer-
ing News,* April 25, 1977.

New techniques for studying reactions in the picosecond range are providing insights into biochemistry.

Brill, Winston J., "Biological Nitrogen Fixation." *Scientific American,* March, 1977.

A few bacteria and simple algae are the major suppliers in nature of "fixed" nitrogen.

Safrany, David R., "Nitrogen Fixation." *Scientific American,* October, 1974.

Life requires nitrogen that has been "fixed" through combination with other elements. Thermodynamics limits all possible methods of fixation.

Yates, John T., Jr., "Catalysis: Insights from New Technique and Theory." *Chemical and Engineering News,* August 26, 1974.

Scientists are approaching a more basic understanding of catalysis using new research techniques.

WATER, SOLUTIONS, AND POLLUTION

16-1 INTRODUCTION

Before a chemical reaction can occur, particles of the reacting substances—molecules, atoms, ions—must be brought together. Substances usually do not react when dry, however. Crystalline silver nitrate and sodium chloride, for example, do not react even when they are ground together to an extremely fine state. However, the addition of water to the system is accompanied by an instantaneous reaction and the formation of a white precipitate, silver chloride.

Many of the processes that influence our environment also involve substances in aqueous solution. Sewage, industrial wastes, and runoff from heavily fertilized land pollute our rivers. As rain falls, it picks up and dissolves gaseous impurities from the atmosphere, helping freshen the air we breathe, the air itself being a solution. Solution chemistry is important as well to an understanding of the vital processes that take place in the aqueous environment of living cells. Medicines are usually administered in solution form, and drugs are absorbed from solution. In this chapter on solution chemistry we first discuss our need for water—the most common liquid and the most common solvent in our environment.

16-2 OUR WATER NEEDS

Water seems to be so abundant that we usually take it for granted. Indiscriminate disposal of wastes has resulted in the pollution of many of our lakes,

Table 16-1   WATER CONTENT OF SELECTED FOODS

FOOD	WATER, %
Wheat flakes	4.0
White bread	35.0
Apple pie	48.0
Hamburger	54.0
Tuna	61.0
Spaghetti	72.0
Eggs	74.0
Apples	85.0
Oranges	86.0
Carrots	88.0
Lettuce	95.0

rivers, and streams. Even the oceans are becoming so contaminated that they can no longer support some forms of marine life to the extent that they once could.

All life depends upon water. Although we can survive for weeks without food, we cannot exist for more than a few days without water. Water is essential for such bodily processes as digestion, transporting nutrients and oxygen, removing carbon dioxide, and regulating body temperature. Our bodies are almost two-thirds water by weight, although the water is not evenly distributed. Blood is 90 per cent water; muscle, 75 per cent; and bones, 20 per cent. A loss of only 12 per cent of the body's water content can be fatal.

Water makes up a large part of the food we eat. An average person obtains nearly half of the daily requirement of water in this manner. Bread is about one-third water, and some vegetables contain as much as 95 per cent water. Table 16-1 gives the water content of selected foods.

Table 16-2   WATER USED IN THE PRODUCTION OF VARIOUS ITEMS

ITEM	WATER USED (gallons)
A Sunday newspaper	200
A loaf of bread	300
Gasoline (20 gal)	400
A ton of paper	43 000
An automobile	50 000
A ton of steel	88 000
A ton of aluminum	350 000
A ton of synthetic rubber	600 000

Table 16–3   THE EARTH'S WATER

SOURCE	PERCENTAGE OF TOTAL
Oceans	97.20
Ice caps	2.15
Subsurface	0.63
Atmosphere	0.01
Fresh-water lakes	0.009
Saline lakes	0.008
Streams	0.0001

Huge quantities of water are used in food production: one egg, 130 gallons; a quart of milk, 1000 gallons; a pound of beef, 4000 gallons. Table 16–2 indicates how important water is in maintaining our standard of living.

Sea water accounts for more than 97 per cent of the water on earth, as Table 16–3 shows. If it could be desalted at a reasonable cost, an almost unlimited supply of usable water would be available. At present, hundreds of desalination plants are in operation around the world. With the development of better technological methods, the number should increase considerably. Table 16–4 shows some of the uses of water around the home.

16–3   SOLUTIONS

Liquid water is never found in nature as a pure substance. It dissolves so many other substances that it is known as the universal solvent. Natural waters are classified according to the proportion of dissolved ionic solids in them, as shown in Table 16–5. The major positive ions dissolved in sea water are sodium (Na^+), magnesium (Mg^{2+}), calcium (Ca^{2+}), potassium (K^+), and

Table 16–4   DOMESTIC WATER USE

NATURE OF USE	WATER USED (average in gallons)
Flushing a toilet	4
Washing dishes	10
A washing machine load	25
A shower (5 minutes)	25
A tub bath	35
Watering lawn (1 hour)	300

Table 16–5 NATURAL WATERS CLASSIFIED BY
LEVEL OF DISSOLVED SOLIDS

CLASS	DISSOLVED SOLIDS (parts per million)	EXAMPLES
Fresh	0–1000	Lake Tahoe; Lake Michigan; Missouri River
Brackish	1 000–10 000	Baltic Sea
Salty	10 000–100 000	Oceans
Brine	Above 100 000	Dead Sea; Great Salt Lake

strontium (Sr^{2+}). The major negative ions are chloride (Cl^-), sulfate (SO_4^{2-}), bicarbonate (HCO_3^-), bromide (Br^-), fluoride (F^-), and iodide (I^-) (Fig. 16–1).

How does sodium chloride dissolve in water? The surface of a sodium chloride crystal consists of positive sodium ions and negative chloride ions that attract polar water molecules (Fig. 16–2). The forces of attraction between the ions and the water dipoles are strong enough to pull the ions away from the crystal. In this process, called *solvation,* each ion becomes surrounded by a number of water molecules and the ions are said to be solvated. The solvated ions diffuse through the water by random molecular motion and more of the crystal dissolves. In time, if enough sodium chloride is present, the rate at which ions leave the surface equals the rate at which they return to the crystal, and the solution is saturated at that temperature (Fig. 16–3).

This limit, at which no additional salt goes into solution, is called the *solubility* of sodium chloride in water. It is expressed as grams of sodium chloride per 100 grams of water. A saturated solution of sodium

Figure 16–1 In a cubic mile of sea water, weighing nearly 5 billion tons, there are about 165 million tons of dissolved matter, mostly chlorine (90 million tons) and sodium (50 million tons).

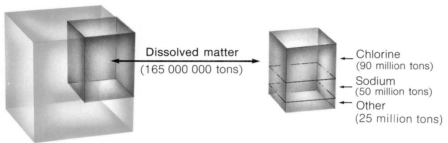

Dissolved matter
(165 000 000 tons)

Chlorine
(90 million tons)
Sodium
(50 million tons)
Other
(25 million tons)

1 cubic mile of sea water
(5 billion tons)

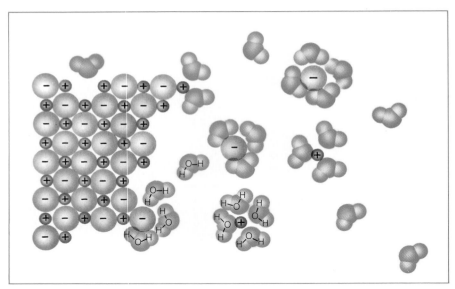

Figure 16–2 Sodium chloride being dissolved in water. The sodium and chloride ions become solvated in the solution.

chloride contains about 35 g NaCl per 100 g water at room temperature. A saturated solution of silver chloride, by comparison, contains 0.00009 g AgCl per 100 g water.

A *solution* is a homogeneous mixture—one that is uniform throughout—of two or more nonreacting substances. In soda pop, a solution of sugar, carbon dioxide, and coloring agents, water is the solvent and the other ingredients the solutes. The component of the solution that is present in the greater amount is regarded as the solvent, and all the other components the solutes. When a solution involves a solid and water, the solid is the solute and the water the solvent.

Solutions that contain a relatively small amount of solute are said to be *dilute*. Solutions that contain a relatively large amount of solute are said to be *concentrated*. Since the composition of solutions can vary between certain limits, the law of definite composition does not apply. The terms "soluble," "slightly solu-

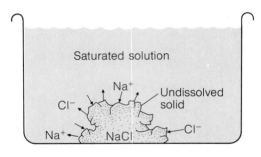

Figure 16–3 In a saturated solution a dynamic equilibrium is established between the undissolved matter and the solution.

Table 16–6   RELATIVE SOLUBILITIES

SUBSTANCE	STRUCTURE	SOLUBILITY IN WATER (Polar Solvent)	SOLUBILITY IN BENZENE (Non-polar Solvent)
Sodium chloride	Ionic	Soluble	Slightly soluble
Hydrogen chloride	Polar	Soluble	Slightly soluble
Sucrose (sugar)	Polar	Soluble	Slightly soluble
Chlorine	Non-polar	Slightly soluble	Soluble
Iodine	Non-polar	Slightly soluble	Soluble
Carbon disulfide	Non-polar	Slightly soluble	Soluble

ble," and "insoluble" are used to describe the tendency of solutes to dissolve in a particular solvent at a given temperature.

It is useful to remember the general rule "like dissolves like" for predicting solubilities. Solvents that have polar molecules, such as water, dissolve both solutes with polar molecules and solutes that are ionic, since the polar molecules and ions exert mutual attractions. Solvents having non-polar molecules dissolve solutes with non-polar molecules. Thus oils, which consist of non-polar hydrocarbons, dissolve in non-polar cleaning solvents but not in a polar solvent such as water. Sodium chloride (ionic) is insoluble in liquid gasoline (non-polar) but soluble in water. Table 16–6 gives the relative solubilities of several substances in a polar solvent, water, and in a non-polar solvent, benzene.

The three states of matter—solid, liquid, and gas —can be combined in a variety of ways to form solutions. Air is a mixture of gases that are soluble in each other, so it is properly called a solution. Liquid mercury dissolves in silver and gold to form solutions called amalgams. Brass is a solid solution of zinc and copper, an alloy. Silver and gold can be alloyed in all proportions. Table 16–7 gives examples of possible types of solutions.

Table 16–7   TYPES OF SOLUTIONS

	SOLID	LIQUID	GAS
Solid	Zinc in copper (brass; alloys)	Salt in water	—
Liquid	Mercury in gold (amalgams)	Alcohol in water	—
Gas	Hydrogen in palladium	CO_2 in water (carbonated beverages)	Oxygen in air

Table 16–8   PARTICLE SIZE OF MIXTURES

	←——————————suspensions——————————→					←——colloids——→		←——solutions
Centimeters	10^0	10^{-1}	10^{-2}	10^{-3}	10^{-4}	10^{-5}	10^{-6}	10^{-7}	10^{-8}
Ångstrom units	10^8	10^7	10^6	10^5	10^4	10^3	10^2	10^1	10^0

16–4   SUSPENSIONS AND COLLOIDAL SYSTEMS

When clay is mixed with water, the particles of clay slowly settle to the bottom. The particles of clay consist of groups of molecules. A mixture of this type is called a *suspension*. There is no clear line of demarcation between a suspension and a solution. A mixture is a suspension if it contains particles large enough to be visible with a high-powered optical microscope. Such particles have a diameter of 10^{-4} cm or greater. A mixture is considered to be a solution if the particles are about the size of ordinary molecules, which have a diameter of approximately 10^{-7} cm. Solutions and suspensions thus stand at the limits of a continuum with respect to particle size, as Table 16–8 indicates.

Mixtures containing groupings of molecules intermediate in size between molecules and the particles in

Figure 16–4   Colloidal particles reflect a beam of light and make its path visible in the Tyndall effect.

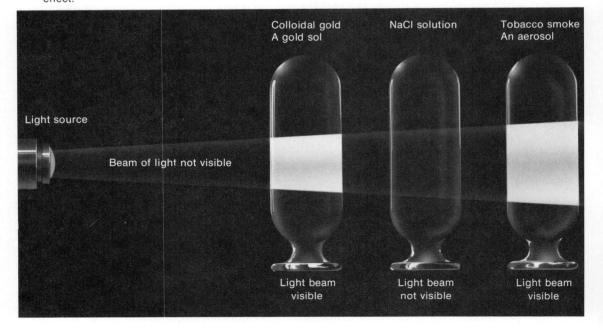

Colloidal gold
A gold sol

NaCl solution

Tobacco smoke
An aerosol

Light source

Beam of light not visible

Light beam
visible

Light beam
not visible

Light beam
visible

Table 16–9 COLLOIDAL SYSYTEMS

DISPERSED PHASE	DISPERSING MEDIUM	EXAMPLE
Solid	Solid	Colored gems, pearls
Solid	Liquid	Paints, jellies
Solid	Gas	Dust, smoke, smog
Liquid	Solid	Butter, cheese
Liquid	Liquid	Milk, cream, mayonnaise
Liquid	Gas	Clouds, mist, sprays
Gas	Solid	Marshmallow, floating soap
Gas	Liquid	Whipped cream, suds, foams

suspensions are called colloidal suspensions, colloidal dispersions, or simply *colloids*. Milk and blood are common examples. Although colloids appear to be homogeneous, individual particles can be seen with an optical or electron microscope and can be detected with a light beam (Fig. 16–4). Colloids are frequently optically opaque, whereas solutions are always transparent. The colloidal particles in a colloidal system are the *dispersed phase,* and the medium in which they occur, the *dispersing medium.* The possible types of colloidal systems are summarized in Table 16–9.

16–5 CONCENTRATION UNITS OF SOLUTIONS (Molarity and Molality)

Concentration refers to any measure of the amount of solute in a given amount of solvent or solution. It may be regarded as a measure of the crowdedness of the particles of solute (Fig. 16–5). The terms concen-

Figure 16–5 A pure solvent and solution compared at the molecular level. (From Brescia, F., et al., *Chemistry: A Modern Introduction,* 2nd ed. Philadelphia: W. B. Saunders Company, 1978, p. 338.)

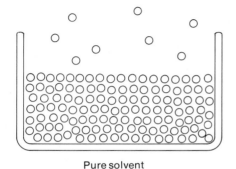

Pure solvent

a

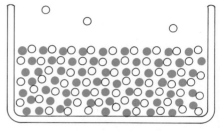

1 mole of solvent
1 mole of solute

b

trated and dilute are non-quantitative descriptions of crowdedness. Two quantitative means of expressing concentrations are molarity and molality, both based on the mole concept. Chemists favor concentration units based on moles, since they interpret chemical phenomena in terms of atoms, molecules, and ions.

　　　　The *molarity (M)* of a solution is the number of moles or gram-formula weights of a solute per liter of solution. Therefore,

Equation 16–1

$$\text{Molarity} = \frac{\text{Moles of solute}}{\text{Liters of solution}}$$

$$M = \frac{\text{Moles of solute}}{V \text{ (volume)}}$$

EXAMPLE 16–1

　　　　What is the concentration in molarity of 1 liter of an antifreeze solution prepared by dissolving 124 grams of ethylene glycol ($C_2H_6O_2$) in water?

SOLUTION

　　　　To express the amount of solute in moles, determine the molecular weight of $C_2H_6O_2$ and, by dimensional analysis, convert 124 g to moles of $C_2H_6O_2$. Then substitute in the equation for molarity, and solve for M.

Volume = 1 liter

$$124 \text{ g } C_2H_6O_2 \times \frac{\text{mole}}{62 \text{ g}} = 2 \text{ moles } C_2H_6O_2$$

$2C = 2 \times 12 = 24$

$6H = 6 \times 1 = 6$

$2O = 2 \times 16 = \underline{32}$

$\text{M.W.}_{C_2H_6O_2} = \overline{62}$

Molarity = ?

$$\text{Molarity} = \frac{\text{moles of solute}}{\text{liters of solution}}$$

$$= \frac{2 \text{ moles } C_2H_6O_2}{1 \text{ liter}}$$

$$= 2 \text{ M}$$

EXAMPLE   16–2

A batch of maple syrup has a concentration of 0.25 M in sucrose, its principal constituent. What amount of sucrose $(C_{12}H_{22}O_{11})$ is present in 1 liter of this maple syrup?

SOLUTION

Using the molarity relationship, determine the number of moles of sucrose present. Then express the number of moles of sucrose in grams.

Molarity = 0.25 M

$$\text{Molarity} = \frac{\text{moles of solute}}{\text{liters of solution}}$$

Volume = 1 liter

moles of sucrose = ?

$$\text{moles of solute} = \text{Molarity} \times \text{liters of solution}$$

$12C = 12 \times 12 = 144$

$$= 0.25 \text{ M} \times 1 \text{ liter}$$

$22H = 22 \times 1 = 22$

$$= 0.25 \text{ mole } C_{12}H_{22}O_{11}$$

$11O = 11 \times 16 = 176$

$\text{M.W.}_{C_{12}H_{22}O_{11}} = 342$

$$0.25 \text{ mole } C_{12}H_{22}O_{11} \times \frac{342 \text{ g}}{\text{mole}} \times = 85.5 \text{ g sucrose}$$

Solutions of known molarity are prepared by weighing the solute and making up the solution in calibrated volumetric glassware (Fig. 16–6). For this reason, molarity units are convenient for laboratory work. However, 1 M solutions of different solutes contain different amounts of solvent, since the amount of solvent needed to make up a given volume of solution depends upon the nature of the solute. Another way of expressing the concentration of solutions is molality, in which the amount of solvent is the same.

The *molality (m)* of a solution is the number of moles of a solute in 1 kilogram of solvent.

$$\text{Molality} = \frac{\text{Moles of solute}}{\text{Kilograms of solvent}} \qquad \text{Equation   16–2}$$

A 1-m solution of ethanol $(C_2H_5OH, \text{M.W.46})$ contains 1 mole of ethanol (46 g) in 1 kilogram of water. Each

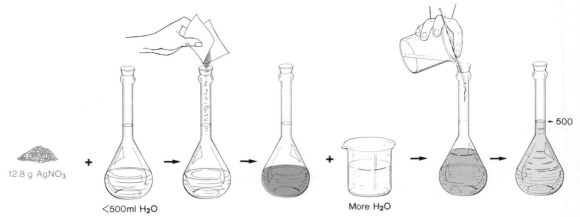

Preparation of 500 ml 0.150 M AgNO₃.

12.8 g AgNO₃

<500ml H₂O

More H₂O

←500

Figure 16-6 Preparing a solution of known concentration. (From Peters, F. I., *Introduction to Chemical Principles,* 2nd ed. Philadelphia: W. B. Saunders Company, 1978, p. 326.)

kilogram of a given solvent contains the same number of moles or molecules of that solvent—for example, 1000 g/18 g/mole, or 55.5 moles, for water. So molality indicates the relative number of moles of both solute and solvent. Thus, a 1-m aqueous solution of ethanol consists of 1 mole of ethanol (6.02×10^{23} ethanol molecules) and 55.5 moles of water ($55.5 \times 6.02 \times 10^{23}$ water molecules). Further, since temperature has no effect on mass, molality does not change with temperature as does molarity.

EXAMPLE 16-3

How many grams of ethanol (C_2H_5OH) must be dissolved in 500 g of water to make a 3-m solution?

SOLUTION

Using Equation 16-2, calculate the number of moles of ethanol required, and convert by dimensional analysis to grams.

molality = 3 m

$$\text{Molality} = \frac{\text{moles of solute}}{\text{kilograms of solvent}}$$

$$500 \text{ g } H_2O \times \frac{\text{kg}}{10^3 \text{ g}} = 0.5 \text{ kg } H_2O$$

moles of solute = m × kg

moles C_2H_5OH = ?

$2C = 2 \times 12 = 24$

$6H = 6 \times 1 = 6$

$1O = 1 \times 16 = \underline{16}$

$M.W._{C_2H_5OH} = 46$

$= 3 \, m \times 0.5 \, kg$

$= 1.5$ moles C_2H_5OH

1.5 moles $C_2H_5OH \times \dfrac{46 \, g}{mole} = 69 \, g \, C_2H_5OH$

16–6 ARRHENIUS' THEORY OF IONIZATION

The electrical conductivity of solutions can be tested with a simple apparatus (Fig. 16–7). The brightness of the light bulb is a measure of the conducting ability of the solution. When the electrodes are placed in distilled water, the bulb does not glow, but when they are placed in salt water the bulb glows brightly.

A solution that conducts electricity is an *electrolyte*. This name is also given to solutes whose solutions are conducting. Acids, bases, and salts are examples. If the electrical conductivity is good, the substance is a strong electrolyte; if it conducts the current poorly, it is a weak electrolyte. Table 16–10 lists some typical electrolytes. Solutes whose solutions are nonconducting are non-electrolytes. Electrolytes occur in nature mostly in sea water and to a lesser extent in fresh water. Most organic substances, such as alcohols, sugars, and starches, are non-electrolytes.

Svante August Arrhenius (1859–1927) proposed a theory to explain the properties of aqueous solutions of electrolytes in his doctoral thesis at the University of Uppsala, Sweden. Though his theory of ionization met with strong opposition at first, it was to revolutionize chemistry and win for Arrhenius a Nobel Prize. Its major points are as follows:

1. Solutions of electrolytes contain ions.
2. Electrolytes dissociate into ions in solvents.
3. The electrical conductivity of solutions of electrolytes is due to the presence of ions.
4. The degree of conductivity depends upon the extent of dissociation of the electrolyte.

According to the Arrhenius theory, solutions of strong electrolytes are nearly completely ionized in dilute aqueous solutions. All ionic compounds are

Figure 16–7 Apparatus for determining the electrical conductivity of solutions. The lamp glows brightly when the electrodes are placed in a strong electrolyte, and glows dimly in a weak electrolyte.

Lamp glows if solution is an electrolyte

Electrodes

Test solution

Source of current

Table 16–10   SOME TYPICAL ELECTROLYTES

STRONG ELECTROLYTES

Hydrochloric acid	HCl
Hydrobromic acid	HBr
Nitric acid	HNO_3
Chloric acid	$HClO_3$
Sulfuric acid	H_2SO_4
Sodium hydroxide	NaOH
Potassium hydroxide	KOH
Calcium hydroxide	$Ca(OH)_2$
Barium hydroxide	$Ba(OH)_2$
Magnesium hydroxide	$Mg(OH)_2$

WEAK ELECTROLYTES

Sulfurous acid	H_2SO_3
Phosphoric acid	H_3PO_4
Nitrous acid	HNO_2
Acetic acid	$HC_2H_3O_2$
Carbonic acid	H_2CO_3
Boric acid	H_3BO_3
Hydrogen sulfide	H_2S
Aqueous ammonia	NH_3

strong electrolytes, as are some covalent compounds that react completely with the solvent to produce ions. A sodium chloride solution is actually a solution of sodium and chloride ions; there are no "molecules" of sodium chloride present.

Equation 16–3

$$NaCl \longrightarrow Na^+ + Cl^-$$

In water solution, the covalent compound HCl is also a strong electrolyte; measurements show that a 0.1 M solution of HCl is 92 per cent dissociated at room temperature. Arrhenius assumed that most HCl molecules dissociate into hydrogen and chloride ions, which are in equilibrium with undissociated molecules:

Equation 16–4

$$\underset{\text{undissociated molecules}}{\text{HCl}} \rightleftharpoons \underset{\text{ions}}{H^+} + \underset{\text{ions}}{Cl^-}$$

Since a hydrogen ion, H^+, is a proton, and cannot exist isolated in water, we now believe that the reaction of HCl in water involves reaction with water molecules in which hydronium ions, H_3O^+, are formed:

Equation 16–5

$$HCl + H_2O \rightleftharpoons H_3O^+ + Cl^-$$

16–7 · COLLIGATIVE PROPERTIES

Properties of solutions that depend only on the number of particles present and not on their chemical nature are called colligative properties. The vapor pressure, freezing point, and boiling point are examples.

The *vapor pressure* of a liquid or solid is the pressure at which the substance in the gaseous state is in equilibrium with the liquid or solid at a given temperature (Fig. 16–8). That is, the number of molecules in the space above the liquid increases until there are as many molecules returning to the liquid surface as are escaping from it. Each substance has a characteristic vapor pressure at that temperature. The boiling point of a substance is the temperature at which the vapor pressure of a liquid equals the atmospheric pressure. For pure water, the boiling point is 100°C when the atmospheric pressure is 760 torr.

When the temperature of an aqueous solution of sugar is 100°C, the vapor pressure of the water is less than 760 torr. Because of the presence of sugar molecules, the concentration of water molecules is less in the solution than it is in pure water. In order for the solution to boil, the sugar solution must be heated to a temperature higher than 100°C before the vapor pressure reaches 760 torr. A surprising result is that 1 mole of any non-electrolyte produces the same vapor-pressure lowering in a given solvent and, therefore, the same boiling-point elevation. So, water solutions of

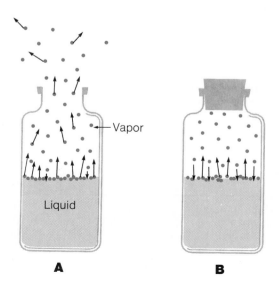

Liquid

Vapor

Figure 16–8 Vapor pressure. *A,* In the open vessel, the molecules escape as the liquid evaporates. *B,* In the closed vessel, an equilibrium is established between the escaping and returning molecules.

A   **B**

non-electrolytes of the same concentration boil at the same temperature.

François Marie Raoult (1830–1901), a physicist and chemist, showed that the lowering of the vapor pressure is proportional to the molal concentration of the solute. It follows from Raoult's law that the boiling point elevation of a solution, as compared with the pure solvent, is proportional to the molal concentration of the solute:

Equation 16–6

$$\Delta T_{B.P.} = \text{Molality} \times B.P._{E.C.}$$

where $\Delta T_{B.P.}$ is the change in the boiling point and $B.P._{E.C.}$ is a constant for a particular solvent, the *molal boiling-point elevation constant*. For water, this constant is 0.51°C. Thus, a 1 m water solution of sugar boils at 100.51°C. The constants have been determined for common solvents, as shown in Table 16–11. In dilute solution, also, the lowering of the freezing point is proportional to the molal concentration of the solute:

Equation 16–7

$$\Delta T_{F.P.} = \text{Molality} \times F.P._{D.C.}$$

where $\Delta T_{F.P.}$ is the change in the freezing point and $F.P._{D.C.}$ is the *molal freezing-point depression constant* for the solvent: 1.86°C for water.

Freezing-point depression is important to automobile owners in colder regions. To prevent the freezing of the water in an automobile radiator as soon as the temperature reaches 0°C, they add a solution called an antifreeze. The active ingredient of "permanent" antifreezes is often ethylene glycol ($C_2H_6O_2$), which also serves as a coolant in the summertime.

Table 16–11 SOME MOLAL FREEZING- AND BOILING-POINT CONSTANTS (°C)

SOLVENT	FREEZING POINT (F.P.)	MOLAL FREEZING-POINT DEPRESSION CONSTANT ($F.P._{D.C.}$)	BOILING POINT (B.P.)	MOLAL BOILING-POINT ELEVATION CONSTANT ($B.P._{E.C.}$)
Acetic acid	16.7	3.9	118.5	3.07
Benzene	5.50	5.12	80.1	2.53
Camphor	180	40	208.3	6.0
Ethyl alcohol	−115	1.99	78.4	1.19
Water	0	1.86	100.00	0.51

Table 16–12 FREEZING POINT OF WATER—
ETHYLENE GLYCOL MIXTURES

PER CENT (by Weight) OF ETHYLENE GLYCOL	FREEZING POINT OF MIXTURE °C	°F
0	0	32
10	−3.5	25.6
20	−8	17
30	−15	5
40	−24	−11
50	−36	−32

Table 16–12 gives the freezing point of water-ethylene glycol mixtures. Another practical application of this principle is the spreading of salt for the melting of ice on highways or walks.

EXAMPLE 16–4

What must the concentration of an anti-freeze solution be (in molality) in order to lower the freezing point of a kilogram of water to −20°C?

SOLUTION

From Raoult's law, we know that the change in the freezing point is proportional to the molality.

$\Delta T_{F.P.} = 20°C$

$F.P._{D.C._{H_2O}} = 1.86°C$

$m = ?$

$\Delta T_{F.P.} = m \times F.P._{D.C.}$

$\therefore m = \dfrac{\Delta T_{F.P.}}{F.P._{D.C.}}$

$= \dfrac{20°C}{1.86°C}$

$= 10.8 \ m$

16–8 ACIDS AND BASES

Acids are compounds that, in water solution, taste sour (*acidus*, sour), turn the color of litmus from blue to red, neutralize bases, and react with certain metals to give free hydrogen. Although many acids, such as sulfuric and nitric, have corrosive properties, others, such as citric and tartaric acids, are common constituents of our food. According to the Arrhenius theory, an acid is a substance that in water solution increases the concentration of hydrogen (hydronium) ions. Although more recent theories define acids differently, the Arrhenius theory is still applicable to common acids in water solution.

The strength of acids depends upon their degree of dissociation. An acid dissociates in a reversible reaction as follows:

Equation 16–8

$$HA \rightleftarrows H^+ + A^-$$

undissociated molecules   ions

Applying the law of chemical equilibrium to this reaction, we get

Equation 16–9

$$K_\alpha = \frac{[H^+][A^-]}{[HA]}$$

in which K_α is the acid constant, ionization constant, or dissociation constant. If an acid dissociates extensively, the hydronium ion concentration and the dissociation constant are large, and the acid is a strong acid. If dissociation is limited, the acid exists largely as undissociated molecules at equilibrium; K_α is small, and the acid is a weak acid. Table 16–13 gives some acid constants.

According to the Arrhenius theory of ionization, a base is a substance that increases the hydroxide ion concentration of water. When dissolved in water, bases have a bitter taste and soapy feel, turn the color of litmus from red to blue, and neutralize acids. Some com-

Table 16–13 DISSOCIATION CONSTANTS OF SOME ACIDS

ACID	REACTION	K_α (0.1 M at 25°C)	PERCENT DISSOCIATED	STRENGTH
Acetic	$HC_2H_3O_2 \rightleftarrows H^+ + C_2H_3O_2^-$	1.8×10^{-5}	1.3	Weak
Nitrous	$HNO_2 \rightleftarrows H^+ + NO_2^-$	4.5×10^{-4}	1.5	Weak
Sulfurous	$H_2SO_3 \rightleftarrows H^+ + HSO_3^-$	1.3×10^{-2}	20	Moderate
Phosphoric	$H_3PO_4 \rightleftarrows H^+ + H_2PO_4^-$	$7\text{-}5 \times 10^{-3}$	27	Moderate
Sulfuric	$H_2SO_4 \rightleftarrows H^+ + HSO_4^-$	$\approx 10^3$	61	Strong
Hydrochloric	$HCl \rightleftarrows H^+ + Cl^-$	$\approx 10^3$	92	Strong

mon bases are NaOH, KOH, NH_4OH, and $Ca(OH)_2$. Neutralization involves the combination of hydrogen ions with hydroxide ions to form water molecules. The dissociation of ammonium hydroxide in water is represented as follows:

$$NH_4OH \rightleftharpoons NH_4^+ + OH^-$$
Equation 16–10

At equilibrium, the dissociation constant is:

$$K_\beta = \frac{[NH_4^+][OH^-]}{[NH_4OH]} = 1.8 \times 10^{-5}$$
Equation 16–11

where K_β is the basic constant, ionization constant, or dissociation constant. The small value of K_β shows that ammonium hydroxide is a weak base. A neutralization reaction is represented as follows:

$$H^+ + OH^- \rightleftharpoons H_2O$$
Equation 16–12

Water is a weak electrolyte, dissociating as follows:

$$H_2O \rightleftharpoons H^+ + OH^-$$
Equation 16–13

In pure water and aqueous solutions, the ionization constant for water, K_I, at equilibrium is:

$$K_I = \frac{[H^+][OH^-]}{[H_2O]}$$
Equation 16–14

Since the concentration of undissociated water, $[H_2O]$, can be considered constant in pure water and dilute solutions, it is combined with the constant K_I to give a new constant, K_W, called the dissociation constant or ion product of water.

$$K_I[H_2O] = K_W = [H^+][OH^-]$$
Equation 16–15

Water is neutral, therefore, since all the hydrogen ions and hydroxide ions come from the dissociation of water molecules. For each water molecule that dissociates, one hydrogen ion and one hydroxide ion are formed. So, in pure water, the concentration of hydrogen ions equals the concentration of hydroxide ions.

Table 16–14   RELATIONSHIP BETWEEN [H⁺]
AND [OH⁻] IN AQUEOUS SOLUTIONS

$[H^+]$ (moles/liter)	$[OH^-]$ (moles/liter)	$[H^+][OH^-]$	CHARACTER
$1 = 10^0$	10^{-14}	10^{-14}	Acidic
10^{-1}	10^{-13}	10^{-14}	,,
10^{-2}	10^{-12}	10^{-14}	,,
10^{-3}	10^{-11}	10^{-14}	,,
10^{-4}	10^{-10}	10^{-14}	,,
10^{-5}	10^{-9}	10^{-14}	,,
10^{-6}	10^{-8}	10^{-14}	,,
10^{-7}	10^{-7}	10^{-14}	Neutral
10^{-8}	10^{-6}	10^{-14}	Basic
10^{-9}	10^{-5}	10^{-14}	,,
10^{-10}	10^{-4}	10^{-14}	,,
10^{-11}	10^{-3}	10^{-14}	,,
10^{-12}	10^{-2}	10^{-14}	,,
10^{-13}	10^{-1}	10^{-14}	,,
10^{-14}	$1 = 10^0$	10^{-14}	,,

At 25°C, these concentrations are 1×10^{-7} mole/liter. Thus,

Equation   16–16

$$K_W = [H^+][OH^-] = (1 \times 10^{-7} \text{ mole/liter})$$
$$\times (1 \times 10^{-7} \text{ mole/liter})$$
$$= 1 \times 10^{-14} \text{ mole}^2/\text{liter}^2$$

Acid solutions have H^+ concentrations greater than 1×10^{-7} mole/liter, and basic solutions have OH^- concentrations greater than 1×10^{-7} mole/liter, as Table 16–14 shows.

EXAMPLE   16–5

An aqueous solution has a hydrogen-ion concentration of 0.01 mole/liter. (a) What is the concentration of hydroxide ions? (b) Is the solution acidic, basic, or neutral?

SOLUTION

The ion product for water is 1×10^{-14} in water solutions. Knowing $[H^+]$, we can solve for $[OH^-]$.

$$[H^+] = 0.01 \frac{\text{mole}}{\text{liter}} = 1 \times 10^{-2} \frac{\text{mole}}{\text{liter}}$$

$K_W = 1 \times 10^{-14} \text{ mole}^2/\text{liter}^2$ $\qquad\qquad [OH^-] = ?$

(a) $\qquad K_W = [H^+][OH^-]$

$\qquad \therefore [OH^-] = \dfrac{K_W}{[H^+]}$

$\qquad\qquad = \dfrac{1 \times 10^{-14} \text{ mole}^2/\text{liter}^2}{1 \times 10^{-2} \text{ mole}/\text{liter}}$

$\qquad\qquad = 1 \times 10^{-12} \text{ mole}/\text{liter}$

(b) The solution is acidic, since $[H^+] = 1 \times 10^{-2}$ mole/liter, which is greater than 1×10^{-7} mole/liter.

16–9 THE pH SCALE

The acidities of water solutions are expressed on a special scale. The symbol *pH*, introduced by the biochemist Sven P. L. Sörensen (1868–1939), was intended originally as an abbreviation for the phrase "potential of hydrogen ion." The pH scale has the advantage of expressing small concentrations of H_3O^+ ions more conveniently than a decimal scale, and also eliminates negative powers of 10. A solution that is neutral has a pH of 7. Acidic solutions have pH values less than 7, and basic solutions pH values greater than 7.

The solutions of our common experience cover a considerable range of pH values (Fig. 16–9). The water content of the soil may cause it to have a value anywhere from pH 3 (very acid) to pH 10 (very alkaline), although these represent extremes; most agriculture proceeds in soil of a pH close to 6. The pH values of soil determine the solubility and thus the availability of necessary minerals to crops. Soil bacteria are also sensitive to variations in pH. Sea water has an average pH of 8. Many marine organisms die below pH 7.5; their eggs are particularly vulnerable.

The pH values of body fluids must be maintained within certain narrow limits. The normal variation of the pH of blood is from 7.3 to 7.5. If the pH goes below 7.0 or above 7.8, death follows. Enzymes are also sensitive to changes in pH. The pH values of most foods range between 2 and 8; however, an acid food may end up as an alkaline residue when metabolized in the body, and vice versa.

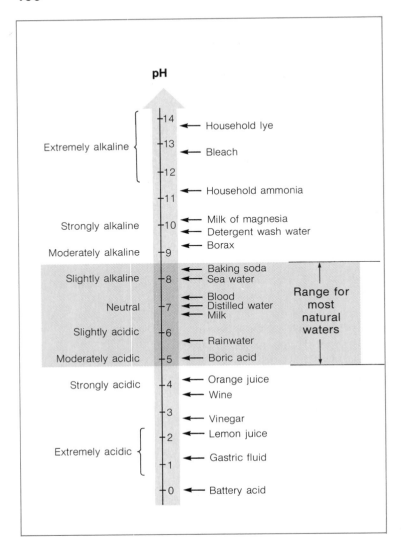

Figure 16-9   The pH scale and pH values of some common solutions.

Rainwater is normally slightly acidic, with a pH of about 5.7, as a result of passing through atmospheric carbon dioxide. Wherever the air is polluted with oxides of sulfur and nitrogen, rainfall also produces sulfuric and nitric acids. The pH of these acid rains may be as low as 3, acidic enough to corrode sturctures and threaten life by lowering the pH of forests, fields, lakes, and ponds.

16-10   SOAP AND WATER HARDNESS

Water contains dissolved impurities in the form of metal ions. It is not necessary to remove these for

drinking purposes, but they may make washing difficult and cause scale to be deposited in pipes and eventually clog them. The chief problems involve calcium
(Ca^{2+}) and magnesium (Mg^{2+}) ions, along with ferrous
(Fe^{2+}) ions. Water containing these ions is said to be
hard. In temporary hardness, water contains bicarbonate ions (HCO_3^-) as well, and can be softened by
heating, causing calcium carbonate to precipitate out of
a solution. When chloride (Cl^-) and sulfate (SO_4^{2-}) ions
are present with Ca^{2+} and Mg^{2+}, the water is said to
have permanent hardness, but may be softened by
other processes: ion-exchange minerals (zeolites), ion-
exchange resins, and the addition of washing soda
(Na_2CO_3). In the last case, Ca^{2+} and Mg^{2+} ions react
with CO_3^{2-} ions and form insoluble precipitates of calcium and magnesium carbonates.

Ordinary soap is a mixture of the sodium compounds (salts) of various organic acids obtained from
natural fats or oils. Perfumes, antiseptics, and other ingredients are added to produce toilet soaps and de-
odorizing soaps. A floating soap is produced if air is
blown through molten soap. Potassium soaps produce
a softer lather and are more soluble in water, and so
are used in liquid soaps and shaving creams. A soap
molecule such as sodium stearate consists of a large
non-polar hydrocarbon portion (hydrophobic–repelled
by water) and an ionic end (hydrophilic–water soluble).

When soap is added to water, the hydrophilic ends
of the molecules are attracted to the water and dissolve
in it. The hydrophobic portions are repelled by the
water molecules and form a thin film called a *mono-
layer* on the surface of the water (Fig. 16–10):

sodium stearate

$$CH_3CH_2CH_2CH_2CH_2CH_2CH_2CH_2CH_2CH_2CH_2CH_2CH_2CH_2CH_2CH_2CH_2-C\begin{matrix} \nearrow O \\ \searrow O^-Na^+ \end{matrix}$$

hydrophobic   hydrophilic

$$CH_3(CH_2)_{16}C\overset{O}{\underset{}{\parallel}}-O^-Na^+$$

The cleansing action of soap involves the lowering
of the surface tension of water by the monolayer of
soap, and emulsification. Dirt is held to clothes by a
thin film of grease or oil (Fig. 16–11). The hydrophobic
portions of soap molecules dissolve in the grease or oil,

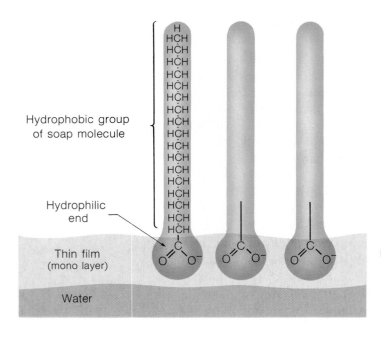

Hydrophobic group
of soap molecule

Hydrophilic
end

Thin film
(mono layer)

Water

Figure 16–10 The addition of soap to water causes a thin film, a mono-layer, to form on the surface of the water. Each soap molecule has a hydrophilic group and a hydro-phobic group.

Figure 16–11 Action of a detergent in removing dirt from a piece of cloth. (From Brescia, F., et al., *Chemistry: A Modern Introduction,* 2nd ed. Philadelphia: W. B. Saunders Company, 1978, p. 346.)

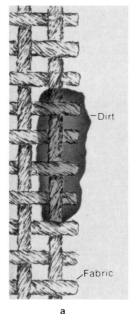  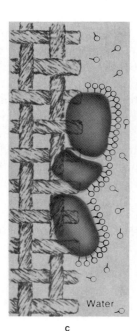 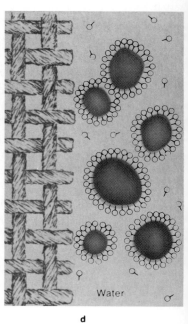

a

A dirty piece
of cloth

b

The cloth in water.
The dirt
strongly adheres
to the cloth.

c

The action of the
detergent lowers
the adhesion of
the dirt to the
cloth, making it
easier to detach
the dirt from
the fabric.

d

The dirt
remains
suspended
in the
solution
and is easily
washed away

468

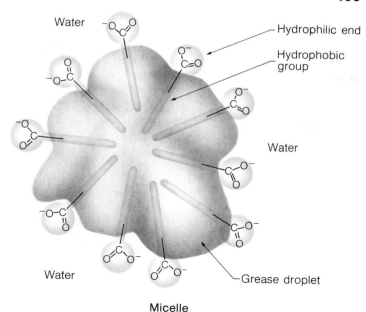

Water — Hydrophilic end

Hydrophobic group

Water

Water

Grease droplet

Micelle

Figure 16–12 Formation of micelles when soap dissolves in grease or oil. The charged micelles are soluble in water and can be washed away.

and the hydrophilic ends remain in the water phase. Scrubbing causes the oil or grease to disperse into tiny droplets, a process called *emulsification*. The oil or grease droplets become surrounded by soap molecules and are known as *micelles* (Fig. 16–12). Since the ionic ends of the soap molecules project outward, the surface of each drop is negatively charged, and because of mutual repulsion, the drops do not coalesce. The entire micelle becomes water soluble and can be washed away by a stream of water.

Soap is an excellent cleansing agent, and it is *biodegradable*—that is, it is capable of being broken down biologically into harmless end products such as CO_2 and water by microorganisms in the soil or in sewage treatment plants—and relatively non-toxic. Its main drawback is that in hard water it reacts to form a scum or precipitate with the Ca^{2+} and Mg^{2+} ions:

Equation 16–18

$$2CH_3(CH_2)_{16}\overset{O}{\overset{\|}{C}}\!-\!O^-Na^+ + Ca^{2+} \rightarrow [CH_3(CH_2)_{16}\overset{O}{\overset{\|}{C}}\!-\!O^-]_2Ca^{++} + 2Na^+$$

Sodium stearate (soluble)   Calcium stearate (insoluble)

This reaction consumes a large amount of the soap, making it unavailable for the washing process and thereby increasing the cost. At the same time, the gummy precipitate remains behind, producing stiffness in clothes, accounting for bathtub ring, and clogging washing machines. The effect of some softening agents is to precipitate the "hard" ions to form harmless granules that settle out, or to exchange the hard ions for Na^+, allowing the use of soap in hard water.

16–11   SYNTHETIC DETERGENTS

The term *detergent* denotes any cleansing agent; soaps fall under this broad category. In popular usage, however, the word refers to synthetic detergents, or *syndets,* of which close to a thousand are available commercially. The new detergents avoid the problems of the curd caused by soap and hard water.

Three related developments ushered in the large-scale use of synthetic detergents: phosphate detergents, synthetic fibers, and automatic washing machines. The most important single factor was the introduction of sodium tripolyphosphate (STPP), $Na_5P_3O_{10}$, as a builder—a component of detergents that softens water and prevents dirt that has already been washed out from being redeposited on clothing. The new automatic washers were redesigned to handle STPP, and synthetic fibers were formulated to be washed by the new detergent in the new washing machines. As a result, easy-to-wash clothing made from synthetic fibers

Figure 16–13 Sudsy detergents polluted many waterways with foam, resulting in fish kills and other adverse effects. (Courtesy of the U.S. Department of Agriculture.)

came into great demand. The more than 40 million washers now in use are designed almost exclusively for detergents. The use of soap, in fact, could seriously damage them.

Society paid a high environmental price when it switched from soap to synthetic detergents (Fig. 16–13). The new products clogged sewage systems and polluted streams with foam, killing fish and wildlife and even getting into city drinking water. Some of the pollution problem was alleviated by replacing the non-biodegradable, highly branched-chain surfactants (surface-active agents that provide the main source of detergent action), such as alkyl benzylsulfonates (ABS), with the more easily biodegradable but sudsless linear alkyl sulfonates (LAS). The metabolism of microorganisms is adapted to the straight-chain alkyl groups found in soaps, natural fats, and LAS; they cannot readily break down the highly branched-chain ABS syndets.

$$CH_3-\underset{\underset{\textstyle CH_3}{|}}{CH}-CH_2-\underset{\underset{\textstyle CH_3}{|}}{CH}-CH_2-\underset{\underset{\textstyle CH_3}{|}}{CH}-CH_2-\underset{\underset{\textstyle CH_3}{|}}{CH}-\langle\bigcirc\rangle-SO_3^-Na^+$$

A sodium alkyl benzene sulfonate (ABS). Not easily biodegradable.

$$CH_3-CH_2-CH_2-CH_2-CH_2-CH_2-CH_2-CH_2-CH_2-\underset{\underset{\textstyle CH^3}{|}}{CH}-\langle\bigcirc\rangle-SO_3^-Na^+$$

A sodium linear alkyl sulfonate (LAS). More easily biodegradable.

16–12 EUTROPHICATION

No other compounds that have been used as detergent builders have been as effective, safe, and cheap as the phosphates. However, these very phosphate builders are at the center of the environmental controversy that surrounds detergents. At stake is the survival of lakes and other bodies of water, some of which are succumbing to *eutrophication.*

Plant nutrients—the elements essential to plant growth—increase in concentration in a lake as streams bring in dissolved compounds of nitrogen, phosphorus, potassium, and sulfur from decaying plant and animal remains. The enrichment of water by nutrients is called eutrophication ("well nourished") and leads to the natural aging of lakes. Ordinarily, eutrophication of a lake takes thousands of years. As nutrients build up in a lake, the plant and animal populations increase. Water weeds become abundant, and thick layers of blue-green algae crowd out other organisms. The oxygen supply is depleted, and the water is unable to support desirable species of fish. When they die, deposits of organic matter build up on the bottom of the lake. Gradually, the lake becomes shallow and evolves into marshland, swamp, and finally dry land.

In the 20th century, man-made pollution has been the cause of the accelerated eutrophication of Lake Zurich in Switzerland, which has gone from youth to old age in less than a century, and of many other lakes. It is estimated that during the past 50 years Lake Erie has "aged" about 15 000 years. The catch of blue pike dropped from nearly 20 million pounds in the 1950's to less than 500 pounds in 1965, and the catch of herring and whitefish also went sharply down. During the same period, the phosphorus concentration in Lake Erie tripled, largely because of the expanded use of phosphate builders in detergents.

Most green plants, including algae, require 15 to 20 elements for their growth. Many of these elements are usually present in natural waters in sufficient concentrations for normal plant growth. However, it appears that phosphorus is a limiting nutrient for algae in many fresh-water lakes—the element in shortest supply with respect to the nutritional requirements of algae. Nitrogen may be a more limiting nutrient in estuaries and other coastal waters. When nitrogen and phosphorus are present in a lake in greater concentrations than normal, several species of algae grow very rapidly and produce an *algal bloom* (Fig. 16–14). This unsightly, foul-smelling green scum sets off the eutrophication process.

Detergents are heavily loaded with phosphate builders (25 to 60 per cent by weight), the source of the hundreds of millions of pounds of detergent phosphorus poured annually into rivers and lakes from municipal sewage. If eutrophication can be halted only by cutting off a critical nutrient, then phosphorus is the best candidate for reduction. Although synthetic detergents are not the only source of phosphorus in natural waters, they are the largest single source. Possible substitutes for phosphates would have to be tested exten-

Figure 16–14 An algal bloom on Lake Tahoe, California. Such contamination causes eutrophication of lakes and streams. (From U.S. Environmental Protection Agency, "Action for Environmental Quality," March, 1973, Photograph by Belinda Rain.)

sively for toxicity and ecological consequences. However, as long as phosphorus compounds continue to be used, the quantities being dumped into waterways should be reduced. Through advanced technology, it may be possible in time to remove phosphates from municipal sewage.

CHECKLIST

Acid
Algal bloom
Alloy
Arrhenius' theory of
    ionization
Base
Biodegradable
Boiling-point elevation
    constant
Colligative properties
Colloid
Concentrated
Detergent
Detergent builder
Dilute
Dispersed phase
Dispersing medium
Electrolyte
Emulsification
Eutrophication
Freezing-point depression
    constant

Hard water
Hydrophilic
Hydrophobic
Ion-product of water
Limiting nutrient
Micelle
Molality, m
Molarity, M
Neutralization
Non-electrolyte
pH
Raoult's law
Soap
Solubility
Solute
Solvation
Solvent
Suspension
Vapor pressure

16–1.　How do a mixture, a solution, and a pure compound differ in composition?

16–2.　Compare the solubility of common table salt, NaCl, with that of naphthalene (moth flakes), $C_{10}H_8$, in (a) water, (b) gasoline.

16–3.　When aqueous solutions of sodium chloride and silver nitrate are mixed, a precipitate forms instantaneously. How does the Arrhenius theory explain this result?

16–4.　Discuss the principle upon which the addition of antifreeze to an automobile radiator exerts its action.

16–5.　Which will freeze faster, a fresh-water pond or a salt-water pond?

16–6.　How does the spreading of salt on ice-covered walks and roads cause the ice to melt?

16–7.　A sample of rainwater has a nitric acid concentration of 3.0×10^{-8} M. How many grams of nitric acid are present in 1 liter?

16–8.　The solution in a car radiator froze at $0°F$. Determine the molality of the solution.

16–9.　A solution contains 684 g of sucrose, $C_{12}H_{22}O_{11}$, per 1000 g of water. Determine the freezing point of the solution.

16–10.　What is the boiling point of the solution in Exercise 16–9?

16–11.　One mole of sugar is added to 1 liter of water. Is the concentration of the solution 1 M? Explain.

16–12.　A solution is made by dissolving 45.0 g of dextrose $(C_6H_{12}O_6)$ in 250.0 g of water. (a) Calculate the molality. (b) At what temperature will the solution freeze?

16–13. Calculate (a) the number of moles, and (b) the number of grams of NaCl in 50 ml of 2 M NaCl.

16–14. Vinegar is an aqueous solution of acetic acid. Would you expect the pH of vinegar to be greater than or less than 7.0?

16–15. How do soaps and detergents cleanse?

16–16. In regions where the water supply is rich in calcium or magnesium compounds, why are detergents better cleaning agents than soaps?

16–17. Of what importance is biodegradability to the formulators of detergents?

16–18. What are the disadvantages of hard water?

16–19. What is the environmental impact of phosphorus compounds in detergents?

16–20. Why is an algal bloom—a thick mat of blue-green algae—on the surface of a lake an undesirable sign?

16–21. How can eutrophication lead to the death of a lake?

16–22. *Multiple Choice*

    A. Which of the following would be most soluble in the polar solvent water?
       (a) I_2 (c) CCl_4
       (b) NaCl   (d) N_2

    B. The freezing point of H_2O is
       (a) lowered by the addition of a solute
       (b) raised by the addition of a solute
       (c) not affected by the addition of a solute
       (d) does not depend upon the amount of solute added

    C. When one mole of NaCl is dissolved in a kilogram of water, the concentration is
       (a) 0.5 molal   (c) 1.0 molal
       (b) 0.5 molar   (d) 1.0 molar

D. A solution that has a pH of 7 is
   (a) slightly acidic   (c) neutral
   (b) strongly acidic   (d) slightly basic

E. K_w
   (a) equals 10^{-14}
   (b) is the ion product of water
   (c) equals $[H_3O^+][OH^-]$
   (d) all of these

SUGGESTIONS FOR FURTHER READING

Buswell, Arthur M., and Worth H. Rodebush, "Water." *Scientific American,* April, 1956.
   We tend to overlook the fact that water is a peculiar substance. Its properties and behavior are unlike those of any other liquid.
Penman, H. L., "The Water Cycle." *Scientific American,* September, 1970.
   Discusses some of the properties of water and their significance in the biosphere.
Rukeyser, William Simon, "Fact and Foam in the Row over Phosphates." *Fortune,* January, 1972.
   What had been a placid backwater of science—limnology, the study of fresh water—erupted into controversy in the 1970's over the role of phosphates in eutrophication.
Snell, Foster Dee, and Cornelia T. Snell, "Syndets and Surfactants." *Journal of Chemical Education,* June, 1958.
   Emphasizes the chemical and industrial aspects of these products.

AN INTRODUCTION
TO ORGANIC
CHEMISTRY

17–1 INTRODUCTION

Excluding carbon, the number of chemical compounds formed by the elements in the periodic table is probably fewer than 200 000. In contrast, carbon is the central element in more than 4 000 000 (20 times as many) compounds, which it forms with a handful of other elements, principally hydrogen, oxygen, and nitrogen. Since the ability of carbon to enter so freely into chemical combination was at first believed to be related to its source in substances of animal or vegetable origin, the term "organic chemistry" came to denote the chemistry of most carbon compounds. Now carbon compounds are synthesized in the laboratory by the thousands every year, and make up a large segment of our industrial production, including plastics, fuels, synthetic fibers, pesticides, detergents, and many other products.

17–2 THE ORGANIC-INORGANIC DUALISM

The belief in the difference between the organic and inorganic realms was a persistent one. Substances that were derived from plant or animal sources—organisms—were known as organic, and all others inorganic, and there was no way to bridge the two worlds. Although the chemical elements were the same in both kinds of substances, there seemed to be two kinds of force that governed them. "Vital force" was supposed to regulate the organic world, and "chemical affinity" the inorganic. An organic substance could only be formed through the vital force of living organisms, and therefore, could never be made in the laboratory.

Friedrich Wöhler (1800–1882), who had been a student of Berzelius, made a discovery in 1828 that led to the eventual abandonment of the organic-inorganic distinction. Wöhler synthesized urea, an organic compound, from ammonium chloride and silver cyanate, which were considered to be inorganic compounds. This reaction was the first evidence that a vital force was not necessary to prepare organic compounds.

Equation 17–1

$$NH_4Cl \quad + \quad AgCNO \quad \rightarrow \quad NH_2-\overset{\overset{\displaystyle O}{\|}}{C}-NH_2 + \quad AgCl \downarrow$$

(ammonium chloride)   (silver cyanate)   (urea)   (silver chloride)

Wöhler wrote to Berzelius in triumph that he could make urea without the use of kidneys. However, the synthesis of urea was still inconclusive, since Wöhler's "inorganic" ammonia was prepared from animal hoofs and horns, not from nitrogen and hydrogen. More decisive tests were the synthesis in 1844 of acetic acid, an organic substance, from completely inorganic materials (the elements carbon, hydrogen, and oxygen) by a student of Wöhler's, Hermann Kolbe (1818–1884), and the synthesis of scores of organic compounds—including methane, acetylene, benzene, methyl alcohol, and ethyl alcohol—by Pierre Berthelot (1827–1907) in the 1850's. It is now recognized that the same principles apply to all molecules regardless of their source, and the term organic chemistry is used to refer to the chemistry of carbon compounds.

17–3 ALKANE HYDROCARBONS: A HOMOLOGOUS SERIES

The simplest organic compounds contain only hydrogen and carbon, for which reason they are designated hydrocarbons. Of these, the alkane hydrocarbons have two kinds of bonds, carbon-to-hydrogen (C—H) and carbon-to-carbon (C—C). Since only single bonds are present, these compounds are also called *saturated* hydrocarbons, and because they are relatively stable toward many reagents, *paraffins* (*parus*, too little; *affinis*, affinity).

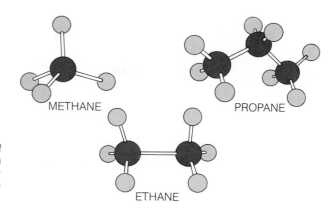

Figure 17–1 Ball-and-stick models of methane, ethane, and propane. (From Brescia, F., et al., *Chemistry: A Modern Introduction,* 2nd ed. Philadelphia: W. B. Saunders Company, 1978, p. 681.)

The alkane hydrocarbons constitute a *homologous series*—a group of related compounds in which the molecular formulas of members differ from one another by a constant increment. The simplest member of the family is one containing only one carbon. Methane, CH_4, is followed by ethane, C_2H_6, then propane, C_3H_8 (Fig. 17–1). Ethane differs from methane by one carbon and two hydrogens, $—CH_2$, and propane differs from ethane by the same $—CH_2$ (methylene) group.

The individual members of a homologous series, called *homologs,* are similar in structure and physical and chemical properties. And they conform to a general formula, in this case, C_nH_{2n+2}. Knowing the number of carbons present, we can predict the number of hydrogens. Thus, octane has 8 carbons and $(2n+2) = 16+2 = 18$ hydrogen atoms per molecule. Some physical properties of the first 12 alkanes are given in Table 17–1.

Table 17–1   PHYSICAL PROPERTIES OF ALKANES

NAME	MOLECULAR FORMULA	USUAL STATE	MELTING POINT, °C	BOILING POINT, °C
Methane	CH_4	Gas	−182	−161
Ethane	C_2H_6	Gas	−183	−89
Propane	C_3H_8	Gas	−188	−45
n-Butane	C_4H_{10}	Gas	−138	−0.5
n-Pentane	C_5H_{12}	Liquid	−130	36
n-Hexane	C_6H_{14}	Liquid	−95	69
n-Heptane	C_7H_{16}	Liquid	−91	98
n-Octane	C_8H_{18}	Liquid	−57	125
n-Nonane	C_9H_{20}	Liquid	−51	151
n-Decane	$C_{10}H_{22}$	Liquid	−32	174
n-Undecane	$C_{11}H_{24}$	Liquid	−25.6	195
n-Dodecane	$C_{12}H_{26}$	Liquid	−9.6	215

Table 17–2   CLASSES OF HYDROCARBONS

CLASS	GENERAL FORMULA	BONDING OF CARBONS	EXAMPLE
Alkane	C_nH_{2n+2}	Single bonds, no rings	CH_4 (methane)
Cycloalkane	C_nH_{2n}	Single bonds, ring	C_6H_{12} (cyclohexane)
Alkene	C_nH_{2n}	One double bond	C_2H_4 (ethylene)
Diene	C_nH_{2n-2}	Two double bonds	$CH_2{=}CH{-}CH{=}CH_2$ (butadiene)
Alkyne	C_nH_{2n-2}	One triple bond	C_2H_2 (acetylene)
Aromatic		At least one benzene ring	C_6H_6 (benzene)

The symbol "n" represents "normal" and refers to a straight unbranched chain of carbon atoms. The names and characteristics of the important classes of hydrocarbons are listed in Table 17–2.

17–4   ISOMERISM; THE NUMBERS GAME

The use of molecular formulas alone to represent organic compounds is not always helpful. Although methane, ethane, and propane occur in just one form, there are two butanes (C_4H_{10}) and three pentanes (C_5H_{12}) (Fig. 17–2). There are 18 known compounds having the molecular formula C_8H_{18}. Each of these has its characteristic boiling and freezing points and other properties. The existence of more than one compound with the same molecular formula is called *isomerism*, and distinct compounds with the same molecular formula are called *isomers*.

There are two reasons for the vast number of organic compounds: (1) the ability of carbon to bond with itself in almost endless succession, and (2) the forma-

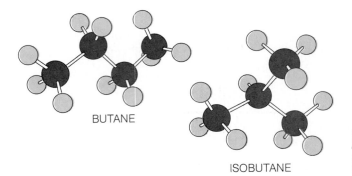

BUTANE

ISOBUTANE

Figure 17–2 Models of the two butanes. (From Brescia, F., et al., *Chemistry: A Modern Introduction*, 2nd ed. Philadelphia: W. B. Saunders Company, 1978, p. 683.)

Table 17–3 NUMBER OF POSSIBLE ISOMERS OF CERTAIN ALKANES

MOLECULAR FORMULA	NUMBER OF POSSIBLE ISOMERS	MOLECULAR FORMULA	NUMBER OF POSSIBLE ISOMERS
CH_4	1	C_9H_{20}	35
C_2H_6	1	$C_{10}H_{22}$	75
C_3H_8	1	$C_{11}H_{24}$	159
C_4H_{10}	2	$C_{12}H_{26}$	355
C_5H_{12}	3	$C_{13}H_{28}$	802
C_6H_{14}	5	$C_{20}H_{42}$	366 000
C_7H_{16}	9	$C_{30}H_{62}$	4 000 000 000
C_8H_{18}	18	$C_{40}H_{82}$	63 000 000 000 000

tion of isomers. Table 17–3 gives the number of isomers possible for certain alkanes. Some properties of
the two butanes and three pentanes are compared in
Table 17–4.

17–5 STRUCTURAL THEORY

Since isomers have the same molecular formulas,
they must differ with respect to the arrangement of the
atoms within the molecules. Kekulé and Couper explained how this is possible.

Friedrich August Kekulé (1829–1896) in 1858 published a paper and within a month, Archibald Scott
Couper (1831–1892) published one of his own, independently. Each presented (1) the idea (already proposed by others) that a carbon atom forms four bonds,
and (2) a new concept (original with Kekulé and
Couper): the self-linking of carbon atoms to form
chains. These two concepts together form the basis

Table 17–4 PHYSICAL PROPERTIES OF THE ISOMERS OF
BUTANE AND PENTANE

COMPOUND	MOLECULAR FORMULA	MELTING POINT, °C	BOILING POINT, °C	DENSITY (g/ml, 20°C)
n-Butane	C_4H_{10}	−135	−0.5	0.62
2-Methylpropane (Isobutane)	C_4H_{10}	−159	−12	0.60
n-Pentane	C_5H_{12}	−130	36	0.63
2-Methylbutane (Isopentane)	C_5H_{12}	−160	28	0.62
2,2-Dimethylpropane (Neopentane)	C_5H_{12}	−20	9.5	0.61

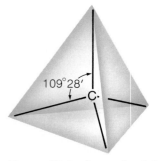

Figure 17-3 A regular tetrahedron. The valence bonds of a carbon atom are directed toward the four vertices.

of a structural theory that provides a simple yet useful interpretation of organic substances and their reactions. The basic assumptions of the structural theory are as follows:

1. Carbon normally forms four bonds; oxygen, two; hydrogen and the halogens, one.
2. A carbon atom may be bonded to atoms of other elements or to other carbon atoms.
3. More than one bond may connect atoms. There may be single, double, or triple bonds between two carbon atoms.

The theory was later expanded to include the concept of the *tetrahedral configuration* of carbon, proposed independently in 1874 by Jacobus Hendricus van't Hoff (1852–1911) and Joseph Achille Le Bel (1847–1930). According to this hypothesis, the four covalent bonds of the carbon atom have definite positions in space, as though the carbon atom were the center of a tetrahedron—a triangular-based pyramid with four faces and four corners (Fig. 17–3). Each bond is equidistant from the remaining three, and the angle between one bond and any of its neighbors is 109°28′.

Methane, CH_4, is a three-dimensional molecule with carbon at the center of a regular tetrahedron, and four bonds directed toward the vertices, where hydrogen atoms are located (Fig. 17–4). Its expanded structural formula is usually represented as a two-dimensional projection.

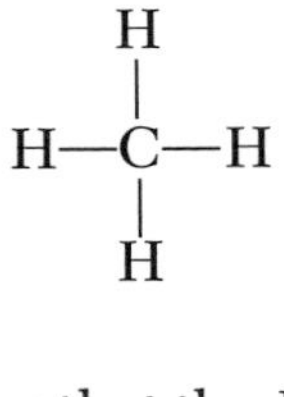

Figure 17-4 Structure of the methane molecule CH_4. The four hydrogen atoms occupy the corners of a regular tetrahedron. The two solid lines are in the plane of the paper, the wedge-shaped line extends out toward you, and the dashed line extends back.

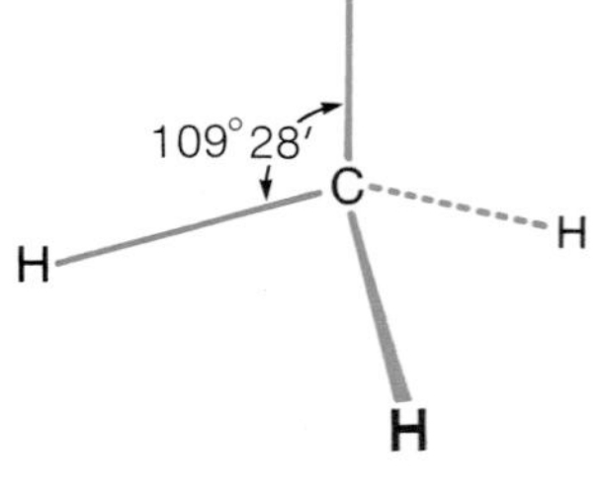

$$H—\overset{\displaystyle H}{\underset{\displaystyle H}{\overset{|}{\underset{|}{C}}}}—H$$

with the understanding that the H—C—H bond angles are 109°28′ (not 90°) and that each dash represents one pair of electrons.

The expanded structural formulas of ethane, C_2H_6, and propane, C_3H_8, are:

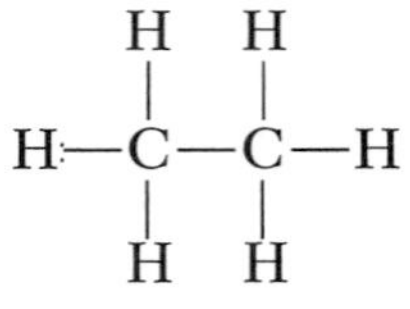

$$H—\overset{\displaystyle H}{\underset{\displaystyle H}{\overset{|}{\underset{|}{C}}}}—\overset{\displaystyle H}{\underset{\displaystyle H}{\overset{|}{\underset{|}{C}}}}—H$$

ethane

$$H—\overset{\displaystyle H}{\underset{\displaystyle H}{\overset{|}{\underset{|}{C}}}}—\overset{\displaystyle H}{\underset{\displaystyle H}{\overset{|}{\underset{|}{C}}}}—\overset{\displaystyle H}{\underset{\displaystyle H}{\overset{|}{\underset{|}{C}}}}—H$$

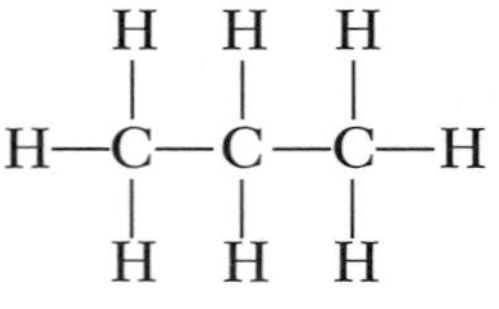

propane

Although chains of three or more carbons are represented as though they were linear, the carbon-to-carbon bond angles are really 109°28', and the carbon skeleton is zig-zag:

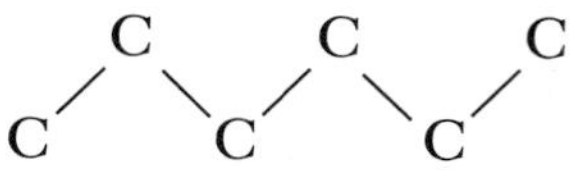

As the chain of carbon atoms lengthens, compounds are represented by condensed structural formulas. Such formulas preserve some idea of the bonding of groups of atoms within a molecule but are more convenient to write out. They may also be written for smaller molecules, such as ethane and propane:

$$CH_3CH_3 \qquad CH_3CH_2CH_3$$

ethane   propane

With butane, it becomes possible to arrange the carbon and hydrogen atoms in more than one way, each representing a structural isomer (Table 17–5). The straight-chain isomer is designated normal, or n-butane. The branched structure is known as isobutane by its common or trivial name.

Table 17–5 STRUCTURAL ISOMERS OF BUTANE

EXPANDED STRUCTURAL FORMULA	CONDENSED STRUCTURAL FORMULA
n-Butane	
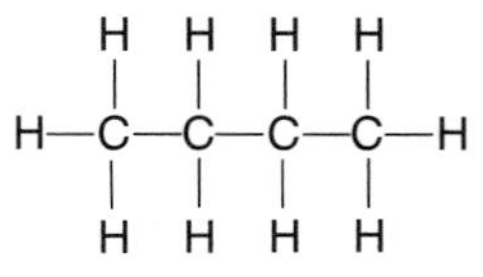	$CH_3CH_2CH_2CH_3$ or $CH_3(CH_2)_2CH_3$
Isobutane	
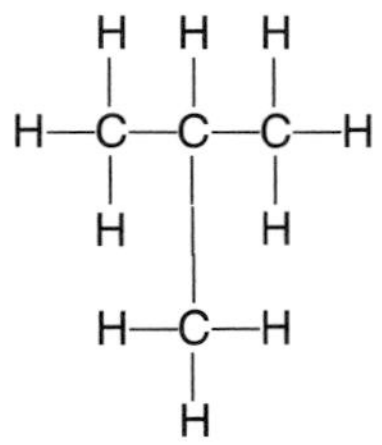	CH_3CHCH_3 $\mid$ CH_3 or $CH_3CH(CH_3)CH_3$

17–6 NOMENCLATURE OF ALKANES

A common suffix or other designation identifies the homologs of a series of organic compounds. Thus, the names of the alkanes all end with the suffix -*ane*. Beginning with pentane, the names indicate the number of carbon atoms in the molecule: *penta*, five; *hexa*, six; *hepta*, seven; *octa*, eight; and so on. To name an unbranched alkane, add the suffix -*ane* to the Greek equivalent of the number of carbons. The compound $C_{10}H_{22}$ is therefore named decane (*deka*, ten).

Compounds are referred to by two names: a common name and a systematic or IUPAC name. The International Union of Pure and Applied Chemistry (IUPAC), a group that meets periodically to consider such matters, recommends that the systematic names of the alkanes be derived by:

1. Determining the longest continuous chain of carbon atoms in the molecule and using the name of the alkane corresponding to this number as the basis for the name of the compound;
2. Numbering the carbon atoms of this chain from one end to the other, beginning at the end nearest the branching;
3. Naming the groups that are attached to this chain;
4. Designating each group by the number of the carbon atom to which it is attached;
5. Using a separate number for each branched group;
6. Numbering the longest continuous chain in such a way that the branched groups have the lowest possible numbers;
7. Using a prefix (*di-, tri-, tetra-*) to number two or more like groups.

EXAMPLE   17–1

What is the systemic (IUPAC) name of the following compound:

$$
\begin{array}{c}
\quad\quad\quad\quad\quad\quad CH_3 \\
\quad\quad\quad\quad\quad\quad | \\
\quad\quad\quad\quad CH_3\ \ CH_2 \\
\quad\quad\quad\quad | \quad | \\
\overset{1}{CH_3}-\overset{2}{CH_2}-\overset{3}{C}-\overset{4}{CH}-\overset{5}{CH_2}-\overset{6}{CH_3} \\
\quad\quad\quad\quad | \\
\quad\quad\quad\quad CH_3
\end{array}
$$

SOLUTION

Since the longest chain contains six carbons, the compound is a hexane. To give the substituents the lowest possible numbers, number the chain from left to right in this case, and assign a number and a name to each branch. The two methyl groups bonded to the third carbon in the chain are designated "3,3-dimethyl." The ethyl group bonded to carbon 4 is "4-ethyl." The complete name is *3,3-dimethyl-4-ethylhexane.* Note that a series of numbers is separated by commas, and numbers are set off from names by hyphens.

The words "methyl" and "ethyl" in the name of the compound in Example 17–1 are derived from the corresponding alkanes minus one hydrogen. They are examples of *alkyl groups.* Some commonly used alkyl groups are listed in Table 17–6. A primary carbon atom is bonded to one other carbon; secondary *(sec-)* to two carbons; and tertiary *(tert-)* to three carbons. In Example 17–1, carbon 1 is a primary carbon; carbon 2 secondary; and carbon 4 tertiary.

17–7 THE STRUCTURE OF SATURATED HYDROCARBONS

What kind of bonding does carbon have in hydrocarbons? The electronic configuration of the ground state of $_6C$, $1s^2 2s^2 2p_x^1 2p_y^1$, shows two unpaired electrons in the 2p subshell (Fig. 17–5). The structure suggests a covalence of 2 for carbon. As has long been known, however, a covalence of 4 for carbon is usual. Through collisions between atoms, the 2s electrons are uncoupled and one is promoted to the vacant $2p_z$ orbital to give an excited atom with four unpaired electrons. But this structure would mean that carbon should form two kinds of bonds with its valence electrons: one with the s electron and three with the p electrons. The experimental fact is that all four C—H bonds in methane, CH_4, are identical in their properties. The four outermost electrons of carbon in this case must produce four equivalent orbitals, each having one electron available for a covalent bond. These four orbitals are said to be *hybridized* and are called sp³ (pro-

Table 17–6 ALKYL GROUPS DERIVED FROM METHANE THROUGH BUTANE

ALKANE	***ALKYL GROUP***	***ALKYL CHLORIDE***
Methane 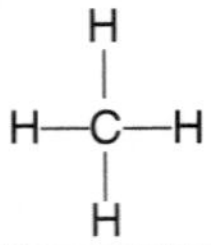	Methyl 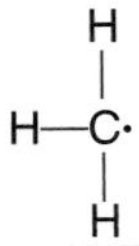	Methyl chloride CH_3Cl
Ethane 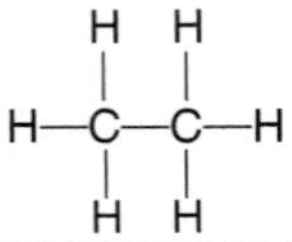	Ethyl 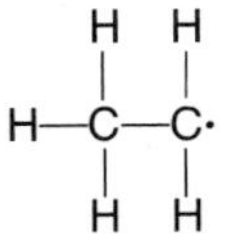	Ethyl chloride CH_3CH_2Cl
Propane 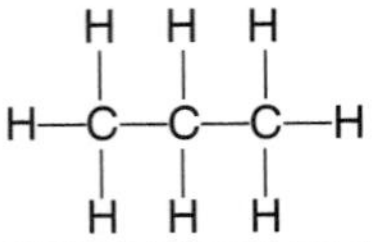	n-Propyl 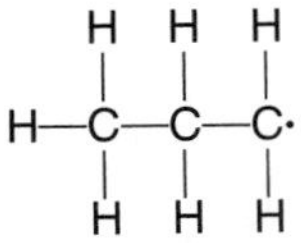	n-Propyl chloride $CH_3CH_2CH_2Cl$
	Isopropyl 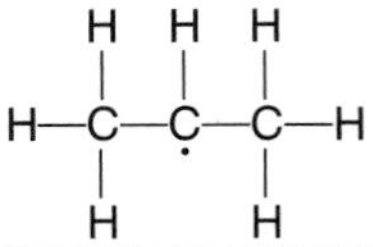	Isopropyl chloride 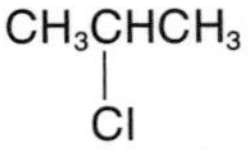
n-Butane 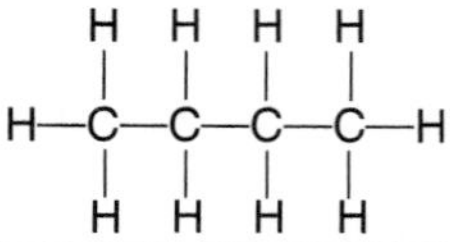	n-Butyl 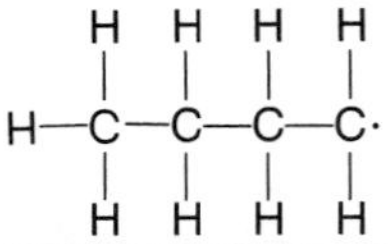	n-Butyl chloride $CH_3CH_2CH_2CH_2Cl$
	sec-Butyl 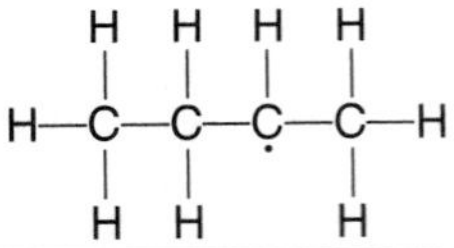	sec-Butyl chloride 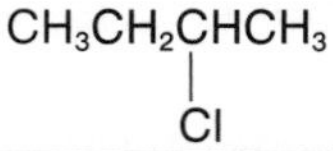
Isobutane 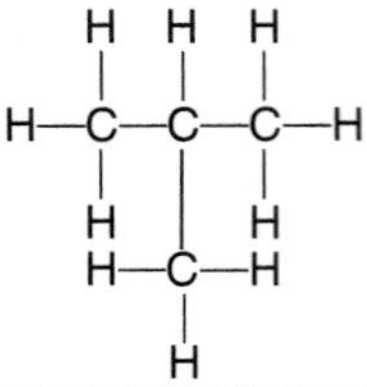	Isobutyl 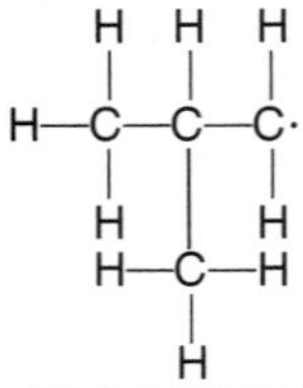	Isobutyl chloride 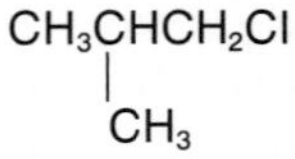
	tert-Butyl 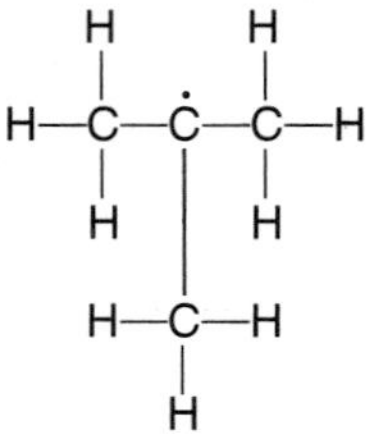	tert-Butyl chloride 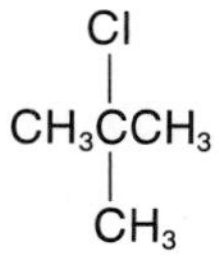

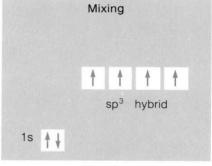

Figure 17–5 Energy-level model of the ground state of carbon.

Figure 17–6 Formation of sp³ hybrid orbital. One of the 2s electrons of carbon is promoted to the vacant $2p_z$ orbital. The remaining 2s electron and the three p electrons then mix and form four sp³ hybrid orbitals.

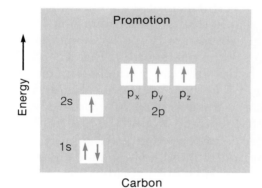

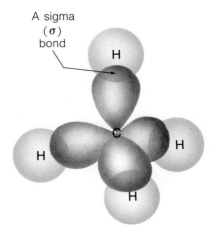

Figure 17–7 Orbital structure for methane, CH_4. The carbon atom is shown with four sp³ hybrid orbitals in a tetrahedral arrangement, and each hydrogen atom with an s orbital.

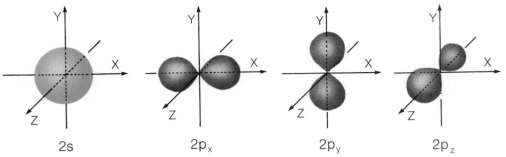

Figure 17–8 Representations of 2s and 2p orbitals.

nounced "s–p–three") orbitals (Fig. 17–6). In the formation of CH_4, each of the four sp^3 hybrid orbitals overlaps the 1s orbital of a hydrogen atom. With the carbon kernel in the center, each bond is directed to one of the four vertices of a regular tetrahedron, making an angle of 109°28' with it (Fig. 17–7).

The shape of the sp^3 hybrid orbitals differs from the s and p orbitals from which they are formed. The 2s orbital is spherical and therefore non-directional. The three 2p orbitals are propeller-shaped and oriented at angles of 90° to one another (Fig. 17–8). The shape of the sp^3 hybrid orbitals produces a much stronger covalent bond. The best overlap is obtained if a hydrogen atom lies on the axis of an sp^3 orbital. The cylindrically symmetrical orbital around the line joining carbon and hydrogen is called a sigma (σ) bond.

The continuous and branched chains of the higher alkanes also have sp^3 hybridization. Ethane, C_2H_6, is a combination of two CH_3— units: CH_3CH_3. The sp^3 hybrid orbitals of each carbon overlap the s orbitals of three hydrogen atoms and an sp^3 hybrid orbital of the

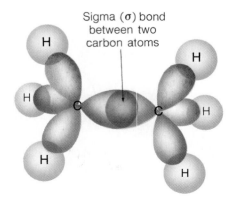

Sigma (σ) bond between two carbon atoms

Figure 17–9 Structure of ethane, C_2H_6.

other carbon, forming sigma bonds in each case (Fig. 17–9).

17–8 THE STRUCTURE OF UNSATURATED HYDROCARBONS

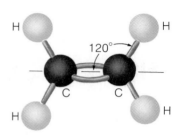

Figure 17–10 Model of ethylene, $H_2C{=}CH_2$.

In unsaturated hydrocarbons there is at least one multiple bond between the carbon atoms. An *alkene* or *olefin* contains a double bond; a *polyene* (*poly*, many) contains more than one double bond; an *alkyne* contains a triple bond. Each carbon atom in an unsaturated hydrocarbon contributes two or three electrons to the formation of a double or triple bond. Since fewer electrons are available for bonding with other atoms, unsaturated hydrocarbons contain fewer hydrogens than the maximum number possible.

The simplest alkene is ethylene, C_2H_4. Experiments show that the bond angles are 120° and that ethylene is a flat, planar molecule (Fig. 17–10). The 2s electrons of carbon are uncoupled and one is promoted to the p level in each carbon atom. Then the s and two of the p orbitals hybridize to form three sp^2 hybrid orbitals, which lie in one plane at angles of 120°. One p orbital is left over (Fig. 17–11). The sp^2 orbitals of each carbon overlap directly and form sigma bonds with two hydrogens and with each other. The remaining propeller-shaped p orbitals of the two carbon atoms are parallel to each other and perpendicular to the plane of the sp^2 hybrid orbitals, the lobes lying above and below the plane. They interact with each other laterally, producing a second, weaker bond between the carbons, called a pi (π) bond (Fig. 17–12). It is easier to disrupt the π portion of a double bond than a single bond.

Acetylene, C_2H_2, is the simplest hydrocarbon having a triple bond. The third bond between the carbons is formed at the expense of two bonds between carbon and hydrogen, so that acetylene is still more unsaturated than ethylene. Experiments indicate that the shape of acetylene is linear—the two carbons and two hydrogens lie in a straight line, with bond angles of 180° (Fig. 17–13).

In each carbon atom of acetylene, sp hybridization occurs between the s and one of the p orbitals after the s electrons have been uncoupled and one promoted to the p_z level. Two sp hybrid orbitals and the two remaining p orbitals are then available for bonding. Each

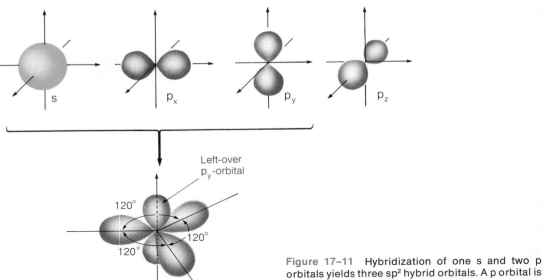

Figure 17–11 Hybridization of one s and two p orbitals yields three sp² hybrid orbitals. A p orbital is left over.

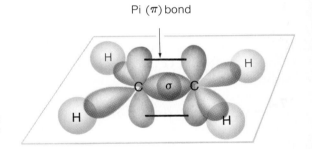

Figure 17–12 The bonding of ethylene. Each sp² hybrid orbital forms three sigma (σ) bonds. The p orbitals form a pi (π) bond.

Figure 17–13 Model of acetylene, HC≡CH.

Figure 17–14 The bonding of acetylene. Each carbon forms two sigma (σ) bonds: one to the other carbon atom, the second to a hydrogen atom. The remaining p orbitals on each carbon combine to form two pi (π) bonds.

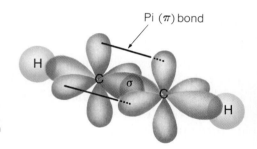

Table 17–7 PROPERTIES OF HYBRID ORBITALS OF CARBON

HYBRID-IZATION	NUMBER OF HYBRID ORBITALS	ORBITALS USED	ORBITALS REMAINING	ANGLE BETWEEN HYBRID ORBITALS	SIGMA (σ) BONDS	PI (π) BONDS
sp^3	4	2s,2p$_x$,2p$_y$,2p$_z$		109°28′	4	0
sp^2	3	2s,2p$_x$,2p$_y$	2p$_z$	120°	3	1
sp	2	2s,2p$_x$	2p$_y$,2p$_z$	180°	2	2

of the sp orbitals is directed along the x-axis, but in opposite directions (180° apart), and each is oriented at 90° to the unhybridized p$_y$ and p$_z$ orbitals (Fig. 17–14). The sp orbitals of each carbon form a sigma bond with the s orbital of one of the hydrogens and with each other. The p orbitals are at right angles to each other and to the sp orbitals, and form two π bonds between the carbon atoms. The third bond between the carbon atoms is another weak point, although the structural formula makes no distinction among the three bonds. A summary of the properties of the hybrid orbitals of carbon is given in Table 17–7, and the chemical bond characteristics of carbon in Table 17–8.

17–9   SOME REACTIONS OF ALKANES

The saturated hydrocarbons are relatively inert. They react very slowly with chlorine and bromine at room temperature, but quite readily at higher temperatures and in the presence of light. One or more hydrogen atoms are replaced by halogen atoms in these *substitution reactions*, and side reactions occur, so that a mixture of products is usually obtained.

$$CH_4 + Cl_2 \rightarrow CH_3Cl + HCl \qquad \text{Equation 17–2}$$

Table 17–8 CHEMICAL BOND CHARACTERISTICS OF CARBON

BOND	FORMULA	SUFFIX	BOND LENGTH (Å)	BOND ENERGY (kcal/mole)
Single	C—C	-ane	1.54	83.1
Double	C=C	-ene	1.34	142.0
Triple	C≡C	-yne	1.20	192.1

Table 17–9   COMPOSITION OF NATURAL GAS
USED AS FUEL

COMPONENT	PER CENT
Methane	78–80
Nitrogen	10–12
Ethane	5.9
Propane	2.9
n-Butane	0.71
Isobutane	0.26
C_5-C_7 Hydrocarbons	0.13

The chlorination of methane yields not only chloromethane, CH_3Cl, but also dichloromethane, trichloromethane, and tetrachloromethane: CH_2Cl_2, $CHCl_3$, CCl_4.

The most important use of the saturated hydrocarbons is as fuels. In complete combustion they form carbon dioxide and water and evolve large amounts of heat, as can be seen from the combusion reactions of methane (CH_4) and octane (C_8H_{18}), a component of gasoline:

Equation   17–3

$$CH_4 + 2O_2 \rightarrow CO_2 + 2H_2O + 210.8 \text{ kcal/mole}$$

Equation   17–4

$$2C_8H_{18} + 25O_2 \rightarrow 16CO_2 + 18H_2O + 1302.7 \text{ kcal/mole}$$

The composition of natural gas sold as fuel is given in Table 17–9; the composition of other hydrocarbon fuels is given in Table 17–10. Propane and butane are gases at room temperature under normal pressure

Table 17–10   COMPOSITION OF HYDROCARBON FUELS

FUEL	CARBON CONTENT OF COMPOUNDS	BOILING POINT RANGE, °C	USES
Acetylene	C_2H_2		Welding
Natural gas	C_1–C_4		Industrial & home fuel
Propane	C_3H_8		Bottled fuel gas
Butane	C_4H_{10}		Bottled fuel gas
Gasoline	C_5–C_{12}	35–225	Motor fuel
Kerosene	C_{12}–C_{16}	200–315	Jet fuel, domestic heating
Fuel oil	C_{15}–C_{18}	250–375	Diesel fuel, industrial fuel
Paraffin wax	C_{20}–C_{30}	50–60 (melting point)	Candles

but are liquefied under pressure and sold as liquefied petroleum gas (LPG).

If there is a lack of oxygen when a hydrocarbon is burned, carbon monoxide, CO, an invisible, deadly gas, is formed.

$$C_8H_{18} + 12O_2 \rightarrow 7CO_2 + CO + 9H_2O \qquad \text{Equation } 17\text{--}5$$

Millions of tons of carbon monoxide are poured into the atmosphere each year, about 80 per cent of it from automobile exhausts. Hydrocarbons from unburned fuel in the exhaust may also react with ozone and atomic oxygen in the air to form chemically reactive aldehydes with extremely irritating odors. Gasoline, a complex mixture of simple hydrocarbons plus slight amounts of additives, is very volatile and escapes into the air. With no emission controls, 60 per cent of the emitted hydrocarbons come from the exhaust, 20 per cent escapes from the crankcase, 10 per cent results from evaporation at the carburetor, and 10 per cent evaporates from the fuel tank. The mixture of products resulting from reactions in the atmosphere—ozone, carbon monoxide, peroxyacylnitrates (PAN), aldehydes, ketones, and alkylnitrates—is called photochemical smog. It causes eye irritation and may interfere with respiration.

The catalytic afterburner uses a catalyst to allow the hydrocarbon oxidation to be completed at a lower temperature. The control of evaporation losses is brought about by a collection system that transports fuel vapors from the fuel tank and carburetor to the fuel system; the vapors are then burned in the engine. Hydrocarbon emissions from crankcase "blow-by" have been eliminated by using a positive crankcase ventilation system (PCV) that recycles hydrocarbons to the engine intake.

17–10 OCTANE RATING OF GASOLINES

The efficiency of the internal combusion engine is related to the octane rating of the fuel and to the compression ratio. With some gasolines, ignition occurs more as an explosion than as a smooth burning. This not only reduces the efficiency but also produces an audible "knock" in the engine. A fuel of pure isooctane produces little knock, while one of pure n-heptane knocks badly.

The octane rating of a gasoline is a measure of the knock produced when the gasoline is used as an automobile fuel. Isooctane and n-heptane have octane ratings of 100 and 0, respectively. A mixture of 90 per cent isooctane and 10 per cent heptane has an octane rating of 90. A gasoline rated 90 produces the same knock as a mixture of 90 per cent isooctane and 10 per cent n-heptane. Fuels have been developed with octane ratings greater than 100, their anti-knock properties being superior to isooctane. Table 17–11 lists the octane numbers required for various engine compression ratios.

The octane rating of gasolines can be improved by various means. Heating a gasoline in the presence of catalysts such as sulfuric acid or aluminum chloride may convert a straight-chain structure into a highly branched isomer with improved burning characteristics and anti-knocking properties. The addition of small amounts of certain substances has a similar effect.

The most common additive is a compound called tetraethyl lead (TEL), $(C_2H_5)_4Pb$. Its antiknock property was discovered by Thomas Midgley (1889–1944), a chemist at General Motors Research Laboratory, in 1922. When added in amounts as small as 3 milliliters per gallon, it often increases the octane number by a large amount, as Table 17–12 shows. To prevent deposits of metallic lead and lead oxides from fouling the valves and spark plugs, another compound called a scavenger, usually ethylenedibromide $(C_2H_2Br_2)$ or ethylenedichloride $(C_2H_2Cl_2)$, is added. This converts the lead to gaseous lead (II) bromide or lead (II) chloride, which escapes through the exhaust. In this way,

Table 17–11   OCTANE NUMBER AND ENGINE COMPRESSION RATIO

ENGINE COMPRESSION RATIO	OCTANE NUMBER REQUIRED
5:1	73
6:1	81
7:1	87
8:1	91
9:1	95
10:1	98
11:1	100
12:1	102

Table 17–12 OCTANE NUMBERS OF SOME HYDROCARBONS, WITH AND WITHOUT TETRAETHYL LEAD (TEL)

HYDROCARBON	MOELCULAR FORMULA	OCTANE NUMBER	
		No TEL	3.0 ml/gallon TEL
n-Butane	C_4H_{10}	93.6	101.6
n-Pentane	C_5H_{12}	61.7	88.7
2-Methylbutane	C_5H_{12}	92.6	102.0
n-Hexane	C_6H_{14}	24.8	65.3
n-Heptane	C_7H_{16}	0.0	43.5
Methylbenzene (Toluene)	C_7H_8	103.2	111.8
n-Octane	C_8H_{18}	−19.0	25.0
2,2,4-Trimethylpentane (Isooctane)	C_8H_{18}	100.0	115.5
Isopropylbenzene (Cumene)	C_9H_{12}	113.0	116.7

thousands of tons of lead go out the tailpipe into our atmosphere.

$$Pb(C_5H_5)_4 + CH_2CH_2 + 16O_2 \rightarrow 10CO_2 + 12H_2O + PbBr_2$$
$$\quad\quad\quad\quad\quad | \quad\ |$$
$$\quad\quad\quad\quad Br \quad Br$$

Equation 17–6

Catalytic converters use platinum and palladium to convert harmful carbon monoxide and hydrocarbons into harmless carbon dioxide and water, as required by the Clean Air Act of 1970, which ordered 90 per cent reduction in these emissions. Since the catalysts can be ruined by lead in gasoline, the Environmental Protection Agency required the addition of unleaded pumps at stations.

17–11 BENZENE; AROMATIC HYDROCARBONS

Many hydrocarbons have a higher ratio of carbon to hydrogen than the alkenes or alkynes, yet have distinctly different properties from these unsaturated hydrocarbons. They are frequently fragrant themselves or can be derived from aromatic substances such as cinnamon, sassafras, and wintergreen.

Michael Faraday discovered one of these compounds in illuminating gas in 1825. Later the same compound was obtained from coal tar, petroleum, and benzoic acid isolated from the aromatic substance gum benzoin, and it became known as benzene. Its molecular formula was determined to be C_6H_6. Josef Loschmidt (1821–1895) proposed that aromatic compounds, such as phenol, aniline, and toluene, could be considered derivatives of benzene, just as the alkanes were considered to be derivatives of methane, CH_4. Since then, the term *aromatic* has been applied to compounds chemically similar to benzene. Aromatic hydrocarbons are used extensively in the manufacture of plastics, dyes, insecticides, and many other materials.

The molecular formula of benzene, C_6H_6, suggests greater unsaturation than is present in alkenes and alkynes. Contrary to expectations, however, benzene does not show the slightest chemical similarity to these compounds. Surprisingly, benzene resembles the alkanes in its inertness toward halogens and other reagents. Its usual reaction is one of substitution:

Equation 17–7

$$C_6H_6 \ + \ Br_2 \ \xrightarrow[\text{catalyst}]{\text{Fe}} \ C_6H_5\!-\!Br \ + \ HBr$$

benzene   bromine   bromobenzene   hydrogen bromide

Equation 17–8

$$C_6H_6 \ + \ HONO_2 \ \xrightarrow[\text{catalyst}]{\text{H}_2\text{SO}_4} \ C_6H_5\!-\!NO_2 \ + \ H_2O$$

benzene   nitric acid   nitrobenzene   water

Benzene can be converted to cyclohexane by reacting with H_2, but under more severe conditions than the hydrogenation of alkenes and alkynes. This reaction suggests a cyclic structure of benzene.

Equation 17–9

$$C_6H_6 + 3H_2 \ \xrightarrow[\text{Ni catalyst, 150°C}]{\text{high pressure}}$$

benzene   cyclohexane

Friedrich August Kekulé in 1865 proposed a ring structure for benzene in which three double bonds alternate with three single bonds.

$$
\begin{array}{cc}
\overset{\displaystyle H}{\underset{\displaystyle }{|}} & \overset{\displaystyle H}{\underset{\displaystyle }{|}} \\
\end{array}
$$

A second Kekulé structure can be drawn with the double bonds alternating the other way. Abbreviated Kekulé formulas are often written as follows:

The Kekulé formula for benzene is supported in broad outline by modern studies. Electron diffraction experiments and x-ray analysis of crystalline benzene derivatives indicate that the six carbon atoms lie in a plane and that they form the corners of a regular hexagon. The hydrogen atoms lie in the same plane. Benzene is a flat, completely symmetrical molecule. However, the Kekulé formula predicts different bond lengths for the single and double bonds, but measurements show that the six carbon-to-carbon bonds are equal in length, 1.39Å, intermediate between the length of a single bond and a double bond.

The geometry of the benzene molecule suggests that the carbon atoms are sp^2 hybridized. Each carbon uses its three sp^2 hybrid orbitals to form sigma bonds with a hydrogen atom and two carbon neighbors. The fourth valence electron of each carbon occupies a p orbital with lobes above and below the plane of the ring; there are thus six p orbitals in the molecule, all parallel to one another (Fig. 17–15). Evidently, the p orbitals do not overlap to form localized pi bonds as in alkenes and alkynes, as shown by the unvarying carbon-to-carbon bond lengths in benzene.

Instead, each p orbital must interact with each of the adjacent carbons rather than with just one. The result is two doughnut-shaped electron clouds, lying above and below the plane of the benzene ring (Fig. 17–16). This participation in the formation of more than

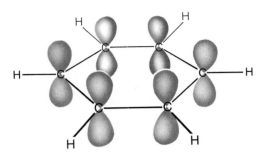

Figure 17–15 Structure of benzene. The carbon and hydrogen atoms all lie in the same plane. The p orbitals are all at right angles to the plane of the ring.

one bond is called delocalization. The six p electrons of benzene are delocalized and form an extended cyclic π bond around all six carbons, rather than three localized π bonds between specific pairs of carbon atoms.

Delocalized bonding is stronger than ordinary covalent bonding between two atoms. Benzene is stable because the extended π bond resists breaking up; however, it is unaffected by substitution reactions. Since the benzene molecule has no definite single or double bonds, it is best represented by a Thiele formula, a circle in the center of a hexagon. The circle represents the π electron cloud of six electrons; the hexagon, the skeleton of six carbon atoms.

Thiele formula for benzene

Some examples of compounds derived from benzene are:

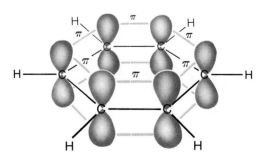

Figure 17–16 Extended bonding in the benzene molecule involves all six p orbitals, which are described as being delocalized. A cyclic charge cloud is formed above and below the plane of the carbon and hydrogen atoms.

para-dichlorobenzene
(moth flakes)

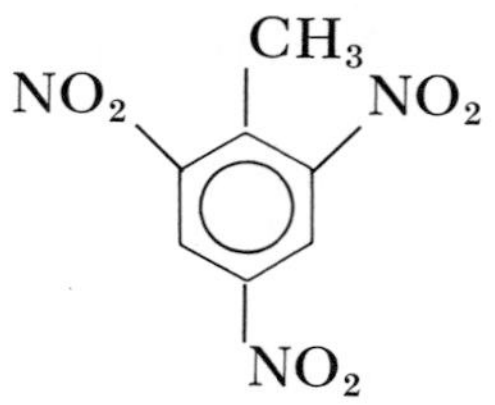

2,4,6-trinitrotoluene
(TNT)

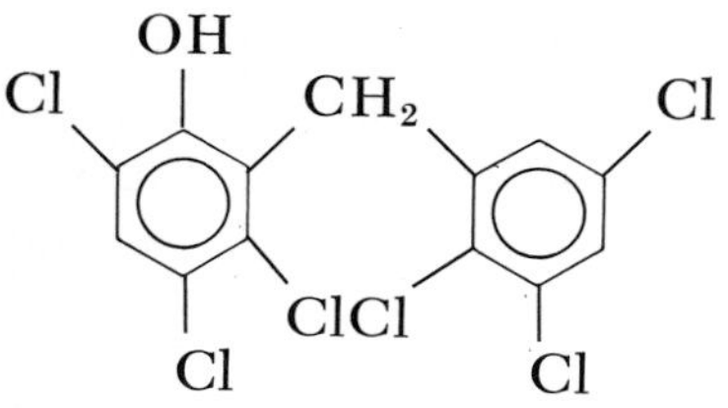

hexachlorophene

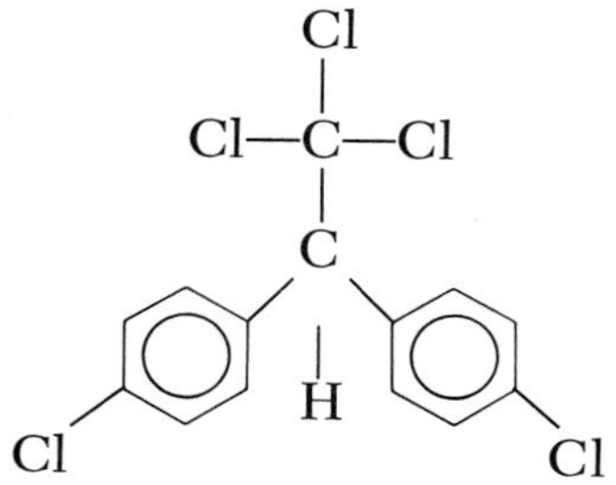

dichlorodiphenyltrichloroethane
(DDT)

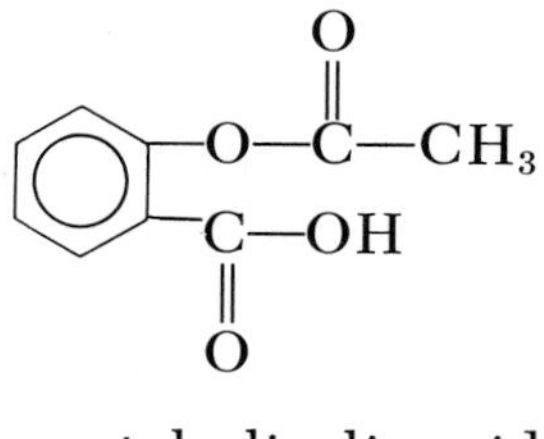

acetylsalicylic acid
(aspirin)

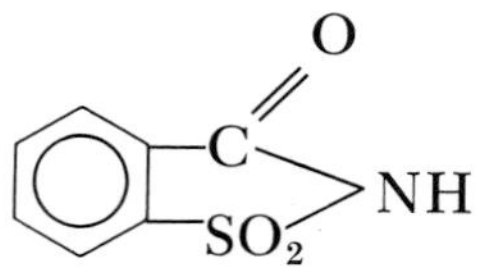

saccharin

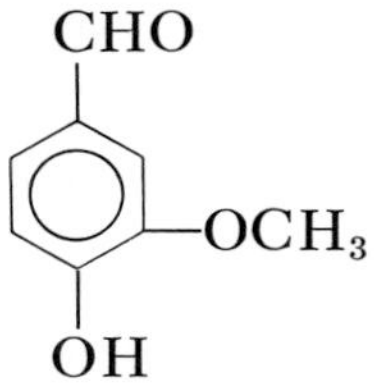

vanillin

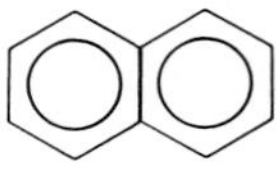

naphthalene

17–12   FUNCTIONAL GROUPS; ALCOHOLS

The replacement of a hydrogen atom of a hydrocarbon by another atom or group of atoms results in compounds called *hydrocarbon derivatives*. The atom or group of atoms is a *functional group*. The chemistry of these derivatives is mainly that of the functional group, which imparts certain common chemical and physical properties to all members of the family. Thus, the hydroxyl group, —OH, is found in all alcohols, represented by the general formula R—OH. The symbol R represents an alkyl group, the radical that remains following the removal of a hydrogen from an alkane, such as methyl, CH_3—, from methane, CH_4. Methyl alcohol is CH_3OH; ethyl alcohol, C_2H_5OH.

The many thousands of organic compounds can be organized on the basis of a relatively small number of functional groups. Some of the more common ones are listed in Table 17–13. It is not unusual for more than one such group to occur in a compound.

Table 17–13   COMMON FUNCTIONAL GROUPS IN ORGANIC COMPOUNDS

TYPE OF COMPOUND	FUNCTIONAL GROUP	GENERAL FORMULA	EXAMPLE	NAME (IUPAC)
Alkane	H	R—H	CH_4	Methane
Alkene	$\diagdown$C=C$\diagup$	$R_2C{=}CR_2$	CH_3, CH_3 on C=C on CH_3, CH_3	2,2-Dimethylbutene
Alkyne	—C≡C—	RC≡CR	CH_3—C≡C—CH_3	2-Butyne
Halide	X:F,Cl,Br,I	R—X	CH_3Br	Bromomethane
Alcohol	—OH	R—OH	CH_3CH_2OH	Ethanol
Ether	—O—	R—O—R'	CH_3—O—CH_3	Methoxymethane
Aldehyde	$-\overset{\displaystyle O}{\overset{\|}{C}}-H$	$R-\overset{\displaystyle O}{\overset{\|}{C}}-H$	$CH_3\overset{\displaystyle O}{\overset{\|}{C}}H$	Ethanal
Ketone	$-\overset{\displaystyle O}{\overset{\|}{C}}-R'$	$R-\overset{\displaystyle O}{\overset{\|}{C}}-R'$	$CH_3\overset{\displaystyle O}{\overset{\|}{C}}CH_3$	Propanone
Carboxylic acid	$-\overset{\displaystyle O}{\overset{\|}{C}}-OH$	$R-\overset{\displaystyle O}{\overset{\|}{C}}-OH$	$CH_3\overset{\displaystyle O}{\overset{\|}{C}}OH$	Ethanoic acid
Ester	$-\overset{\displaystyle O}{\overset{\|}{C}}-OR'$	$RC\overset{\displaystyle O}{\overset{\|}{}}-OR'$	$CH_3\overset{\displaystyle O}{\overset{\|}{C}}-OCH_3$	Methyl ethanoate
Amine	—NH_2	R—NH_2	CH_3NH_2	Methylamine

Each functional group is assigned a suffix in the IUPAC system: *-ol* for alcohols; *-one* for ketones; and so on. The basis of the name of a compound is the longest continuous alkane to which the functional group is attached. That alkane is considered to be the parent hydrocarbon, and the functional groups are derivatives of it.

EXAMPLE 17–2

Give the systematic (IUPAC) name for the following compound.

$$\overset{\displaystyle CH_3}{\underset{\displaystyle \underset{\textstyle CH_3OH}{|}}{\overset{|}{\underset{1}{CH_3}\underset{2}{C}}}}\quad \underset{3}{CH}\underset{4}{CH_2}\underset{5}{CH_2}\underset{6}{CH_3}$$

SOLUTION

Number the carbon atoms of the parent alkane starting at the end nearest the functional group, the —OH group, and designate the position of the functional group by the number of the carbon atom to which it is attached. This compound is an alcohol, specifically, a 3-hexanol. The names of any additional groups attached to the parent chain and the numbers of the carbons to which they are attached are indicated, followed by the name of the parent. Thus, the complete name is *2,2-dimethyl-3-hexanol*.

Alcohols are in a sense alkyl derivatives of water, in which one of the H atoms of water is replaced by an R group, as much as they are hydroxy derivatives of alkanes, in which a hydrogen atom of an alkane is replaced by an —OH radical. The lower homologs are similar to water in structure:

$$\underset{\text{water}}{\overset{\displaystyle H \qquad H}{\diagdown\!\underset{O}{}\!\diagup}} \qquad \underset{\text{an alcohol}}{\overset{\displaystyle R \qquad H}{\diagdown\!\underset{O}{}\!\diagup}} \qquad \underset{\text{methanol}}{\overset{\displaystyle CH_3 \qquad H}{\diagdown\!\underset{O}{}\!\diagup}}$$

As in water, there are strong intermolecular attractions between alcohol molecules; these attractions arise from dipole-dipole interactions and from hydrogen bonding, an intermolecular force between the hydrogen of one molecule and the more electronegative oxygen in a nearby molecule (Fig. 17–17). As a result, methyl alcohol, CH_3OH, is a liquid at room temperature. Alcohols of low molecular weight are soluble in water, but as the alkyl group lengthens, the higher alcohols resemble the corresponding hydrocarbons more and more. Polyhydric alcohols exist in which two or more —OH groups are present. Ethylene glycol, commonly used as an antifreeze, is a dialcohol:

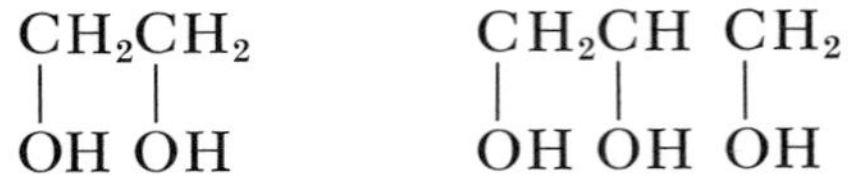

$$\begin{array}{cc} \underset{|}{CH_2CH_2} & \underset{|}{CH_2CH\ CH_2} \\ OH\ OH & OH\ OH\ OH \end{array}$$

ethylene glycol   glycerol

and glycerol (or glycerine) is a tri-hydroxyalcohol. The formulas and names of some alcohols are listed in Table 17–14.

Ethyl or grain alcohol (CH_3CH_2OH) is the most common alcohol. It is present in varying concentrations in such beverages as cider, beer, wine, champagne, and gin. Even the strongest alcoholic beverages are seldom more than 45 per cent ethyl alcohol, or 90 proof. (Proof is twice the alcohol percentage by volume.)

Grain alcohol is often produced by fermenting sugars from various sources—corn, rice, potatoes, sugar cane—in one of the oldest chemical processes known. Fermentation takes place through certain en-

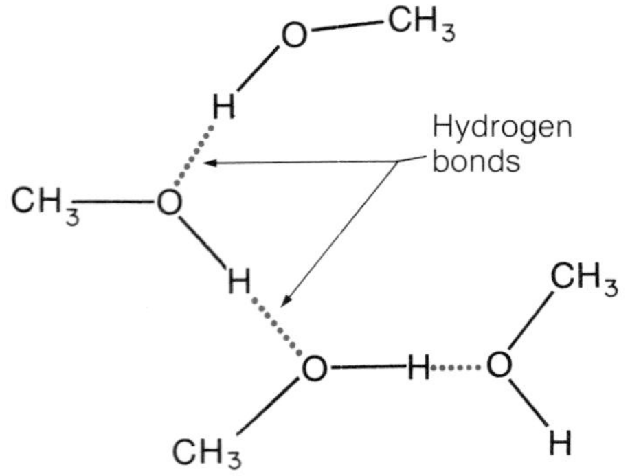

Figure 17–17 Hydrogen bonding in methyl alcohol, CH_3OH.

Table 17–14 SOME COMMON ALCOHOLS

ALCOHOL	IUPAC NAME	COMMON NAME	SOLUBILITY IN WATER (g/100g H_2O, 20°C)
CH_3OH	Methanol	Methyl alcohol	Miscible
CH_3CH_2OH	Ethanol	Ethyl alcohol	Miscible
$CH_3CH_2CH_2OH$	1-Propanol	n-Propyl alcohol	Miscible
$CH_3CH(OH)CH_3$	2-Propanol	Isopropyl alcohol	Miscible
$CH_3CH_2CH_2CH_2OH$	1-Butanol	n-Butyl alcohol	7.9
$CH_3CH_2CH(OH)CH_3$	2-Butanol	sec-Butyl alcohol	12.5
$CH_3CH(CH_3)CH_2OH$	2-Methyl-1-propanol	Isobutyl alcohol	10
$CH_3C(OH)(CH_3)CH_3$	2-Methyl-2-propanol	t-Butyl alcohol	Miscible

zymes found in yeast *(Saccharomyces cerevisiae)* that convert sugars to ethyl alcohol and carbon dioxide.

$$C_{12}H_{22}O_{11} + H_2O \xrightarrow{\text{yeast}} 4CH_3CH_2OH + 4CO_2$$
$$\text{sugar} \qquad\qquad\qquad \text{ethyl alcohol}$$

Equation 17–10

Most industrial alcohol is produced by reacting ethylene with water in the presence of sulfuric acid. An addition reaction takes place.

$$CH_2{=}CH_2 + H_2O \xrightarrow[\substack{100\ \text{atm}}]{\substack{300°C \\ H_2SO_4}} CH_3CH_2OH$$

Equation 17–11

Since alcohol for drinking purposes is heavily taxed, industrial ethyl alcohol is denatured by adding poisonous compounds such as methyl alcohol (wood alcohol) to it, making it unfit to drink. Blindness can result from methanol poisoning, the optic nerve being particularly susceptible. Ethyl alcohol is used in large quantities as an industrial solvent and as a starting material for other chemical products.

Alcohols react with carboxylic acids—organic compounds containing a carboxyl group (—COOH—) —to form the class of compounds known as *esters.* Animal and vegetable fats, oils, and waxes are largely made up of esters. Fruit aromas are composed of many different components, of which esters are a dominant one.

Table 17–15   SOME ESTERS: THEIR AROMAS AND FLAVORS

ESTER	FORMULA	AROMA OR FLAVOR
Ethyl acetate	$CH_3COOCH_2CH_3$	Apple
Ethyl butyrate	$CH_3CH_2CH_2COOCH_2CH_3$	Apricot
Ethyl formate	$HCOOCH_2CH_3$	Rum
Ethyl heptanoate	$CH_3(CH_2)_5COOCH_2CH_3$	Cognac
Isoamylacetate	$CH_3COOCH_2CH_2CH(CH_3)CH_3$	Banana
Methyl-n-valerate	$CH_3(CH_2)_3COOCH_3$	Pineapple
Octylacetate	$CH_3COO(CH_2)_7CH_3$	Orange

Equation 17–12

$$\underset{\text{acetic acid}}{CH_3\overset{O}{\overset{\|}{C}}\boxed{-OH + H}}\;\;\underset{\text{n-butyl alcohol}}{O\;\;CH_2CH_2CH_2CH_3} \rightleftarrows \underset{\text{n-butyl acetate}}{CH_3\overset{O}{\overset{\|}{C}}-O-CH_2CH_2CH_2CH_3} + \underset{\text{water}}{H_2O}$$

Some common esters used in perfumes and flavors are listed in Table 17–15. Natural fats are solid esters; oils are liquid esters of glycerol.

17–13 POLYMERS

Among the oldest and most familiar materials are natural high polymers (*poly*, many; *meros*, part)—namely, silk, cotton, wool, and fur. Until the early decades of the 20th century, polymers were a chemical mystery because of their complexity and fragility. Almost every aspect of our contemporary life, on the other hand, reflects in some way the advances made in the laboratory and industrial synthesis of giant molecules. Much of our clothing is made from synthetic fibers. We walk on carpets and sit on furniture made from synthetic polymers.

Cellulose, rubber, starch, and proteins, like all organic compounds, are composed mainly of carbon, hydrogen, and oxygen. Their main distinguishing feature is the large size of their molecules. Since these macromolecules are made of simpler building blocks —isoprene in the case of rubber, amino acids in proteins—they are called polymers. Molecular weights of polymers may range from 100 000 to several million. Any substance that can serve as a building unit is a *monomer*. More than 40 organic monomers derived from coal and oil can be combined in an almost limit-less number of ways to form new materials.

Addition polymerization is one way by which small molecules are assembled into large ones. Many olefins polymerize in the presence of free radicals. Styrene is an ethylene derivative in which a benzene ring replaces one of the hydrogens of ethylene. In the first step of the process, a catalyst is dissociated by heat or light into free radicals.

$$R\text{—}R \xrightarrow[\text{or light}]{\text{heat}} 2R\cdot$$

catalyst   free radicals

Equation 17–13

benzoyl peroxide   benzoyloxy radicals

In the propagation step, a free radical attacks a molecule of the monomer and reacts with its double bond by adding to it on one side and freeing an electron on the other. The new free radical is also capable of attacking a styrene molecule and maintaining the free radical character of the chain end. A large number of monomer units are added in this way to the growing chain, and a macromolecule is built up. The process terminates when two of the giant molecules happen to collide with each other and react to form a normal stable molecule, completing the chain reaction.

Equation 17–14

styrene   a new radical

Polystyrene is used widely in packaging materials, food containers, electrical insulation, and molded objects. As styrofoam it is used to make picnic coolers and other insulated containers. It has many other uses as well, and can be polymerized with other materials. Polymers are so versatile in their applications that world production is millions of tons annually, and rising.

The radical formed by the loss of one hydrogen from ethylene is known as the vinyl group. Styrene is the name given to vinyl benzene.

$$H_2C{=}CH_2 \qquad H_2C{=}CH \qquad H_2C{=}CH$$

    ethylene   vinyl group   vinyl benzene
                                        (styrene)

Other vinyl-type addition polymers and some of their uses are listed in Table 17–16.

Another type of polymerization was discovered by Leo Hendrick Baekeland (1863–1944). An organic chemist in search of a substitute for shellac, Baekeland became interested in the gummy, tarlike liquid material that fouled up his glassware when in 1909 he carried out a reaction between phenol and formaldehyde. He found that by applying heat and pressure he could turn the liquid into a hard, transparent resin that had interesting properties: electrical insulation and resistance to heat, moisture, chemicals, and mechanical wear. The phenol and formaldehyde monomers polymerize to form a large three-dimensional network as heat "splits out" water molecules, the hydrogen coming from the benzene ring and oxygen from formaldehyde. In effect, the phenolic rings are joined together by CH_2 units.

Equation 17–15

phenol formaldehyde

Table 17–16 REPRESENTATIVE VINYL
ADDITION POLYMERS

MONOMER	POLYMER	USES
H H | | C═C | | H H Ethylene	H H H H H H | | | | | | —C—C—C—C—C—C— | | | | | | H H H H H H Polyethylene	Wrapping film, tubing, molded objects
H H | | C═C | | H Cl Vinyl chloride	H H H H H H | | | | | | —C—C—C—C—C—C— | | | | | | H Cl H Cl H Cl Polyvinyl chloride	Phonograph records, raincoats, bottles
F F | | C═C | | F F Tetrafluoroethylene	F F F F F F | | | | | | —C—C—C—C—C—C— | | | | | | F F F F F F Polytetrafluoroethylene	Teflon, chemically resistant coatings
H Cl | | C═C | | H Cl Vinylidene chloride	H Cl H Cl H Cl | | | | | | —C—C—C—C—C—C— | | | | | | H Cl H Cl H Cl Polyvinylidene chloride	Saran
H H | | C═C | | H C≡N Acrylonitrile	H H H H H H | | | | | | —C—C——C—C——C—C— | | | | | | H C≡N H C≡N H C≡N Polyacrylonitrile	Orlon, Acrilan
H H | | C═C | | H CH_3 Propylene	H H H H H H | | | | | | —C—C——C—C——C—C— | | | | | | H CH_3 H CH_3 H CH_3 Polypropylene	Fibers, films, carpets, heart valves, steering wheels
H CH_3 | | C═C | | H CH ‖ CH_2 Isoprene	H CH_3 H H | | | | —C—C═C—C— | | H H Polyisoprene	Synthetic rubber

The invention of the material, called Bakelite, gave birth to the synthetic plastics industry. It is used in molded objects such as telephone housings, insulators, varnishes, and in many other ways.

Wallace Hume Carothers (1896–1937), a research chemist at DuPont, in 1935 prepared the first completely synthetic fiber—nylon—which has since replaced silk in almost all its uses. The most common type, Nylon 66, is formed by the condensation polymerization of a mixture of two 6-member monomers, hexamethylenediamine $(NH_2-(CH_2)_6-NH_2)$ and adipic acid $(HOOC-(CH_2)_4-COOH)$, in which molecules of water are split out:

Equation 17–16

$$HOOC(CH_2)_4COOH + NH_2(CH_2)_6NH_2 \longrightarrow$$

adipic acid　　　hexamethylenediamine

$$\left[\underset{}{-\overset{\overset{O}{\parallel}}{C}-(CH_2)_4-\overset{\overset{O}{\parallel}}{C}-NH(CH_2)_6NH-} \right]_n + 2nH_2O$$

Nylon is an example of a polyamide, a polymer having the repeating unit —CONH—. It comes in sheet, fiber, and bristle form, and has many uses: hosiery, shirts, molded items, brushes, tires, and parachutes.

The discovery of nylon laid the foundation for the development of all synthetic fibers. The easy-care, permanent-press polyster used in clothing, carpeting, and draperies is obtained from the condensation of ethylene glycol and terephthalic acid:

$$2\ \underset{\underset{OH}{|}}{CH_2}\ \underset{\underset{OH}{|}}{CH_2} + HO-\overset{\overset{O}{\parallel}}{C}-\underset{}{\bigcirc}-\overset{\overset{O}{\parallel}}{C}-OH \longrightarrow$$

ethylene glycol　　　　　terephthalic acid

Equation 17–17

$$\left(-O-CH_2-CH_2-O-\overset{\overset{O}{\parallel}}{C}-\underset{}{\bigcirc}-\overset{\overset{O}{\parallel}}{C}- \right)_n + nHOCH_2CH_2OH$$

poly(ethylene terephthalate)

(PET)

Poly(ethylene terephthalate), or PET, is sold as Dacron, Kodel, and Fortrel. The fabrics can be set or stabilized so that the resulting configuration is maintained during wear or use with little or no ironing required after laundering. Trouser creases or skirt pleats are there for the life of the garment. For these purposes, PET fiber is blended with cotton in a 50:50 or 65:35 ratio of PET to cotton. The consumption of synthetic fibers now exceeds that of the natural fibers cotton and wool, and polyester is one of the most versatile available.

17–14 ISOMERISM REVISITED

The type of isomerism discussed in Section 17–4 is *chain isomerism*, a type of structural isomerism in which there is a difference in the sequence of carbon atoms. Examples are n-butane and isobutane. Consider these two halides:

$$
\begin{array}{cc}
\underset{\displaystyle \text{1,1-dichloroethane}}{
H-\overset{\displaystyle \overset{H}{|}}{C}-\overset{\displaystyle \overset{H}{|}}{\underset{\displaystyle \underset{Cl}{|}}{C}}-Cl}
& \text{and} \qquad
\underset{\displaystyle \text{1,2-dichloroethane}}{
H-\overset{\displaystyle \overset{H}{|}}{\underset{\displaystyle \underset{Cl}{|}}{C}}-\overset{\displaystyle \overset{H}{|}}{\underset{\displaystyle \underset{Cl}{|}}{C}}-H}
\end{array}
$$

Their molecular formulas are identical: $C_2H_4Cl_2$. They contain the same number and the same types of bonds, and the same carbon skeleton. But they differ in the point on the chain where some type of substitution occurs, and for this reason they are positional isomers. Another example are the alcohols, 1-propanol and 2-propanol. The carbon skeleton is the same in both molecules, which differ only in the position of the —OH group.

$$
\begin{array}{cc}
CH_3CH_2CH_2OH & CH_3\overset{\displaystyle \underset{\displaystyle \underset{OH}{|}}{C}}{H}CH_3 \\
\text{1-propanol} & \text{2-propanol}
\end{array}
$$

A third type of structural isomerism is found in C_2H_6O. When the atoms are arranged in the sequence CH_3-O-CH_3, the substance is an ether; in the sequence CH_3CH_2OH, the substance is an alcohol. The different sequences result in different functional groups. The properties of functional group isomers differ.

Dimethyl ether	*Ethyl alcohol*
(CH_3-O-CH_3)	(C_2H_5OH)
1. A gas	1. A liquid
2. Practically insoluble in H_2O	2. Miscible in all proportions in H_2O

There are compounds that have the same molecular formula (the same number and kind of atoms) and the same structural formula (the same sequence of atoms), but which differ in the spatial arrangement of the atoms. They are *stereoisomers*. In geometrical or *cis-* and *trans-*isomers, the compounds differ in their configuration because of lack of free rotation, most commonly about a double bond. For example, 2-butene exists in two isomeric forms that can be represented only by their configurational formulas. *Cis-* denotes that the groups that are part of the parent chain lie on the same side of an axis drawn through the double bond; *trans-* indicates that they lie on opposite sides.

cis-2-butene   *trans*-2-butene

Note the differences in some of the physical properties of the *cis-* and *trans-*isomers listed in Table 17–17, which reflect the differences in the shapes of the molecules and intermolecular attractions. The chemical properties of *cis-* and *trans-*isomers are similar, since the same functional group is present, but there are differences in the rates of their reactions.

A more subtle type of isomerism occurs in many compounds in which one or more asymmetric carbon

Table 17–17 SOME PHYSICAL PROPERTIES OF CIS- AND TRANS-2-BUTENE

	BOILING POINT, °C	MELTING POINT, °C	DENSITY (g/ml, 20°C)	HEAT OF FUSION (kcal/mole)
cis-2-Butene	+3.7	−139	0.621	1.75
trans-2-Butene	+0.9	−106	0.604	2.33

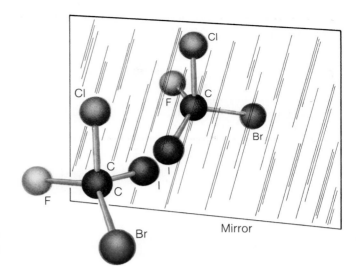

Figure 17–18 Optical isomers. Mirror images of an asymmetric carbon atom cannot be superimposed.

atoms are present. An asymmetric carbon is one that is bonded to four different groups, like the carbon in CFClBrI, fluorochlorobromoiodomethane. The compound and its mirror-image isomer are optical isomers (Fig. 17–18). An estimate of the maximum possible isomers for a given structural formula is given by van't Hoff's Rule: $\# = 2^n$, in which n is the number of asymmetric carbons. If n = 4 in a compound, then the maximum number of optical isomers is $2^4 = 16$. The combined effect of the various kinds of isomerism is to increase the possible number of carbon compounds enormously.

A classic illustration of optical isomerism is that of two forms of tartaric acid, which are mirror images of each other:

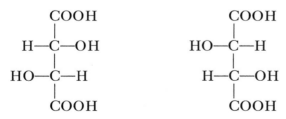

As Table 17–18 indicates, levo- and dextro-tartaric acids have identical physical properties, with one exception—their effect on polarized light, denoted by $[\alpha]_D^{20}$, the specific rotation. One isomer rotates the plane of polarized light 12° to the right (dextro, d); the other 12° to the left (levo, l).

Table 17–18   SOME PHYSICAL PROPERTIES OF D- AND
L-TARTARIC ACIDS

ACID	MELTING POINT, °C	SOLUBILITY (g/100g H_2O)	DENSITY (g/ml)	SPECIFIC ROTATION
d-Tartaric acid	170	139	1.76	+12°
l-Tartaric acid	170	139	1.76	−12°

CHECKLIST

Addition reaction	Octane number
Alcohol	Olefin
Alkane	Organic compound
Alkene	Paraffin
Alkyl group	Pi (π) bond
Aromatic hydrocarbons	Polymer
Bond angle	Primary carbon
Bond energy	Saturated
Conjugated system	Secondary carbon
Double bond	Sigma (σ) bond
Ester	sp hybrid orbitals
Ether	sp^2 hybrid orbitals
Free radical	sp^3 hybrid orbitals
Functional group	Substitution reaction
Homologous series	Tertiary carbon
Homologs	Tetrahedral configuration
Hydrocarbon	Tetravalence
Hydrogen bonding	Triple bond
Isomerism	Unsaturated
Macromolecule	Vital force
Monomer	

EXERCISES

17–1.   What is meant by a homologous series of compounds?

17–2.   Explain the statement that the octane number of a gasoline is 94.

17–3.   Write condensed structural formulas for each of the following compounds:
(a) 2,3-dibromo-3-heptene
(b) 1,4,5-trichloro-2-hexyne
(c) 3-bromo-2-methyl-2-pentanol
(d) methyl ethyl ether
(e) 3-bromopentanoic acid

17–4. What characteristics of carbon make possible the formation of the great number of organic compounds?

17–5. Write structural formulas for an example of each of the following:
(a) alkene
(b) alcohol
(c) ether
(d) ester
(e) aldehyde

17–6. How do (a) benzene and (b) ethylene differ in their reactions with bromine?

17–7. Why are molecular formulas alone inadequate to represent organic compounds? Illustrate with an example.

17–8. (a) What is meant by an asymmetric carbon? (b) What unique property does it have?

17–9. Contrast the physical properties of the ionic compound table salt, sodium chloride, with those of the hydrocarbon naphthalene ($C_{10}H_8$), moth flakes.

17–10. Why has the structure of benzene been a problem to scientists?

17–11. Draw sections of the following polymers and their monomers:
(a) polyethylene
(b) polyacrylonitrile
(c) polystyrene
(d) polyisoprene

17–12. Identify the functional groups present in each of the following:

(a) $\underset{\displaystyle \text{OH} \quad \text{OH}}{\text{C}\,\text{H}_2\text{CH}_2}$

(b) $C_2H_5\text{—O—}C_2H_5$

(c) $CH_3 \overset{\displaystyle O}{\overset{\|}{\text{C}}}\text{—O—}CH_3$

(d) $\overset{\displaystyle O}{\overset{\|}{\text{C}}}\text{—H}$

(e) $\underset{\displaystyle \text{O—}\overset{\displaystyle O}{\overset{\|}{\text{C}}}\text{—CH}_3}{\text{C OOH}}$

17–13.   Write condensed structural formulas for the following compounds:
(a) acetylene
(b) di-n-propyl ether
(c) 2,2-dimethyl-3-chlorobutane
(d) paradibromobenzene
(e) methyl acetate

17–14.   Why is methyl alcohol not used as a solvent for medicines to be taken internally?

17–15.   Identify the functional groups present in:
(a) vitamin C   (b) aspirin

(c) vanillin

17–16.   Discuss the main source of the man-made atmospheric pollutant, hydrocarbons, and the prospects for minimizing it.

17–17.   Discuss the role of π bonding in benzene.

17–18.   Draw the structures of all the dichlorobenzene isomers and name them.

17–19.   If the annual consumption of gasoline is 100 billion gallons and lead is added at a maximum rate of 1.5 g/gallon, how many tons of lead are needed for this purpose?

17–20.   Define and illustrate the following terms:
(a) functional group
(b) delocalization energy
(c) unsaturation
(d) aromaticity
(e) geometrical isomerism

17–21. Name each of the following compounds by the IUPAC system.

(a) $CHCl_3$

(b) $CH_3CH_2—O—CH_2CH_3$

(d) $HC≡CH$

(e) CH_3CHCH_3
　　　|
　　　OH

(c)

17–22. *Multiple Choice*

A. Which is an isomer of 2-methylbutane?
   (a) propane
   (b) pentane
   (c) 2-methylpropane
   (d) 2-methylpentane

B. The type of hybridization present in CH_4 is
   (a) sp
   (b) sp^2
   (c) sp^3
   (d) p-p

C. A functional group is
   (a) one of the vertical columns in the periodic table
   (b) a horizontal row in the periodic table
   (c) exemplified by R in ROH
   (d) exemplified by OH in ROH

D. A double bond between two carbon atoms consists of
   (a) two sigma bonds
   (b) two pi bonds
   (c) one sigma and one pi bond
   (d) sp hybridized carbon atoms

E. Polyethylene is an example of
   (a) a monomer
   (b) an addition polymer
   (c) a radical
   (d) a rubber

Breslow, Ronald, "The Nature of Aromatic Molecules." *Scientific American,* August, 1972.
　　The surprising stability of the benzene ring is referred to as aromaticity, a property that depends upon the presence of delocalized electrons. The exploration of aromaticity has both practical and theoretical significance.
Lambert, Joseph B., "The Shapes of Organic Molecules." *Scientific American,* January, 1970.
　　A knowledge of the shapes of molecules, which influence chemical reactivity, is valuable in the synthesis of compounds such as antibiotics of the penicillin family.

SUGGESTIONS FOR FURTHER READING

Lieber, Charles S., "The Metabolism of Alcohol." *Scientific American,* March, 1976.

> Alcoholism can cause cirrhosis and death because alcohol disturbs liver metabolism and damages the liver cells.

Lipman, Timothy O., "Wöhler's Preparation of Urea and the Fate of Vitalism." *Journal of Chemical Education,* August, 1964.

> Organic chemistry was not established immediately with Wöhler's preparation of urea. That, however, was a major step forward in the evolution of a unified scheme of chemistry.

"Nylon—From Test Tube to Counter," *Chemistry,* September, 1964.

> The story of Wallace H. Carothers' discovery of nylon.

Westheimer, Frank H., "The Structural Theory of Organic Chemistry." *Chemistry,* June, 1965.

> Structural theory accounts for the existence of more than one substance with a given molecular formula and explains why more than 2 million organic compounds have already been isolated.

NUCLEAR REACTIONS AND STRUCTURE

18–1 INTRODUCTION

The dual aspect of science as a boon and a threat to mankind is nowhere more evident than in the field of nuclear science. On the one hand, as the sources of fossil fuel—particularly oil and gas—are becoming depleted, society is more and more dependent upon alternative energy forms, including nuclear energy. On the other hand, the possibility of nuclear holocaust underlies the foreign policy of the great nations of the world. In this chapter we examine the nature of the atomic nucleus, the source of this awesome energy.

18–2 NUCLEAR STABILITY

As we explore the nucleus, it will be convenient to refer to specific nuclides. A *nuclide* is any nuclear species that has a specific atomic number and mass number. For example, uranium-235, $^{235}_{92}U$, is a nuclide; magnesium-27, $^{27}_{12}Mg$, is another nuclide. The *atomic number* is characteristic of the element, while the *mass number* identifies a particular nuclide of that element. The various nuclides of an element are isotopes of that element. Thus, other nuclides of uranium exist, including $^{234}_{92}U$ and $^{238}_{92}U$ (Fig. 18–1).

There are nuclides like $^{20}_{10}Ne$ that have equal numbers of protons and neutrons; in this case, 10 of each. In most nuclides, however, the proton and neutron numbers are different, the neutron-to-proton ratio increasing to about 1.5 for heavier elements such as $^{184}_{74}W$. A plot of the number of neutrons, N, versus the num-

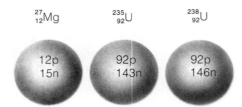

$^{27}_{12}\text{Mg}$ $\quad$ $^{235}_{92}\text{U}$ $\quad$ $^{238}_{92}\text{U}$

12p 15n $\quad$ 92p 143n $\quad$ 92p 146n

Figure 18-1 Examples of nuclides.

ber of protons, Z, shows that the stable (non-radioactive) nuclides fall within a narrow belt (Fig. 18–2).

A nuclide outside the stability band tends to stabilize itself by spontaneously transforming or "decaying" to a nuclide closer to or within the stability region. If the nuclide is to the left of the stability region, it has an excess of neutrons and can achieve stability through a process that lowers the neutron-to-proton ratio. One to the right of the stability band has an excess of protons; it can achieve stability by a process that lowers the number of protons or increases the number of neutrons, thus increasing the neutron-to-proton ratio. These processes are treated in greater detail in the following section.

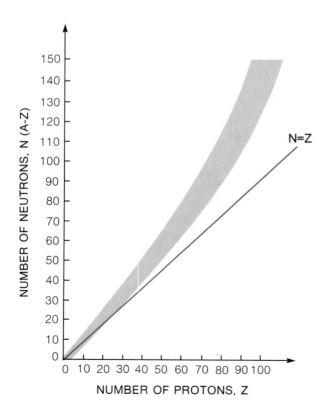

N=Z

NUMBER OF NEUTRONS, N (A-Z)

NUMBER OF PROTONS, Z

Figure 18-2 Stability band of nuclides. The stable nuclides fall approximately within the area indicated. The sloping line represents nuclides with equal numbers of protons and neutrons. Nuclides having more neutrons than protons lie above this line.

18-3   NUCLEAR TRANSMUTATIONS

The processes by which unstable nuclides achieve stability are called *transmutations,* since they usually involve a change in atomic number and mass number. Transmutations may be either spontaneous (as in natural radioactivity) or artificially induced in various ways to achieve stability. As a result, one element is changed, or transmuted, into another.

Like chemical reactions, nuclear reactions can be represented by equations. The same symbols are used for atoms and nuclei, although the atomic number and mass number are added to the symbols in the case of nuclei. Just as atoms and molecules are not created or destroyed in ordinary chemical changes, so nucleons (protons and neutrons) are not created or destroyed in nuclear reactions, although they may interchange. Both mass numbers and electric charge are conserved in nuclear reactions. In a balanced nuclear equation the sum of the atomic numbers (electric charges) of the reacting nuclei and particles will be equal to the sum of the atomic numbers of the products, and the mass numbers (total number of nucleons) of the reactants and products will balance.

EXAMPLE   18-1

When $^{82}_{35}Br$ emits a beta particle, it is transmuted into another element. Identify the new element.

SOLUTION

Write a balanced equation for this transmutation, noting that there will be conservation of electric charge and conservation of mass numbers. Recall that a beta particle is an electron with a charge of -1; in nuclear equations, its mass is considered to be zero since it is thousands of times less massive than either a proton or a neutron.

$$^{82}_{35}Br \rightarrow\ ^{82}_{36}X +\ ^{0}_{-1}e$$

A beta particle can be thought of as formed from a neutron that divides into a proton and an electron (beta particle). Because the proton remains in the nucleus, the positive nuclear charge is increased by one unit, from 35 to 36, but there is no change in the mass of the nucleus (Fig.18–3). The atomic number of the new element, therefore, must be 36, and its mass number 82. The element represented by X has to be krypton, Kr.

EXAMPLE 18–2

In the first artificial transmutation ever carried out, Ernest Rutherford (1919) bombarded nitrogen with alpha particles and identified hydrogen nuclei among the products. What other nuclide was produced?

SOLUTION

Recalling that alpha particles are helium nuclei, ^4_2He, consider them to be reactants along with $^{14}_7\text{N}$.

$$^{14}_7\text{N} + {}^4_2\text{He} \rightarrow {}^1_1\text{H} + {}^{17}_8\text{X}$$

The atomic number of the other product nuclide, represented by X, is 8; thus the newly produced element must have been oxygen. In order for the mass numbers to balance, the oxygen nuclide must have had a mass number of 17.

EXAMPLE 18–3

Frederic (1900–1958) and Irène Joliot-Curie (1897–1956) were the first to produce an artificially radioactive element. In 1933, they bombarded $^{27}_{13}\text{Al}$ with α-particles, and found that neutrons and $^{30}_{15}\text{P}$ were produced. (The phosphorus, however, was radioactive, and decayed into $^{30}_{14}\text{Si}$ by positron emission.) Write the equation for the first reaction.

SOLUTION

$$^{27}_{13}\text{Al} + {}^4_2\text{He} \rightarrow {}^{30}_{15}\text{P} + {}^1_0\text{n}$$

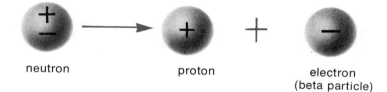

Figure 18-3 Beta emission. A neutron divides into a proton and a beta particle.

neutron   proton   electron (beta particle)

Several modes of natural transmutation, or radio-activity, are known, two of which are discussed here. Alpha emission, or α-decay, is common for elements of atomic number greater than 82 and mass number greater than 209. Such nuclides are found to the right of the region of stability and have too many protons. The emission of α-particles reduces the number of protons by two and the number of neutrons by two, bringing the nuclear composition closer to the stability band. An example is

$$^{254}_{102}No \rightarrow\; ^{4}_{2}He + \;^{250}_{100}Fm \qquad \text{Equation 18–1}$$

Nuclides that have a high neutron-to-proton ratio and lie to the left of the stabilty band may be stabilized by β-decay. The net effect is a decrease in the number of neutrons by one and an increase in the number of protons by one, thus decreasing the neutron-to-proton ratio. An example is

$$^{14}_{6}C \rightarrow\; ^{0}_{-1}e + \;^{14}_{7}N \qquad \text{Equation 18–2}$$

18-4 HALF-LIFE; CARBON-14 DATING

Consider a sample of radioactive material. Its activity, or rate of decay, is the number of transmutations or disintegrations per unit time. Suppose that the decay rate at the start is 1000 disintegrations per minute (dpm) and that we measure the decay rate over a period of time. Assume that after 10 minutes the activity has dropped to 500 dpm; after 20 minutes, to 250 dpm; and after 30 minutes, to 125 dpm (Fig. 18–4). Every 10 minutes the activity is reduced exactly one-half. We say that the half-life ($t_{1/2}$) of the sample is 10 minutes.

The half-life is a useful property for identifying nuclides, since each radioactive nuclide has a characteristic half-life. If this is short, the sample loses its radioactivity rapidly; if it is long, the sample retains much of its radioactivity (Fig. 18–5). The half-lives of some radionuclides are listed in Table 18–1.

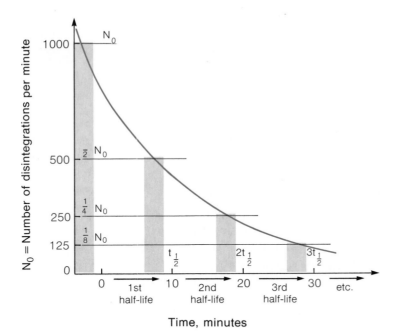

Figure 18–4 A radioactivity decay curve. A half-life is the time required for the activity to drop to one-half of any value and is a constant for a given radio-nuclide.

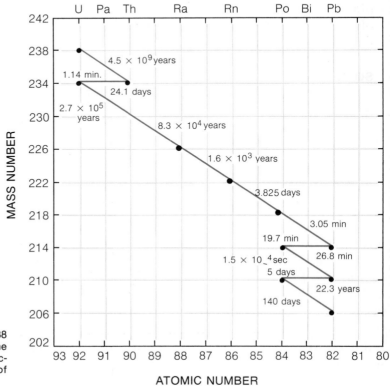

Figure 18–5 The uranium-238 radioactive decay series. The range of half-lives is from fractions of a second to billions of years.

Table 18–1 HALF-LIVES OF SELECTED
RADIONUCLIDES

ELEMENT	RADIONUCLIDE	HALF-LIFE, $t_{1/2}$	RADIATION
Hydrogen	^3_1H	12.4 years	β
Carbon	$^{14}_6C$	5570 years	β
Sulfur	$^{35}_{16}S$	87.1 days	β
Cobalt	$^{60}_{27}Co$	5.3 years	α
Strontium	$^{90}_{38}Sr$	27.7 years	β
Radium	$^{226}_{88}Ra$	1620 years	α
Uranium	$^{235}_{92}U$	7.1×10^8 years	α

EXAMPLE 18–4

Iodine-131, used in the treatment of diseases of the thyroid, decays by β-emission according to the following equation:

$$^{131}_{53}I \rightarrow {}^{131}_{54}Xe + {}^{0}_{-1}e$$

The half-life for the reaction is approximately eight days. Assuming that we start with 10.00g of of ^{131}I, how much will remain in 32 days (approximately one month)?

SOLUTION

For each half-life, the activity and therefore the amount of ^{131}I is reduced by one-half.

Number of Half-Lives	Number of Days	Quantity of ^{131}I Remaining (grams)
0	0	10.00
1	8	5.00
2	16	2.50
3	24	1.25
4	32	0.612

In 32 days, 0.612g of ^{131}I will remain.

Willard F. Libby (b. 1908) at the University of Chicago devised a method based on the half-life of carbon-14 by which archeological discoveries can be dated. Charcoal from the Lascaux caves in France, carbon from the ashes of a prehistoric Indian campfire, the wood from a Mayan temple, or a sandal from Oregon can be dated with considerable accuracy using the carbon-14 clock.

Carbon-14 is produced through cosmic-ray bombardment of nitrogen in the atmosphere:

Equation   18–3

$$^{14}_{7}N + ^{1}_{0}n \rightarrow ^{14}_{6}C + ^{1}_{1}H$$

It is radioactive and decays by beta-emission:

Equation   18–4

$$^{14}_{6}C \rightarrow ^{14}_{7}N + ^{0}_{-1}e \qquad\qquad t_{1/2} = 5570 \text{ years}$$

The concentration of carbon-14 in the atmosphere, as radioactive carbon dioxide, is approximately constant, since it is replenished at the same rate that it disintegrates. Deviations from constancy are detectable by dating tree rings.

Plants absorb radioactive carbon dioxide from the atmosphere and convert it to carbohydrates in the process of photosynthesis. The carbon-14 content of living plants becomes stabilized at approximately one carbon-14 atom for every 10^{12} carbon-12 atoms. The same ratio becomes established in animals and human beings who eat the plant materials. When a plant or animal dies, the consumption of carbon-14 ceases and the ratio of carbon-14 to carbon-12 starts to decrease. To determine its age, a sample is burned to carbon dioxide, which is then analyzed for the $^{14}C/^{12}C$ ratio. After 5570 years, the $^{14}C/^{12}C$ ratio would be only one-half the normal ratio. In this way, dates can be determined quite accurately over several half-lives.

18–5   THE GEIGER-MUELLER COUNTER

Although nuclear particles are too small to be seen directly, there are ways to study their properties. The methods are based on one of the following features: (1) the ability of nuclear particles to cause ionization in matter through which they pass; (2) their ability to cause flashes of light (scintillations) when they strike certain crystals or liquids; and (3) their ability to produce tracks in specially prepared photographic emulsions. The traces or tracks that were left in various in-

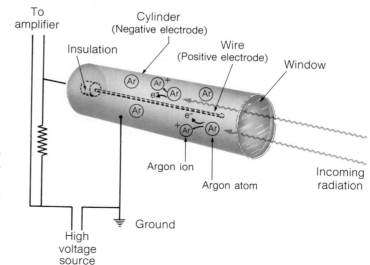

Figure 18–6 A Geiger-Mueller radiation counter detects radioactivity by ionization of argon atoms resulting from collisions with a particle. The positive ions and negative electrons conduct an electric current between the oppositely charged electrodes, giving rise to a pulse in the external circuit that can be amplified.

struments led scientists to conclude that they were observing particles in action.

The Geiger-Mueller counter consists of a metal cylinder that contains within it a fine tungsten wire stretched along its axis and insulated from it, and a small amount of gas (argon or helium) at low pressure (Fig. 18–6). A potential difference of about 1000 volts is maintained across the cylinder and wire (electrodes). When a particle from a radioactive source enters the cylinder, it collides with some of the atoms, knocking off electrons and creating positively charged gaseous ions. Under the accelerating influence of the electric field, the electrons and ions formed in the original collisions gain enough energy to produce further (secondary) ionization as they collide with neutral atoms. The process may be repeated again and again until there is an avalanche of a million or more electrons upon the wire. The discharge between cylinder and wire produces a pulse that can be amplified, activate a clicker and be heard, or be recorded electronically and counted.

18–6   ENERGY IN NUCLEAR PROCESSES

Imagine assembling a carbon-12 atom from its components: six protons, six neutrons, and six electrons. The rest masses of these particles in atomic mass units (amu) are given in Table 18–2. What mass do we anticipate for the carbon-12 atom?

Table 18–2   MASSES OF ELECTRON, PROTON, AND NEUTRON

PARTICLE	MASS (amu)
Electron	0.000549
Proton	1.007277
Neutron	1.008655

For carbon-12:

$$6 \text{ protons} = 6 \times 1.007277 = \quad 6.043662 \text{ amu}$$

$$6 \text{ neutrons} = 6 \times 1.008665 = \quad 6.051990 \quad \text{''}$$

$$6 \text{ electrons} = 6 \times 0.000549 = \quad \underline{0.003294} \quad \text{''}$$

$$\text{Mass of particles} = 12.098946 \quad \text{''}$$

$$\text{Mass of carbon-12} = \underline{12.000000} \quad \text{''}$$

$$\text{Mass difference} = \quad 0.098946 \text{ amu}$$

Since it is reasonable to assume that the whole is equal to the sum of its parts, we find it surprising that the actual mass of a carbon-12 atom is less than the combined masses of the electrons, protons, and neutrons that compose it (Fig. 18–7). The difference, known as the *mass defect,* exists to varying degrees for each nuclide.

We account for the "missing" mass by assuming that it is converted into energy. Since mass and energy are related by the expression $E = mc^2$, we can calculate the energy equivalent of 1 atomic mass unit.

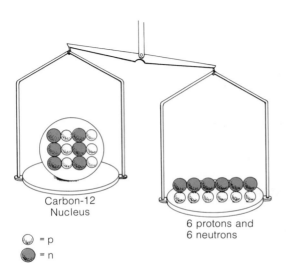

Carbon-12
Nucleus

6 protons and
6 neutrons

◔ = p

● = n

Figure 18–7   A carbon-12 nucleus compared to its components. The whole in this case is not equal to the sum of its parts.

1 amu = 1.66×10^{-27} kg $E = mc^2$

   c = velocity of light

$$= (1.66 \times 10^{-27}\,\text{kg}) \left(3 \times 10^8 \frac{m}{sec}\right)^2$$

$$= 3 \times 10^8 \frac{m}{sec}$$

$$= (1.66 \times 10^{-27}\,\text{kg}) \left(9 \times 10^{16} \frac{m^2}{sec^2}\right)$$

   E = ?

$$= 14.9 \times 10^{-11}\,\text{joule}$$

It is customary to express the energy of nuclear processes in units of millions of electron volts (MeV), so we convert joules to MeV. The conversion factors are:

1 electron volt = 1 eV = 1.60×10^{-19} joule
1 million electron volts = 1 MeV = 1.60×10^{-13} joule

Therefore,

$$14.9 \times 10^{-11}\,\text{joules} \times \frac{1\ \text{MeV}}{1.6 \times 10^{-13}\,\text{joule}} = 931\ \text{MeV}$$

Equation 18–5

    or: 1 amu = 931 MeV

   The mass defect in carbon-12 is equivalent to 0.098946 amu $\times 931 \frac{\text{MeV}}{\text{amu}} = 92.1$ MeV. This energy is called the *binding energy*—the energy required to separate a nucleus into its individual nucleons. The binding energy per nucleon of carbon-12 is 92.1 MeV/ 12 nucleons = 7.67 MeV/nucleon.

   A plot of the binding energy/nucleon of the elements shows a maximum in the vicinity of iron (8.790 MeV/nucleon), and a gradual decrease toward the heavier elements such as uranium (7.591 MeV/nucleon) (Fig. 18–8). Nuclei that have the greatest binding energy/nucleon are the most stable. It appears from the curve that if heavy nuclei, such as those of uranium, could be split into lighter nuclei, energy would be released in the process. The nucleus has been split— *fission*—and its considerable energy harnessed. Further, the combination of lighter nuclei, such as hydrogen and lithium, to form heavier nuclei should also be accompanied by the release of energy. This process— *fusion*—has also been achieved.

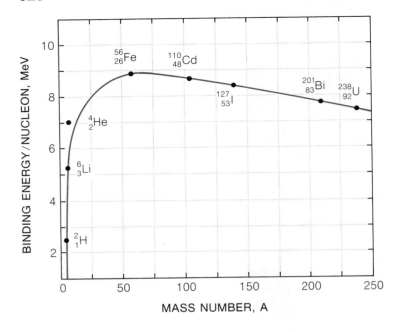

Figure 18–8 Nuclear binding energy curve. The most stable nuclei are located near iron at the maximum of the curve, where the binding energy per nucleon is greatest.

18–7 PARTICLE ACCELERATORS; THE CYCLOTRON

Investigators of nuclear reactions soon realized that they would not get very far without the aid of suitable "hardware." To produce a nuclear reaction with a proton or alpha particle, the Coulomb barrier of the positively charged nucleus must be penetrated. Energies in the range of millions of electron volts and higher are required.

There are a few natural sources of energetic particles. Radioactive nuclei emit alpha, beta, and gamma rays with energies of about 1 MeV. Rutherford used alpha particles emitted from polonium in his early research. High-energy nuclear particles known as cosmic rays rain in on us partly from the sun, partly from outside the solar system, especially from the center of our galaxy, with energies as high as 10^{13} MeV, but they are not available for controlled experiments.

Rutherford and others established that alpha-particle bombardment transmuted nearly all the lighter elements through potassium. The alpha particles from naturally radioactive sources, however, did not have high enough energies to penetrate the nuclei of heavier atoms with their considerable positive charge. Since nature is limited in the range of particles that can be used as projectiles, some means of accelerating charged particles to any desired energy had to be found. That is the reason for developing particle *accelerators*.

Accelerators were not really new either in principle or in practice. The gas discharge tube and the X-ray tube, both widely used, accelerate electrons through potential differences of thousands of volts. The problem is to accelerate particles much heavier than electrons through voltages in the MeV range and greater. Although they differ widely in design, accelerators share the same purpose: to study the effect of bombarding atomic nuclei with particles traveling at speeds approaching that of light. These high-speed particles penetrate atomic nuclei and serve as probes in the investigation of matter.

To produce a proton with 10 MeV of energy, two approaches can be used. In the first, the proton can be accelerated through one single and large potential difference of 10 million volts to acquire the desired high energy. The Van de Graaff generator is based on this principle. The voltage is controllable to better than 0.1 per cent so that all the particles in the beam have very nearly the same energy, a favorable feature for many kinds of precision work in research, industry, and medicine.

A second approach to producing high-energy par-

Figure 18–9 Principle of the cyclotron. (From Masterton, W. L., and Slowinski, E. J., *Chemical Principles,* 4th ed. Philadelphia: W. B. Saunders Company, 1977, p. 584.)

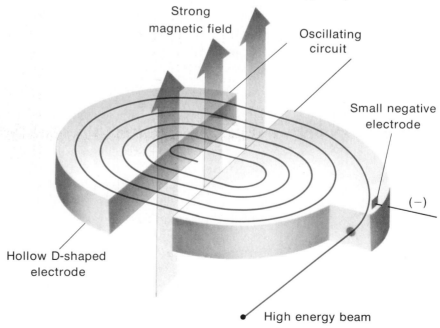

Strong magnetic field

Oscillating circuit

Small negative electrode

(−)

Hollow D-shaped electrode

High energy beam

Figure 18–10 The main accelerator tunnel at the Fermi national accelerator laboratory, Batavia, Illinois. A view of the bending magnets is shown. Protons are accelerated to high energies in this accelerator. (Fermilab photo.)

ticles was introduced by Ernest O. Lawrence (1901–1958) at the University of California in 1931. It is based on the concept of accelerating a particle many times in succession through a relatively low potential difference, in such a way that it picks up energy at each acceleration. Finally, the particle arrives at the desired energy. Lawrence's first cyclotron was barely 1 foot across and produced 1 200 000-volt protons. It employed a circular track and a succession of small pushes activated by a magnetic field (Fig. 18–9). Each time the particle went around it was accelerated a slight amount. Accelerators have since gone to the million- and billion-volt class (Fig. 18–10).

18–8 NEUTRON REACTIONS

A drawback in the use of charged particles such as electrons, protons, and alpha particles is that they interact strongly with the planetary electrons of atoms and ionize atoms or molecules of the material through which they pass. In the process they lose energy rapidly and are brought to rest and neutralized after passing through only a few centimeters of air or through thin foils. Since neutrons carry no electric charge and interact very little with planetary electrons, they are more penetrating than charged particles of comparable energy. They are easily obtained by bombarding light elements such as beryllium, boron, and lithium with alpha particles:

$$^{9}_{4}\text{Be} + {}^{4}_{2}\text{He} \rightarrow {}^{12}_{6}\text{C} + {}^{1}_{0}\text{n}$$

Equation 18–6

At the University of Rome, Enrico Fermi (1901–1954) and his co-workers used neutrons to bombard various elements. Since then, neutrons of low, intermediate, and high energy have been used to bombard light and heavy nuclei throughout the periodic table. When a neutron collides with a nucleus and is captured by it, the resulting compound nucleus is unstable. The nucleus may achieve stability through radioactivity. In the case of $^{238}_{92}\text{U}$, radioactivity was induced, but the product nucleus was a beta-emitter, whereas uranium-238 is an alpha-emitter. Fermi's interpretation of the results was that neutrons had penetrated uranium nuclei, forming an unstable isotope, $^{239}_{92}\text{U}$, which decayed by beta-emission:

$$^{238}_{92}\text{U} + {}^{1}_{0}\text{n} \rightarrow ({}^{239}_{92}\text{U}) \rightarrow {}_{-1}^{0}\text{e} + {}^{239}_{93}\text{X}$$

Equation 18–7

If this were true, it would mean the production of an element beyond the limits of the periodic table known at that time. In the same way, elements 94, 95, and other "trans-uranium" elements, that is elements beyond uranium, could be synthesized. At the University of California, Edwin M. McMillan (b. 1907) identified element 93, called neptunium, in 1940. A year later, Glenn T. Seaborg (b. 1912), at the same university, discovered element 94, plutonium. The elements were named for the planets beyond Uranus in the solar system.

18–9 NUCLEAR FISSION

When Otto Hahn (1879–1968) and his co-worker Fritz Strassmann (b. 1902) at the Kaiser Wilhelm Institute repeated Fermi's neutron bombardment of uranium in 1938, they expected to find uranium isotopes with greater atomic weights than ordinary uranium (assuming that the uranium nuclei captured neutrons) or evidence of trans-uranium elements. Instead, they found radioactive isotopes of much lighter elements, such as barium, krypton, strontium, and cerium. Since no particles larger than alpha particles had previously been found ejected from nuclei, it was difficult to believe that barium or krypton nuclei were emitted from uranium following neutron bombardment. Puzzled, they sent their results to physicist Lise Meitner (1878–1969), a former co-worker, for her interpretation, and published their work in a scientific journal in January, 1939.

Meitner and her nephew, Otto Frisch (b. 1904) came to the conclusion, soon confirmed by others, that in Hahn's and Strassmann's experiments, nuclei of uranium-238 were being split by captured neutrons into two nearly equal parts. Frisch coined the term "fission" for this process, comparing it to a cellular process, and Meitner calculated that about 200 MeV of energy is released from each uranium nucleus that undergoes fission. The energy appears as kinetic energy in the fission products as they fly apart, and is nearly 100 million times as much energy per atom as any common explosive. Meitner and Frisch also published their results in a scientific journal in January, 1939.

The energy released in a nuclear fission comes from the conversion of mass into energy. In a typical event, uranium-235 captures a neutron, and fissions into barium-141 and krypton-92, with the release of three neutrons (Fig. 18–11):

Equation 18–8
$$^{235}_{92}\text{U} + {}^{1}_{0}\text{n} \rightarrow {}^{141}_{56}\text{Ba} + {}^{92}_{36}\text{Kr} + 3{}^{1}_{0}\text{n}$$

A comparison of the masses of the reactants and products shows a mass difference. The combined masses of the products after fission is less than the sum of the masses of the reactants before fission. The mass difference appears as energy, approximately 200 MeV.

Before Fission

$^{235}_{92}\text{U}$ =	235.0439 amu
$^{1}_{0}\text{n}$ =	1.0087 ”
	236.0526 ”
	−235.8373 ”
Mass difference	0.2153 amu

After Fission

$^{141}_{56}\text{Ba}$ =	140.9139 amu
$^{92}_{36}\text{Kr}$ =	91.8973 ”
$3{}^{1}_{0}\text{n}$ =	3.0261 ”
	235.8373 amu

$$0.2153 \text{ amu} \times 931\frac{\text{MeV}}{\text{amu}} = 200 \text{ MeV}$$

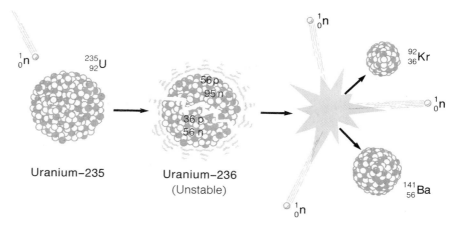

Uranium–235

Uranium–236
(Unstable)

$^{1}_{0}n$

$^{92}_{36}Kr$

$^{1}_{0}n$

$^{141}_{56}Ba$

$^{1}_{0}n$

Fission fragments
and neutrons
+ ENERGY

Figure 18–11 Fission of uranium-235 following capture of a neutron. Note the formation of an intermediate unstable nuclide, ^{236}U.

18–10 THE CHAIN REACTION

On the average, approximately three neutrons are released during each fission of uranium-235. When uranium-235 is present in an amount exceeding a critical size, so that the neutrons are not lost, these neutrons may initiate fission in other nuclei, releasing more neutrons, which in turn can cause fission in still other nuclei. This process is known as a chain reaction (Fig. 18–12). In nuclear reactors, the chain reaction is controlled and the energy is put to useful purposes. In a bomb, the chain reaction, once started, is uncontrolled.

Enrico Fermi and Leo Szilard (1898–1964) discovered that "slow" neutrons were effective in causing uranium-235 to undergo fission. They can be obtained when neutrons collide with atoms without combining, thus losing some of their kinetic energy. The first large-scale chain reaction with fissionable uranium was realized under Fermi's direction in 1941. Rods containing uranium metal and uranium oxide were embedded in a pile of graphite (carbon) blocks. The chain reaction was started by neutrons escaping from a beryllium source placed at the bottom of the pile. However, the reaction was not self-sustaining, since neutrons released by fission were lost by non-fission capture in uranium-238, or by diffusion, and the external neutron source was necessary to inject neutrons into the pile to keep the reaction going. The first self-sustaining chain reaction was carried out on December 2, 1942, in a pile constructed under the squash courts in the stadium of the

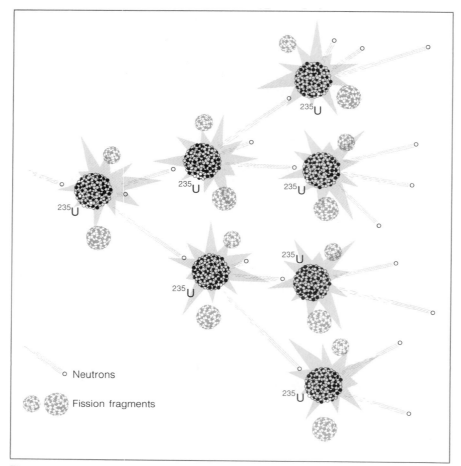

Figure 18–12 A nuclear chain reaction. The neutrons released when ²³⁵U fissions induce fission reactions in other ²³⁵U nuclei, thereby sustaining the process.

○ Neutrons

Fission fragments

University of Chicago. Refinements in the nuclear pile included the introduction of cadmium rods, which could control the speed of the reaction by their ability to absorb neutrons. Today fission reactors are used for many purposes: the generation of electricity; medical, industrial, and agricultural research; and as an energy source to drive submarines and surface ships.

A fission bomb is essentially a reactor, so designed that the chain reaction grows at as high a rate as possible and releases the maximum energy in the minimum time. The critical mass of uranium-235 is approximately 4 kilograms, or less than a cupful because of its great density (19.05 g/cm³). The fissionable material in a bomb is separated into subcritical sections (Fig. 18–13). Triggering causes these sections to coalesce into a mass greater than the critical size, and a chain reaction is initiated by a stray neutron from cosmic radiation or from a neutron source. The devastating effects of nuclear explosions are due to the intense heat, which produces fires, the strong mechanical shock waves, and the radioactivity of the fission products.

18–11 NUCLEAR POWER REACTORS

The consumption of power in the United States has doubled every 10 years for the past 30 years. If this doubling rate continues to the end of the century, the question will be not whether energy production should be increased, but how to do it with a minimum of harmful side effects.

Nuclear energy was rapidly becoming a national energy source in the 1970's. More than 70 nuclear power plants were in operation, producing about 12 per cent of the electric power supply in the United States. Nearly 100 others were being built or designed (Fig. 18–14). They were expected to furnish about 25 per cent of the total in the 1980's, and even more by the year 2000.

Nuclear power plants are a source of thermal pollution (as are fossil-fuel plants) and low-level radiation.

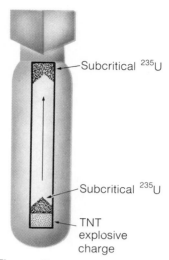

Subcritical ^{235}U

Subcritical ^{235}U

TNT explosive charge

Figure 18–13 Diagram of a nuclear bomb. A high explosive (TNT) drives ^{235}U pieces together. When the mass of ^{235}U becomes supercritical, it explodes.

Figure 18–14 Nuclear power reactors in the United States. (Courtesy of the Office of Information Services, U.S. Atomic Energy Commission.)

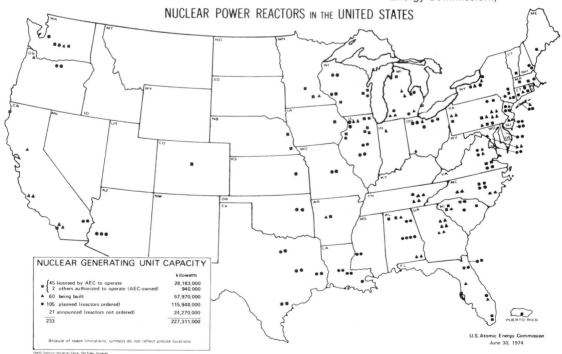

NUCLEAR POWER REACTORS in the UNITED STATES

NUCLEAR GENERATING UNIT CAPACITY	
	kilowatts
■ { 45 licensed by AEC to operate	28,183,000
{ 2 others authorized to operate (AEC-owned)	940,000
▲ 60 being built	57,970,000
● 105 planned (reactors ordered)	115,948,000
21 announced (reactors not ordered)	24,270,000
233	227,311,000

Because of space limitations, symbols do not reflect precise locations.

USAEC Technical Information Center, Oak Ridge, Tennessee

U.S. Atomic Energy Commission
June 30, 1974

PUERTO RICO

Courtesy of the Office of Information Services, U.S. Atomic Energy Commission.

The advantages, according to some people, are low-cost electricity; conservation of coal, oil, and gas, which are more useful as sources of organic molecules than as sources of heat; and reduction of pollution. Once the energy has been produced to generate heat, the operation of a nuclear plant is practically the same as that of a conventional plant (Fig. 18–15).

Nuclear reactors currently in use employ fuel rods containing uranium-238 and small amounts of uranium-235 and plutonium-239, the last two isotopes used as fuel (Fig. 18–16). About 99.3 per cent of natural uranium is in the form of non-fissionable uranium-238, which can be used as a fuel if it is converted to plutonium by capturing a fast neutron.

Equation 18–9

$$^{238}_{92}U + {}^{1}_{0}n \rightarrow {}^{239}_{92}U \xrightarrow{{}^{0}_{-1}e} {}^{239}_{93}Np \xrightarrow{{}^{0}_{-1}e} {}^{239}_{94}Pu$$

This does not occur to a great extent in light water reactors, in which water is used as the coolant, because the fast neutrons that are necessary for the reaction are slowed down by the water. The use of a liquid sodium or some other coolant could make available a high enough concentration of fast neutrons so that more nu-

Figure 18–15 A nuclear power plant. (From Brescia, F., et al., *Chemistry: A Modern Introduction,* 2nd ed. Philadelphia: W. B. Saunders Company, 1978, p. 546.)

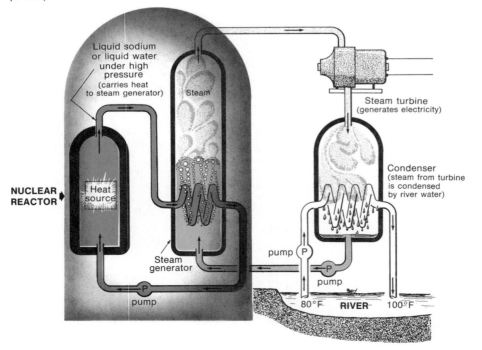

clear fuel is produced than consumed. Such a reactor is designated a *light-metal-cooled-fast-breeder-reactor* (LMFBR), and non-fissionable uranium-238 that can be so transformed is described as "fertile." A relatively abundant non-fissionable material is thus transformed into fissionable fuel.

Fissionable fuels are exhaustible, and the supply of inexpensive ores containing uranium-235 could be used up in the near future. More costly fuel would then be used, but the known reserves could be consumed in a few decades by the continued deployment of inefficient non-breeding reactors. In breeder reactors, the fuel life expectancy could be extended by hundreds of years.

18–12   REACTOR SAFETY

A major issue involving nuclear power plants is concern over safety standards and equipment. There is some evidence that current standards for radioactive emissions are not strict enough and should be revised, since radioactivity leakage is a possibility.

The most serious safety hazard is possible fuel melting. Although a nuclear reactor cannot explode like a bomb, since there is no way to bring sufficient fissionable material together, excessive heat could melt the core and discharge large amounts of radioactive materials into the environment. This possibility exists because of the way in which pressurized water reactors are designed. Uranium atoms fission, releasing energy in the form of heat. Steam circulates around the reactor's radioactive core and absorbs this heat. The heat expands the steam, which is channeled into turbines that produce electricity. If the pressurized steam system ruptures, the heat within the reactor core cannot be dissipated immediately and could cause the reactor, which might contain 50 to 100 tons of fuel, to melt, generating more heat.

A cooling system has been designed to prevent the sequence of events leading to a fuel melt. Called the Emergency Core Cooling System, it would operate if a reactor's primary cooling system fails. It consists of a water supply with pumps and pipes, which would flood an overheating reactor with cooling water long enough to give the reactor operators time to shut it down. But

Figure 18–16   This assembly is fueled with approximately 43 000 uranium pellets stacked in sealed rods. After about three years, the assembly is replaced. (Courtesy of Westinghouse Electric Corporation.)

doubts have been expressed that the system will perform as designed. Initial tests on an experimental reactor in the Idaho National Engineering Laboratory in the late 1970's did show that the emergency cooling system functioned generally as expected. But, a chain of equipment malfunctions led to a partial core meltdown at the Three Mile Island nuclear power plant near Harrisburg, Pennsylvania, in 1979, the most serious accident in the history of commercial nuclear power plants.

18–13   DISPOSAL OF RADIOACTIVE WASTES

The wastes produced by reactors are gaseous, liquid, and solid. If fission reactors supplied all the power needs of the United States, the safe disposal of radioactive ashes and gases would be a staggering problem. Each year would require the disposal of radioactive waste equivalent to that from the explosion of hundreds of thousands of nuclear bombs. Even at the present rate, the disposal problem is serious.

The waste gases are mostly radioactive isotopes of the noble gases. The most difficult problem involves krypton-85, which is now released to the atmosphere. It is long-lasting ($t_{1/2} = 10.76$ years) and unreactive chemically. Its concentration in the atmosphere is increasing, and it contributes to the general background radiation. The basic technology for trapping krypton— by cryogenic cooling, for instance—is available, but is a very complicated process.

Tritium ($^3_1$H) enters the liquid discharge of power and fuel reprocessing plants. It is produced from lithium and boron in the coolant and combines with oxygen to form water, becoming part of the effluent. In reprocessing plants, tritium is released to the atmosphere as water vapor. In both cases, it is extremely difficult to remove. As part of a water molecule, tritium could become incorporated into biological systems. With a half-life of 12.26 years, the radiation produced by its decay would be quite damaging. No mechanism is known, however, by which tritium can concentrate in food chains, and there is no evidence to indicate that it does.

Liquid wastes are generally reduced to solid form through precipitation, and added to the solid waste. In the process, the volume is reduced to one tenth of the original liquid material. Solid waste, consisting of spent fuel rods, is first stored at the reactor site while the radioactivity is reduced. It is then shipped to fuel reprocessing plants, where it is treated and stored for 10 years before being shipped to a repository. Deep geological formations, salt formations in particular, are favored by some scientists. The ultimate disposal problem, however, has not been solved.

18-14 FUSION ENERGY

If a way could be found to harness nuclear fusion, another source of useful energy would be available. In this process, light nuclei are fused together into heavier nuclei and energy is released (Fig. 18-17). The binding energies in the resulting nuclei are greater than in the nuclei undergoing fusion. The advantages of fusion are illustrated by the almost limitless supplies of hydrogen and other light elements as compared with uranium. The radioactive waste problem would be almost nonexistent for a system relying upon fusion.

The sun and the stars are believed to owe their energy to nuclear fusion. Hans Bethe (b. 1906) at Cornell University proposed a mechanism, called the carbon cycle, which could generate energy of this magnitude. Carbon acts as a catalyst as four protons are fused by successive capture into one alpha particle (helium nucleus). Two positrons (particles with one positive charge and zero mass) and 24.7 MeV of energy are released for each alpha particle created. In fusion as in fission, mass is consumed and converted into energy. In effect, hydrogen functions as a nuclear fuel in the fiery furnace of the sun, and is consumed at the rate of millions of tons each second. The mass of the sun is

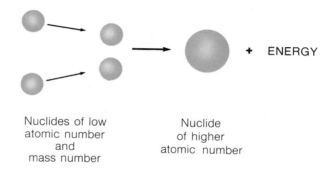

Nuclides of low
atomic number
and
mass number

Nuclide
of higher
atomic number

+ ENERGY

Figure 18-17 Nuclear fusion.

so great, however, that hydrogen will likely remain for billions of years. The carbon cycle is as follows:

Equation 18–10

$$^{12}_{6}C + ^{1}_{1}H \longrightarrow ^{13}_{7}N$$

$$^{13}_{7}N \longrightarrow ^{13}_{6}C + ^{0}_{1}e$$

$$^{13}_{6}C + ^{1}_{1}H \longrightarrow ^{14}_{7}N$$

$$^{14}_{7}N + ^{1}_{1}H \longrightarrow ^{15}_{8}O$$

$$^{15}_{8}O \longrightarrow ^{15}_{7}N + ^{0}_{1}e$$

$$^{15}_{7}N + ^{1}_{1}H \longrightarrow ^{12}_{6}C + ^{4}_{2}He$$

$$4\,^{1}_{1}H \longrightarrow ^{4}_{2}He + 2^{0}_{1}e + 24.7 \text{ MeV}$$

Fusion reactions occur at temperatures of millions of degrees and for this reason are called *thermonuclear reactions*. In order to fuse, (1) the nuclei must approach closely (about 10^{-15}m), and (2) they must possess enough kinetic energy to overcome the Coulomb repulsion of the nuclei. Since reactions between ordinary nuclei of hydrogen proceed more slowly than those between the nuclei of the heavier isotopes, deuterium or tritium, the latter are preferred (Fig. 18–18). Deuterium occurs in nature as a rare isotope of hydrogen; tritium is practically nonexistent, and must be manufactured in nuclear reactions. A reaction between a deuteron and a triton is energetically favorable and is a key reaction in the hydrogen bomb.

Equation 18–11

$$^{2}_{1}H + ^{3}_{1}H \rightarrow ^{4}_{2}He + ^{1}_{0}n + 17.6 \text{ MeV}$$

The ignition temperature of about 60 million degrees is provided by the detonation of a fission bomb. The first H-bomb was exploded at Eniwetok Proving Ground in the Pacific in 1952 with energy in the megaton range (10^6 tons of TNT). Since then, fission and fusion have become linked. Hydrogen, or fusion, bombs do not have a critical mass and can be made on any scale, but they depend upon the explosion of a fission bomb to get started.

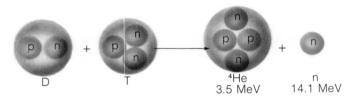

D   T ^{4}He   n   **Figure 18–18** A fusion reaction with
3.5 MeV   14.1 MeV   deuterium (D) and tritium (T) nuclei.

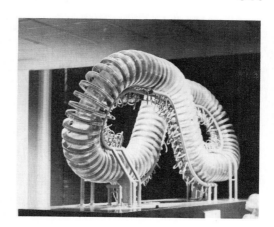

Figure 18–19 A figure-8 shaped stellarator. Plasma (ionized gas) is confined by a twisted magnetic field, then heated.

Thus far, temperatures of millions of degrees for periods long enough for controlled fusion reactions to occur have not been obtainable except in the case of fission bombs. Research is being carried out on methods of holding a "plasma"—a gas in which atoms are reduced to their nuclei and electrons—in a container by means of magnetic fields. If deuterium gas in a tube is completely ionized, it will exist as electrons and nuclei, not as atoms. A strong magnetic field could prevent the plasma from touching the sides of the container; this would be necessary because of the conductive loss of heat. A number of configurations have been tested, such as the tokamak and the stellarator (Fig. 18–19). Collisons between deuterons could result in fusion reactions with the release of nuclear energy in these "magnetic bottles."

Work on the problem of controlling thermonuclear reactions is continuing in many laboratories throughout the world. The Plasma Physics Laboratory at Princeton University in 1978 achieved a 60 million degree Celsius tokamak plasma temperature. This level is well above the 44 million degree Celsius temperature estimated to be needed for sustained fusion reactions, but it lasted for only 15 thousandths of a second. Indications are that an effective fusion reactor may become scientifically feasible before the end of the 20th century. If it is ever developed, its environmental and other advantages will be great. For these reasons, it is being called the "downtown" reactor. The fuel supply problem would be solved, as would the problem of radioactive wastes. There is an inexhaustible supply of the basic fusion fuel, deuterium, in the oceans, and there are no significant amounts of radioactive byproducts produced in fusion. The quantity of fuel present would be so small that there would be no possibility of a critical mass forming to create a nuclear explosion.

18–15　NUCLEAR STRUCTURE

Two models of nuclear structure, the liquid-drop model and the nuclear shell model, are discussed here.

The *liquid-drop model* portrays the nucleus as a configuration of closely packed protons and neutrons, each interacting only with its nearest neighbors. There is a striking resemblance in this model to atoms in a liquid drop, in which the forces of attraction and repulsion between particles are balanced.

Unless it is radioactive, a nucleus is stable despite the presence in it of positively charged protons, which tend to disrupt it. Just as in a liquid drop particles evaporate from the surface if they have enough energy to overcome the surface tension, so nucleons may be emitted from a nucleus by picking up enough speed through chance collisions with other nucleons and breaking through the barrier of cohesive nuclear forces. On the other hand, a particle may enter a nucleus from the outside and by its presence and the kinetic energy which it imparts to the nucleons, permit the escape of a proton, a neutron, or an alpha particle. The liquid-drop model describes successfully certain nuclear phenomena—spontaneous alpha-particle emission, nuclear reaction, and particularly nuclear fission. But nulcei are incredibly complex, and this model yields to others for a better explanation of some effects.

As early as 1934 it was observed that nuclei having 50, 82, and 126 neutrons are unusually stable. Since then, evidence has been found that nuclei having 2, 8, 20, 28, 50, 82, or 126 nucleons of the same kind, either protons or neutrons, also have exceptional properties. These numbers are known as "magic numbers." Helium has 2 protons and 2 neutrons, and is extremely stable. So is oxygen-16, with 8 protons and 8 neutrons. Tin, with 50 protons, has more stable isotopes than any other element. Lead-206 has 82 protons and is the end product of several radioactive series.

A shell structure, analogous to that which is used to explain the electron structure of atoms, has been proposed to explain nuclear stability. In the Rutherford-Bohr model of the atom, the electrons are arranged in concentric shells, each capable of accommodating a definite number of electrons, two for the innermost shell, eight for the next, and so on. In the shell model of the nucleus, the basic assumption is that in the magic-number nuclei the shells are filled (Fig. 18–20). The excited states of nuclei are explained by nucleons being raised from one energy level to a higher energy level. The nuclear shell model successfully explains alpha, beta, and gamma emission, and the nature of the electric and magnetic fields that surround nuclei. The puzzling aspect that led to the designation "magic numbers" comes logically out of the theory, and the mystery surrounding these numbers is removed.

18–16 BIOLOGICAL EFFECTS OF RADIATION

Radiation, whether in the form of X rays or gamma rays, or particles such as neutrons or beta particles,

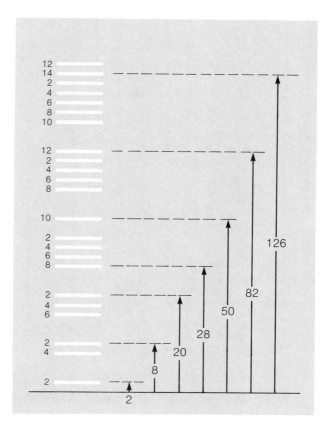

Figure 18–20 Nuclear magic numbers and energy levels. The magic numbers correspond to filled levels of nucleons.

is potentially damaging to cells. The radiation from nuclear reactors and fallout, therefore, is hazardous to living things. Most nuclear radiation has very high energies that can be transferred to any matter through which it passes. A single alpha particle can cause the ionization of millions of atoms before its energy is dissipated. The chemistry of cells is disrupted and cells may die.

Radiation damage to the bone marrow may cause it to lose its capacity to produce enough red blood cells. When white blood cells are destroyed by radiation, the body loses its resistance to infection. The lining of the gastrointestinal tract is readily damaged, with ulcers developing in the stomach and intestines. Radiation-induced changes in the chromosomes of cells in the reproductive organs may result in genetic mutations. The lymph nodes, spleen, and tonsils are most vulnerable. The skin loses its hair when exposed to heavy radiation, and the walls of the blood capillaries may be so weakened that hemorrhages occur throughout the body. Concern for such consequences underlies the reservations that some have for the further development of nuclear power.

CHECKLIST

Accelerator
Binding energy
Breeder reactor
Carbon-14 dating
Chain reaction
Critical size
Cyclotron
Fission
Fuel melt
Fusion
Geiger-Mueller counter
Half-life

LMFBR
"Magic numbers"
Mass defect
Moderator
Nuclide
Pile
Plasma
Shell model
Thermonuclear reaction
Transmutation
Trans-uranium elements

EXERCISES

18-1. Complete the following nuclear equations:

(a) $^2_1H + {}^3_1H \rightarrow (\ ) + {}^1_0n$

(b) $(\ ) + {}^4_2He \rightarrow {}^{42}_{20}Ca + {}^1_1H$

(c) $^{27}_{13}Al + {}^1_0n \rightarrow {}^{28}_{13}Al + (\ )$

(d) $(\ ) + {}^4_2He \rightarrow {}^{35}_{17}Cl + {}^1_1H$

(e) $(\ ) \rightarrow {}^{14}_7N + {}^0_{-1}e$

18–2.   Identify nuclide X in the following uranium-235 fission:

$$^{235}_{92}U + ^{1}_{0}n \rightarrow ^{141}_{54}Xe + X + 3^{1}_{0}n$$

18–3.   How does beta-emission from a nucleus affect (a) the number of neutrons? (b) protons?

18–4.   Copper-66 decays by beta emission. Write the nuclear equation.

18–5.   Nobelium-254 has been produced by the bombardment of curium-246 with carbon-12 nuclei. Four neutrons are released in the process. Write the balanced nuclear equation.

18–6.   Write equations for the following induced nuclear reactions. The first symbol stands for the target nuclide. In parentheses the symbol of the projectile particle is followed by the symbol of the ejected particle. The last symbol stands for the product nuclide.

(a) $^{6}_{3}Li$ (n, α) _____

(b) _____ (α, $^{3}_{1}H$) $^{250}_{99}Es$

(c) $^{45}_{21}Sc$ (α, p) _____

(d) _____ ($^{10}_{5}B$, $3^{1}_{0}n$) $^{257}_{103}Lr$

18–7.   The half-life of strontium-90 is 25 years. What does this mean?

18–8.   Carbon from an ax handle discovered in an archeological "dig" is only one-half as radioactive as the carbon from the handle of a new ax. How old is the artifact?

18–9.   Discuss some advantages and disadvantages of fission reactors as an energy source.

18–10.   In what form is energy obtained from a nuclear reactor?

18–11.   Why is it that chain reactions do not occur in natural deposits of uranium?

18–12. Why is the chance of a nuclear reactor's exploding like a bomb remote? Explain.

18–13. What becomes of the "mass defect" in fission and fusion?

18–14. Beryllium-9 has a mass of 9.01219 amu. Calculate its binding energy per nucleon.

18–15. What does a "breeder" reactor breed? Why is this significant?

18–16. In a nuclear reaction, tritium ($_1^3H$) and deuterium ($_1^2H$) fuse into helium-4 with the release of a neutron and 17.6 MeV as follows:
$$_1^3H + _1^2H \rightarrow _2^4He + _0^1n + 17.6 \text{ MeV}$$
Calculate the mass defect of this reaction in atomic mass units (amu).

18–17. *Multiple Choice*

    A. Radioactivity is caused by
        (a) too many isotopes
        (b) short half-lives
        (c) gamma radiation
        (d) unstable nuclei

    B. Which of the following detects radioactivity?
        (a) atomic pile
        (b) Geiger-Mueller counter
        (c) cylcotron
        (d) plasma

    C. The smallest amount of fissionable material that will support a self-sustaining chain reaction is called the
        (a) isotopic weight
        (b) atomic weight
        (c) critical mass
        (d) mass number

    D. The fission of uranium-235 nuclei is initiated by
        (a) electrons
        (b) protons
        (c) neutrons
        (d) gamma rays

E. The energy obtained from nuclear reactions comes from
   (a) explosions
   (b) helium being converted to hydrogen
   (c) radioactive decay of uranium
   (d) conversion of mass to energy

Blomeke, John O., Jere P. Nichols, and William C. McClain, "Managing Radioactive Wastes." *Physics Today,* August, 1973.

> Although underground salt deposits are still the most favored, the arctic ice cap, deep ocean trenches, and solar orbit have been considered as storage areas.

Bump, T. R., "A Third Generation of Breeder Reactors." *Scientific American,* May, 1967.

> A number of organizations throughout the world are working on the development of a third generation of breeder reactors, which may be commercially operational in the 1980's.

Emmett, John L., John Nuckolls, and Lowell Wood, "Fusion Power by Laser Implosion." *Scientific American,* June, 1974.

> Fusion reactions can be initiated without a magnetic field by focusing a powerful laser pulse on a frozen pellet of fuel. Laser-fusion research is comparable in scale to the magnetic-confinement approach.

Schoenborn, Benno P., "Neutron Scattering and Biological Structures." *Chemical and Engineering News,* January 24, 1977.

> The way molecules diffract neutrons discloses details of the structure of biological materials that neither X-ray scattering nor electron microscopy can attain.

Sparberg, Esther B., "A Study of the Discovery of Fission." *American Journal of Physics,* January, 1964.

> Neither an accident nor a case of serendipity, nuclear fission represented the climax of activity in nuclear science during the 1930's.

Yonas, Gerold, "Fusion Power with Particle Beams." *Scientific American,* November, 1978.

> This dark horse in the race to fusion power uses intense beams of electrons, instead of laser beams, to implode tiny pellets of deuterium and tritium.

19

ASTRONOMY TODAY

19-1 INTRODUCTION

Astronomy is today one of the most active and exciting fields of physical science. The excitement extends not only to astronomers but to the public at large, through such events as the launching of artificial satellites, moon landings, orbiting observatories, space labs, planetary probes, and the search for extraterrestrial life. Pulsars, black holes, quasars, and scores of interstellar molecules are just a few recent discoveries that have changed some of our ideas of the solar system and the universe.

Almost everything we know about planets and stars comes to us from the light they send us. Until 1610, knowledge of the universe was based on observations made with the naked eye. When Galileo pointed his telescope to the moon, sun, and stars, he changed the course of astronomy dramatically. The telescope has since become the astronomer's single most important tool, gathering light, magnifying, and resolving. It is basically a large and powerful eye.

Light energy originates in nuclear reactions in the interior of stars where matter is transformed into energy according to Einstein's equation $E = mc^2$. Each second, for example, 600 million tons of hydrogen in our sun are changed into 595½ million tons of helium, and 4½ million tons of matter are transformed into energy. This energy appears first in the star's interior as gamma rays with very short wavelengths and high energy. As the gamma rays interact with matter in the star, they are gradually transformed into photons of longer wavelengths, and work their way to the surface to appear as starlight.

The astronomer uses a telescope to collect light,

the raw material of astronomy. Instruments attached to the telescope, such as the camera and the spectroscope or spectrograph, record and analyze the information received. The astronomer's second most important tool, the spectrograph, takes starlight from a telescope, passes it through a prism, and records the resulting spectrum on a photographic plate. The amount of information that spectral analysis gives is nothing less than amazing. We discussed spectra previously, and in this chapter we shall see their applications to astronomy.

The Doppler effect has become another powerful tool in astronomy. We discussed this effect earlier with reference to wave motion, mainly in the realm of sound. Light also behaves as a wave and therefore also exhibits the Doppler effect. Unlike sound, the Doppler effect in light is too slight for us to notice in our daily lives. When we are dealing with the great velocities encountered in astronomy, however, changes in wavelength and frequency can be measured. If a star is moving toward us, the light waves become shorter, shifting toward the blue end of the spectrum. If the star is receding, the wavelength is lengthened, and there is a red shift. Such evidence is also important to any theory of the universe, as we shall see, because it is quantitatively observable.

In the 1970's, astronomers for the first time were not limited to what their eyes could see or their telescopes could gather, that is, the visible region of the electromagnetic spectrum. It became possible to observe the heavens and collect radiation from nearly every region of the spectrum. Radio astronomy uses the waves in the radio region of the spectrum, which are about one million times longer than light waves and requires different types of instruments to record and analyze data. Infrared, ultraviolet, and X-ray astronomy deal with wavelengths in those regions of the spectrum. Observations at these various wavelengths are being brought together, and during the last quarter of the twentieth century may give the first complete picture of the universe.

19–2 THE CONSTELLATIONS

To the early observers, all objects in the sky were regarded as stars. The "fixed stars" were distinguished from the "wandering stars" or planets, and with the sun and moon were believed to revolve around the earth in their respective "spheres." About 3000 stars are visible to the naked eye on any clear, moonless night, but they are not evenly distributed. The brightest are grouped into constellations to which the names of mythological characters have been given. There is often no apparent resemblance, however, between the shape of the mythological figure and the star grouping. There are eighty-eight constellations, some of which are listed in Table 19–1 with their meanings.

The constellations are used in astronomy mainly to indicate the region in the sky in which a celestial object may be found. The first discrete source of radio waves outside our solar system, for example, was discovered in the constellation Cygnus and designated Cygnus A. Many objects have been located in the directions of other constellations and mapped, the constellations serving as area markers.

A band of 12 constellations, called the zodiac, lies in the same plane as the earth's orbit around the sun (Fig. 19–1). These constellations are considered by some to be important in influencing human affairs, although there is no scientific basis for this belief. Table 19–2 give the signs of the zodiac in sequence.

Table 19–1   NAMES OF SELECTED CONSTELLATIONS

NAME OF CONSTELLATION	MEANING
Andromeda	Chained maiden
Aquarius	Water bearer
Aries	Ram
Cancer	Crab
Capricorn	Sea goat
Cassiopeia	Lady in a chair
Cygnus	Swan
Gemini	Twins (Castor and Pollux)
Hydra	Sea serpent
Leo	Lion
Libra	Balance
Orion	Hunter
Pegasus	Winged horse
Pisces	Fishes
Sagittarius	Archer
Scorpius	Scorpion
Taurus	Bull
Ursa Major	Big bear
Ursa Minor	Little bear
Virgo	Virgin

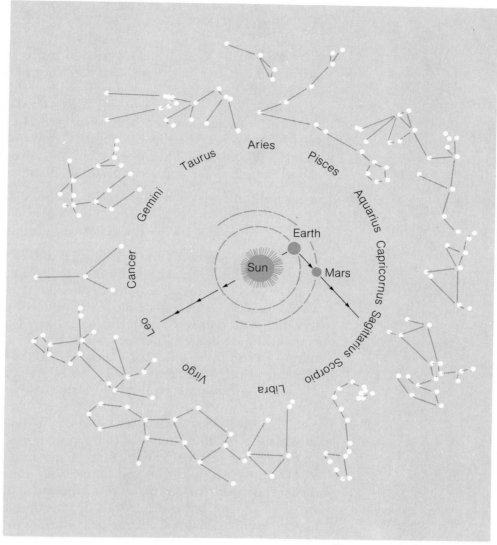

Figure 19–1 The zodiac. This band of 12 constellations lies in the same plane as the earth's orbit around the sun. (From *Astronomy One,* by Hynek and Apfel, W. A. Benjamin, Inc. © 1972. Reprinted by permission of the Benjamin/Cummings Publishing Co., Inc.)

Table 19–2 SIGNS OF THE ZODIAC

1. Aries	7. Libra
2. Taurus	8. Scorpius
3. Gemini	9. Sagittarius
4. Cancer	10. Capricornus
5. Leo	11. Aquarius
6. Virgo	12. Pisces

19-3   CELESTIAL DISTANCES AND POSITIONS

The star 61 Cygni is 64.8 trillion miles from the earth, and it is one of the nearest stars. For expressing distances to the stars, we need units more convenient than miles. Even the astronomical unit (AU), 93 million miles—the mean distance between the earth and the sun—is too short; the distance of 61 Cygni is approximately 700 000 AU. The speed of light, however, is a useful standard for expressing stellar distances. In 1 year, light travels 6 trillion miles, $\left(186\ 000\ \dfrac{\text{mi}}{\text{sec}} \times \right.$

$\left. 3.2 \times 10^7 \dfrac{\text{sec}}{\text{yr}} = 6 \times 10^{12} \dfrac{\text{mi}}{\text{yr}} \right)$, a distance called the light-year (ly). The distance to 61 Cygni is 10.8 ly.

Although no star is close enough to have a parallax as large as 1 second of arc, the distance to such a star would be 3.26 ly. This distance is defined as 1 parallax-second, or 1 parsec (pc). If the parallax of a star is known, the distance in parsecs is its reciprocal. Alpha Lyrae, for example, has a parallax of 0.1″; its distance is therefore 1/0.1 = 10 parsecs, or 10 × 3.26 = 32.6 ly. The distances of several well-known stars are given in Table 19-3.

The positions of stars and other celestial objects are specified in the same way that a position is defined on the earth's surface. The location of any spot on earth is given by its latitude and longitude. Latitude is measured north or south of the equator, assigned a value of 0°; the poles are 90° to the north and south. Longitude is measured east or west of the Greenwich Meridian, a line running from the north pole to the south pole through the former site of the Royal Greenwich Observatory in England, and is expressed in two

Table 19-3   PARALLAXES AND DISTANCES OF SELECTED STARS

STAR	PARALLAX (arc-seconds)	DISTANCE (parsecs)	DISTANCE (light-years)
Proxima Centauri	0.76	1.3	4.2
Sirius	0.38	2.6	8.5
61 Cygni	0.30	3.3	10.8
Aldebaran	0.06	16.7	54.4

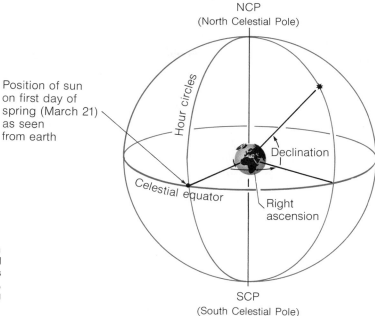

NCP
(North Celestial Pole)

Position of sun
on first day of
spring (March 21)
as seen
from earth

Hour circles

Declination

Celestial equator

Right
ascension

SCP
(South Celestial Pole)

Figure 19-2 Right Ascension
(RA)-Declination (DEC) Celestial
Coordinate System. Coordinates
give RA in hours around equator,
and DEC in degrees along
meridian.

ways: angular measure (degrees, minutes, seconds)
and time (hours, minutes, seconds). Thus, a place on
earth may have a latitude of +42°43′ (a positive
latitude means that the location is north of the equator),
and 73°12′ western longitude (measured west from
Greenwich). Declination (DEC) is the celestial
equivalent of latitude and is specified in degrees, positive to the north and negative to the south of the extension of the earth's equator, the celestial equator, at
0°. Right ascension (RA) is similar to longitude except
that astronomers measure only in the eastward direction; it is the distance measured eastward along the
celestial equator from the location of the sun on the
first day of spring and is usually expressed in hours,
minutes, and seconds (Fig. 19-2). An object identified
as PSR 0531 +21 (a pulsar) is located at 5^h31^mRA and
21°DEC—that is, 5 hours and 31 minutes right ascension, and +21° declination.

19-4 CHARACTERISTICS OF STARS

Astronomers depend on the analysis of light
emitted by a star for their knowledge of its physical
state. A spectrum is an indicator of temperature,
luminosity, chemical composition, and size. Light from

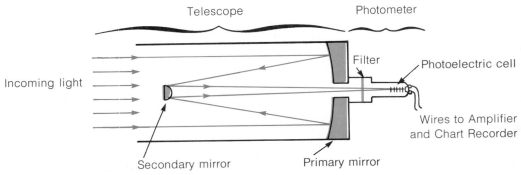

Telescope   Photometer

Incoming light

Filter

Photoelectric cell

Wires to Amplifier
and Chart Recorder

Secondary mirror   Primary mirror

Figure 19–3 Photoelectric photometer. A photoelectric cell converts starlight into an electric current that can be amplified and recorded. (Adapted from Kaufmann, William J., III, *Astronomy: The Structure of the Universe.* New York: Macmillan Publishing Co., 1977, p. 166.)

faint stars is measured by a photoelectric photometer, a device containing a sensitive photoelectric cell. Starlight is converted in the cell to an electric current proportional to the amount of light received; the current is then amplified and recorded electronically (Fig. 19–3).

Most spectra of stars are similar to the solar spectrum—a bright continuum crossed by dark absorption Fraunhofer lines—but there are some stars that have only bright-line spectra and no dark absorption lines. From the hottest to the coolest stars, the spectra fall into groups designated O — B — A — F — G — K — M, in the Henry Draper (1837–1882) sequence, a sequence easily remembered with the aid of the mnemonic device: *Oh Be A Friendly Girl and Kiss Me*. Table 19–4 lists this sequence and some of its characteristics. Hundreds of thousands of stars were classified in this way by Annie Jump Cannon (1863–1941), an astronomer at the Harvard College Observatory, whose catalogue is called the Henry Draper catalogue after the benefactor who made the investigation possible. Many stars are still known by their HD (Henry Draper catalogue) numbers, such as HD 42581.

Table 19–4 CLASSIFICATION OF STARS

CLASS	TEMPERATURE RANGE, °C	COLOR	SPECTRUM	EXAMPLE
O	25000 and up	Blue-white	He^+	Lambda Cephei
B	11 000–25 000	Blue-white	H^0, He^0	Beta Centauri
A	7500–11000	White	H^0, Mg^{2+}	Sirius A
F	6000–7500	Yellow-white	Fe^+, Ti^+	Procyon
G	5100–6000	Yellow	Ca^+, Ca^{2+}	Our sun
K	3600–5100	Yellow	Na^+	Alpha Centauri B
M	3000–3600	Reddish	AlO, TiO	Betelgeuse

The surface temperatures of stars vary from a relatively cool 2500° C or less to over 50 000° C. Some stars, like the sun, are rich in metals. Stars vary considerably in density and size, ranging from subdwarfs to supergiants. Their diameters range from a fraction of the diameter of the sun to a diameter 500 times as large. Stellar masses range from a hundredth the mass of the sun to about a hundred solar masses.

The Doppler effect (discussed in Chapter 8) gives information about the motion of stars. If a star is moving away, the whole spectrum of its light is shifted toward the red or long-wave end. Most stars, for example, show strong absorption lines by calcium atoms at wavelengths of 3933.664Å and 3968.470Å. In the spectrum of the star delta Leporis, one of the lines of calcium is displaced 1.298Å toward the red. Assuming that the displacement is due to the Doppler effect, the speed of the star's receding motion is calculated to be 99 kilometers per second.

19–5 THE LIFE AND DEATH OF STARS

Despite their number and diversity, stars fall into a kind of periodic table on the basis of (1) brightness, and (2) temperature, or spectral class. When brightness is plotted along the vertical axis and spectral class along the horizontal axis, the resulting diagram is known as a *Hertzsprung-Russell (H-R) diagram,* after Ejnar Hertzsprung (1873–1967) and Henry N. Russell (1877–1957), who proposed it independently (Fig. 19–4). It has been called the most important diagram in all of astronomy, showing a pattern in the size and temperature of stars that indicates that there must be an underlying physical cause.

The most prominent region of the H-R diagram is the *main sequence*. This is a long narrow band running diagonally across the diagram, in which over 90 per cent of all stars near the sun fall. The more luminous stars on the main sequence are hot and bluish-white in color; the dimmest stars are cool and reddish. The sun, with its predominantly yellow color, falls about three-fourths of the way along the band.

A star spends most of its lifetime on the main sequence. During this time hydrogen is converted into helium through nuclear reactions. When the supply of hydrogen is depleted, the star moves off the main sequence into a region lying above it and to the right,

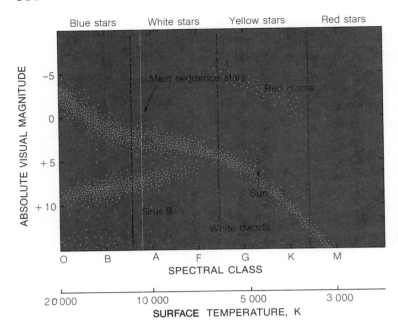

Figure 19–14 Hertzsprung Russell (H-R) diagram. The luminosity of stars is plotted along the vertical axis, and the spectral class along the horizontal axis. Most stars in the Milky Way occupy the "main sequence." Numerically smaller magnitudes are associated with the brighter stars.

the *red giant* branch, the only other well-defined sequence. There the star enters the "helium-burning" stage of development, during which nuclear reactions involving helium take place. The majority of stars in the solar neighborhood lie in the main sequence and red giant regions.

The lifetime of the sun as a main sequence star may be 10 billion years, half of which it has already spent. In another 5 billion years its hydrogen will have been exhausted and it may evolve into a red giant. Then its volume will expand far beyond its present limits to the distance of the earth's orbit.

Once they have passed the helium-burning stage, stars evolve in one of three directions: *white dwarfs, neutron stars,* or *black holes.* By then they have virtually exhausted the nuclear fuel that enabled them to shine for billions of years. White dwarfs occupy the lower left region of the H-R diagram and represent more than 10 per cent of all the stars in our galaxy. Stars lighter than about 1.2 solar masses can die as white dwarfs. They are relatively hot stars, but their low luminosity indicates that they must be very small. Sirius B is a white dwarf. Although it is about the size of the earth, its mass is nearly that of the sun. The average density of a white dwarf is about a million times greater than the sun's density; a cubic inch of matter from Sirius B would weigh an amazing 1 ton on earth.

Neutron stars were first predicted theoretically in the 1930's. Their internal pressure is so great that the plasma electrons combine with protons to form neutrons. Such stars are much denser than white dwarfs; a cubic inch of a neutron star would weigh a billion tons on earth. Only a few kilometers in diameter, a neutron star should spin very rapidly as it contracts from the size of a normal star, at rates between 30 times per second and once every few seconds, because of the conservation of angular momentum (much the same as skaters spin more rapidly as their arms are brought closer to the body).

Astronomers now largely agree that pulsars, discovered in 1968 as a class of objects that emit bursts of radio noise with periods ranging from 0.033 to 3.75 seconds, are rapidly rotating neutron stars. Only neutron stars can rotate at pulsar frequencies without disruption. It is estimated that as many as 100 million of the 100 billion stars in our galaxy have burned themselves out and collapsed into neutron stars.

There has been speculation that certain stars (in which the mass considerably exceeds the mass of the sun in their final stage of evolution) may collapse to densities even greater than those found in neutron stars. The gravitational field around such a star, called a "black hole," would be so great that neither matter nor light could ever leave it. The density of a black hole would be greater than 10^{16} g/cm^3, the densest matter known in the universe. Black holes are the subject of considerable research. Astronomers in 1973 detected X-ray evidence of a black hole in the constellation Cygnus. The black hole seemed to be a kind of invisible companion to a visible supergiant star called HDE-226868. Every 56 days the black hole passes behind the visible star. As it does so, it blocks X rays that are created as gases spiral toward the black hole and are compressed and heated to 100 million degrees. A suspected black hole was also discovered in the constellation Scorpio in 1978 and named Scorpio V-861.

19–6 NOVAE AND SUPERNOVAE

Stars that at one time appeared very faint may suddenly become extremely brilliant, with a light output thousands of times greater than before. Such stars are called *novae* ("new stars"). Novae bright enough to be visible to the naked eye occur only once every few years in our galaxy. Those that are observable by telescope occur more frequently. A nova may reach its maximum brightness in 1 or 2 days, or it may take as long as several days or weeks. Novae are exploding stars that blow off parts of their atmospheres. The optical radiation decreases gradually after several days or weeks, and after a few years the star seems to return to its former state.

More rare than novae, but considerably more energy-rich, are the *supernovae*. Supernovae are major explosions, flaring up to millions of times brighter than their original state. Only two have been recorded within our galaxy by Western astronomers, the first discovered by Tycho Brahe in 1572 in the constellation Cassiopeia, and the second in the constellation Serpens discovered by Johannes Kepler in 1604. At its peak, Tycho's supernova was several times brighter than Venus and could be seen even in the daytime; Kepler's was as bright as Jupiter. With powerful modern telescopes, supernovae are regularly found in other galaxies.

Chinese astronomers have recorded at least four other supernovae, including one in 1054 in the constellation Taurus, the remains of which are now recognized as the Crab Nebula (Fig. 19–5). Why Western chroniclers failed to mention a star which had been invisible and suddenly became (on July 4, 1054) one of the brightest in the sky—so bright that it could be seen in full daylight—is a mystery. The event must have been one of the most spectacular in history. Although ignored by the West, in China and Japan the "guest star" was described in detail. After glowing brilliantly for 3 weeks, it gradually faded and disappeared after 2 years. The Veil Nebula in Cygnus is believed to be another visual remnant of a supernova.

Figure 19–5   The Crab Nebula, photographed with the 200-inch (5-meter) telescope. The gaseous debris is believed to have come from a supernova explosion. (Courtesy of the Hale Observatories.)

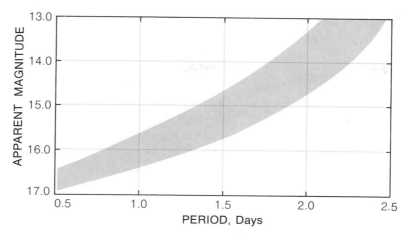

Figure 19-6 Period-luminosity relation for cepheids in the Clouds of Magellan. The longer the period, the brighter the star.

19-7   PULSATING STARS

Pulsating stars wax and wane in brightness in periods ranging from 1 to 100 days. The most familiar of these is Polaris, the pole star, which brightens and fades to a slight extent in a period of 3.97 days. One class of such variable stars is called the *Cepheid variables,* named after a pulsating star discovered in the constellation Cepheus and named delta Cephei. Because of their changing internal structure, pulsating stars are unstable. As they increase and decrease in size, their surface temperature also changes, and their light emission varies.

The Cepheid variables act as a "celestial lighthouse," and as a means of measuring distances deep into the universe where other methods of measurement are inadequate. The Cepheids were first investigated by Solon I. Bailey (1854–1931), a Harvard astronomer, who studied them in aggregations of stars called *globular clusters* within our own galaxy. Henrietta S. Leavitt (1868–1921) of the Harvard College Observatory later investigated the Cepheids in the Clouds of Magellan, two small companion galaxies of our own, and determined that most had periods of more than a day. At the same time, she discovered that the average apparent brightness of these Cepheids was related to their periods of pulsation. She established a *period-luminosity relation,* which became a potential yardstick for measuring great astronomical distances (Fig. 19–6).

Harlow Shapley (1885–1972), Director of the Harvard College Observatory, applied the period-luminosity relation to the determination of the distances of the globular clusters in our galaxy. Edwin P. Hubble (1889–1953) at the Mount Wilson Observatory then used it in measuring the distances to the Magellanic Clouds, to the Great Nebula in Andromeda, and to more distant galaxies up to 20 million light years away. The relation was further used to calibrate the red-shift, making possible measurement to still more distant galaxies in which the Cepheids cannot be resolved even by the most powerful telescopes available. From such data, the size of the universe, and, if the universe is expanding, its age from the origin of expansion could be estimated.

19-8 GALAXIES, NEBULAE, AND STAR CLUSTERS

When the sky is observed on a clear, moonless night, a faint milky band of light is visible across certain constellations. Ancient astronomers called it appropriately the Milky Way, or *galaxy* (Greek *gala*, milk). All the stars that can be seen with the naked eye, and most of those that can be distinguished by the most powerful telescopes, are members of the Milky Way system. Even with a pair of field glasses or a small telescope, the milky appearance can be seen to be the effect of countless stars merged together to form a nebulous mist. There are more than 100 billion stars in the Milky Way, the majority of them so distant that we cannot see them as individual points of light. Aside from stars, a galaxy includes such components as star clusters, clouds of gas and dust, and interstellar molecules.

Our knowledge of our place in the Milky Way and of the distribution of galaxies in the universe has been gained only in the present century. The early telescopes revealed many diffuse patches of light, which looked nearly alike and were classified as "nebulae" (Fig. 19–7). These objects were believed to lie within the Milky Way, but they did not seem to be at all like stars. Many of them are beautiful spirals and are seen in all different orientations. As early as the 18th century, the astronomer Sir Wiliam Herschel (1738–1822) and the philosopher Immanuel Kant (1724–1804) suggested that these nebulae were actually "island universes" or galaxies lying far beyond the Milky Way, but this hypothesis was not generally accepted. As re-

Figure 19-7   The Orion Nebula.
(Lick Observatory photograph.)

cently as the 1920's, the nature of nebulae was being
argued.

The matter was settled in 1924 when Edwin P.
Hubble, using the Cepheid variable period-luminosity
relation, succeeded in measuring the distance to the
Great Nebula in Andromeda and to a number of other
spiral nebulae with the 100-inch telescope at Mount
Wilson, then the most powerful in the world. (Fig.
19–8). Hubble established that the nearest spiral nebulae were galaxies like our own Milky Way—vast systems of stars, clusters, gas, and dust—situated a million
or more light years outside our own galaxy. The universe is populated by billions of galaxies that appear to
be the building blocks of the universe. A single photograph with a telescope may easily show 1000 galaxies.

The term *nebula* now refers to a cloud of gas and
dust, one of the many components of a galaxy. The Ring
Nebula in Lyra is an example of a planetary nebula, a
small shell centered on a star that is thought to have
ejected it, which shines by fluorescence stimulated by
ultraviolet radiation from the central star (Fig. 19–9).
The Horsehead nebula in Orion is a dark cloud of dust
which hides background stars (Fig. 19–10). Some
nebulae are supernova remnants produced by the ex-

Figure 19-8 The Andromeda Galaxy, M31, is a spiral galaxy similar to our own. (Lick Observatory photograph.)

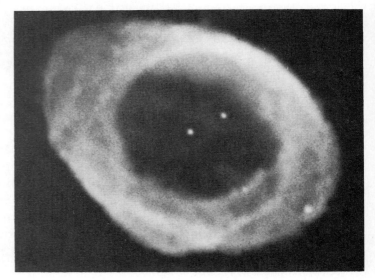

Figure 19-9 The Ring Nebula in the constellation Lyra. (Courtesy of the Hale Observatories.)

Figure 19-10 The Horsehead Nebula in Orion. (Courtesy of Hale Observatories.)

plosion of a massive star; they expand at high velocities and radiate as a result of collision with interstellar gas.

If viewed from the outside, the Milky Way would probably appear like a flat disk with tightly wound spiral arms very much like the spiral Andromeda galaxy. It resembles a giant pinwheel made up of more than 100 billion stars concentrated toward a central hub and rotating slowly around a compact, brilliant nucleus. The spiral arms consist of concentrations of stars, dust, and gas. At the speed of light it would take 100 000 years to pass from one end of the Milky Way to the other; it is 5000–10 000 ly in thickness (Fig. 19–11). The sun is 30 000 ly from the center of the Milky Way, orbiting around it once about every 200 million years.

The space above and below the pinwheel is filled with billions of much fainter stars, and some 200 fuzzy globular *star clusters*. A star cluster is a group of stars moving together through the galaxy in relative proximity. It is made up of hundreds of thousands of stars concentrated toward the center so that they appear to fuse together and suggest a continuous area of light. These globular clusters are scattered around the main nucleus of the galaxy, forming a sort of halo around the main body (Fig. 19–12). None is near our sun; the closest is about 20 000 ly away. The slightly ellipsoidal shape of many of them suggests that they probably rotate.

Many of the stars in globular clusters are red giants, known as Population II stars. They are apparently very ancient, approximately 10 billion years old, suggesting that they may have been formed before the Milky Way evolved into its present spiral structure. Hundreds of "open" star clusters, most consisting of a few hundred stars held together by their mutual gravitation, have also been found, usually in or near the spiral arms. Stars in open clusters, called Population I stars, are relatively young, from about 1 million to 1 billion years old, and are often associated with interstellar matter from which new stars are believed to be forming. Little, if any, interstellar matter is found in the globular clusters.

With modern instruments, a billion or more galaxies are detectable, and several thousand are close enough for detailed inspection. They vary in size, shape, mass, and luminosity. Galaxies are usually identified by number, prefaced by initials standing for the reference in which they are catalogued. The

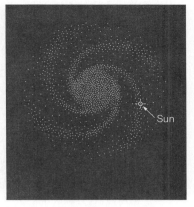

Figure 19–11 Pinwheel appearance of the Milky Way viewed from the outside.

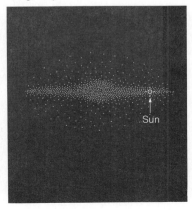

Figure 19–12 Globular star clusters around the nucleus of the galaxy.

Figure 19–13 Spiral galaxy in Ursa Major, designated NGC 3031 and M81. (Courtesy of Hale Observatories.)

brightest have the initial M (such as M51) for Charles Messier (1730–1817), an astronomer who listed the 103 brightest "nebulae" in 1784. Many of the fainter ones are designated NGC for the New General Catalogue of John L. E. Dreyer (1852–1926), based largely on discoveries of nebulae by the pioneers of galaxy studies, William and his son John Herschel (1792–1871) (Fig. 19–13).

Galaxies, in turn, are themselves generally found in clusters that vary in size and in the number of member galaxies. The Milky Way is a member of a cluster of 20 or so galaxies known as the *Local Group*. Three of the members of this cluster were discovered in 1971, suggesting that there may be others as yet undiscovered. Some clusters, such as the Virgo and Coma clusters, contain thousands of members. The galaxies within a cluster are not crowded, being separated by distances several times their individual diameters. Clusters of galaxies apparently form the largest structural units in the universe.

19–9 RADIO ASTRONOMY

The recent revolution in astronomy is due in large part to a serendipitous discovery made in 1932 by Karl G. Jansky (1905–1950) of the Bell Telephone Laboratory which opened a new window on the universe. Until then, our only window into space was the visible region of the electromagnetic spectrum. In order to understand better the causes of interfering static, so that improved transoceanic short-wave radio-telephone systems could be designed, Jansky built a rotating radio antenna 100 feet long. He found that thunderstorms and man-made electrical interferences accounted for one kind of static, and learned enough to solve the immediate problem at hand. At the same time, Jansky discovered a second kind of static, which sounded like a steady radio hiss on the earphones and did not appear to come from any obvious source. After months of experiment, he proved that the hiss could not possibly originate on earth and that it must come from outer space, although not from the sun, moon, or planets. The source, he established, was in the direction of the center of the Milky Way. Without knowing it, Jansky had built the world's first radio telescope.

Grote Reber (b. 1911) a radio engineer and amateur astronomer, having read about Jansky's discovery, built a radio telescope in his backyard. It was movable, 31.4 feet in diameter, and resembled a giant parabolic dish, with the radio receiver held out in front on four legs. With this telescope, Reber mapped the radio sky for several years. His paper containing the first maps seemed so fantastic that it encountered resistance when it arrived at the *Astrophysical Journal*. The editor decided to publish it nevertheless, believing that it was the beginning of a new kind of astronomy. Today, radio telescopes are an indispensable tool not only for exploring the far reaches of the universe but also for providing detailed information of space closer at hand (Fig. 19–14).

If our eyes were sensitive to radio waves instead of

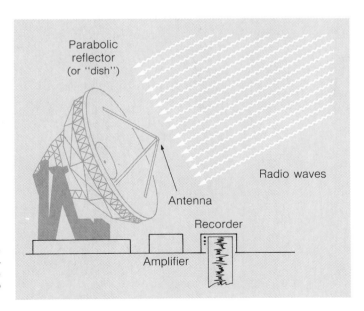

Figure 19–14 Radio telescope (schematic). The parabolic "dish" reflects radio waves to a focus at the antenna, which feeds the signals into an amplifier and recording device.

ELECTROMAGNETIC
RADIATION

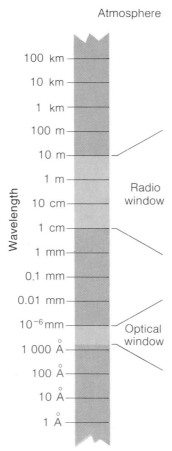

Figure 19–15 The two windows in the atmosphere that allow radiation to reach the earth: the radio window and the optical window. (From *Astronomy One,* by Hynek and Apfel, W. A. Benjamin, Inc. © 1972. Reprinted with permission of the Benjamin/Cummings Publishing Co., Inc.)

to light, the sky would appear very different. The sun would be much less bright, while the Milky Way would shine brilliantly. Hundreds of new "stars" would cover the sky, forming unfamiliar constellations. Clouds of interstellar gas in space would become visible, as would the gaseous remnants of supernovae. One of the brightest objects would be Cygnus A.

A radio telescope, unlike an optical one, can receive from outer space waves between 1 centimeter and 20 meters in length (Fig. 19–15), which are able to penetrate the earth's atmosphere. (Wavelengths less than 1 cm are absorbed by the atmosphere; those above 20 m are reflected back into space.) Observations can be made with a radio telescope throughout a 24-hour period, since radio signals are not affected by the sunlit sky. An important advantage of a radio telescope is that it can explore behind the dust clouds of space, which are opaque to visible light. It has been estimated that this dust must hide 90 per cent of the stars in the Milky Way from visual detection, obscuring the regions beyond about 6000 ly. Radio waves, however, penetrate this material without absorption. Further, radio has an inherent advantage over light in probing to great distances, at which the red-shift is substantial. For example, the light from galaxies at a distance of 5 billion ly, moving away from us at half the speed of light, would be shifted so far to the red that only part of their spectrum in the visible range could be recorded on photographic plates; the rest of their light would be lost. With radio waves, the loss due to a wavelength shift is relatively small, and galaxies can be detected at considerably greater distances.

A basic problem of radio astronomy, on the other hand, is the much greater length of radio waves as compared to light waves, making resolution of the source, that is, the ability to discriminate fine detail, more difficult. Secondly, celestial radio signals are weak, the faintest having only a hundred-millionth of the power of a television signal. One reason for this is that the energy of a photon is inversely proportional to its wavelength, and radio waves are very long. Larger antennas have been the solution to both problems. A large antenna has superior resolution, enabling radio sources to be narrowed down to smaller and smaller regions, and it collects more radio energy from the source, just as a larger mirror in an optical telescope collects more light.

There are many kinds of radio telescopes, each with its special use. The two principal forms of construction are (a) fixed or steerable metal-mesh or solid metal parabolic dishes, which collect radio signals and focus them to a radio receiver where they are amplified and then fed to recording devices, and (b) linear arrays of antennas arranged in various patterns.

A large radio telescope built at the Jodrell Bank station of the University of Manchester, England, is similar in shape to the one used by Reber, but is 250 feet in diameter. Other giant radio telescopes have been built throughout the world. The largest movable, steerable paraboloid is 300 feet in diameter. The largest parabola of any kind, a dish of metal mesh 1000 feet in diameter, is built into a hole carved out of the mountains of Arecibo, Puerto Rico (Fig. 19–16).

Another way to obtain better resolution of radio sources is through *interferometry*. Two antennas are placed some distance apart and connected to a single radio receiver. Radio waves from space strike one slightly before the other and produce interference patterns. As the earth rotates, this radio interferometer sweeps the sky. Through interferometry, the position of a strong radio source can be determined with the

Figure 19–16 The 1000-ft. (305-meter) diameter telescope at Arecibo, Puerto Rico. (Courtesy of the Arecibo Observatory.)

same accuracy that could be obtained with a single antenna as large as the distance between the two small antennas. One of the most sensitive radio telescope systems is the National Radio Astronomy Observatory's (NRAO) three-element interferometer at Green Bank, West Virginia.

Thousands of discrete sources of radio emissions have been discovered and mapped. In many cases no optically visible object could be identified as the source. Radio astronomers have catalogued and numbered the sources. The catalogue to which reference is made most frequently is the Third Cambridge Catalogue. The entry designated 3C 48, for example, refers to the forty-eighth entry in this catalogue.

The first discrete source of radio waves outside our solar system was discovered in 1946. It was a spot of such intensity in the constellation Cygnus, designated Cygnus A, that it startled astronomers. Additional sources were soon found in other constellations. The first identification of a radio source with a visual object also involved Cygnus A. Rudolph Minkowski (b. 1895) and Walter Baade (1893–1960), working with photographs made with the 200-inch telescope on Palomar Mountain, then the world's largest optical telescope, showed that Cygnus A coincided with the position of a galaxy now estimated to be 700 million ly away. It was originally thought that Cygnus A represented the collision of two galaxies and that the strong radio emission came from the energy released. Astronomers no longer accept the colliding-galaxy hypothesis, and the origin of the radio energy of this and other strong sources continues to be a mystery. Many other radio sources have since been identified with visible galaxies. As resolution improved, it was shown that many individual objects were emitting radio waves. One of the brightest sources is the sun, and the planets also broadcast. Some of the nebulae within our galaxy, the remnants of supernovae, are also radio sources, as are certain classes of stars: red supergiants, red dwarfs, blue dwarfs, novae, pulsars, and X-ray stars.

19–10 HYDROGEN 21-CENTIMETER WAVES

A new dimension was added to astronomy with the discovery of radio waves emitted by hydrogen in space.

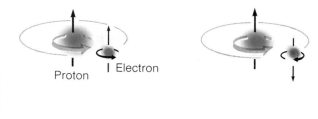

Figure 19–17 Hydrogen 21-cm radiation. When the proton and electron spins are parallel (*A*), the atom contains more energy than when the spins are opposed (*B*). When the spin of the electron flips from *A* to *B*, radiation of $\lambda = 21$ cm is emitted.

It had been suggested that neutral hydrogen atoms in interstellar space might undergo a change in energy state and emit radio energy. The reasoning is as follows: A hydrogen atom in space can exist in only one of two energy states (Fig. 19–17). In the lower energy state, the electron and proton are spinning in opposite directions; in the slightly higher state, they are spinning in the same direction. Every few million years, the spin axis of the electron flips over, and a hydrogen atom in the higher states passes into the lower state. Simultaneously, a photon is emitted and carries off the energy lost in the transition.

The wavelength of the emitted photon is 21 centimeters, thousands of times longer than the wavelengths of visible light, and equivalent to a frequency of 1420 megahertz. Equipment sensitive enough to detect this radiation was not available until 1951. In that year, Edward M. Purcell (b. 1912), a physicist at Harvard University, detected 21-centimeter waves from hydrogen. For the first time, astronomers had a specific spectral line to work with in the radio spectrum.

The 21-centimeter radiation has become a potent astronomical tool. It can penetrate interstellar dust, because the waves are about a million times longer than the dimensions of a dust particle. Light waves tend to be scattered, their length being about the same as a dust particle. Otherwise, the two kinds of radiation behave in the same way. The discovery of this radiation made possible a complete picture of the Milky Way. The spiral shape is revealed by maps using the distribution of neutral hydrogen, which is concentrated in the spiral arms of the galaxy. The Doppler shifts in the line, which can be measured far more accurately than the Doppler shift of light, have revealed the motions of turbulent hydrogen clouds. Studies of the proportion of hydrogen in other galaxies, made possible by this radiation, provide clues to their evolution.

19–11 QUASARS

The strong radio source known as 3C 48 engaged the attention of a young astronomer, Allan Sandage (b. 1926), in 1960. With the 200-inch Palomar optical telescope, he took photographs of its position in the sky. What he saw was one of the most puzzling objects in the universe. Previously, the sources of radio emissions had been gas clouds or galaxies, but never stars. The object Sandage photographed was not a galaxy; it resembled a faint star with a wispy line to one side. Since it was apparently not a normal star, judging from the streak of light, 3C 48 was called a quasi-stellar radio source, or *quasar*. Using a photoelectric photometer, Sandage further found that 3C 48 had an abnormal color for a star, being too bright in blue and ultraviolet. Moreover, 3C 48 had a brightline spectrum, never before found in a star, rather than a continuous spectrum.

The puzzle was resolved when astronomers measured an accurate position for another radio source, 3C 273. This source coincided with another unusual star, fairly bright, with unusual colors and a bright-line spectrum. Maarten Schmidt (b. 1929) established that the lines were mostly hydrogen lines that were shifted to the red end of the spectrum. The red-shift corresponded to a velocity away from the observer of approximately 50 000 km/sec. When Sandage's spectrum of the first quasar, 3C 48, was reexamined, it was found that the bright lines were essentially the same lines observed in 3C 273: normal hydrogen lines, but shifted even more to the red, corresponding to a velocity of recession of 110 000 km/sec.

Only a small fraction of all quasars are strong radio sources. Some quasars have little or no emission at radio wavelengths. The term quasar is now applied to starlike objects with large red-shifts, regardless of their radio emission. Over 200 quasars have been identified, although some astronomers have estimated that millions of quasars exist in the universe.

According to one hypothesis, quasars are very remote objects that provide a direct glimpse of the universe as it existed 8 or 9 billion years ago. According to another hypothesis, quasars are relatively nearby objects. The tremendous energy associated with quasars has been attibuted to various sources: supernova explosions; gravitational collapse of a massive superstar; matter-antimatter annihilation; galaxies at an early stage of formation; and condensed central cores of galaxies.

19–12   THE INTERSTELLAR MEDIUM

Discoveries in spectroscopy and radio astronomy have shed considerable light on the composition of the space between the stars, once believed to be largely empty. The most abundant constituent is hydrogen in various forms. Neutral atomic hydrogen (H) emits and absorbs radiation at the radio wavelength of 21 cm; ionized hydrogen (H^+) and molecular hydrogen (H_2) have been detected in a variety of ways, including Balmer lines and radio astronomy. The elements helium, carbon, nitrogen, oxygen, sulfur, iron, chlorine, and many others are also present, as indicated by their spectral lines in nebulae.

Signals from hydroxyl radicals ($\cdot OH$) were received following a search by Sander Weinreb (b. 1936) of the Massachusetts Institute of Technology and others. Unlike the 21-cm signal of hydrogen, these signals consist of four different frequencies. When the dish of an 84-foot radio telescope was point toward Cassiopeia A, absorption was obtained at the 1665 and 1667 megahertz lines of the $\cdot OH$ radical. Later, two additional lines were detected in the direction of Sagittarius, at 1612.2 and 1720.6 megahertz.

Many molecules, as well as atoms and ions, have been detected in intersellar space since 1968, when molecular astronomy was born with the discovery of molecules consisting of more than two atoms. Until then, most astronomers believed that the density of the interstellar medium was too low to get more than two atoms to combine. Such ideas needed drastic revision when Charles H. Townes (b. 1915) of Stanford University and others found ammonia (NH_3) in several interstellar clouds near the center of the Milky Way. Then Townes and coworkers detected emission signals from water vapor (H_2O) in several regions of the galaxy. Many other molecules have been detected since. They have probably been formed by collisions between atoms of the interstellar gas or by combining on the surface of particles of interstellar dust.

The discovery of some very complex molecules in interstellar space, such as formaldehyde (H_2CO) and methyl alcohol (CH_3OH), shown in Table 19–5, has

Table 19–5   Iɴᴛᴇʀsᴛᴇʟʟᴀʀ Mᴏʟᴇᴄᴜʟᴇs

YEAR DETECTED	MOLECULE	FORMULA
1937	Methylidyne (ionized)	CH^+
1937	Methylidyne	CH
1939	Cyanogen radical	CN
1963	Hydroxyl radical	OH
1968	Ammonia	NH_3
1968	Water	H_2O
1969	Formaldehyde	H_2CO
1970	Carbon monoxide	CO
1970	Hydrogen cyanide	HCN
1970	X-ogen	?
1970	Cyanoacetylene	HC_3N
1970	Hydrogen (molecular)	H_2
1970	Methyl alcohol	CH_3OH
1970	Formic acid	$HCOOH$
1971	Carbon monosulfide	CS
1971	Formamide	NH_2CHO
1971	Carbonyl sulfide	OCS
1971	Silicon monoxide	SiO
1971	Methyl cyanide	CH_3CN
1971	Isocyanic acid	$HNCO$
1971	Hydrogen isocyanide	HNC
1971	Methylacetylene	CH_3CCH
1971	Acetaldehyde	CH_3CHO
1972	Thioformaldehyde	H_2CS
1972	Hydrogen sulfide	H_2S
1972	Methanimine	CH_2NH
1973	Sulfur monoxide	SO
1974	Ethyl alcohol	C_2H_5OH
1974	Silicon monosulfide	SiS
1975	Sulfur dioxide	SO_2
1975	Acrylonitrile	H_2CCHCN
1975	Methyl formate	$HCOOCH_3$
1975	Nitrogen sulfide radical	NS
1975	Cyanamide	NH_2CN
1976	Cyanodiacetylene	HC_5N
1976	Formyl radical	HCO
1976	Acetylene	C_2H_2

led to interesting speculation. According to some theories, life may occur when certain substances in a planetary atmosphere—ammonia, water, formaldehyde, hydrogen—are irradiated by ultraviolet radiation or by lightning to produce amino acids. It appears from recent discoveries in molecular astronomy that the conditions for organized life may exist elsewhere in the universe. The discovery of hydrogen cyanide (HCN) in 1970, important in the laboratory synthesis of amino acids, and the occurrence of so many biochemical compounds lend support to this possibility.

This subject is discussed in greater detail in Chapter 24.

19–13 THE EXPANDING UNIVERSE

Cosmology is concerned with such general questions of the nature of the universe as: How did it originate? What is it like now? What is it likely to be like in the future? Most contemporary cosmologies stem from *Hubble's law* (derived from red-shift studies) that other galaxies are moving away from ours, and are doing so at speeds that are proportional to the distance of the galaxy. Table 19–6 gives the recession speeds of a few galaxies from among the hundreds that have been tested. Note that the greater the distance of the galaxy, the greater its speed away from us.

Hubble's law and Einstein's general theory of relativity laid the foundation for the theory of the expanding universe. According to this theory, the average density of matter in the past was greater than it is now. As we go back in time, the density was enormously high, possibly 10^{14} times as dense as water. The universe started from a highly compressed and extremely hot state. About 20 billion years ago, according to the most recent estimates, the matter erupted in a vast explosion—the "big bang"—and the expansion of the universe began. As the matter expanded in all directions and thinned out, it cooled down and condensed to form stars and galaxies. The red-shifts in the spectra of distant galaxies indicate that the universe continues to expand at the present time.

The "big bang" theory began with Vesto Melvin Slipher's (1875–1969) discovery in 1913 at the Lowell Observatory that about a dozen galaxies are moving

Table 19–6 RECESSION SPEEDS OF SOME GALAXIES

POSITION	DISTANCE (light-years)	SPEED (kilometers/second)
Virgo	6 000 000	1 120
Pegasus	23 000 000	5 600
Ursa Major	85 000 000	15 700
Leo	105 000 000	20 000
Gemini	135 000 000	24 500
Hydra	360 000 000	62 000

away from the earth at speeds ranging up to about one million miles an hour. This was the first hint that the universe is expanding. When Slipher reported his finding at a meeting of the American Astronomical Society in 1914, his slides clearly revealed the red shift that indicated an enormously rapid motion of these galaxies. The audience stood up and cheered, something that almost never happens at a scientific meeting. One of the people in Slipher's audience was Edwin Hubble.

Meanwhile, Einstein published his equations of general relativity in 1917. The astronomer Willem de Sitter (1872–1934), Director of the Leiden Observatory, found a solution to them that predicted an exploding universe in which the galaxies moved rapidly away from one another. This was just what Slipher had observed. After World War I, de Sitter's theoretical prediction of an expanding universe made a great impression on astronomers. Hubble said that it was mainly de Sitter's result that influenced him to take up the study of the moving galaxies where Slipher had left off.

In the 1920's, Hubble and Milton Humason (b. 1891) followed up on Slipher's work, first with the 60-inch telescope at Mount Wilson, then with the 100-inch, the world's largest at that time. They measured the speeds and distances of many galaxies too faint to be seen by Slipher, and confirmed Slipher's discovery. Some of the galaxies were moving away from us at 100 million miles an hour. In 1919, Hubble discovered that the farther away a galaxy is, the faster it moves, the relationship known as Hubble's Law and one predicted by Einstein's theory of relativity.

The agreement of both theory and observation pointing to an expanding universe and a beginning in time made a tremendous impression on astronomers. Even Einstein, who resisted the new developments and held on to the idea of a static, unchanging universe, was convinced following a visit to Hubble in 1930 that the structure of the universe is not static. At the present time, the big-bang theory has no competitors.

The latest evidence makes it almost certain that the "big bang" really did occur many billions of years ago. When the material of the universe was enormously hot—billions of degrees—and compressed, there occurred a "primeval fireball" of radiation, as described by John Wheeler (b. 1911) of Princeton University. Starting out as enormously energetic gamma rays, the radiation was "cooled" by expansion. Its wavelength increased, and it now appears mostly in the radio and microwave bands. Following a suggestion of Robert H. Dicke (b. 1916) of the same university that radiation surviving from the primeval fireball ought to be detectable, Arno A. Penzias (b. 1933) and Robert W. Wilson (b. 1936), both of the Bell Telephone Laboratory, discovered such radiation in 1965. Adherents to the big bang theory claim that the background microwave radiation, corresponding to the spectrum of radiation of an object at 3K, represents the remnant signals from near the origin of the "big bang."

EXERCISES

19–1. (a) What information does a red-shift in its spectrum yield about a star? (b) a blue-shift?

19–2. What role has the radio telescope played in the recent revolution in astronomy?

19–3. What are some problems posed by the discovery of quasars?

19–4. How are the positions of celestial objects specified?

19–5. What is the distinction between a nebula and a galaxy?

19–6.　If the Andromeda spiral galaxy is approaching us at a velocity of 290 km/sec, how long will it take to reach us?

19–7.　What evidence have we that the universe is expanding?

19–8.　Why is the Hertzprung-Russell diagram considered by some astronomers to be the most important diagram in astronomy?

19–9.　Galaxies have been referred to as the structural units of the universe. Explain.

19–10.　Distinguish between pulsars and pulsating stars.

19–11.　How have the 21-cm radio waves emitted by neutral hydrogen extended our knowledge of the universe?

19–12.　The parallax of a star is 0.001″. How far away is it?

19–13.　Why is the name Milky Way appropriate for our galaxy?

19–14.　Indicate the close relationship of other physical sciences to astronomy.

19–15.　A star is 45 ly away. Traveling at a speed of 100 000 mph, how long would it take to reach this star?

19–16.　How do we know the composition of stars?

19–17.　*Multiple Choice*

　　A. The average distance from the earth to the sun is
　　　(a) a light-year
　　　(b) an astronomical unit
　　　(c) an angstrom unit
　　　(d) a parsec

　　B. A nova is
　　　(a) a star that suddenly increases in brightness
　　　(b) a pulsar

(c) a quasar

(d) a receding galaxy

C. The remnant of the big bang is also known as
   (a) a neutron star
   (b) a black hole
   (c) 3K background radiation
   (d) the interstellar medium

D. The energy of a main sequence star comes from
   (a) gravitation
   (b) nuclear fission
   (c) nuclear fusion
   (d) helium burning

E. A Doppler red shift for a galaxy indicates that it is
   (a) approaching   (c) slowing down
   (b) receding   (d) speeding up

SUGGESTIONS FOR FURTHER READING

Bethe, Hans A., "Energy Production in Stars." *Physics Today,* September, 1968.
   Stars have a life cycle: they are born, they grow to maturity, and they die.

Bok, Bart J., "The Birth of Stars." *Scientific American,* August, 1972.
   Conditions for the formation of stars exist in the central disk of the galaxy, where interstellar gas and dust mix with young stars.

Chaisson, Eric J., "Gaseous Nebulas." *Scientific American,* December, 1978.
   These clouds within our galaxy are sites of star formation.

DeVorkin, David H., "Steps toward the Hertzsprung-Russell Diagram." *Physics Today,* March, 1978.
   The story of the discovery of the temperature-luminosity plot that revolutionized astronomy.

Gott, J. Richard III, James E. Gunn, David N. Schramm, and Beatrice M. Tinsley. "Will the Universe Expand Forever?" *Scientific American,* March, 1976.
   Their answer: Yes.

Ostriker, Jeremiah P., "The Nature of Pulsars." *Scientific American,* January, 1971.
   Pulsars are believed to be neutron stars in rapid rotation. Their discovery has opened new fields and illuminated old ones.

Schmidt, Maarten, and Francis Bello, "The Evolution of Quasars." *Scientific American,* May, 1971.
   According to the view presented, quasars provide a glimpse of the universe as it existed 8 or 9 billion years ago, only 1 or 2 billion years after the "big bang."

Solomon, Philip M., "Interstellar Molecules." *Physics Today,* March, 1973.
   Molecules have been discovered in interstellar clouds. Their abundance and chemical complexity were totally unexpected.

20

THE SOLAR SYSTEM

20-1 INTRODUCTION

Explorations of earth's close neighbors in space—the moon, Mercury, Venus, Mars, and Jupiter—have given us considerable information about our small corner of the universe. And what a fascinating corner it is, containing a central star, planets and their satellites, meteors, asteroids, and comets. Although countless systems like the solar system are presumed to exist, ours is the only one known at present. Within it there are clues to the formation and evolution of the universe as a whole. The findings thus far support the uniqueness of the earth in many respects, above all, the presence of intelligent life.

20-2 THE SOLAR SYSTEM

The solar system is composed of the sun, nine known planets, at least 35 moons, thousands of "minor planets" (asteroids), meteoroids, and perhaps a billion comets. The sun, an average star, radiates its own light and differs in this respect from a planet or a moon, which shines by reflected light only. A planet and a moon have many similarities; they differ chiefly in that a *planet* revolves around the sun, whereas a *moon* revolves around a planet.

Viewed from a point far above the Northern Hemisphere of the earth, all of the planets revolve around the sun in elliptical orbits in a counterclockwise direction (Fig. 20–1). The orbits of the planets have been defined with such precision that the position of the earth with respect to any planet may be predicted with-

Figure 20-1 The solar system (not to scale). (From Pasachoff, J. M., *Contemporary Astronomy.* Philadelphia: W. B. Saunders Company, 1977, pp. 406–407.)

in hundredths of a second of arc for many years ahead. The orbits of Mars, Jupiter, Uranus, and Neptune are all tilted less than 2° out of the plane of the *ecliptic*, the plane of the earth's orbit (Fig. 20–2). The orbit of Saturn is tilted slightly more than 2°, and Venus slightly more than 3°. The tilt of Mercury's orbit is 7°, and Pluto's, 17°. All of the planets rotate on their axes, although Venus rotates in a direction opposite to that of the other planets. Most of the planets have some form of atmosphere, except Mercury and possibly Pluto. Only one moon—Saturn's Titan—is known to have an atmosphere.

The fact that all the planets lie nearly in the plane of the ecliptic is a puzzle, although it can be accounted for by a modern version of the nebular theory proposed by Immanuel Kant (1724–1804). Like our galaxy, the solar system is thought to have evolved slowly out of a "nebula," which in time would form aggregations and then condense into solid planets. According to the modern view, now also called the "dust-cloud" theory,

Figure 20-2 The plane of the ecliptic is the plane of the earth's orbit around the sun. (From Pasachoff, J. M., *Contemporary Astronomy.* Philadelphia: W. B. Saunders Company, 1977, p. 281.)

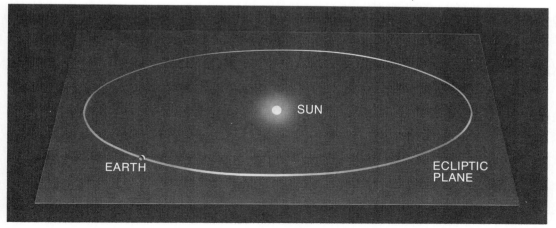

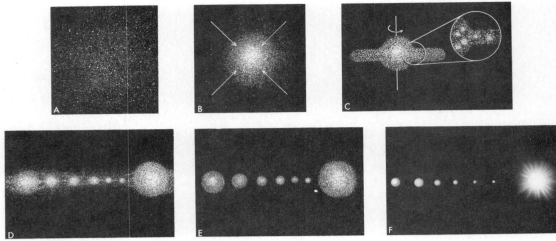

Figure 20–3 Formation of the solar system. (From Pasachoff, J. M., *Contemporary Astronomy*. Philadelphia: W. B. Saunders Company, 1977, p. 286.)

the planets and the sun began by the condensation of a cold cloud of dust and gas (Fig. 20–3). Inertial effects set up by the rotation of the cloud opposed gravitational collapse, with the result that the cloud was flattened into a rotating disk. Further contraction of the cloud caused it to increase in density, and local instabilities broke it up into individual units. At the center was a protostar, and toward the periphery a series of protoplanets. When the density of the protoplanets exceeded a certain limit, they condensed into planets. It now appears that the cold planetary bodies crystallized out at about the same time, some 4.5 billion years ago.

Radiation from the protosun, glowing from the release of gravitational energy, caused the evaporation of most (perhaps 99 per cent) of the original mass of the protoplanets. Dissipation of material from the atmosphere into space is still going on, though at a reduced rate. Hydrogen and helium, which must have constituted the major part of the nebula and the protoplanets, have almost entirely escaped from the earth and have left behind the heavier elements. The stronger gravitational attraction of the more massive planets, Jupiter and Saturn, have enabled them to retain more of the light elements, accounting for their lower density.

Although each of the planets is unique, they fall into two groups—with the exception of Pluto, about which little is known. The terrestrial planets are quite small and located relatively close to the sun, and have a composition quite similar to that of the earth; they include Mercury, Venus, Earth, and Mars. They are composed mainly of metals and are rather dense. The "Jovian" or major planets, located in the outer regions of the solar system, are Jupiter, Saturn, Uranus, and Neptune. They are much less dense than the earth, and are enveloped in atmospheres several thousand miles deep. The nearest Jovian planet, Jupiter, has been most thoroughly studied. All the satellites (moons) are very small in comparison to their planets, with the exception of that of the earth. However, two of the moons—Jupiter's Ganymede and Saturn's Titan—are larger than the planet Mercury.

Estimates of the age of the solar system were little better than guesses before the discovery of radioactivity, although geologists were beginning to speak in terms of millions or hundreds of millions of years, based upon studies of the earth's rocks. When it had been proved that the radioactive elements uranium and thorium decay ultimately into lead and helium, Bertram Boltwood (1870–1927), a radiochemist working with Rutherford, proposed that this fact could be used to calculate the ages of uranium- and thorium-containing minerals.

It is possible to tell how long ago the mineral had been formed by measuring the amount of lead or helium and comparing it with the content of uranium and thorium. Since the rate of decay is known, time can be measured by comparing the relative abundance of the parent elements and their noble-gas daughters present in a sample of matter. Various methods—including the decay of potassium to argon and the decay of rubidium to strontium—indicate that the cold planetary bodies of the solar system all crystallized at about the same time, 4.5 or 4.6 billion years ago. Table 20–1 gives some of the radioactive elements used to determine the

Table 20–1 RADIOACTIVE "CLOCKS" USED TO DETERMINE THE AGES OF MINERALS

ISOTOPE	HALF-LIFE (millions of years)	DECAY PRODUCTS
Potassium-40	1310	Calcium-40, Argon-40
Rubidium-87	50000	Strontium-87
Thorium-232	13900	Lead-208, Helium
Uranium-235	710	Lead-207, Helium
Uranium-238	4510	Lead-206, Helium

Table 20-2 CHARACTERISTICS OF THE SUN AND PLANETS*

	SUN	MERCURY	VENUS	EARTH	MARS	JUPITER	SATURN	URANUS	NEPTUNE
Distance from Sun (millions of miles)	333000	36	67	93	140	480	890	1800	2800
Mass (earth = 1)		0.06	0.82	1.0	.11	480	95.2	14.5	17.2
Diameter (thousands of miles)	867	3.0	7.6	7.9	4.2	89	76	30	28
Density (water = 1)	1.4	5.4	5.1	5.5	3.9	1.3	0.7	1.3	1.6
Rotation Period (earth days)	27	59	243	1	1	0.4	0.4	0.4	0.6
Length of Year†		88ᵈ	225ᵈ	1ʸʳ	687ᵈ	12ʸʳ	30ʸʳ	84ʸʳ	165ʸʳ
Velocity of Escape (miles/second)		2.2	6.2	7.0	3.1	37	22	13	14
Number of Moons		0	0	1	2	14	10	5	2
Average Surface Temperature (K)	5800	600	750	285	240	128	105	70	55
Surface Gravity (earth = 1)	28	0.37	0.89	1.00	0.38	2.74	1.14	.96	1.15

*Excluding Pluto
†d = earth days; yr = earth years

ages of minerals. These isotopes were apparently made at the time the elements originated and are not being formed to a significant extent on earth today.

Some characteristics of the solar system are summarized in Table 20–2.

20–3 THE SUN

The dominant member of the solar system is the sun, which accounts for nearly the entire mass (more than 99 per cent). It has a diameter of 865 000 miles, 109 times the earth's diameter, and a volume over 1 million times as great as the earth's. Its average density is a little greater than that of water, 1 g/cm³. The density increases rapidly toward the interior, where it is approximately 150 g/cm³. The average distance between the earth and the sun is 93 000 000 miles and is called 1 astronomical unit (AU).

The sun is a typical star on the main sequence of the Hertzsprung-Russell diagram. Like other G-class stars, it is composed mainly of hydrogen and helium, and its energy is generated by nuclear reactions deep in its interior. The outer, visible layers of the sun are collectively known as its atmosphere. It consists of three layers—the *photosphere, chromosphere,* and *corona*—which blend into each other and have no sharp boundaries (Fig. 20–4). The chromosphere and corona can normally be seen briefly only during a total eclipse of the sun, but can be observed at other times with instruments such as the coronagraph.

The Photosphere. The most conspicuous feature of the photosphere are irregular dark spots. With the telescope, Galileo became convinced that these areas, known as *sunspots*, were actually on the surface of the sun itself and, since they apparently move across its face, he used them to demonstrate the sun's rotation about its axis. At the equator the period of rotation is about 25 days, while near the poles it is 34 days. The spots are regions where the gases, though hotter than the surfaces of many stars, are somewhat cooler than those of the surrounding areas and appear dark only in contrast. The largest spots may be over 125 000 miles across; the smallest, 1000 miles. They continually change in appearance and size. Although their nature is not fully understood, they may be due to magnetic phenomena originating in the sun's interior.

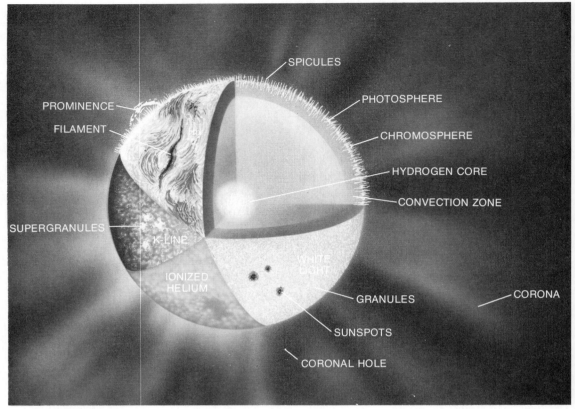

Figure 20-4 Regions of the sun. (From Pasachoff, J. M., *Contemporary Astronomy.* Philadelphia: W. B. Saunders Company, 1977, p. 120.)

Sunspots follow an 11-year cycle during which they increase in number toward a maximum and then fall off to a minimum. Most of the spots appear, then disappear, within a day or so (Fig. 20–5). A sunspot cycle begins with the appearance of small spots at latitudes of 40 to 45° in both solar hemispheres. As the cycle progresses, new and larger groups of spots form closer to the solar equator, the closest appearing 5 to 10° from the equator at the close of the cycle. As these die out, small spots reappear at the higher latitudes. Correlations have been noted between events on earth and the sunspot cycle. One link is a long-period weather cycle; total annual rainfall varies approximately in an 11-year period, the years of maximum rainfall corresponding to the times of sunspot maxima.

The Chromosphere. Above the photosphere is the chromosphere, a relatively thin layer 2000 to 6000 miles deep, with a temperature of 50 000°C. During an eclipse it is visible for only a few seconds, just after the moon has covered the photosphere. The brightest eruptive activity on the sun—*solar flares*—is asso-

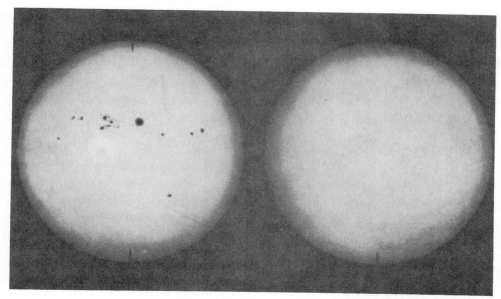

Figure 20–5 Photographs of the sun. At left, near a time of sunspot maximum; at right, near sunspot minimum. (Courtesy of the Hale Observatories.)

ciated with this layer. These sudden short-lived increases in the light intensity of the chromosphere release great amounts of energy, especially in the ultraviolet and X-ray regions, producing radio fade-outs and brilliant auroral displays in the earth's higher latitudes.

Figure 20–6 The solar corona during a solar eclipse. (From Pasachoff, J. M., *Contemporary Astronomy.* Philadelphia: W. B. Saunders Company, 1977, p. 147.)

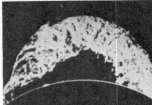
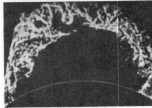
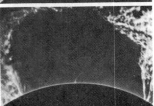

Figure 20-7 Photographs of a solar prominence. The explosion took one hour from beginning to end. (Courtesy of High Altitude Observatory National Center for Atmospheric Research; sponsored by the National Science Foundation.)

Solar flares are related to the sunspot cycle, occurring mostly in the regions of sunspot groups.

The Corona. The chromosphere blends gradually into the outermost layer, the corona. In parts of the corona, the temperature may rise as high as 3 million °C. This layer is seen most spectacularly during a total solar eclipse (Fig. 20–6). It extends more than a million miles above the photosphere. Solar prominences appear as red flamelike protuberances rising from the edge of the sun during a total eclipse. The prominences are cool, denser accumulations of hydrogen gas that form in the inner corona and may extend tens of thousands or even a million miles above the sun, then gracefully fall back. They are often geyser-like eruptions; others form huge arches (Fig. 20–7).

20-4 THE MOON

Our closest permanent neighbor in space, the moon lies at an average distance from the earth of 239 000 miles. It never comes closer than 222 000 miles, or gets farther away than 253 000 miles. The moon goes around the earth once every 27 days 7 hours 43 minutes. The force of gravity on the moon's surface is one-sixth as strong as that on the earth's surface—so feeble that the moon has been unable to hold on to an atmosphere. The moon's gravitational pull is much weaker than earth's with an escape velocity of only 2.38 km/sec, compared to 11.2 km/sec on earth; thus, any atmosphere that may have been there at one time would have leaked away into space long ago. The lack of an atmosphere to retain the sun's heat results in sudden and extreme temperature changes, from more than 90°C during the day to −180°C or less when the sun sets. Any water that may have been present on the moon would have boiled away and the vapor escaped into space along with other gases. There are, therefore, no seas, lakes, rivers, or glaciers on the moon; nor are there clouds, fog, rainstorms, or snowfalls.

The moon has been known by various names, among them Diana, "she who hunts the clouds"; Selene, the sister of Helios (the sun); and Luna, the name which is often used to distinguish it from the moons of other planets. From Selene is derived the term "selenography," the study of the same aspects of the moon that geography is of the earth, while the supposed effect of the full moon upon human behavior is reflected in the word "lunacy." The dream of reaching the moon was made a goal of national policy for the United States during the 1960's and came to fruition in 1969.

Except for the earth, the planets that have moons are considerably larger than their satellites. The diameter of Jupiter is 28 times as great as that of its largest moon, and Saturn's is 21 times as great. The earth's diameter is only 3.6 times as great as the moon's, the two forming a sort of double planet system.

The question of the moon's origin has not yet been satisfactorily answered. There are few supporters today of the once popular planetary-capture theory, which said that the moon had once been a planet in its own right orbiting the sun and was somehow "captured" by the gravitational pull of the earth. Nor are there many adherents to the theory that the moon had once been part of the earth and broke off by centrifugal force, orbiting around the earth and leaving behind a huge basin now filled by the Pacific Ocean. The evidence today favors the theory that both the earth and the moon were formed near each other at about the same time, by the process of material accretion. Investigations by potassium-argon dating of rocks brought back from the moon show them to be equal to the earth's estimated age, 4.6 billion years.

Light and dark areas are easily distinguished even when the moon is observed without a telescope. Through a telescope, these features are resolved in much greater detail. The largest features on the moon are dark areas, or *maria*, so called because they were originally thought to be seas (*mare*, sea) (Fig. 20–8). The maria are not seas at all, but large, roughly circular, smooth, flat, dry basins up to 700 miles in diameter, which appear dark because of their low reflectivity. When the moon is full, the dark maria suggest the appearance of the "man in the moon." Many 17th century names for these features are still in use: Sea of Serenity, Sea of Tranquillity, Sea of Fertility, and others. The biggest of the maria is named the Ocean of Storms. Smaller areas of darkness have such names as the Bay of Rainbows and the Lake of Dreams.

The brighter areas are the highlands—crater-covered mountains and hills. The names of the mountain chains are drawn from those of the earth: Lunar Alps, Lunar Caucasus, and so on. Some of the mountain peaks reach heights of 25 000 feet or more, as do those on earth.

MARE NUBIUM

MARE HUMORUM

MARE NECTARIS

MARE FOECUNDITATIS

MARE TRANQUILLITATIS

Copernicus

Kepler

MARE VAPORUM

OCEANUS PROCELLARUM

MARE CRISIUM

MARE SERENITATIS

MARE IMBRIUM

MARE FRIGORIS

Plato

Figure 20–8 The moon. (From Pasachoff, J. M., *Contemporary Astronomy.* Philadelphia: W. B. Saunders Company, 1977, p. 292.)

In a map of the moon published in 1651, Giovanni Riccioli (1598–1671) began the custom of naming the lunar craters after famous men, mostly scientists and philosophers. He named the most conspicuous lunar crater after Tycho Brahe, and honored Copernicus with a crater nearby. Craters are also named for Plato, Aristotle, Ptolemy, Archimedes, Kepler, Newton, and many others. This practice continues to be followed. An international conference was called in 1970 to consider names for craters previously identified only by number. The number of craters great and small is in the millions.

The lunar surface has been mapped in detail by

several methods. Television scanning techniques were employed with the Ranger and Surveyor spacecraft, and the pictures transmitted to earth. Photographs of higher quality were obtained when spacecraft carrying astronauts orbited the moon. Finally, extremely detailed photographs were taken by the Apollo astronauts, who opened the moon as a scientific laboratory (Fig. 20–9). The smaller features of the moon are similar to the larger ones. Tiny craters have essentially the same appearance as large ones, and miniature ridges resemble the larger ranges.

The problem of the origins of the lunar craters and other features is still far from being resolved. Some selenologists have favored a volcanic origin for the craters and maria, while others favor an impact origin; there is evidence that both processes have contributed. Craters range in size from tiny pits to several miles, and some appear to be of more recent origin than others. The majority of the moon's craters, particularly the larger ones, may have been formed by the impact of meteorites; they closely resemble the craters left by bombs, in that their walls are much steeper on the inside than on the outside. Many of the smaller ones may have been caused by volcanic activity; in some cases, the volcanoes may have been triggered by meteoritic

Figure 20–9 An astronaut on the moon. St. George crater is about three miles in the distance. (Courtesy of NASA.)

impact. Lunar rocks collected by the Apollo astronauts show evidence of both forms of activity. The astronauts did resolve one question, the nature of the moon's surface. They found it to be covered with a fine, dustlike sand, which varied in depth from a fraction of an inch to several inches and clung to their space suits.

One of the surprises of the Lunar Orbiter photographic survey on the moon was the discovery of regions of dense material below the surface, called *mascons*, from the two words *mass concentration*. They were detected from the strong gravitational pull they exerted on satellites, which distorted the orbits of the satellites over the maria. It has been determined that the mascons are thin, circular disks of high-density, (lavalike) materials extending below the surfaces of the maria.

The orbital movement of the moon changes the visible portion of its surface as viewed from the earth. In a one-month period, the moon completes a cycle of phases (Fig. 20–10). New moon occurs when the moon is exactly in line between the earth and the sun, so that none of its illuminated surface is visible. First quarter occurs when half the moon's visible surface is illumi-

Figure 20–10 The moon's monthly cycle of phases. Different parts of the lunar surface are visible from earth as the moon moves in its orbit around the earth.

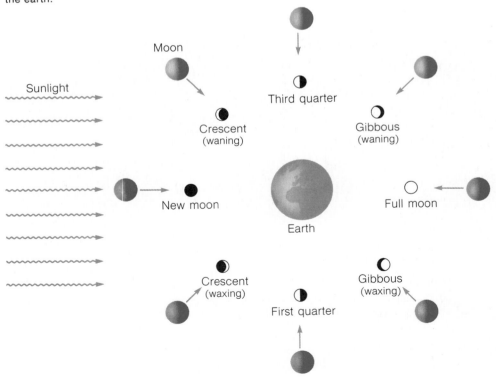

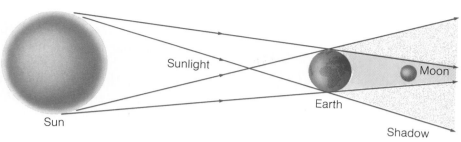

Figure 20–11 Lunar eclipse. When the moon is completely within the darkest part of the earth's shadow (the umbra), a total eclipse occurs.

nated, full moon when the whole disk is illuminated, and last quarter when, again, only half the moon's surface is illuminated.

By a remarkable coincidence, the apparent size of the full moon equals that of the sun, even though the moon is a much smaller body. A solar eclipse (*ekleipsis,* disappearance) occurs when the moon covers the sun's disk. The most spectacular kind is a total solar eclipse experienced on only a very narrow path on the earth, 150 to 170 miles wide. Although the period of totality lasts only a few minutes, a total solar eclipse has been an awesome spectacle throughout history. In modern times, total solar eclipses have provided so much information about the sun that teams of astronomers journey to remote places on the earth to observe them.

A lunar eclipse is much less spectacular than a solar eclipse and of less scientific value. Lunar eclipses occur when the full moon enters the shadow of the earth (Fig. 20–11). Since the earth's shadow is larger in diameter than the moon, totality can last as long as 1 hour 40 minutes. Although there are more solar than lunar eclipses, the latter can be observed simultaneously by many more observers, being visible from the earth's hemisphere facing the moon.

20–5 MERCURY

The smallest planet, Mercury, is also the closest to the sun and makes its circuit around the sun in only 88 days. The Greeks named it for Hermes, the fast-moving messenger of the gods, who was represented with winged sandals and a winged hat, while the Romans named it for their messenger, Mercury. Mercury's highly eccentric orbit takes it to within 28.3 million miles of the sun at perihelion and 43.2 million miles at aphelion; its mean distance is about 36 million miles. Its period of rotation was not known until it was determined through radar measurements in 1965. By bouncing radar waves off the surface of Mercury and catching the echoes with the radio telescope at Arecibo, Puerto Rico, Gordon Pettengill and Rolf Dyce found that Mercury rotates on its axis once in 58.65 days, or essentially two-thirds the period of revolution.

Mercury's orbit lies within that of the earth, making the planet invisible at night, when astronomical work is usually done. The side of the earth that is in darkness faces away from the sun, toward the planets whose orbits are outside the earth's. Only the side of the earth that is in daylight, facing toward the center of the solar system, can face Mercury (Fig. 20–12). The sun's strong glare prevents observation of Mercury with the naked eye, except for short periods of about an hour before sunrise and an hour after sunset at certain times of the year. When it is visible at all, Mercury is a very bright object low on the horizon, and through a telescope exhibits the same phases as the moon.

From infrared studies of its surface, the temperature of Mercury is known to range from 350°C to −160°C. If an atmosphere exists (none is detectable), it must be extremely thin, since the velocity of escape on Mercury is only 4.2 km/sec and, since it is so close to the sun, the heat is intense. Mercury's density, 5.4 g/cm³, suggests that it contains mainly rock and metals and that its interior may contain compounds of iron or other heavy elements. Its surface appears to be made of the same kind of rock as that found on the moon.

The spacecraft Mariner 10 completed a 5-month

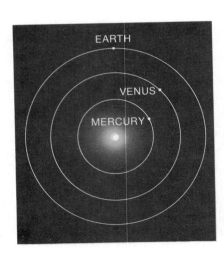

Figure 20–12 Orbits of Mercury and Venus lie within the earth's orbit. (From Pasachoff, J. M., *Contemporary Astronomy*. Philadelphia: W. B. Saunders Company, 1977, p. 309.)

Figure 20-13 Mercury as photographed from Mariner 10 spacecraft. Cratered surface is strikingly similar to the moon's. (Courtesy of NASA.)

journey to Mercury in March, 1974. It swept over the planet at a distance of 465 miles, and transmitted more than 2000 television pictures across 80 million miles of space. The photographs revealed the surface of the planet to be extraordinarily similar to that of the moon, on a regional as well as on a global basis (Fig. 20–13). Craters are the predominant surface feature, their structure being similar to lunar craters of the same size. The plains strongly resemble lunar maria. As on the moon, a heavily cratered terrain has been preserved on Mercury, without modification by volcanism or atmospheric erosion. Some of the features are probably 4 to 4.5 billion years old.

20-6 VENUS

After the sun and moon, Venus is the brightest object in the sky, reflecting more than 75 per cent of the sunlight that strikes the dense cloud cover. Venus can often be seen in full daylight with the naked eye. It is the body most frequently referred to as the "evening star" or "morning star" when it is visible shortly after sunset or just before sunrise. It was not realized at first that both "stars" were the same object; the Greeks called it Hesperus ("west") in the evening, and

Phosphorus or Lucifer ("bearer of light") in the morning. The planet was almost always identified by ancient peoples as a goddess of beauty, Venus to the Romans.

Venus is 67 million miles from the sun. Its orbit is one of the least eccentric of the planets, so that its perihelion and aphelion distances are similar. At its closest approach, it is 24 million miles from the earth. Galileo was the first to observe through a telescope that Venus, like the moon, exhibits a full set of phases (Fig. 20–14). Radar and the radio telescope at Arecibo (optical telescopes can probe no deeper than the outer cloud layers) established that Venus has a rotation period of 243 days, the slowest for a planet in the entire solar system, and that it rotates in a direction opposite to the typical motion of the other planets. Its period of revolution is 225 days. It has little or no magnetic field.

A number of space probes aimed at Venus in the 1960's and 1970's obtained more data about Venus than astronomers had been able to gather in the

Figure 20–14 Phases of Venus. Venus exhibits the same phases as the moon. (Lowell Observatory photograph.)

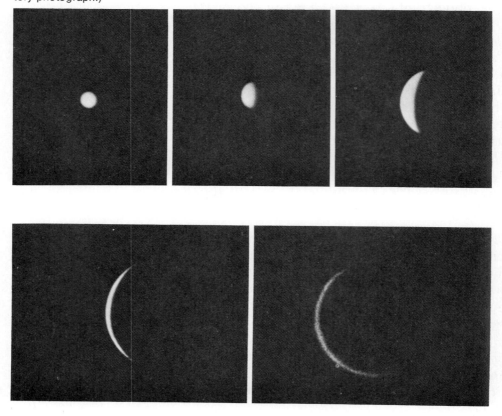

previous 400 years. Early Russian probes were apparently crushed by the pressure of the Venusian atmosphere, although Venera 3 evidently made a crash landing in 1965—the first time a space vehicle had reached another planet. In 1972, high-resolution radar probed through the thick clouds of Venus and for the first time resolved features on the planet's surface. The first map of a part of Venus along its equator showing discrete features—a landscape of huge, shallow craters from 20 to 100 miles wide and less than a quarter of a mile deep—was made in 1973 using a computer technique. Mariner 10, equipped with television cameras and radar, flew by Venus in 1974 and sent back 3400 television pictures of the cloud cover, which appeared to be moving along the equator at 350 km/hr. Venus' clouds are something like terrestrial smog, but they are believed to be composed mainly of sulfuric acid droplets.

The planet's atmosphere was the object of two Pioneer flights in December, 1978, the first devoted primarily to a study of the atmosphere of another planet on a global scale. Pioneer Venus 1 was to send daily reports on the atmosphere as well as pictures and radar maps of the surface topography. It was intended to operate for up to a year in orbit around Venus. Pioneer Venus 2, called the transporterbus, scattered five separate probes into the atmosphere over different areas of the planet. They gathered a wide range of data, including temperatures, composition, density, and distribution of the atmosphere. The data were to be translated into sketches of clouds, winds, and other features, a process that would take months or years to complete.

The atmosphere of Venus apparently consists largely of carbon dioxide (75 to 90 per cent) and smaller amounts of nitrogen and the noble gases (5 per cent); there is no oxygen. The cloud layer, 13 miles thick, is penetrated by some sunlight. The solar radiation heats the lower regions of the atmosphere and is then trapped by the "greenhouse effect" due to the carbon dioxide (Fig. 20–15). Although the temperature at the top of the cloud layer is −50°C, at the base of the clouds 45 miles above the surface it is 100°C, and on the surface it is a furnace-like 500°C. The once popular idea of Venus as a planet having lush vegetation and being an abode for life as we know it now seems impossible. Not only does its surface seem far too inhospitable, but no great amount of water vapor has been detected.

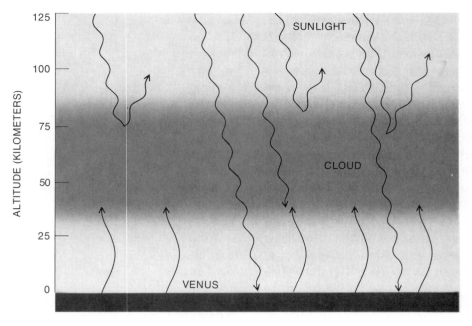

Figure 20–15 Greenhouse effect on Venus. (From Pasachoff, J. M., *Contemporary Astronomy*. Philadelphia: W. B. Saunders Company, 1977, p. 329.)

20–7 MARS

Long known as the "red planet," Mars was associated with the Roman god of war because its reddish hue seemed to suggest images of war, blood, and flames. It is a conspicuous object in the sky, rather bright, and can easily be seen with the naked eye. Except for the finer details, its surface can be seen with a telescope; it is the only planet whose surface is observable from the earth. Many color changes, both seasonal and non-seasonal, have been observed.

Mars has excited more interest over the years than any other planet, because it was at one time seriously considered to harbor intelligent life. The astronomer Giovanni Schiaparelli (1835–1910) in 1877 observed what appeared to be very thin lines on the surface, which he called "canali." When the term was translated into "canals," it created the impression that intelligent life existed on Mars. Percival Lowell (1855–1916), who built an astronomical observatory near Flagstaff, Arizona, conceived the idea that canals had been built to irrigate the reddish desert areas of Mars by drawing water from the polar caps. But there was little resemblance between his drawings of the canals and what could be observed through the telescope or photographed.

Only in the past decade have the conditions that prevail on the planet become known. Mariner 4 made the first close-up pictures of Mars in 1965, which revealed a surface covered with large craters reminiscent of those on the moon. Mariners 6 and 7 in 1969 sent back pictures that showed that the Martian surface also had large areas of terrain unlike anything ever seen on the moon or the earth.

Mariner 9, however, was one of the high points in the exploration of Mars and decisively changed our view of the planet that generations had thought most closely resembled the earth (Fig. 20–16). Photographs (7000 of them) sent back by Mariner 9 in 1972 revealed surprising evidence of four enormous volcanic mountains, larger than any on earth. Perhaps Mars is just beginning to "boil" inside, as illustrated by the well advanced process in the Nix Olympica–Tharsis Ridge, but vulcanism has not yet spread to the planet as a whole.

The photographs also show a vast system of canyons, gullies, and channels. The huge canyons that stretch east and west along the equator of Mars are 50 to 75 miles wide and 3 to 5 miles deep, much larger than

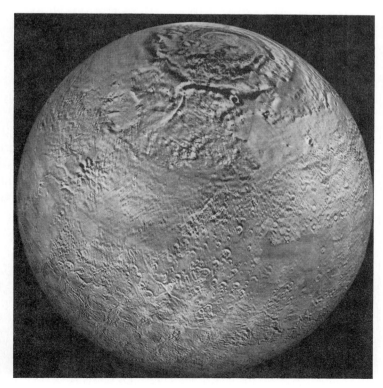

Figure 20–16 The North Pole of Mars. Photographed by Mariner 9. (Courtesy of the Jet Propulsion Laboratory.)

any found on earth. The largest corresponds to a feature known as Coprates, which sometimes changes in appearance with the seasons. These variable markings may have nothing to do with the surface of Coprates. The canyon is so deep that dust persists in the atmosphere between the canyon walls and makes the canyon look brighter than the surrounding landscape after the atmosphere above is comparatively free of dust. Once the canyon atmosphere is clear, there is little contrast between the interior of the canyon and the surrounding area.

Since the Mariner flights, it has been realized that the conditions for life on Mars are extreme by terrestrial standards. The diameter of the planet is only one-half that of the earth, and its mass only one-ninth. Its surface gravity, less than half that of the earth, accounts for the thin Martian atmosphere, resembling the thin air found 100 000 feet above the earth. The Martian atmosphere differs considerably from the earth's in composition, being rich in carbon dioxide and containing much less nitrogen. The thin atmosphere affords little protection against ultraviolet solar radiation, which reaches the Martian surface at an intesity that would be fatal to most forms of life. No liquid water has ever been detected on the surface, which appears to be drier than the earth's most arid deserts, but there is water vapor in the atmosphere. The temperature at the equator ranges from a daytime high of 28°C to a nighttime low of −100°C. At the poles the temperature is −215°C, compared to −89°C in Antarctica in the winter.

Two Viking landers touched down on the surface of Mars in 1976, while two Viking orbiters circled and photographed the planet with unprecedented resolution and clarity. They have provided evidence that the cratered terrain has been modified by volcanic activity, that water was a significant agent in shaping its features early in its history, and that since then surface material

Figure 20–17 View of the surface of Mars taken by Viking I lander. This view resembles deserts in the southwest United States. (From Pasachoff, J. M. *Contemporary Astronomy.* Philadelphia: W. B. Saunders Company, 1972, pp. 358–359.)

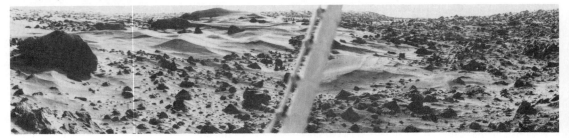

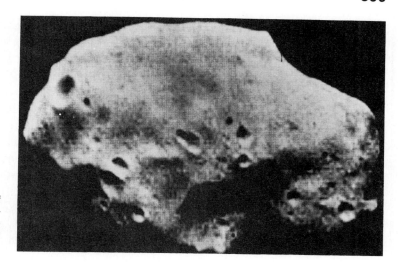

Figure 20–18 Phobos, one of the two moons of Mars, photographed by Mariner 9. (Courtesy of the Jet Propulsion Laboratory.)

has been redistributed by high-velocity wind, though causing little erosion. The surface of Mars is more like rocky volcanic deserts on the earth than the highly cratered surface of the moon (Fig. 20–17). Yet Mars seems to have little sand. The residual polar caps, it now appears, consist of water ice, not carbon dioxide ice. This discovery has given hypotheses involving water as an active agent in Mars' past greater acceptance. The channels on Mars are even more abundant than the Mariner 9 photographs showed.

With a diameter of 4200 miles, Mars is midway in size between the earth and the moon. In two respects it is remarkably similar to the earth: the Martian rotation period is 24 hours 37 minutes 23 seconds, and its axis is tilted 23°59' to the plane of the orbit, which produces seasons similar to those on earth. It receives less radiation from the sun than does the earth. When it is closest to the earth, Mars is about 34 500 000 miles away. Mars has two small moons, Phobos and Deimos. The Mariner flights show Phobos to be an elongated object with a maximum diameter of about 14 miles (Fig. 20–18). Deimos is probably not larger than about 4 miles. It is likely that both are captured asteroids that approached Mars too closely and were perturbed into new orbits.

20–8 JUPITER

Jupiter is by far the largest of all the planets and the most massive, more than twice as massive as the rest of the planets combined and 318 times as massive as the earth. Aside from the sun, it is the only other significant mass in the solar system. In spite of its great mass, Jupiter's density is only about 1.3 g/cm³, less than one-fourth the earth's and about the same as the sun's. Its equatorial diameter is 88 700 miles and it exhibits marked flattening at the poles, with a diameter measured through the poles of 83 000 miles. Jupiter is the most rapidly rotating planet; its day is 9 hours 50 minutes at its equator and 9 hours 55 minutes near the poles. Its average distance from the sun is 484 million miles, and it takes 11.86 terrestrial years to complete a circuit of the sun.

The structure of Jupiter is a matter of considerable debate. Although the center of the planet probably contains a dense solid core, the outer layer appears to be composed of various gases. Methane and ammonia have been detected in its atmosphere. Water is probably present, but hydrogen and helium are very likely the predominant constituents of the planet, as they are of the universe, since Jupiter is massive enough to hold the lightest gases. The presence of helium in Jupiter's atmosphere was confirmed in 1973.

Jupiter generally appears as a very bright object, nearly equaling Venus as a reflector of sunlight. As seen through a telescope, Jupiter's most striking feature is a complex system of bands of colors. These spectacular bands, representing the outer region of the planet's atmosphere, come and go, fade and darken, widen and become narrow, and move up and down in latitude. The colors that show up in the bands—red, brown, yellow, green, and blue—may be caused by various free radicals, or molecular fragments. A conspicuous atmospheric feature called the Red Spot (although it is often anything but red), appears as a huge oval shape in the southeastern quadrant. Measurements made by Pioneer 10, a spacecraft which passed within 81 000 miles of Jupiter in 1973, indicate that the Red Spot is a huge hurricane (Fig. 20–19). The red color may be due to a mix of organic compounds that form ice crystals at the tops of the clouds.

The four Jovian satellites that Galileo discovered are comparatively large bodies: Ganymede and Callisto are both larger than the moon, and Europa and Io are about the same size as the moon. The other satellites are much smaller in diameter; the 12th was discovered in 1951 and is only visible in the largest telescopes. The 13th and 14th were discovered in 1974 and 1975.

Figure 20–19 Jupiter, its Great Red Spot, and Io (the innermost large moon) are visible in this photograph taken February 5, 1979, by Voyager 1. (Courtesy of NASA.)

Jupiter's outer satellites, it is believed, represent captured asteroids scooped up by the massive planet during the course of its history.

Jupiter is a source of powerful radio emissions in the decameter range. Although their origin has not yet been determined, they are probably related to Jupiter's magnetic field, which is 10 times stronger than the earth's, and to its radiation belts, 10 000 to 1 million times as intense as those surrounding the earth. In some ways Jupiter is like a small star, but it is not quite massive enough to shine on its own. Its mass is just below the mass that would have made it a star. It is the only such object that we know.

Two Voyager spacecraft were launched in 1977 to explore the Jovian planets of the solar system. They were expected to make their closest approach to Jupiter in 1979. Their mission: to send back pictures of Jupiter, Saturn, and their satellites, and to measure the atmosphere, magnetic field, cosmic rays, and density of the planets. By studying the outer planets, it may be possible to find out more about how the earth was formed. The cold outer planets are more like the way they were when the solar system was formed.

The analysis of data and photographs from the first Voyager in 1979 showed that Io is covered with volcanoes, except for the polar caps, making it the only body in the solar system beside earth that is known to have active volcanoes. Europa has a surface of ice and water and miles of long, linear structures crisscrossing it. Ganymede, the largest of Jupiter's moons, resembles our moon and may be composed of water-containing rocks. Callisto has huge dents, probably caused by meteor impacts, and is marked with bright-colored ridges. The spacecraft also discovered a ring around Jupiter, making it the third planet in the solar system (after Saturn and Uranus) known to have such a feature. There is a very narrow outer ring and an inner ring of thinner matter that goes all the way to the surface.

Figure 20–20 Trajectory of a Voyager space craft. Following encounters with Jupiter and Saturn, there could be a rendezvous with Uranus. (Courtesy of NASA.)

Jupiter's atmosphere is swirling and turbulent, rather than stable, and has well-ordered circulation patterns as had been thought. The Red Spot's counterclockwise vortical motion is one of the most striking features of the planet.

The trajectories of the Voyagers were chosen so that the gravity and orbital motion of Jupiter could act as a slingshot to send them to Saturn, one arriving there in 1980, the other in 1981 (Fig. 20–20). Measurements would be made of Saturn, its satellite Titan, and its spectacular ring system. One of the Voyagers might then be sent to Uranus, arriving there in 1986. Eventually, the Voyagers would pass beyond the solar system.

20–9   SATURN

Saturn ranges between 839 and 937 million miles from the sun, twice as far as Jupiter. It moves so slowly along its gigantic orbit, taking 29.5 years to make one complete circuit of the sun, that the Romans named it for their god of time, Saturn. Although it is nearly the same size as Jupiter, it has less than one-third as much mass; its density is only 0.7 g/cm^3. Given a large enough body of water, it would float.

Saturn rotates once every 10 hours 14 minutes at the equator and once every 10 hours 38 minutes at the poles, indicating that, like Jupiter, it is not a solid surface but a dense atmosphere the different parts of which rotate at different speeds. Any solid matter within must be very near the center, as with Jupiter. It appears to be composed largely of hydrogen and helium, followed by methane and, to a lesser extent, ammonia. Its atmosphere is deep and violent.

The most striking feature of Saturn as seen through the telescope is its ring system. Galileo was the first to view the rings, although not clearly—they appeared to him more like appendages of the planet than rings. Christiaan Huygens (1629–1695) saw that the rings went completely around the planet. The ring system, 171 000 miles in diameter, consists of four separate rings, all in the plane of Saturn's equator. The two brightest are separated by the Cassini Division, named after one of its first observers, G. D. Cassini (1625–1712). In 1850 a third, dark ring lying closer to the planet was discovered. Called the Crepe Ring, because its dark appearance suggests to some a crinkly fabric, it casts its shadow across the face of the planet. In 1961 a fourth ring was discovered, which reaches almost to the planet itself.

It is generally agreed that Saturn's rings are made up of individual particles or satellites each in its own gravitationally stable orbit around Saturn. A solid ring or "halo" of material is not physically possible. The solid appearance of the rings is an effect created by the close proximity of the tiny particles when viewed from the great distance of the earth. The particles were once believed to consist of frozen ammonia, but they are now thought to be chunks of water-ice or bits of rock coated with ice. The belt of rings is so thin, approximately 12 miles, that it disappears from sight when viewed exactly edge-on (Fig. 20–21). One theory is that the rim represents material that was left over and did not accrete to form Saturn during its planet-building state.

Well beyond the ring system, Saturn has 10 known moons. Titan is the largest, being larger than Mercury and nearly as large as Mars. Titan has another distinction—it is the only moon in the solar system known to have an atmosphere, consisting of a thin envelope of methane and possibly hydrogen. Clouds are probably present, and a greenhouse effect in the atmosphere may result in Titan's absorbing more of the sun's warmth that it radiates back into space. Titan could be

Figure 20–21 The rings of Saturn seen from various angles. (Courtesy of New Mexico State University Observatory.)

a miniature of what the primitive earth was like at the dawn of life.

20–10 URANUS

This cold, remote planet was unknown to early astronomers, since it is rarely visible to the naked eye. Its discovery by William Herschel in 1781 added a seventh planet to the solar system and doubled the size of the solar system, since Uranus is farther from Saturn than Saturn is from the sun. Named for the father of Saturn, Uranus varies in distance from 1608 to 1956 million miles from the sun, a distance so great that the sun is probably only a faint gleam in the sky when viewed from Uranus. It takes Uranus 84.02 years to complete a circuit around the sun, and each of its seasons is 21 years long. It is believed to be covered by clouds, and its surface temperature is below −200°C. Its rotation period is 10 hours 49 minutes. Uranus appears greenish.

A most unusual feature of Uranus is its tilt. All of the other planets spin on their axes like tops as they travel their orbits. Mercury and Jupiter stand virtually straight; Venus, Earth, Mars, Neptune, and Pluto (perhaps) lean over at angles ranging from 23° to 29°. Uranus, however, is tilted at an angle of 98°. Instead of spinning like a top, it lies on its side and rolls along. Uranus has five known moons; the largest is Titania, with a diameter of 625 miles.

The first major structural discovery in the solar system in nearly 50 years involved Uranus. The discovery of Pluto in 1930 held claim to that distinction until 1977 when rings were discovered around Uranus (Fig. 20–22). Until then, Saturn was the only one of the nine known planets encircled by rings. A team of astronomers spotted the five rings around Uranus while working in an airborne observatory 41 000 feet aloft. Other observatories confirmed their existence, and they have been named alpha, beta, gamma, delta, and epsilon after the first five letters of the Greek alphabet. The rings are vertical to the earth's equatorial plane and were not seen earlier because light reflected from Uranus obscures the rings' lesser reflections. The rings lie in a 7000-km-wide (4400-mile) belt. Four of the flattened rings are about 10 km (6 miles) across, while the outermost one is 100 km (60 miles) wide.

Figure 20–22   Rings around Uranus. (From Pasachoff, J. M., *Contemporary Astronomy*. Philadelphia: W. B. Saunders Company, 1977, p. 389.)

20–11   NEPTUNE

Neptune is a twin of Uranus, which it physically resembles except for the rings. It too is greenish. But it is a billion miles farther from the sun than Uranus and is therefore much colder, −215°C or less. Light from the sun takes 4 hours to reach it. It is never visible to the naked eye, but can be seen with a good pair of binoculars. Its orbital period is 164.8 years, so great that since its discovery in 1846 it has not yet made a complete revolution around the sun; that will not be until the year 2011. Although its mass is 173 times that of the earth, its density is only about half as great. It has a fast rotation period of 15 hours 30 minutes.

Neptune, we recall, was the first planet whose position was predicted before its discovery. John Couch Adams and Urbain Leverrier independently calculated its position from perturbations in Uranus' orbit, and Johann Galle made the actual discovery of the planet. It was then found that many observers had previously seen Neptune just as they had seen Uranus, but had mistaken it for a fixed star. Neptune has two satellites. William Lassell (1799–1880) found Triton, the brighter one, only a few weeks after the discovery of the planet; it is larger than the moon and revolves clockwise (retrograde). More than a hundred years later, in 1949, Gerard Kuiper (b. 1905) discovered Nereid, which has the most eccentric of all satellite orbits in the solar system, ranging from 830 000 miles to more than 6 000 000 miles from the planet.

20–12 PLUTO

The story of the decades-long search for a "Planet X" which was known to perturb the orbits of Uranus and Neptune has been told in a previous section. The search ended on March 13, 1930—50 years after astronomers began to think seriously about it; the 75th anniversary of the birth of Percival Lowell and the 149th anniversary of Herschel's discovery of Uranus—with the announcement at the Lowell Observatory of the discovery of an object beyond Neptune that was apparently a planet. Other observatories soon confirmed the discovery of the planet, which was located not far from the spot originally predicted by Lowell. The name Pluto, brother of Jupiter and Neptune, was suggested.

Pluto is over 3.5 billion miles from the sun, revolving in an orbit that takes 248 years to complete. It is so faint that it can be seen only with the most powerful telescopes. Its orbit is the most eccentric of all the planets; at one point it even comes inside the orbit of Neptune, and therefore comes closer to the sun than Neptune does. For this reason it is believed by some to be an escaped satellite of Neptune. There is no chance of collision between the two planets, because Pluto's orbit is at a relatively steep inclination of 17°.

Pluto appears to be a small body, 4000 to 6000 miles in diameter. Its rotation period as derived from variations in brightness is 6 days 9 hours and 17 minutes. Pluto's surface temperatures is less than −240°C. Since even methane would freeze on Pluto, it probably has no gaseous atmosphere. It was discovered in 1976 that the surface of Pluto is frozen methane. Pluto may be the only planet that has survived in a pristine state under conditions at which it was formed 4.6 billion years ago. Pluto could also have

Table 20-3 PLANETARY DISTANCES IN AU USING THE TITIUS-BODE LAW

	MERCURY	VENUS	EARTH	MARS	—	JUPITER	SATURN	URANUS	NEPTUNE	PLUTO
	0	3	6	12	24	48	96	192	384	768
Add:	4	4	4	4	4	4	4	4	4	4
Sum:	4	7	10	16	28	52	100	196	388	772
Divide by 10:	.4	.7	1.0	1.6	2.8	5.2	10.0	19.6	38.8	77.2

a more complex structure: a core of water ice, followed by layers of methane and ammonia, and finally pure methane ice. In 1978 a moon in a 12 000-mile high orbit was discovered around Pluto. The great distance from the earth had prevented astronomers from resolving the planet and its moon into two separate specks of light. Tentatively named Charon, it circles Pluto once every 6 days 9 hours and 17 minutes, an interval identical to Pluto's own period of rotation. An observer on one side of Pluto would always see the moon in the same position in the sky. On the other side, it would never be visible.

20-13 ASTEROIDS

Johannes Titius (1729–1796), a mathematician, made an interesting discovery in 1766 related to the distances from the sun of the six planets known in his time. He set down the names of the planets in order, and under each name he wrote a number: 0 for mercury, 3 for Venus, and a number twice that of the planet before it for the planets beyond Venus. Titius then added 4 to each number, and divided each sum of two numbers by 10. The resulting series gave the approximate distances of the planets from the sun in astronomical units, as shown in Table 20-3.

Titius showed his calculations to the astronomer Johann Bode (1747–1826), who published them. The formula became known as Bode's law, but it is also called the *Titius-Bode law*. It works quite well for the planets then known, the distances in astronomical units coming out very close to those observed, *if* a planet is slipped in between Mars and Jupiter, as shown in Table 20-4. Further, it could be predicted from the Titius-Bode law that an imaginary planet beyond Saturn would have a distance of 19.6 AU. When Uranus was discovered, its distance turned out to be 19.19 AU, in fairly close agreement.

The "missing" planet between Mars and Jupiter seemed to be discovered on New Year's Day, 1801, when Guiseppe Piazzi (1746–1826) found a small body in orbit around the sun between the two planets. He sent data on the object to a number of scientists, including Bode and Karl Friedrich Gauss (1777–1855), who later calculated its orbit, showing that it was

Table 20–4   PLANETARY DISTANCES FROM SUN

PLANET	DISTANCE CALCULATED FROM TITIUS-BODE LAW (AU)	ACTUAL DISTANCE (AU)
Mercury	0.4	0.39
Venus	0.7	0.72
Earth	1.0	1.0
Mars	1.6	1.52
—	2.8	—
Jupiter	5.2	5.20
Saturn	10.0	9.55
Uranus	19.6	19.2
Neptune	38.8	30.1
Pluto	77.2	39.5

nearly circular, like that of a planet. Bode pointed out that the body's average distance from the sun was 2.77 AU, remarkably close to the predicted distance of the "missing" planet of 2.8 AU.

The "planet" was very small, only 480 miles in diameter, not much more than an orbiting chunk of rock, but a planet by definition nevertheless. Piazzi named the new planet Ceres, after the Roman goddess of agriculture. A year later, Wilhelm Olbers (1758–1840) found another "planet" in the same distance range; it was even smaller, and he named it Pallas. By 1807, two others had been discovered, Juno and Vesta. The four objects were at first called "planetoids," meaning planet-like bodies, and were later referred to as minor planets. For some reason, however, the name "asteroids," meaning starlike bodies, became attached to them, and the name stuck.

For about 30 years it was assumed that Ceres, Pallas, Juno, and Vesta were the only asteroids. Then each year saw the discovery of five or six more. With a new technique involving the use of a camera as well as a telescope, astronomers found hundreds of asteroids. Several thousand are now catalogued, and the total number is estimated to be about 100 000. Ceres is still the largest, while the great majority are much less than 1 mile in diameter. Between them, Ceres and Pallas probably account for half the total mass of all asteroids. The asteroid belt between Mars and Jupiter is at an average distance from the sun of 250 million miles. It is not known whether asteroids are the debris of a former planet or planets that disintegrated for some reason, or whether they are part of the original matter of the solar system that never became a planet.

20–14   METEORITES

On February 12, 1947, thousands of people in the environs of the Siberian village of Novoprovka were eyewitnesses to a "rain of iron," one of the rare occasions when the earth has collided with another member of the solar system. They saw, against the blue of the sky, a ball of light as brilliant as the sun and about the size of the full moon. On the slopes of the nearby mountains, explorers later found more than 100 craters, some of them 30 to 40 feet deep and as wide as 75 feet at the top. The ground over an area of several square miles was strewn with pieces of iron, ranging from several hundred pounds in weight to tiny specks. Before it entered the earth's atmosphere, it is estimated that the meteorite weighed a thousand tons and had a diameter of 30 feet. After it broke up into millions of pieces, each piece flew through the air at supersonic speed. The shock waves felled trees and completely shattered the rocks of the mountain slopes.

In prehistoric times a vastly more destructive collision caused the meteor crater near Flagstaff, Arizona, a huge hole about a mile across and several hundred feet deep (Fig. 20–23). Countless pieces of iron, the largest of which weigh hundreds of pounds, are embedded in the ground for a distance of several miles from the crater. On September 28, 1969, an object exploded over the town of Murchison in Australia, showering frag-

Figure 20–23 The Barringer meteor crater in Arizona. (From Pasachoff, J. M., *Contemporary Astronomy*. Philadelphia: W. B. Saunders Company, 1977, p. 412.)

ments over an area 7 miles long and 2 miles wide. The
noise lasted about a minute, and sounded like thunder
or a series of sonic booms. At least 82 kilograms of
meteorite fragments from the Murchison event are now
in public collections.

The idea that meteorites were of extraterrestrial
origin enjoyed roughly the same reputation in the year
1800 that UFO's do today. Primitive peoples had made
objects of stones believed to have fallen from heaven,
but in more sophisticated times this source was put
down as an old wives' tale. In 1803, however, after
the village of L'Aigle in France was pelted by a dense
shower of falling stones, the Academy of Sciences appointed a commission headed by Jean Baptiste Biot
(1774–1862) to investigate the event. The commission's
report eliminated the possibility that the stone shower
was a terrestrial phenomenon, and produced evidence
that satisfied the Academy that the stones came from
outside our planet.

Subsequently, museums began to collect meteor-
ites. The largest in "captivity," found by Eskimos in
Greenland and named Ahnigito, was brought to the
United States by the explorer Robert E. Peary (1856–
1920) in 1897. Weighing 33 tons, it is housed in the
Hayden Planetarium in New York City (Fig. 20–24).

There seems to be little doubt that meteorites are
small asteroids. When the Pribram meteorite fell in
Czechoslovakia in 1959, and the Lost City meteorite in
the United States in 1970, their orbital trajectories
were photographed in precise detail. There was

enough information to show that they were typical asteroids with paths overlapping that of the earth. Further studies indicated that they were very old, at least comparable to the age of the earth.

There are two principal types of meteorites, *stony* and *metallic*, the latter consisting mainly of iron and some nickel. Because the iron meteorites survive the trip through the atmosphere better than the stony ones, and are more easily distinguished from terrestrial rocks, more of them are found. The chemist Edward C. Howard seems to have been the first to examine the internal structure of stony meteorites (around 1800), and he found small masses ranging in size from a pinhead to a small pea. Later, the mineralogist Gustav Rose (1798–1873) established a classification of meteorites, naming this class *chondrites*, after their internal structure (*chondros*, grain of seed). The small rounded bodies in chondrites came to be called *chondrules*. There is evidence that chondrites are pieces of planetary matter in a primitive state, and that the sun and chondrites were made from the same parent material, the metal content being the same in both. This has opened the possibility of analyzing in detail the stuff of which the planets are made, since even on our own planet we can examine only the crust and know very little about the interior.

As early as 1834, Berzelius extracted complex organic substances from a meteorite that had fallen in a village in France. By 1962, it had been clearly shown that meteorites contain hydrocarbons, and by 1970 compounds such as alcohols, carboxylic acids, sugars, and amino acids were reported in meteorites. There has been some reluctance to accept many such reports, since some of the material in carbonaceous chondrites—those with an appreciable amount of carbonaceous material other than free carbon—could be biological contaminants that entered the meteorites from the air or ground. A number of fragments of the Murchison meteorite, however, were collected soon after impact and are believed to have been exposed to very little contamination. The Murchison meteorite contains a number of organic compounds never before encountered in meteorites; the compounds seem to have been synthesized by nonbiological processes and are of extraterrestrial origin. It has been suggested that because of the environment of their formation, the process of chemical evolution stopped rather than proceeded on to life.

20–15   COMETS

Unlike the planets, which move across the sky with great regularity, new comets appear and disappear unpredictably. They are surrounded by an air of mystery partly for this reason, and also because the visible ones often develop a long spectacular luminous tail (*cometes*, hairy, from resemblance to female tresses) and were regarded as an omen of war and pestilence. Although comets have not yet been fully explained, it is believed that they are members of the solar system rather than interlopers from outer space.

Very little conclusive evidence is available concerning the nature or origin of comets. Normally, a comet is an aggregation of solids that has a nucleus a mile to 50 miles in diameter. According to a model nicknamed the "dirty snowball," the nucleus consists mainly of solid water (H_2O), methane (CH_4), and ammonia (NH_3), with molecules and particles of heavier elements interspersed among them. There may be an enormous number of comets, possibly 100 billion. A comet moves around the sun in a huge orbit, one circuit of which may take millions of years. Gravitational attractions of the planets may distort the orbit so that the comet makes a complete turn around the sun in as little as a few years.

When the comet is distant from the sun, it remains frozen, but when it approaches the sun it begins to vaporize. The escaping gases surround the nucleus, forming an envelope called the head, or coma, ranging from 10 000 to more than 100 000 miles in diameter. Ultimately, depending on how close the comet comes to the sun and the chemical composition of the comet, these gases (and some dust) may form the comet's tail, which in some cases extends 100 million miles into space. It is believed that meteors, or "shooting stars," come from the trails of comets. The periods of maximum intensity of certain meteor showers coincide with the times when certain comets pass closest to the earth.

The vast majority of comets observed are not visible to the naked eye. A few are large and brilliant, and can often be seen in broad daylight. When a comet is discovered, it is identified by the name of its discoverer and by an alphabetical letter designating the sequence of comets discovered in any particular year. Comet Ikeya-Seki 1965 f designates the comet discovered by these amateur astronomers, and it was the sixth discovered in the year 1965 (Fig. 20–25).

Figure 20–25 The comet Ileya-Seki 1965 f. (Courtesy of the Smithsonian Astrophysical Observatory.)

CHECKLIST

Asteroid	Planet
Chondrite	"Red planet"
Chondrule	Red spot
Chromosphere	Revolution
Comet	Rotation
Corona	Satellite
Crater	Selenography
Ecliptic	Solar eclipse
Lunar eclipse	Solar flare
Maria	Solar prominence
Mascon	Sunspot cycle
Meteorite	Titius-Bode law
Nebular theory	Totality
Photosphere	

EXERCISES

20–1. What significance may meteorites have concerning the composition of the earth's interior?

20–2. Discuss the nature of sunspots and the chronology of a sunspot cycle. What influence, if any, has a sunspot cycle on human affairs?

20–3. What causes light to be emitted by a meteorite?

20–4. From the information now available, discuss the possibility of life as we know it on other planets in our solar system.

20–5. What information has been obtained about the moon, first by means of improved instrumentation such as the telescope, camera, and lunar orbiter, and finally through astronaut landings, which was not available previously?

20–6. In your opinion, has the price-tag in the billion-dollar range for lunar exploration been worth the return?

20–7. Discuss the structure of a comet.

20–8. Discuss the structure and possible origin of Saturn's rings and their relation to the structure and origin of the solar system.

20–9. List several differences between the terrestrial planets and the Jovian planets.

20–10. What are some essential differences between planets and stars?

20–11. What is the connection between meteorites and asteroids?

20–12. Why are the "seas," "oceans," and "bays" on the moon misnomers?

20–13. Of all the natural moons, what is unique about Saturn's Titan that makes it of particular interest?

20–14. Which of the planets would appear to be most suitable for human exploration? In your opinion, should such exploration be encouraged?

20–15. Why does Mars vary greatly in brightness as seen from the earth?

20–16. What is the basis for the statement that our solar system is not unique in the universe?

20–17. Discuss the changes that occur as a comet approaches the sun.

20–18. Why is there not a solar eclipse at every new moon?

20–19. On which planet do we see the surface rather than the atmosphere?

20–20. *Multiple Choice*

    A. Sunspots appear dark because they are
       (a) hotter than the surrounding surface
       (b) cooler than the surrounding surface
       (c) flares
       (d) coronal streamers

B. Galileo discovered all but
   (a) sunspots
   (b) the phases of Venus
   (c) the moons of Mars
   (d) the moons of Jupiter

C. The lunar maria are presumed younger than the highlands because
   (a) they are lighter
   (b) they are darker
   (c) they have water
   (d) they have fewer craters

D. The greenhouse effect heats a planet because
   (a) infrared radiation is trapped
   (b) more sunlight gets in
   (c) the winds don't blow too fast
   (d) it removes CO_2 from the atmosphere

E. The Great Red Spot of Jupiter is probably
   (a) a continent
   (b) a storm
   (c) a hole in the clouds
   (d) a thick cloud layer

SUGGESTIONS FOR FURTHER READING

Arvidson, Raymond E., Alan B. Binder, and Kenneth L. Jones, "The Surface of Mars." *Scientific American,* March, 1978.
   The Viking spacecraft have viewed Mars from the ground and from orbit, adding evidence on how it has been shaped by volcano, meteorite impact, water and wind.

El-Baz, Farouk, "Naming Moon's Features Created 'Ocean of Storms,'" *Smithsonian,* January, 1979.
   A glimpse of a Space Age twist in a centuries–old problem: classifying and naming the features of the moon.

Grossman, Lawrence, "The Most Primitive Objects in the Solar System." *Scientific American,* February, 1975.
   There is evidence that the meteorites known as carbonaceous chondrites are mixtures of minerals that have survived unaltered since they condensed out of the nebula that gave rise to the sun and planets.

Lewis, John S., "The Chemistry of the Solar System." *Scientific American,* March, 1974.
   The exploration of the moon, and the fly-by missions to Mercury, Venus, Mars, and Jupiter, have greatly enlarged our knowledge of the chemistry of the solar system.

Mammana, Dennis L., "Now We Have Two Ringed Planets in the Solar System." *Smithsonian,* December, 1978.
   An account of the discovery in 1977 of the rings of Uranus.

Murray, Bruce C., "Mars from Mariner 9." *Scientific American,* January, 1973.
   Mariner 9 provided about 100 times the amount of information accumulated by all previous flights to Mars, decisively changing our view of the planet.

Pasachoff, Jay M., "The Solar Corona." *Scientific American,* October, 1973.

Until this century, the nature of the corona was a mystery. Recently, some of the mystery has begun to disappear, thanks to telescopes, computers, and satellites.

"The Solar System." *Scientific American,* September, 1975.

An entire issue devoted to the sun and the bodies in orbit around it, with an emphasis on the findings from space probes.

Whipple, Fred L., "The Nature of Comets." *Scientific American,* February, 1974.

Comets may be the remains of the cloud from which the sun and the planets formed.

THE EARTH'S CRUST AND INTERIOR—INVITATION TO EARTH SCIENCES

21–1 INTRODUCTION

Recent discoveries have changed the way in which most geologists view the evolution of the earth's surface. Concepts are changing, and as old puzzles are solved new ones replace them.

The central idea of a major new theory—that the earth's surface consists of a small number of rigid plates in motion relative to each other—has gained acceptance. According to *plate tectonics,* as the new theory is called, plate motions are responsible for the present positions of the continents, for the formation of many mountain ranges, and for essentially all major earthquakes. Data gathered over the past 30 years in marine geology—the sounding of the deep ocean, the samples and photographs of the ocean floor, the measurements of heat flow and magnetism—are being reinterpreted according to the new concepts.

Plate tectonic theory has unified widely separated fields in the earth sciences, from volcanology and seismology to sedimentary geology. The forces that cause the plates to move, however, are not yet understood. The main difficulty is a lack of knowledge about the earth's mantle, the shell of rock separating the core from the thin surface crust, where the driving mechanisms are believed to originate. Nevertheless, many long-standing problems within the field of geology are being reconsidered in the light of this revolution in fundamental concepts.

21–2 STRUCTURE OF THE SOLID EARTH

It seems paradoxical that at a time when the universe is yielding so many of its secrets to boundless human curiosity, knowledge of our own earth remains inaccessible. For example, the world's deepest mine (in South Africa) goes down only about 3.4 km. The deepest hole ever drilled in the earth (in a Texas oil field) was 7.7 km, about one tenth of 1 per cent of the earth's radius (which is 6400 km). Although the mantle comprises about 67 per cent of the total mass of the earth, no one has yet obtained an uncontaminated sample of it.

Nonetheless, the basic divisions of the earth have been known for some time (Table 21–1). From earthquake (seismic) waves, geophysicists have inferred that the solid earth consists of three principal concentric zones (Fig. 21–1): (1) a thin *crust* varying from 5 km in ocean basins to 60 km in thickness beneath mountain ranges; (2) a *mantle* about 2900 km thick; and (3) a *core* about 3400 km thick that accounts for more than half of the earth's radius and nearly a third of its mass.

The mantle, in turn, is subdivided into an upper mantle, a transition zone, and a lower mantle, while the core is divided into a liquid outer core and a solid inner core (Fig. 21–2). The crust and part of the upper mantle, a zone 50 to 200 km thick, together comprise the *lithosphere*, which is thicker beneath continents

Table 21–1 THE EARTH'S STRUCTURE

	DISTANCE FROM SURFACE OF THE EARTH (km)	THICKNESS (km)	DENSITY (g/cm³)
Crust	0–60		
Continental		32	2.8
Oceanic		5	3.0
Upper mantle	400		
Transition zone	1000		
Lower mantle	2900	2900	3.3 to 5.6
Outer core	5000	2100	9.5 to 12
Inner core	6371	1400	13+

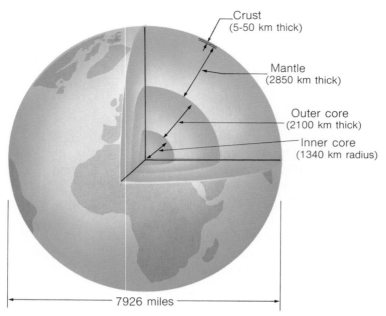

Figure 21–1 Interior of the earth. The major zones are the crust, mantle, outer core, and inner core. (Adapted from Flint, R. F., and Skinner, B. J., *Physical Geology*. New York: John Wiley & Sons, 1974, p. 5.)

than beneath oceans. The lithosphere in turn, overlies the *asthenosphere*, a layer 100 km in thickness. A partially molten zone separates the lithosphere from the asthenosphere. The percentage of molten material is very small, probably less than 2 per cent. This is

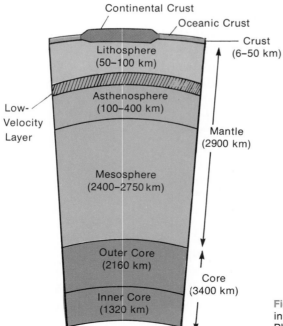

Figure 21–2 Details of the structure of the earth's interior. (From Levin, H. L., *The Earth Through Time*. Philadelphia: W. B. Saunders Company, 1978, p. 213.)

enough, however, to reduce the strength of rocks in the zone and allow the lithosphere to slide over the asthenosphere.

21-3 SEISMIC WAVES

More has been learned about the earth's interior by studying seismic waves than by any other means. The velocities of earthquake waves and of waves produced by controlled underground explosions depend on the rigidity and density of materials, which in turn depend on their chemical composition and structure. Depending upon the medium through which they pass, the seismic waves are bent, speeded up, slowed down, and sometimes stopped altogether. By studying the transmission of waves through the earth, seismologists learn much about the physical properties of the interior.

The faint echoes emerging at the earth's surface actuate the sensitive pendulum recorders known as seismographs (Fig. 21-3). When the earth shakes, the recording drum and support of the inertial unit move relative to the weight. The relative motion is shown on the drum. Modern seismographs employ electronic devices to pick up, amplify, and record the motions of the earth, and digital processing greatly speeds analysis and interpretation.

The major recent improvement in seismology has been the deployment of seismographic arrays in various localities, such as the Large Aperture Seismic Array in Montana, which covers 200 km and contains 525 seismographs. Seismic arrays have become re-

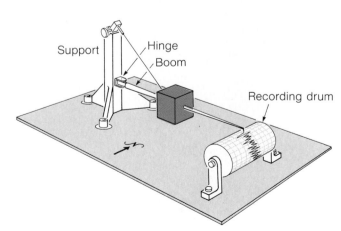

Figure 21-3 A pendulum seismograph. Earth motions are greatly amplified by the use of accessories. (Redrawn from Donn, W. L., *The Earth.* New York: John Wiley & Son, 1972, p. 354.)

markably effective probes for studying the earth's interior. Just as radio telescopes have revealed celestial objects that were once invisible, the new generation of seismic instruments have detected fine details of the earth's structure that were once unobservable. Considerable refinement has also been possible with the advent of the worldwide network of standardized seismic stations recording the effects of large underground explosions.

Four basic types of seismic waves are analyzed—two kinds of body waves (P, S) and two corresponding surface waves (L, R). The first signal on a conventional seismograph is usually due to a fast-moving P (for primary) wave. The P waves are compressional or longitudinal waves (like sound waves; see Chapter 8) in which the displacement of the particles in the ground is along the waves' direction of travel. They are the fastest of the seismic waves, traveling at about 6.0 to 6.7 km/sec in the crust and 8.0 to 8.5 km/sec in the upper mantle.

The other type of body waves are called S (for secondary) waves. They are transverse waves (like light waves; see Chapter 9) in which the direction of ground motion is perpendicular to the direction of travel. The velocity of S waves is lower than that of P waves: 3.0 to 4.0 km/sec in the crust and 4.4 to 4.6 km/sec in the upper mantle. Unlike P waves, S waves cannot be transmitted through liquids. Since S waves do not travel through the outer core, this region is believed to be molten (Fig. 21–4).

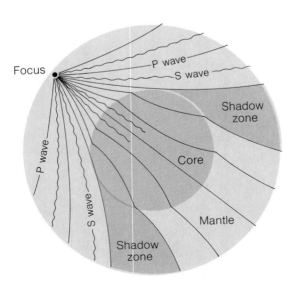

Figure 21–4   Effect of the earth's core on seismic waves. The core bends and concentrates the P waves, but does not transmit S waves. Since a liquid medium would have such an effect, the core is believed to be molten. (Redrawn from Mears, B., Jr., *The Changing Earth.* New York: Van Nostrand Reinhold, 1970, p. 158.)

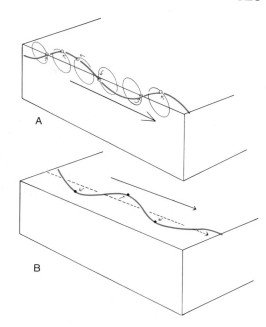

Figure 21–5 (A) Rayleigh surface waves. (B) Particle motion in Love wave. (From Levin, H. L., *The Earth Through Time.* Philadelphia: W. B. Saunders Company, 1978, p. 178.)

The two types of surface waves are L (Love) and R (Rayleigh) waves (Fig. 21–5). They travel near the earth's surface at speeds of about 3.5 to 5.2 km/sec, but are not confined to the crust. L waves cause sideways motion along the surface; R waves cause the particles to execute elliptical motion. (It has been found that the ratio of body-wave magnitude to surface-wave magnitude can be used to distinguish an earthquake from an explosion, such as a nuclear detonation, eliminating in principle the need for on-site inspection to monitor underground testing of nuclear weapons.)

21–4 THE CRUST: ROCKS AND MINERALS

The part of the solid earth with which we are most familiar, the crust, constitutes less than 1 per cent of the earth's total mass. It consists of material with an average density of only about 2.8 g/cm³, compared to 4.5 g/cm³ for the mantle and 11.0 g/cm³ for the core (see Table 21–1). Its overall average thickness is 17 km, but under the oceans it averages 5 km, under the continents about 35 km, and under some mountains, 60 km (Fig. 21–6). Both oceanic and continental crust "float" in mantle rocks in much the same way that icebergs float in water. Continents stand higher because they are thick and are composed of relatively low density rocks.

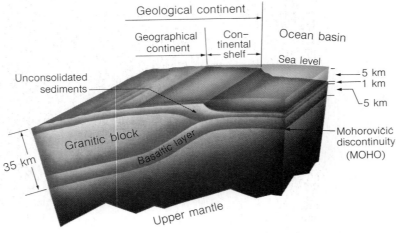

Figure 21–6 Cross-sectional view of the earth's crust. The oceanic and continental portions vary in thickness and composition. (Redrawn from Zumberge, J. H., and Nelson, C. A., *Elements of Geology*, 3rd ed. New York: John Wiley & Sons, 1972, p. 21.)

The thinner, heavier oceanic crust does not stand as high. Table 21–2 gives the 10 most abundant elements of the earth's crust, comprising more than 99 per cent by weight.

The crust contains a great variety of *rocks*, heterogenous aggregates of minerals in varying proportions. A *mineral* is a homogeneous substance with a definite chemical composition and crystal structure, and characteristic physical properties. Some common rock-forming minerals and their chemical composition are listed in Table 21–3. Rocks are subdivided into three categories, depending on their origin: *igneous, sedimentary,* and *metamorphic* (Fig. 21–7).

Igneous rocks are by far the most commonly found rocks in the earth's crust. Among them are granite, basalt, gabbro, and pumice. They have been formed

Table 21–2 THE 10 MOST ABUNDANT ELEMENTS OF THE EARTH'S CRUST

ELEMENT	WEIGHT, %
Oxygen	46.60
Silicon	27.72
Aluminum	8.13
Iron	5.00
Calcium	3.63
Sodium	2.83
Potassium	2.59
Magnesium	2.09
Titanium	0.44
Hydrogen	0.14

Table 21–3   COMMON ROCK-FORMING MINERALS

CHEMICAL GROUP	MINERAL	CHEMICAL COMPOSITION
Carbonates	Calcite	$CaCO_3$
	Dolomite	$CaMg(CO_3)_2$
Halides	Fluorite	CaF_2
	Halite	$NaCl$
Oxides	Hematite	Fe_2O_3
	Magnetite	Fe_3O_4
	Quartz	SiO_2
Sulfates	Gypsum	$Ca(SO_4) \cdot 2H_2O$
Sulfides	Galena	PbS
	Pyrite	FeS_2
	Chalcopyrite	$CuFeS_2$
Silicates	Quartz	SiO_2
	Olivine	$(Mg,Fe)_2SiO_4$
Amphiboles	Hornblende	$Ca_4Na_2(MgFe)_8$ $(AlFe)_2(Al_4Si_{12}O_{44})$ $(OHF)_4$
Feldspars	Albite	$Na(AlSi_3O_8)$
	Anorthite	$Ca(Al_2Si_2O_8)$
	Microcline	$K(AlSi_3O_8)$
	Orthoclase	$K(AlSi_3O_8)$
	Plagioclase	Mixture of albite and anorthite
Micas	Biotite	$K(Mg,Fe)_3(AlSi_3O_{10})(OH)_2$
	Chlorite	$(MgFeAl)_6(AlSi_4O_{10})(OH)_8$
	Kaolinite	$Al_2(Si_2O_5)(OH)_4$
	Muscovite	$KAl_2(AlSi_3O_{10})(OH)_2$
	Talc	$Mg_3(Si_4O_{10})(OH)_2$
Pyroxenes	Augite	$Ca(MgFeAl)(AlSi)_2O_6$

Figure 21–7 The rock cycle. (From Levin, H. L., *The Earth Through Time*. Philadelphia: W. B. Saunders Company, 1978, p. 35.)

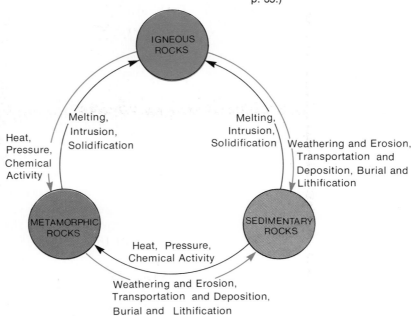

from molten matter (*ignis,* fire), possibly when magma (fluid rock beneath the earth's surface) poured forth from a volcano as lava and then solidified. Or they may have solidified within the earth's interior through the process of crystallization, with widely differing chemical compositions. Among the most common minerals in continental igneous rocks are quartz, feldspars, and micas. Pyroxenes and amphiboles are also quite common.

Sandstone, limestone, and shale are examples of *sedimentary rocks,* products formed from the deposition of weathering and erosion of older rocks at the earth's surface. These sediments may be carried away by water, wind, or ice and deposited elsewhere. Or the material may be dissolved in water and later precipitated elsewhere. The loose material is eventually compressed into sedimentary rocks by the weight of the overlying mass; it may also be cemented by material dissolved and reprecipitated between the grains. Some of the common sedimentary materials are gypsum, calcite, halite, clays, quartz, and dolomite.

Metamorphic rocks are igneous, sedimentary, or metamorphic rocks that have been changed or recrystallized by increased pressure and temperature, usually without melting. In this way, limestone is converted to marble, shale to slate, and sandstone to quartzite.

21-5 THE MANTLE AND CORE

A sharp boundary between the crust and the mantle, discovered by Andrija Mohorovičić (1857–1936), a seismologist at the University of Zagreb, Yugoslavia, provides a means of defining the depth of the earth's crust. This boundary is known as the *Mohorovičić discontinuity,* or the Moho for short, and constitutes the transition between the crust and the mantle. The speed of P waves here increases abruptly from approximately 6.7 km/sec, typical of the bottom of the lower crust, to about 8.0 km/sec, characteristic of the top of the upper mantle. The sudden change in seismic velocity indicates that the material above the Moho is different from the material below.

It is generally accepted that the mantle is made up predominantly of denser silicate materials that are rich in magnesium and iron. By comparison, stony meteorites of the chondrite variety, which are believed to have been formed at about the same time as the earth, contain 80 to 90 per cent silicates, mainly magnesium-iron silicates. Compounds of four elements—magnesium, iron, silicon, and oxygen—are believed to comprise more than 90 per cent of the mantle. In the upper mantle, these elements are mainly in the form of silicates. The important minerals of the upper mantle are olivine, pyroxene, and garnet.

The mantle constitutes about 85 per cent of the earth's total volume. It consists of a relatively thin (50 km) high-velocity layer; a second layer, 100 km thick, the asthenosphere; a transition zone, 400 to 1000 km in thickness; and the lower mantle, 1000 to 2900 km thick. The transition region is characterized by several abrupt increases in velocity, which laboratory studies indicate can be attributed to changes in crystal structure induced by high pressures. The main component of the mantle, olivine, is transformed at pressures corresponding to a depth of about 400 km to a structure in which the atoms rearrange themselves into a tighter packing arrangement, thereby increasing the density of the material.

Like the crust, the mantle is believed to be solid, with the exception of isolated regions containing small pockets of magma. However, large regions of the mantle may be very close to their melting point at the prevailing high temperatures and pressures and may have a fluid-like character. The lower mantle is believed to be relatively homogeneous, consisting of mixtures of such high-density silicates and oxides as enstatite ($MgSiO_3$), stishovite (SiO_2), and periclase (MgO).

The outer core is believed to be liquid, since S waves cannot penetrate it. It has the properties of molten iron (80 to 85 per cent) mixed with nickel, cobalt, sulfur, or silicon. Although the inner core is believed to be solid, there is evidence from wave velocities suggesting that it is near the melting point or is partly molten. At the center of the earth, the density is estimated to be 12.86 g/cm³, the pressure about 3.5 million atmospheres, and the temperature between 3500°C and 6000°C.

21-6 VOLCANOES

The dynamic nature of the apparently solid earth is evident in the violent eruptions of volcanoes. More than 500 volcanoes have erupted in the past four centuries alone. The eruptions at the surface are only small manifestations of great events going on in the earth's interior. So far, there is only speculation about these events, for no single theory has yet embraced all of the details of volcanic behavior.

A volcanic eruption can be an awesome spectacle. Krakatoa, a small volcanic island in Indonesia, had been dormant for 200 years when it suddenly exploded in 1883. Detonations were heard in Australia, nearly 5000 km away, as the volcano ejected 17 cubic kilometers of rock and pumice into the air and destroyed two thirds of the island. A catastrophic tidal wave that followed drowned 36 000 people on the adjacent coasts of Sumatra and Java. For years, the volcanic ash circled the earth, creating spectacular blood-red sunsets. One clue to the origin of volcanoes is their location (Fig. 21–8). Active volcanoes are concentrated in parts of the world where earthquakes are most common. Furthermore, volcanoes are commonly found in young mountain belts (although there are young mountains, such as the Himalayas, which do not have them).

Volcanic activity has been most widespread during the periods of adjustment that follow the formation of mountain ranges. Extremely hot material

Figure 21–8 Distribution of volcanic activity. Locations of some active and recently active volcanoes are shown. (Redrawn from Donn, W. L., *The Earth.* New York: John Wiley & Sons, 1972, p. 297.)

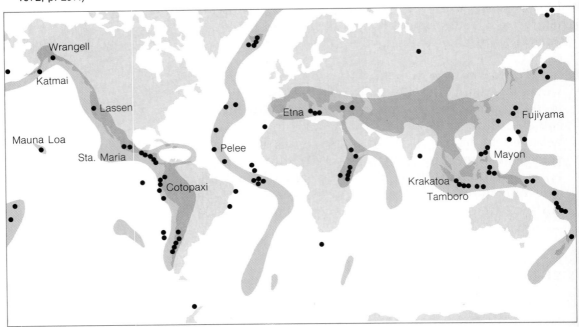

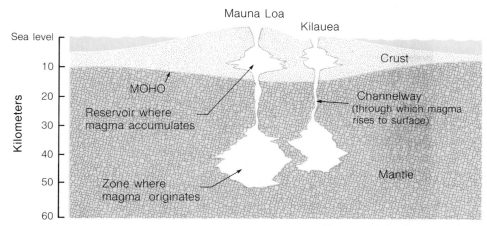

Figure 21–9 Cross-sectional view of two Hawaiian volcanoes; Mauna Loa and Kilauea. (Redrawn form Zumberge, J. H., and Nelson, C. A., *Elements of Geology*. 3rd ed. New York: John Wiley & Sons, 1972, p. 70.)

located tens of kilometers below the surface becomes liquefied if the pressure is reduced or the temperature rises. The pressure may be reduced by the bending or cracking of the rocks above, or the temperature may be increased by radioactive heating. The liquefied material, called magma, is lighter than the rocks lying above it and tends to rise wherever it finds an opening (Fig. 21–9). If it rises directly to the surface of the earth through fractures in the rock, it may come out quietly as fluid lava or as cinders, glowing ash, and partially solid plugs. In the event that it reaches a roof of solid rock some distance below the surface, it may spread sideways and eventually erupt into one or more volcanoes where there are cracks in the roof.

Lava is the molten material with its associated gases that flows out of volcanic vents and fissures; the term is also applied to the rock that solidifies from this material. *Magma* is the name for this material when it occurs below the surface, where the gas content tends to be greater. One type of magma is represented by basalt—a dark, heavy, relatively fluid material rich in iron and magnesium minerals and low in silica. Basalt is typical of oceanic volcanoes and some continental volcanoes (Table 21–4). Rhyolite, a lighter material at the other end of the spectrum, is low in iron and magnesium minerals and high in silica; it is extrusive granite and relatively viscous. Lavas of intermediate composition are typical of island arcs and young mountain ranges. Lavas range in temperature from 600 to 1200°C. Magma temperatures below the surface must be higher than those for surface lavas, because of the greater pressures. The basaltic magma probably forms

Table 21-4   COMMON IGNEOUS ROCKS

COARSELY CRYSTALLINE INTRUSIVE IGNEOUS ROCKS	APPROXIMATE EXTRUSIVE EQUIVALENTS	COMMON SILICATE MINERAL COMPONENTS	AVERAGE SPECIFIC GRAVITY
Granite	Rhyolite	Quartz Potash Feldspar Sodium Plagioclase (Minor Biotite, Amphibole, Magnetite)	2.7
Diorite	Andesite	Sodium-Calcium Plagioclase Amphibole, Biotite (Minor Pyroxene)	2.8
Gabbro	Basalt	Calcium Plagioclase Pyroxene Olivine (Minor Amphibole, Ilmenite)	3.0

Source: Levin, H. L., *The Earth Through Time.* Philadelphia: W. B. Saunders Company, 1978, p. 37.

because localized regions of the upper mantle are somehow heated, and melt. Most granitic magma may involve the melting of the lower part of the earth's continental crust.

Volcanoes are found in a great variety of shapes

Figure 21–10 Lassen Peak in California. This type of volcano results from viscous lava. (Courtesy of U.S. Air Force Information Office.)

Figure 21–11 Mount Fujiyama, Japan.

and sizes. Not all are explosive; some, like many of the volcanoes of Hawaii, are sluggish and gentle. Basaltic lavas are fluid and tend to flow; viscous "granitic" lavas tend to erupt explosively. When the flow of lava is rapid, the structure of the volcano tends to be flat, like an inverted saucer, and explosive eruptions are few and weak. If the lava is squeezed out in a very viscous condition, like toothpaste from a tube, and moves at little more than a snail's pace, a very steep-sided mountain, such as Lassen Peak in California (Fig. 21–10), is produced. Volcanoes such as Mount Rainier in the United States and Fujiyama in Japan have a graceful profile, rising from a wide base to a tall, slender peak (Fig. 21–11). A million years may be required to build a composite volcano such as these, involving in part the outpouring of lava and in part explosive discharges. The explosive, viscous varieties, on the other hand, may grow spectacularly. In just 1 year, the Mexican volcano of Paricutin, born in 1943 in a corn field, grew to a height of 1200 feet (Fig. 21–12).

21-7 CONTINENTAL DRIFT

Until recently, most geologists conceived the earth's crust as being a fairly stable layer enveloping the fluid mantle and core. The only kind of motion

Figure 21–12 Paricutin volcano, Mexico. This volcano erupted in a farmyard in 1943. (Photograph by Tad Nichols.)

perceived in this framework was the tendency of the crust as a rigid shell of varying rock density to float on a plastic layer of greater density. However, a scientific revolution involving a change in view from a static earth to a dynamic one is profoundly changing our understanding of the earth. The concept of a static earth held that the continents and ocean basins were permanent features that had existed in their present form since early in the earth's history. The basis for a dynamic earth is the idea that continents are constantly moving and ocean basins are opening and closing.

The theory of *continental drift* proposes that the continents did not always exist where they are today but were part of a supercontinent, which the geophysicist Alfred L. Wegener (1880–1930) called *Pangaea* (all-earth). About 200 to 250 million years ago, this single continent began splitting apart at what are now the rifts in the oceanic ridges, and the pieces (continents) drifted to their present locations. Until the 1960's, the idea that the continents had drifted apart was regarded with considerable skepticism. Since then, as a result of many new findings, the theory has gained much support. The evidence today favors the temporary formation of two large land masses: *Gondwanaland* in the Southern Hemisphere and *Laurasia* in the Northern Hemisphere (Fig. 21–13).

The geological and paleontological evidence for the theory of continental drift is strong. Wegener himself said that he began taking the idea of drifting continents seriously after learning of the fossil evidence for a former land connection between Brazil and Africa. The similarities and differences between fossils in various parts of the world not only support the continental drift concept but also provide a reasonably precise timetable for a number of the key events before and after the breakup. Following the separation of the continents, evolution proceeded on different paths leading to the present biological diversity observed on the various continents.

On the geological side, Patrick M. Hurley (b. 1912) at the Massachusetts Institute of Technology studied rocks in Africa and Brazil which were 2 billion and 550 million years old. When the maps of Africa and South America were fitted together, the regions consisting of rocks of the same age appeared to be extensions of one another on both continents. An additional correlation

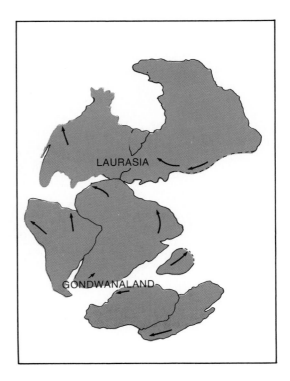

Figure 21–13 Pangaea breaking up into Gondwanaland and Laurasia. (From Levin, H. L., *The Earth Through Time*. Philadelphia: W. B. Saunders Company, 1978, p. 372.)

in the Southern Hemisphere was found in a number of glaciations that left a distinctive record in the southern parts of South America, Africa, Australia, Antarctica, India, and Madagascar, which suggested that they were joined into a single unit—Gondwanaland—until about 200 million years ago, when it began to break up.

Many attempts have been made to guess precisely how South America, Africa, Australia, Antarctica, and India were once joined as Gondwanaland, but there is as yet no general agreement. The fit between the coast of South America and Africa is excellent, and the fit between Australia and Antarctica is good (Fig. 21–14). But the fit of all five major units is not settled, and the original position of Madagascar is unknown. Unlike

Figure 21–14 Fit of the continents. The fit between South America and Africa is excellent. (Adapted from Spencer, E. W., *The Dynamics of the Earth.* New York, Thomas Y. Crowell Co., 1972, p. 558; after Wegener, 1915.)

the original Wegener idea that the continents alone are drifting across the ocean floor, the current view is that the continents, the ocean floor, and the upper part of the upper mantle are drifting.

21–8 SEA-FLOOR SPREADING

Opposition to the theory of continental drift was centered on the belief that the earth's crust and mantle were too rigid to permit such motions. It seemed unlikely that a large continent could cross an entire ocean basin and leave no trace of its passage. On the other hand, the existence of great young mountain ranges, deep ocean trenches, and volcano and earthquake belts that in some cases are continuous over distances of several thousand kilometers indicates the large-scale motion of material forming the earth's crust. Furthermore, the relatively thin layers of sediment on the sea floor suggest that it has been regularly rejuvenated, and the oceanic lands and submerged volcanoes are all relatively young.

When oceanographers explored and mapped the oceanic ridges and rifts in meticulous detail, it became clear that the ocean floor was relatively young compared with the continents. Arthur Holmes (1890–1965) and Harry H. Hess (1906–1969) then proposed sea-floor spreading as a mechanism for continental drift. According to this hypothesis, the continents do not plough their way through the ocean floor, but move with it, much like material on a conveyor belt. Molten rock from the mantle below rises up along a crack-like valley on the crest of the mid-oceanic ridge, a mountain range which runs for thousands of kilometers through the major ocean basins, then spreads outward and quickly hardens into new ocean floor (Fig. 21–15). In this way the sea floor is continuously rejuvenated, sweeping along with it the layer of sedimentary material so that no part of it remains truly ancient. Elsewhere on the earth the ocean floor plunges into a trench and then sinks deep into the mantle (Fig. 21–16).

Support for the hypothesis of sea-floor spreading has come from the rocks of the ocean floor, which carry a permanent record of their own history. The oceanic crust is not uniformly magnetized. Magnetometer surveys of the ocean floor reveal a remarkably regular pattern of symmetric magnetic strips on either side of

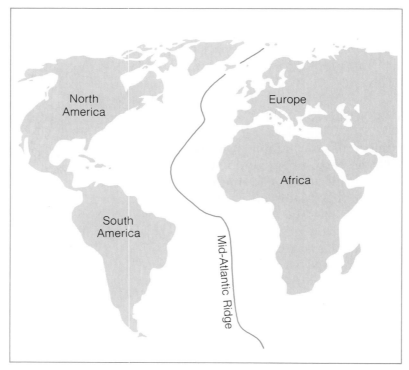

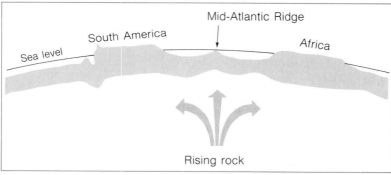

Figure 21–15 Mid-Atlantic ridge. This ridge is slowly expanding and pushing the continents apart.

the ocean ridges. Strips of material with normal polarization, with north pointing in the north direction, alternate with strips of reverse or anti-normal polarization. The entire ocean floor is covered with long strips of alternating magnetic polarity, each strip at least 10 km wide (Fig. 21–17).

To account for the symmetrical pairs of parallel strips that have the same direction of magnetization, it has been proposed that when lava wells up in the ocean ridge, it is magnetized as it cools in the direction of the earth's magnetic field. The alternations of polarity indicate that the earth's magnetic field reverses at widely

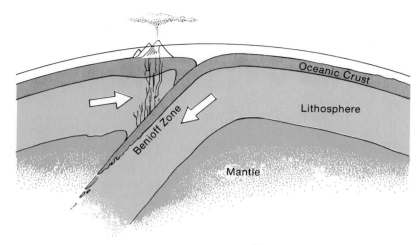

Figure 21–16 Convergence of two plates. (From Levin, H. L., *The Earth Through Time.* Philadelphia: W. B. Saunders Company, 1978, p. 217.)

varying intervals, from 50 000 years to 20 million years. Unless this magnetism is disturbed by reheating or physical distortion, it becomes a permanent record of the direction and polarity of field at the time the rock was formed.

Verification of sea-floor spreading has changed the status of continental drift from speculative conjecture to a generally accepted fact. The speed of spreading on each side of a mid-ocean ridge varies from less than a centimeter to as much as 18 centimeters per year. Such rates are geologically fast and can account for the formation of entire ocean floors. At 15 cm/year, the floor of the Pacific Ocean (about 15 000 km wide) could be produced in 100 million years. The directions and rates of motion for both sea-floor spreading and continental drift are in complete agreement. It is now apparent that virtually all of the present area of the oceans has been created by sea-floor spreading during the past 200 million years.

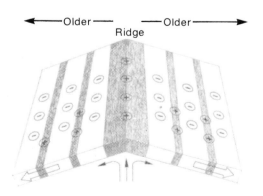

Figure 21–17 The normal (+) and reversed (−) magnetizations of the sea floor. (From Levin, H. L., *The Earth Through Time.* Philadelphia: W. B. Saunders Company, 1978, p. 225.)

21-9   PLATE TECTONICS

As a result of marine magnetic and earthquake studies, the theory of plate tectonics has gained increasing acceptance among geologists. Plate tectonics unites the concept of sea-floor spreading with the earlier idea of continental drift in a single unifying theory. It is believed that relative plate motions are responsible for the positions of the continents, for the formation of many mountain ranges, and for essentially all major earthquakes.

"Tectonics" is a geological term that refers to earth movements. Plate tectonics is the theory that the surface of the earth is divided into a number of large, rigid, platelike regions that move with respect to one another. North America is on one such plate; Africa is on another (Fig. 21–18). The plates are believed to be 50 to 100 km thick and to slide on the warmer, less rigid material of the upper mantle. The boundaries of the plates are not limited to continental boundaries; plates often include both oceans and continents.

Three types of plate boundaries are recognized: spreading or divergent, in which new crust is created as two adjacent plates move away from each other; convergent, when two plates converge and one plate overrides the other, pushing it into the mantle where it is heated and assimilated; and transcurrent fault boun-

Figure 21–18   Six major plates. These plates account for the pattern of continental drift. The African plate shown here is assumed to be stationary. Arrows indicate the direction of motion of the five other large plates. (Redrawn from Hamblin, W. K., *The Earth's Dynamic Systems.* Minneapolis: Burgess Publishing Co., 1975, p. 57.)

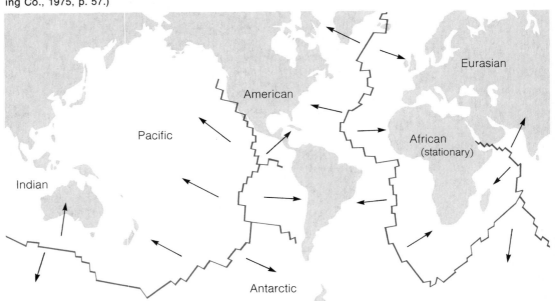

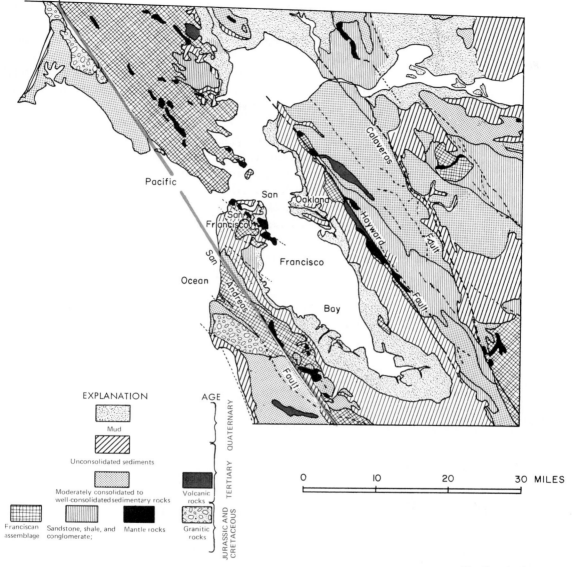

EXPLANATION

Mud

Unconsolidated sediments

Moderately consolidated to
well-consolidated sedimentary rocks

Volcanic
rocks

Franciscan   Sandstone, shale, and   Mantle rocks   Granitic
assemblage   conglomerate;   rocks

AGE

QUATERNARY

TERTIARY

JURASSIC AND
CRETACEOUS

0   10   20   30 MILES

Figure 21–19 The San Andreas fault is the boundary between two plates of the earth's surface. The earth's crust on the right is moving in the opposite direction to that on the left. (From Levin, H. L., *The Earth Through Time*. Philadelphia: W. B. Saunders Company, 1978, p. 218.)

daries, which occur when two adjacent plates slide past each other. Mid-ocean ridges are the most common example of divergent boundaries. Trenches and young mountain ranges mark convergent zones. Long linear faults, such as California's San Andreas fault, mark the contact where plates slide side by side (Fig. 21–19). In the San Andreas case, the edge of the Pacific plate is moving northwesterly at the rate of a few centimeters per year, carrying with it past the mainland a narrow strip of the California coast.

According to plate tectonics, the vast Himalayan range was created when a plate of the earth's crust carrying the land mass of India collided with the plate carrying Asia some 45 million years ago, having traveled 5000 km across what is now the Indian Ocean. The northern edge of the Indian plate crumpled up the many layers of shallow-water sediments laid down over millions of years on the continental shelf that bordered the southern edge of Asia, and formed the Himalayas.

It is believed that plate tectonics has been an active mechanism for much longer than 200 million years. During this time, virtually all the present oceans were created and others destroyed. Where the plates floated apart, the continents embedded in them also drifted away from one another. The processes we see today may have been in action all through geological time; continents may have split many times and formed new oceans, and collided and been welded together. Although the geometry and kinematics of plates are fairly well known, the dynamics—the driving mechanisms of the plate motions—is a subject of con-

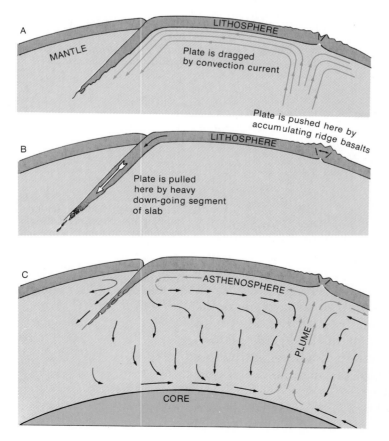

Figure 21-20 Three models of possible driving mechanisms for plate movements. (From Levin, H. L., *The Earth Through Time.* Philadelphia: W. B. Saunders Company, 1978, p. 220.)

Figure 21–21 Rock arches. Weathering of sandstone caused these structures. Arches National Monument, Utah. (Photograph by Arthur N. Strahler, Santa Barbara, Calif., 93108.)

siderable research. Most models assume some form of thermal convection within the mantle, but there is as yet little quantitative evidence available on the composition and properties of the mantle (Fig. 21–20).

21–10 EARTH SCULPTURING

Ever since it was formed, the earth's crust has been subjected to processes leading to the destruction of land forms through erosion, the removal of parts of the crust to new locations by various agents. Most of the features of the topography are the results of sculpting by erosional forces, which have their origin in the sun, the oceans, and the atmosphere.

The preliminary or static phase of erosion involves *weathering*, of which two broad types are identified: *chemical* and *mechanical*. Wherever rocks are exposed to air, water, and organisms, weathering reduces them to a soft, loosened debris that can be carried away by active agents of erosion, such as running water, wind, ice, and waves (Fig. 21–21).

Chemical weathering results in the decomposition of rock. Through the chemical activity of water, oxygen, or carbonic acid, new minerals, usually softer and of lower density, are produced from the original ones; for example, feldspar is reduced to clay. Since it usually involves water, chemical weathering proceeds more rapidly in moist than in arid climates. A striking example is a monument that stood nearly unscathed for 3500 years in Egypt, but deteriorated badly in a relatively few years when brought to the more humid and polluted environment of New York City.

Mechanical weathering causes rocks to disintegrate into smaller pieces, as when quartzite breaks into smaller fragments. In regions that experience frequent alternation of freezing and thawing, ice wedgework becomes a dominant disintegration process. Water seeps into rock openings and expands as it freezes; the resultant pressure on the surrounding rock often splits it apart. Root weathering also cause rocks to disintegrate. The roots and trunks of trees growing in rock crevices develop enormous splitting pressures on the surrounding rocks.

When weathering has caused rocks to disintegrate or decay, gravity may force the debris to move downslope from higher to lower ground (Fig. 21–22). This process, called mass-wasting, may be carried out in several ways: rockfalls, landslides, mudflows, and creep. Important results of weathering are the formation of soil and mineral ore deposits.

Figure 21–22 Piles of debris at foot of cliffs. Snowy Range, Lake Marie, Wyoming. (From Mears,, B., Jr., *The Changing Earth.* New York: Van Nostrand Reinhold, 1970, p. 69. Photograph by Brainerd Mears, Jr.)

Figure 21-23 The Canyon of the Yellowstone River. Note the V-shaped valley, the mark of a youthful stream. (Photograph from Geological Survey. Photograph #57-HS-87 in the National Archive.)

Ultimately, streams transport the rock material to the sea. Incidental to their flow to the sea, streams are by far the dominant agent of erosion on the earth's surface today, more important than wind and ice. In the process, they are responsible for producing many landforms, consuming uplands by eroding their valleys (Fig. 21-23). The most important erosional process of streams involves hydraulic action—the removal and transportation of loose material from the channel sides and bottom. The swifter the current, the larger the size of the particles that can be transported. The solvent action of water is less significant as an erosive force.

As the velocity of the stream decreases, the stream deposits the denser particles first, and then the finer materials. The sediments accumulate and may form a delta near the junction of the river with a lake or the ocean. Some deltas, such as those of the Mississippi and Nile rivers, extend their coastal areas for hundreds of square miles into the sea (Fig. 21-24). The flat, fertile alluvium, the material deposited by the running water, makes excellent farm soil.

Figure 21-24   The fertile Nile Delta, about 500,000 square miles in area, surrounded by light-colored deserts. (Courtesy of NASA.)

21-11   MATERIALS OF THE EARTH'S CRUST

The new ideas of plate tectonics have had an impact on economic geology on a global scale. Many ore deposits occur at the boundaries, present or past, of the crustal plates. What ores are formed and where they are placed in the crust may depend on the tectonic history of a region. This can be important in a resource-hungry world.

Many major classes of mineral deposits fit the new conceptual framework. They occur in volcanic or igneous rocks and were formed at the same time as these rocks. According to plate tectonics, volcanism occurs at diverging plate boundaries such as mid-ocean ridges, where mantle material rises to form new crust; at converging boundaries, where plates descend into the mantle through subduction, leading to volcanism that forms chains of mountains, or island arcs; and over hot spots caused by ascending plumes of mantle material. Each process may give rise to characteristic types of ore deposits.

The mineral deposits of the island of Cyprus, a rich source of copper, are a prime example of the first type. The copper sulfide ore occurs in a distinctive progression of rock types known as an ophiolitic sequence. The sequence is exactly that which should be formed at a mid-ocean ridge; on top, sediments of a type formed on the ocean floor; beneath the sediments, pillow lavas formed when molten volcanic material erupts into sea water; farther down, sheets of basalt formed as cracks in the ocean floor are filled from below with volcanic material; and on the bottom, rocks rich in iron and magnesium that are believed to be characteristic of the mantle. The minerals include sulfides of copper, iron, and sometimes zinc, chromium ore, and asbestos deposits. Mineral deposits of the Cyprus type are found in many parts of the world, including the northeastern United States and eastern Canada.

A second type of mineral deposit, ores known as porphyry coppers, is associated with converging plate boundaries. The copper deposits in the Andes are an example. There the eastward moving oceanic crust of the Pacific plunges under the lighter material of the westward moving South American continent. Partial melting of the oceanic plate produces magmas that rise through the overlying continental rocks, sometimes forming volcanoes. The upper parts of these magma intrusions often contain copper and molybdenum, and sometimes gold and silver as well.

Plate tectonic mechanisms explain the source of much of the world's mineral wealth—iron ores, gold deposits, copper, zinc, lead, and silver ores. Still, the new models of ore formation are far from complete and not entirely accepted. Mineral exploration can be guided by an understanding of how, and perhaps where, ores are formed and deposited but far more exhaustive studies are required.

CHECKLIST

Continental drift
Core
Crust
Gondwanaland
Igneous
Laurasia
Lava
Lithosphere
Magma
Mantle

Metamorphic
Mineral
Mohorovičić discontinuity
Pangaea
Plate tectonics
Rock
Sea-floor spreading
Sedimentary
Seismic waves
Seismograph

EXERCISES

21-1. Discuss some evidence in support of the concept of continental drift.

21-2. How is sea-floor spreading related to plate tectonics?

21-3. How is information concerning the earth's interior obtained?

21-4. Explain the relationship between magma and lava.

21-5. Discuss the varieties of volcanic eruptions and their relation to the shapes of volcanoes.

21-6. Discuss the structure of the earth's interior.

21-7. Why is the ocean floor youthful in relation to the age of the earth?

21-8. What evidence exists for the idea that the earth's magnetic field has reversed at intervals over the ages?

21-9. In what way is the comparison of continents to floating icebergs valid?

21-10. Distinguish between rocks and minerals.

21-11. How are regions of active volcanoes related to earthquake belts?

21-12. What is the Mohorovičić discontinuity? How was it discovered?

21-13. Why is none of the original crust of the earth now visible?

21-14. Discuss the kinematics of three types of plate boundaries.

21–15. Why is the rate of weathering greater in a humid than in an arid climate?

21–16. *Multiple Choice*

    A. The oldest rock on the earth's surface is
      (a) sedimentary   (c) tectonic
      (b) igneous   (d) metamorphic
    B. An instrument that measures earthquake activity is a
      (a) polarimeter   (c) seismograph
      (b) refractometer   (d) micrometer
    C. The theory that continents float on layers of mantle rock is called
      (a) seafloor spreading   (c) mass wasting
      (b) plate tectonics   (d) volcanism
    D. The mid-Atlantic ridge is
      (a) a region where tectonic plates are colliding
      (b) bordered by young igneous rocks
      (c) bordered by old sedimentary rocks
      (d) a region where earthquakes are common
    E. An agent of significant erosion is
      (a) oxidation of minerals in air
      (b) falling snow
      (c) wave action
      (d) radioactive decay

SUGGESTIONS FOR FURTHER READING

Bullard, Sir Edward, "The Origin of the Oceans." *Scientific American,* September, 1969.
    The ocean floor is surprisingly young. Growing outward from mid-ocean ridges, it is pushing continents apart.
Dewey, John F., "Plate Tectonics." *Scientific American,* May, 1972.
    As the plates making up the earth's surface shift, new crust and mountains are formed and continents drift.
Hallam, A., "Alfred Wegener and the Hypothesis of Continental Drift." *Scientific American,* February, 1975.
    The most important points in Wegener's hypothesis of 1912, neglected or scorned for 50 years, have been substantiated by discoveries in geophysics and oceanography.
Hurley, Patrick M., "The Confirmation of Continental Drift." *Scientific American,* April, 1968.
    The evidence now favors the idea that the continents were once combined into two land masses: Gondwanaland and Laurasia.
Jordan, Thomas H., "The Deep Structure of the Continents." *Scientific American,* January, 1979.
    There are new insights into processes that control continental evolution and tectonics.
Molnar, Peter and Paul Tapponnier, "The Collision between India and Eurasia." *Scientific American,* August, 1977.
    For 40 million years, the Indian subcontinent has been pushing northward against the Eurasian land mass.

648

Pollack, Henry N., and David S. Chapman, "The Flow of Heat from the Earth's Interior." *Scientific American,* April, 1977.

The heat-flow pattern of the earth is interpreted in terms of plate tectonics.

Wyllie, Peter J., "The Earth's Mantle." *Scientific American,* March, 1975.

Although the surface of the earth is shaped by the action of the mantle, the mantle is inaccessible. Nevertheless, much has been learned about it by indirect means.

THE EARTH'S HYDROSPHERE— OCEANOGRAPHY

22–1 INTRODUCTION

Earth's abundant supply of water make it unique among the planets. The large planets—Jupiter, Saturn, Uranus, and Neptune—have only a small, solid core surrounded by enormous atmospheres. Among the terrestrial planets, Mercury has practically no atmosphere and the side facing the sun is hot enough to melt lead. Venus has a thick atmosphere and a surface that may be even hotter than the surface of Mercury. Mars and the moon have a surface marked with craters formed by meteorites and perhaps by volcanoes. Since much of the surface of the earth is covered by a liquid shell, or *hydrosphere,* Earth is called the "water planet." Without the oceans the earth would probably be as desolate as the surface of the moon. *Oceanography* is the application of techniques and principles from many sciences—chemical, geological, physical, biological, and meteorological—to the study of the sea.

22–2 THE WORLD OCEAN

The ocean waters cover 4/5 of the Southern Hemisphere and more than 3/5 of the Northern Hemisphere, in all about 71 percent of the earth's surface. They are joined together in essentially a world ocean containing about 1350 million cubic kilometers of water at an average depth of 4 kilometers. There are 3 square kilometers of water for each square kilometer of land of the earth's surface area. The continents project like islands above the vast surrounding sea, at an average elevation of close to 1 kilometer (875 meters).

This continuous body of water is subdivided, sometimes arbitrarily, into oceans and seas (Fig. 22–1). The relatively circular Pacific Ocean, covering about a third of the earth's surface, is by far the largest ocean. The S-shaped Atlantic Ocean is about half the size of the Pacific but has more coastline than the Pacific and Indian Oceans combined. The Atlantic and Pacific Oceans are often subdivided at the equator into northern and southern parts. The third main ocean is the Indian Ocean. The Arctic and Antarctic Oceans are sometimes regarded as separate oceans.

Seas are smaller bodies than oceans. Deep seas such as the Mediterranean and the Gulf of Mexico occur between continents. Shallow seas such as the North Sea and Hudson Bay occur within depressed parts of a continent. Marginal seas such as the Bering Sea may be deep or shallow and are separated from the main oceans by island arcs.

Figure 22–1   The hydrosphere.

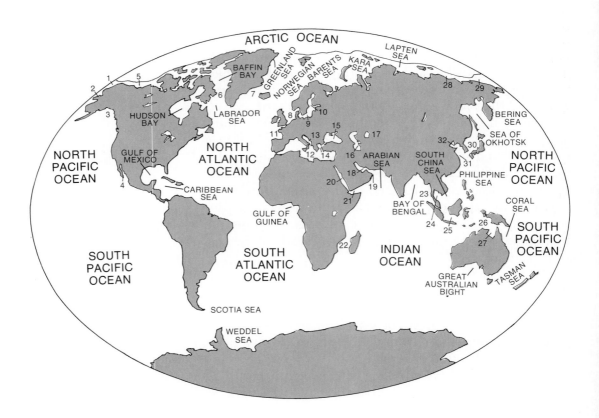

22–3   ORIGIN OF SEAWATER

Unlike the moon or Mercury, which were too small gravitationally to retain their hydrospheres or atmospheres, or Mars, which held on to only a trace of each, the earth was able to keep its mantle of water and air. Yet gases such as neon and argon were almost completely lost. The earth's water was bound up in more stable compounds. Several mechanisms have been proposed for the origin of water and other "excess volatiles"—substances such as carbon dioxide, chlorine, nitrogen, and sulfur—that are more abundant in the ocean and atmosphere than can be accounted for by the weathering of rocks alone.

Most scientists consider the degassing of the earth's interior, as proposed by William Rubey (b. 1898), of the United States Geological Survey and the University of California, to have been the best mechanism for producing the excess volatiles, and thereby the atmosphere and the ocean. Soon after it was formed, the earth cooled and solidified. When molten magma cools, solid minerals are formed. During the crystallization process, some of the gaseous constituents are freed. Gaseous and volatile substances, including water vapor, escaped to the surface in a process known as degassing. The rate at which this process occurred is not known. One possibility involves rapid degassing followed by reaction between the ocean and the earth's crust. Or, degassing may have been slow, with the ocean evolving over most of the earth's history.

The water vapor released from the surface of the earth was caught in the gaseous envelope surrounding it. As the earth's crust continued to cool, water condensed and fell to the earth's irregular surface as a liquid, accumulating in depressions and forming the oceans. The release of water into the oceans was gradual during the first billion years of earth's history. In addition, oceans received water from volcanoes and geysers. For the past 1.5 billion years, the oceans have essentially maintained their present chemical composition.

22–4   SALT IN THE SEA; DESALINATION

For billions of years chemical matter dissolved from the land has been deposited in the sea. The

Table 22–1 MAJOR CONSTITUENTS OF
DISSOLVED SOLIDS IN SEA WATER

ION		PERCENTAGE BY WEIGHT
Chloride	Cl^-	55.06
Sodium	Na^+	30.58
Sulfate	$SO_4^=$	7.68
Magnesium	Mg^{++}	3.74
Calcium	Ca^{++}	1.20
Potassium	K^+	1.12
Bicarbonate	HCO_3^-	0.41
Bromide	Br^-	0.19

evaporation of 100 kg of sea water yields about 96.5 kg water. The rest, 3.5 kg, consists of dissolved solids such as common salt (NaCl) and small amounts of dissolved gases such as oxygen (O_2). If all of the oceans could be evaporated into space, enough salt would remain to cover the earth with a layer close to 70 m in thickness.

Certain elements are more abundant than others in seawater because some substances are more soluble than others. For example, sodium chloride (NaCl) is very soluble. But calcium chloride ($CaCl_2$), as in limestone and clam shells, is much less soluble, and silicon dioxide (SiO_2), as in quartz and glass, is almost insoluble.

From 1872 to 1876, the ship H.M.S. *Challenger* carried scientists on a round-the-world cruise to study the biology of the sea, collect water samples and bottom samples, and measure water temperatures. Over a nine-year period, William Dittmar (1833–1892), a chemist at the University of Glasgow, analyzed 77 water samples taken on this voyage. He determined the percentage composition of the dissolved solids, such as chloride, sulfate, and sodium, in each of the samples. In another study, Georg Forchhammer (1794–1865), a geologist at the University of Copenhagen, spent 20 years analyzing 200 samples of seawater collected for him by friends, ship captains, and naval officers.

These analysts and others reached essentially the same conclusion: about 95 per cent of the dissolved solids are made up of only six major elements (Cl, Na, Mg, S, Ca, K) (Table 22–1). In fact, 86 per cent of all dissolved salts consist of only two elements, sodium and chlorine. At least 35 other elements have also been detected in seawater. Although actual concentrations of dissolved substances may vary from place to place, the ratio of any one of the major elements to the total dissolved solids is nearly constant. This illustrates the effectiveness of the mixing processes of elements within the oceans on a worldwide basis.

The increasing demand for fresh water is focusing attention on the sea. Hundreds of desalination plants are in operation throughout the world, producing hundreds of millions of gallons of pure water per day. The oldest desalination technique is distillation. Seawater is evaporated using solar energy, and fresh water is condensed. More elaborate methods than solar stills are also used, but they are more energy-intensive.

22–5 WHY IS THE OCEAN BLUE?

Although intense sunlight strikes the ocean surface, the amount of light that enters the water depends upon such factors as the sun angle, the surface condition, and the clarity of the water. Light traveling through the water becomes progressively dimmer through absorption and scattering by particles in the water and the seawater itself (Fig. 22–2). Less than 50 percent penetrates below 1 meter, and only 2 percent may be left at a depth of 30 m. Thus the sea is a dark world.

Most of the red light is filtered out in the upper few

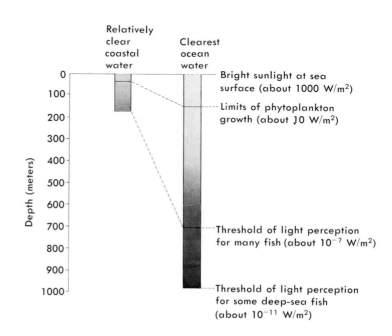

Figure 22–2 Light penetration in clear coastal water and the clearest ocean water on a bright sunny day. (From McCormick, J. M., and Thiruvathukal, J. V., *Elements of Oceanography*. Philadelphia: W. B. Saunders Company, 1976, p. 85.)

Relatively clear coastal water
Clearest ocean water

Bright sunlight at sea surface (about 1000 W/m^2)

Limits of phytoplankton growth (about 10 W/m^2)

Threshold of light perception for many fish (about 10^{-7} W/m^2)

Threshold of light perception for some deep-sea fish (about 10^{-11} W/m^2)

Depth (meters)

meters. Blue light penetrates the deepest, more than 500 m in the clearest ocean water, followed by green light. The color of seawater is due to the light that is reflected from the surface and the light that is scattered within the water. Since blue and green penetrate the deepest, they are scattered the most. The color of the sea is deep blue in the clearest open ocean water. It is green or yellowish in turbid coastal water that contains a considerable amount of suspended particulate matter.

22-6   HEAT IN THE SEA

On the whole, the ocean is a tremendous reservoir of cold water that is not much above freezing. More than 99 per cent of the heat entering the sea comes from the absorption of sunlight and its conversion into heat. Since most light is absorbed near the surface, the greatest warming occurs here. Warm water tends to stay at the surface, being less dense than cold water, resulting in a stratification of warm surface water over cold deeper water.

Sea-surface temperatures are highest near the equator and lowest near the poles. They are measured directly using an ordinary thermometer in a sample of water scooped in a bucket or a Nansen bottle (Fig. 22–3), named for Arctic explorer and oceanographer Fridtjof Nansen (1861–1930). Another method is used with artificial weather satellites. Photographs taken with infrared-sensitive film show differences in ocean shading that are directly related to the sea-surface temperature.

The temperature of a deep-water sample is taken with a special thermometer attached to a Nansen bottle. The thermometer in the bottle is lowered into the water upside down and then reversed at the desired depth, registering the temperature at that depth. The mercury column breaks, leaving a column of mercury which is proportional to the temperature. After the temperature is read on board ship the thermometer is tipped back to the original position. The mercury column is rejoined and can be used again. Another instrument, the bathythermograph, (bathys, deep), using a coated glass plate, records the temperature and pressure continuously from the surface to the desired depth (Fig. 22–4).

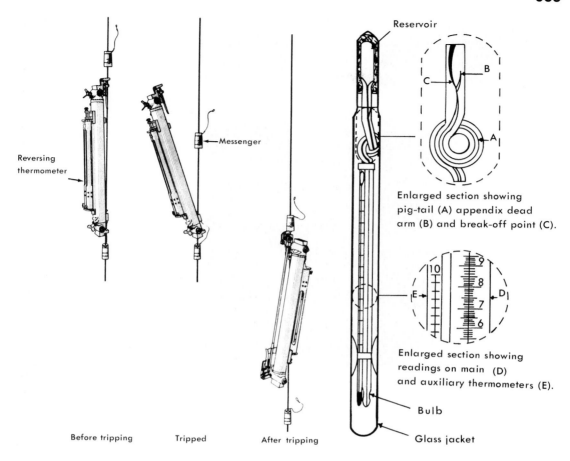

Reservoir

←Messenger

Reversing thermometer

Enlarged section showing pig-tail (A) appendix dead arm (B) and break-off point (C).

Enlarged section showing readings on main (D) and auxiliary thermometers (E).

Bulb

Glass jacket

Before tripping   Tripped   After tripping

(in reversed position)

Figure 22–3 Nansen bottle with reversing thermometer to record temperature of a deep-water sample. (From McCormick, J. M., and Thiruvathukal, J. V., *Elements of Oceanography.* Philadelphia: W. B. Saunders Company, 1976, p. 98.)

Sea-surface temperatures vary from about −2°C (ice) around Antarctica and Greenland to about 30°C (85°F) at the equator. The average is about 16°C (60°F). The temperature of the uppermost layer is relatively constant, changing little between day and night and between summer and winter. It is called the mixed layer, from the effects of wind stirring and radiation cooling, and may be up to 250 m deep (Fig. 22–5). The *thermocline*, a zone of rapid decrease in temperature with depth, extends to a depth of 1.2 to 1.4 km. Below this level the temperature decreases only slightly to the bottom. At a depth of 1.6 km and lower, the temperature of all ocean water is less than 4°C (39°F). Thus, most seawater is only slightly above freezing, even the bottom water in tropical latitudes.

The amount of heat entering the sea, most of it from solar radiation with minute amounts from the

A

B

Figure 22–4 The bathy-thermograph plots temperature changes with depth. (From McCormick, J. M., and Thiruvathukal, J. V., *Elements of Oceanography.* Philadelphia: W. B. Saunders Company, 1976, p. 99; photo courtesy of NOAA.)

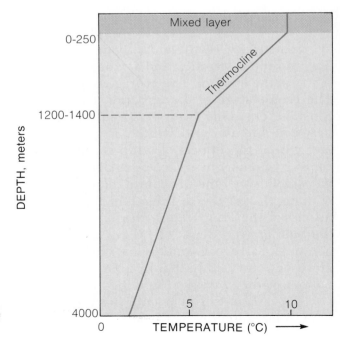

Figure 22–5 Temperature distribution in the oceans. The oceans are a reservoir of cold water.

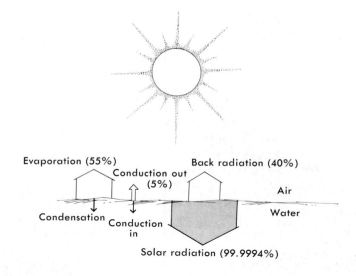

Figure 22–6 Heat budget of the ocean. (From McCormick, J. M., and Thiruva-thukal, *Elements of Oceanography*. Philadelphia: W. B. Saunders Company, 1976, p. 90.)

earth's interior, is in balance with that leaving the sea (Fig. 22–6). Evaporation accounts for 55 per cent of the outgoing heat, radiation 40 per cent, and conduction 5 per cent. The heat budget equation for the oceans relates these aspects of heat transfer as follows:

Solar Radiation = Evaporation + Re-radiation + Conduction   Equation 22–1

22–7 SOME TOOLS OF THE OCEANOGRAPHER

Recent advances in oceanography have come from the use of advanced instrumentation. Echo-sounding instruments—sonar—give a continuous trace of ocean depths. Topographic features of the sea floors, and even materials or structures several kilometers below them, are recorded in detail (Fig. 22–7).

Probing devices lowered from oceanographic ships photograph and sample the sea floors. Long "cores" of sediment are removed from depths of several kilometers, revealing details of sea-surface temperature and salinity spanning millions of years.

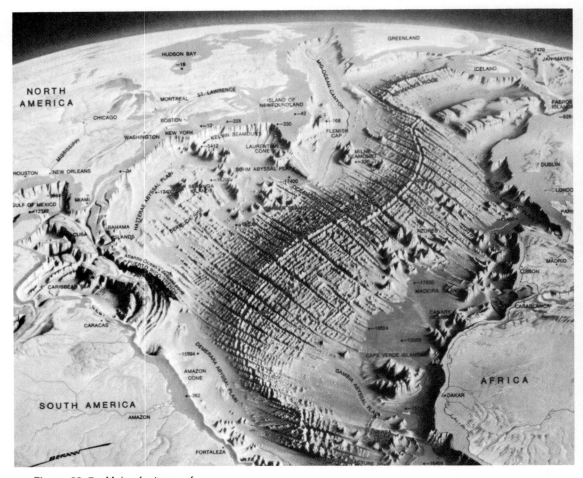

Figure 22-7 Major features of the Atlantic Ocean. (From McCormick, J. M., and Thiruvathukal, J. V., *Elements of Oceanography*. Philadelphia: W. B. Saunders Company, 1976, p. 47; painting by Heinrich Berann, courtesy of ALCOA.)

Radioactive decay methods applied to the dating of lavas indicate the age of different parts of the sea floor.

More recently, direct human observation of the sea has yielded valuable information about its composition and deep-sea life. Diving operations range from simple scuba equipment used to explore coral reefs to the deep-sea diving saucers developed by explorer Jacques-Yves Cousteau (b. 1910). People have been lowered into some of the greatest depths in the seas. Others have drifted along while embedded in the flow of deep ocean currents.

22-8 THE CONTINENTAL MARGINS

The edges of the continents are covered by seawater. There are two zones that vary in slope, depth, and width: (1) the gently sloping continental shelf, and (2) the steeper continental slope descending to the deep-sea floor.

During the last glacial period, the continental shelves were largely dry land. About 70 percent of the area, as a result, is covered by terrestrial deposits laid down on river and coastal plains 10 000 to 25 000 years ago. Thirty percent is covered with marine deposits— muds, silts, biological sediments such as shell debris, and chemical precipitates. Sand deposits occur near the shore zone to depths of 10 to 20 m. The continental shelf varies from zero to 1500 km in width, with an average of 80 km, and 20 to 550 m in depth, with an average of 133 m.

The steeper continental slope extends from the continental shelf to the deep-sea floor, merging with the floor at depths of 3000 to 4500 m. Sixty percent of the material of the slope surface is mud; 25 percent is sands, 10 percent is rock and gravel, and 5 percent is shell beds and organic oozes.

The most conspicuous features of the continental slope are submarine canyons, which are V-shaped, rock-walled, winding valleys. Most of the large canyons are near the mouths of large, ancient rivers. The Monterey Canyon off California, the largest, is deeper and wider than the Grand Canyon. Some submarine canyons owe their origins to turbidity currents—fast-flowing bodies of water loaded with suspended mud and sand. The currents flow downslope along the bottom of the ocean. The suspended particles, under the high speed of the water, can apparently cut canyons in hard rocks. Millions of years are needed for most canyons to form. Although turbidity currents have never been observed directly, there is indirect evidence for their existence.

22–9 UNDER THE DEEP SEAS

Beyond the continental slope, the deep seas account for about 84 percent of ocean area. They were once thought to be entirely floored by flat, featureless surfaces, the sea-floor plains. Instead, the ocean floor is almost as complex as the land surface.

Sea-floor plains are best developed between the continental slope and the mid-ocean ridge at depths of 3000 to 6000 m. Many are flat and featureless and covered with organic oozes and clays. Others form fans from silt and sand derived from rivers with much suspended matter; the largest are beyond the Ganges and Indus deltas. Undersea volcanoes called *sea-mounts* rise abruptly 3 to 4 km above the sea floor. They are present in all oceans, but are most common in the Pacific. Some, like the Hawaiian and the Canary Islands, rise above sea level. Many are found along chains, with the oldest at one end and the youngest at the other.

As the volcanic activity subsides, the peaks are eroded by wave action. Eventually, the flattened sea-mounts sink below sea level and are then called *table-mounts* or *guyots*. The great naturalist Charles Darwin (1809–1882) suggested that coral atolls, ringlike islands made up of coral reefs, are built over guyots (Fig. 22–8). This view is still accepted.

The deepest parts of the ocean, at depths of 7500 to 11 500 m, are oceanic trenches. The Challenger Deep in the Marianas Trench is deeper (11 500 m) than Mount Everest is tall. These long, narrow depressions frequently run parallel to island arcs and to continents. Most trenches are located around the edges of the

Figure 22–8 Origin of an atoll. (From McCormick, J. M., and Thiruvathukal, J. V., *Elements of Oceanography.* Philadelphia: W. B. Saunders Company, 1976, p. 53.)

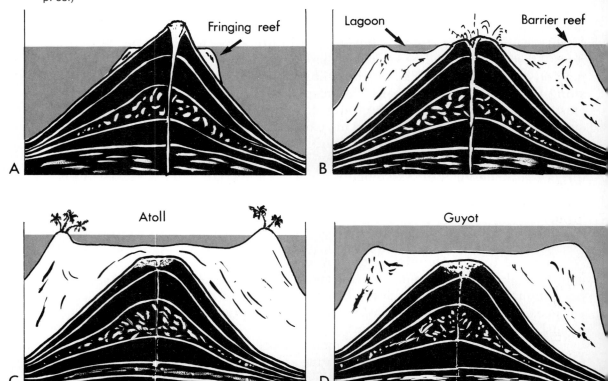

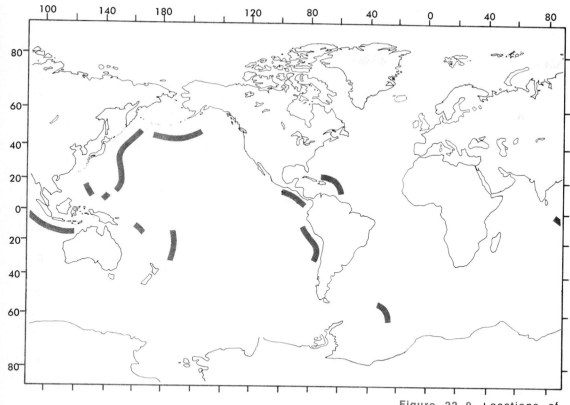

Figure 22–9 Locations of oceanic trenches. (From McCormick, J. M., and Thiruvathukal, J. V., *Elements of Oceanography.* Philadelphia: W. B. Saunders Company, 1976, p. 51.)

Pacific Ocean (Fig. 22–9). Trenches form belts of intense earthquake and volcanic activity, like the mid-ocean ridges, and are closely linked to continental drift.

22–10 OCEAN SEDIMENTS

The most economically important feature of the ocean floor are sediments, matter that settles out of the water onto the floor. Many originate on land and are brought to the oceans by streams or wind. Others are the remains of organisms that live in the oceans.

Ancient sediments are the source of petroleum and various mineral resources. Gold and diamonds are found on continental shelves off land areas rich in these precious minerals. Manganese nodules are just beginning to be mined. They are formed by successive coatings of minerals rich in manganese and other metals precipitated around objects such as rock particles. The world is turning more and more to the oceans as the source of valuable natural resources.

22–11 ENVIRONMENTAL OCEANOGRAPHY

We now have the ability to affect the ocean on a broad scale, either by design or by accident. Industrial and domestic wastes are often dumped directly into the sea. Industrial wastes are often toxic to marine and human life. Domestic sewage provides nutrients for marine plants, including organisms that cause disease. Areas near metropolitan centers become more polluted, but the wastes are not restricted to the area where they are dumped. They are carried by currents often far from the site of intervention. Sewage sludge sinks and pollutes the bottom sediments.

Oil pollution is a particular problem in the coastal environment because oil floats on water and does not mix with it. Thousands of tons of crude oil are released into the sea off the coasts in large-scale oil spills from tankers. They pollute beaches and kill bird and marine life. With the increase in off-shore drilling for oil, rigorous safeguards will have to be instituted if accidents that cause pollution are to be kept at a minimum.

CHECKLIST

Bathythermograph	Nansen bottle
Continental shelf	Oceanic trench
Continental slope	Oceanography
Degassing	Seamount
"Excess volatiles"	Sediment
Guyot	Submarine canyon
Heat budget equation	Tablemount
Hydrosphere	Thermocline
Manganese nodule	Turbidity current

EXERCISES

22–1. Why is earth called the "water planet?"

22–2. List some tools of the oceanographer.

22–3. Discuss the interdisciplinary nature of oceanography.

22–4. What are the oceans used for?

22–5. How does the average depth of the oceans compare with the average elevation of the continents?

22–6. What is meant by the "heat budget" of the ocean?

22–7. Why is the sea cold?

22–8. Why is the sea dark?

22–9. How are submarine canyons formed?

22–10. How are continental shelves different from continental slopes?

22–11. What are some features of the deep sea bottom?

22–12. How are seamounts and tablemounts related?

22–13. Where is the youngest part of the ocean floor?

22–14. What can be done to reduce marine pollution?

22–15. What are the attractions of off-shore drilling and mining?

22–16. *Multiple Choice*

   A. The oceans cover about what percentage of the earth's surface?
   (a) 25   (b) 50   (c) 70   (d) 90

   B. Undersea volcanoes are known as
   (a) submarine canyons   (c) sea-floor plains
   (b) seamounts   (d) trenches

   C. A theory that explains the origin of the ocean is
   (a) degassing   (c) desalination
   (b) oceanography   (d) turbidity current

D. A zone of the ocean in which the temperature decreases rapidly with depth is called the

   (a) continental shelf   (c) thermocline

   (b) continental slope   (d) heat budget

E. Features of the ocean floors can be determined by

   (a) sonar   (c) bathythermo-

   (b) Nansen bottle   graph

                                (d) telescope

SUGGESTIONS FOR FURTHER READING

Bullard, Sir Edward, "The Origin of the Oceans." *Scientific American,* September, 1969.

    The ocean floor is surprisingly young. Growing outward from mid-ocean ridges, it is pushing continents apart.

Heezen, Bruce C., and Ian D. MacGregor, "The Evolution of the Pacific." *Scientific American,* November, 1973.

    Deep-sea drilling shows differences in the floor of the Pacific Ocean. Does the movement of the crust under the basin account for them?

Revelle, Roger, "The Ocean." *Scientific American,* September, 1969.

    Introductory article in an entire issue devoted to the topic.

THE EARTH'S ATMOSPHERE AND BEYOND— METEOROLOGY

23–1 INTRODUCTION

In the most general terms, the earth's atmosphere —the gaseous envelope surrounding the earth—is divided into lower and upper regions. The lower atmosphere is usually considered to extend to the top of the stratosphere, an altitude of about 50 km. Everything above that is the upper atmosphere. The lower and upper atmospheres, in turn, are commonly subdivided on the basis of temperature distribution into the troposphere, tropopause, stratosphere, stratopause, mesosphere, thermosphere, and exosphere (Fig. 23–1).

Weather refers to the state of the atmosphere for a short period of time, a day or so. It is described in terms of temperature, humidity, clouds, precipitation, pressure, and wind. *Meteorology* is the science of the atmosphere and of weather, and depends upon an understanding of basic natural laws. Rockets, balloons, satellites, and ground stations are all used to gather weather information, and sophisticated computers perform in minutes calculations that might otherwise take months. Despite modern instruments and methods, weather forecasting is at best an inexact though improving science.

23–2 THE TROPOSPHERE

The lowest region of the atmosphere extends to a height of about 18 km over the equator and 8 km over

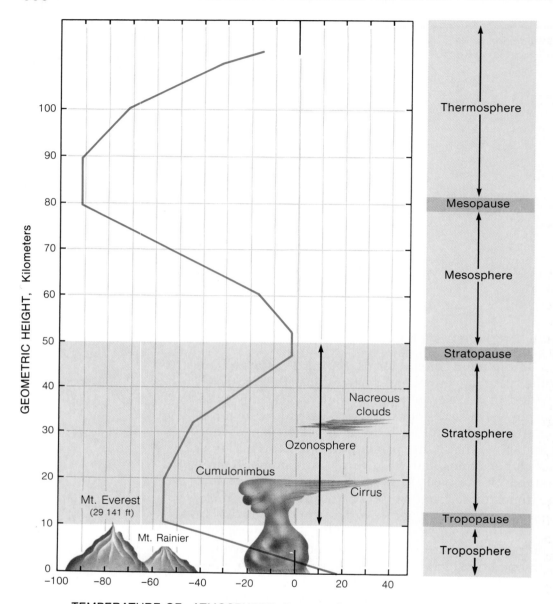

Figure 23–1 Thermal Structure of the atmosphere. (U.S. Navy Research Facility.)

the poles, with an average height of 11 km. It is a turbulent region of swirling air masses, cloud formations, warm and cold fronts, and storms. It is the source of the air we breathe. Two gases, nitrogen and oxygen, account for 99 per cent of dry, clean air, remaining essentially unchanged with increasing altitude (Table 23–1).

The water-vapor content of the troposphere varies from about 20 grams per kilogram of air in the tropics

Table 23–1   COMPOSITION OF UNCONTAMINATED
DRY AIR NEAR SEA LEVEL

COMPONENT	MOLE, %
Nitrogen	78.084
Oxygen	20.947
Argon	0.934
Carbon dioxide	0.0314
Neon	0.00182
Helium	0.000524
Krypton	0.000114
Hydrogen	0.00005
Xenon	0.0000087

to less than 0.5 g/kg at the poles. Dust, smoke, salt, pollen grains, volcanic ash, organic materials, and a variety of other solids are also present. They are involved in cloud and fog formation and rainfall, acting as nuclei around which water droplets and ice crystals develop. Water vapor and carbon dioxide strongly absorb infrared radiation from the sun and earth, and play a key role in maintaining the earth's heat balance by holding in the heat of the earth's lower atmosphere through the "greenhouse" effect. The atmosphere is transparent to incoming visible radiation but is opaque to the infrared radiation (heat) that is emitted by the surface of the earth; such radiation cannot escape back into space.

The temperature in the troposphere decreases with altitude about 6.5 Celsius degrees per kilometer, falling to an average temperature of −56°C at the top. The vertical temperature change or gradient is called the *lapse rate of temperature*. The main reason for the decrease in temperature is the decrease in air density with increasing elevation. The higher one is above sea level, as on a mountain top, the less dense the air. Since the source of heat is the earth's surface, the dense air at lower levels receives heat by conduction and convection more efficiently than less dense air. Further, water vapor and carbon dioxide, the chief infrared absorbers, are more concentrated at lower levels.

23–3   THE STRATOSPHERE;
THE OZONE LAYER

The *stratosphere* extends from an altitude of 11 km to 50 km. It was so named in the belief that it consisted of an arrangement of layers ("strata") that did not mix. The temperature, contrary to popular belief at the turn of the century, stops falling at about 11 km and remains constant with increasing altitude up to 20 km. In the range from 20 km to 50 km the temperature increases progressively, reaching a maximum of about −20°C at 50 km.

Except for *ozone*, the 3-atom molecule of oxygen, O_3, the chemical composition of the stratosphere is essentialy constant. The concentration of ozone, however, increases from about 0.04 parts per million in the troposphere to a maximum of about 10 parts per million at an altitude of 25 km to 30 km in the stratosphere. Although it is only a trace constituent, ozone exerts a profound influence on life on earth by absorbing the sun's ultraviolet radiation from 2100 Å to 2900 Å. Were this radiation not absorbed, many forms of life would cease to exist and others would be imperiled. In the process of absorbing ultraviolet radiation and decomposing into diatomic oxygen molecules and oxygen atoms, ozone releases heat to the stratosphere, enabling the atmosphere to maintain a heat balance. The temperature then decreases progressively from −20°C at 50 km to approximately −92°C at 85 km. In this region, the *mesosphere*, the ozone concentration becomes lower and lower.

$$O_2 \xrightarrow{\text{UV}} O + O \text{ (Dissociation of oxygen molecules)}$$

$$O + O_2 \longrightarrow O_3 \text{ (Formation of ozone)}$$

$$O_3 \xrightarrow{\text{UV}} O_2 + O \text{ (Dissociation of ozone)}$$

Chlorofluorocarbons, such as $CFCl_3$ and CF_2Cl_2, also known as Freons or halocarbons, are gases that were widely used as propellants in aerosol cans and as refrigerants. A theory was advanced in 1974 that these compounds represent a definite hazard to the earth's ozone layer. They are ideal as propellants to spray cosmetic products, cleaning agents, and some medications because they do not react chemically with whatever is being sprayed. But this same chemical inertness causes them to persist in the atmosphere, making their way into the stratosphere where they are broken down by ultraviolet light, releasing free chlorine atoms that react with and deplete the ozone.

$$CFCl_3 \xrightarrow{UV} CFCl_2 + Cl \text{ (Dissociation of Freon-II)}$$

$$Cl + O_3 \longrightarrow ClO + O_2 \text{ (Reaction of chlorine atom with ozone)}$$

A panel of scientists concluded that the continued release of halocarbons would eventually produce a reduction between 2 and 20 percent in stratospheric ozone. Such a reduction would lead to an increase in the amount of damaging ultraviolet radiation reaching the surface of the earth. This would lead, in turn, to a larger increase in all forms of skin cancer, and would also have harmful effects on plants and animals. As a result, the use of halocarbons as propellants was banned, except in cases of special need, as in medicine.

23-4   THE IONOSPHERE; RADIO TRANSMISSION

The *ionosphere* extends from about 50 km to thousands of kilometers beyond the earth's surface. It is a region that has a relatively large concentration of free electrons and positively charged ions, formed by the absorption of the sun's ultraviolet radiation and to a lesser extent by X rays, and by collisions with cosmic rays, which strip electrons from molecules and atoms. The concentrations of charged particles are still so low, however, that they do not recombine to a significant extent.

The existence of the ionosphere was independently proposed by Arthur E. Kennelly (1861–1939), an electrical engineer, and Oliver Heaviside (1850–1925), a physicist, to explain engineer and inventor Guglielmo Marconi's (1874–1937) successful transmission of radio waves over long distances around the earth. Since radio waves travel in straight lines, it was thought that they would tend to be lost in space rather than follow the curvature of the earth. Kennelly and Heaviside explained that long-distance radio transmission occurs because an electrically conducting region in the upper atmosphere reflects radio waves back to earth, and their idea was confirmed experimentally. When a radio wave enters the ionosphere, it induces a current among the charged particles there. The upper part of the wave front, being in a more highly charged area, is speeded up. The wave pivots and thus is bent down-

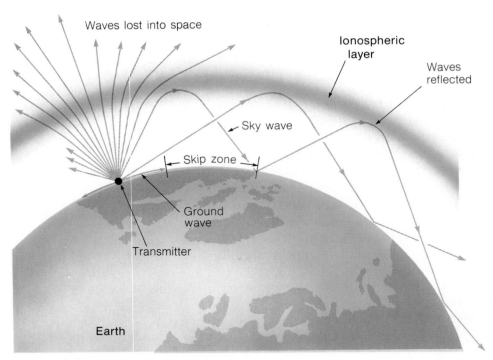

Figure 23–2 Reflection of
radio waves from the ionosphere
(Kennelly-Heaviside layer). (Re-
drawn from Strahler, A. N., *The
Earth Sciences.* New York:
Harper & Row, 1963, p. 197.)

ward. The repeated bouncing of the radio wave between the ionosphere and the ground accounts for the
reception of radio signals around the curvature of the
earth (Fig. 23–2).

The ionosphere is divided into D, E, F_1, and F_2
regions according to increasing altitude and electron
concentration, the electron concentration depending
upon a number of factors, such as time of day, season of
the year, and phase of the solar cycle. The D region,
for example, between 50 km and 85 km has a maximum
concentration at noon of about 1.5×10^4 electrons/cm³,
which almost vanishes at night; this is why radio reception from distant stations is usually better at night than
during the day. The electron concentration reaches a
maximum of about 1×10^6 electrons/cm³ at an altitude
of 250 km, then gradually falls off.

23–5   VAN ALLEN BELTS AND AURORAS

A region called the *thermosphere* overlaps part of
the ionosphere, beginning at an altitude of about 85 km
and extending to an average altitude of about 500 km.
In the lower thermosphere, the temperature increases
rapidly with increasing altitude as molecular oxygen

absorbs solar radiation up to 2000 Å and dissociates with the release of heat. Although the temperature in the upper thermosphere averages about 1300°C, the low concentration of matter in this region does not cause the gas to be hot in the conventional sense of being hot to the touch. With increasing altitude, the number of oxygen atoms exceeds the number of oxygen molecules as a result of dissociation until, at 400 km, less than 1 per cent of the oxygen is in the molecular form. Nitrogen, on the other hand, is not easily dissociated by solar ultraviolet radiation and remains predominantly diatomic throughout the upper atmosphere.

The outermost region of the atmosphere, above an average altitude of 500 km, is known as *exosphere*. Atomic oxygen is the most abundant atom in the lower exosphere, and helium and atomic hydrogen at higher altitudes, although the concentrations of all are very low (Fig. 23–3). Atomic hydrogen, the main constituent of the upper exosphere, is produced mainly by the dissociation of water vapor and methane by ultraviolet light at altitudes below 100 km. The light hydrogen atoms diffuse rapidly upward to the top of the atmosphere, where many are freed from the earth's gravitational pull and escape. The ions of hydrogen, helium, and oxygen are also present in the exosphere, but cannot escape, being confined by the earth's magnetic field. (The problem of how some helium manages to acquire sufficient energy to escape has not yet been solved.)

With the advent of artificial satellites, a region of the atmosphere called the "magnetosphere" has been recognized. Extending outward for thousands of miles, this region consists of a thin, electrified gas of electrons and protons that gyrates along the lines of force of the earth's magnetic field and forms a vast region of radiation around the earth (Fig. 23–4). The Van Allen belts, named for James A. Van Allen (b. 1914), a physicist at the University of Iowa, were discovered and mapped with instrumentation placed aboard the first American satellites. A belt of high-energy protons is trapped in the region of the magnetic field about 3200 km above the earth at the magnetic equator. Another belt, containing high-energy electrons, surrounds the magnetic equator about 16000 km out from earth. The earth's magnetic field acts as a trap for charged particles, deflecting them from their course and causing

Figure 23–3   The composition of the atmosphere. Information is based upon the Explorer satellite program of the National Aeronautics and Space Administration. (Courtesy of NASA.)

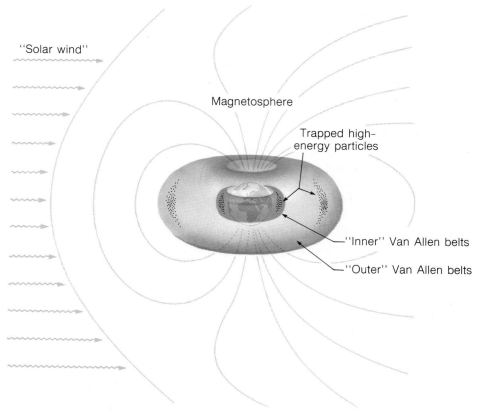

"Solar wind"

Magnetosphere

Trapped high-
energy particles

"Inner" Van Allen belts

"Outer" Van Allen belts

Figure 23–4 Trapped radiation in earth's magnetosphere.
Original "inner" and "outer"
Van Allen belts hold high-energy
protons and electrons, respectively.

them to pursue a spiral path around the lines of force.
According to one theory, cosmic-ray protons collide
with atmospheric nuclei, producing neutrons; the neutrons stream back into the magnetosphere where they
decay into protons, electrons, and neutrinos. The
charged protons and electrons spiral around the magnetic lines of force and move around the earth. The
outer regions of the magnetosphere interact with hot,
ionized gas streaming from the sun, generating auroras
and magnetic storms that disrupt communications.

The spectacular phenomena called *auroras* are
produced directly or indirectly by fast-moving electrons and protons from the sun that are deflected, because of their electrical charge, toward the magnetic
poles by the earth's magnetic field. They are observed
most frequently near the magnetic poles, generally at
magnetic latitudes from 60° to 70°. Auroras take a great
variety of forms and colors, and usually occur at altitudes from 75 to 400 km. The colors are produced when
solar electrons excite the atoms and molecules of the

earth's atmosphere, causing them to emit light. Excited atomic oxygen emitting at 5577 Å accounts for the green; atomic oxygen emitting at 6300 Å and 6364 Å accounts for the red; and excited ionized molecular nitrogen is the source of the blue and violet.

23–6   WEATHER LORE

Long years of human observation of weather patterns have produced a body of weather superstitions that are part of our folklore. Although some have a grain of truth, they are for the most part often broad generalizations that have limited accuracy:

Red sky in the morning, sailor's warning;
Red sky at night, sailor's delight.

Rain long foretold, long last;
Short notice, soon past.

Rain before seven, shine before eleven.

A rainbow in the morning is the shepherd's warning,
A rainbow at night is the shepherd's delight.

But I know ladies by the score
Whose hair, like seaweed, scents the storm;
Long, long before it starts to pour
Their locks assume a baneful form.

23–7   OBSERVING THE WEATHER

A wide variety of instruments has been developed for the study of the atmosphere. *Wind vanes* are the simplest instruments for determining wind direction. They usually consist of a thin metal vane that is moved by the wind and points in the direction from which the wind is blowing. Some produce an electrical current that is converted to a reading on a dial in a weather station.

An *anemometer* is a series of cups attached to a vertical shaft that measures wind velocity on an indicator. The wind strikes the cups, causing them to spin and vary an electrical current that gives wind velocities (Fig. 23–5). The wind vane and anemometer are combined in the *aerovane*.

Pilot balloons are the most common method of gathering upper-air data, to 20 km. These helium-filled bags are carried aloft and moved by winds and air currents. Their motion can be tracked by radar. A *radio-*

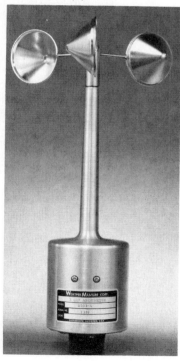

Figure   23–5 Three-cup anemometer. As it rotates, it varies an electric current that drives a wind speed indicator. (From Navarra, J. G., *Atmosphere, Weather, and Climate.* Philadelphia: W. B. Saunders Company, 1979, p. 155; photo courtesy of Weather Measure Corp.)

sonde is an instrument package carried by a balloon that measures temperature, pressure, and humidity with miniature electronic devices (Fig. 23–6). These data are radioed continuously to a receiver at a ground station. At a predetermined level, the balloon bursts and the radiosonde is parachuted to earth. A *dropsonde* is a variation of a radiosonde released by aircraft to record data as it falls through the air.

Rocket-borne instrument packages are used at heights between 30 and 90 km above the earth's surface. The electronic equipment measures such variables as pressure, temperature, wind, and density. *Radar* is used for tracking storms and weather patterns over large areas. *Computers* process vast amounts of data in a fraction of the time required by people and draw synoptic weather maps almost instantaneously.

Air temperature is the most fundamental of the weather elements. Variations in atmospheric heat energy are the ultimate driving force of all weather

Figure 23–6 Balloon and radiosonde for gathering upper-air weather data. The instruments measure temperature, pressure, and humidity. A transmitter sends the data to a ground station. (From Navarra, J. G., *Atmosphere, Weather and Climate.* Philadelphia: W. B. Saunders Company, 1979, p. 41; photo courtesy of NCAA.)

changes. The *thermograph* makes a continuous record of temperature. Changes in temperature cause changes in the shape of a bimetallic element or liquid-filled tube that are magnified by a lever arrangement to which a pen arm is connected. Minute temperature changes of 0.01°C or less can be measured with instruments containing a *thermistor*, a metallic unit whose electrical resistance varies with temperature. They are also involved in instrument packages designed to telemeter to earth the temperatures of the surface of the planets and the moon.

The *sling psychrometer* measures relative humidity, the ratio of the actual amount of water vapor present in the air to the capacity of the air to hold water vapor (Fig. 23–7). It consists of two thermometers, one of which is covered with muslin that is moistened when used. Evaporation from the wet bulb lowers its temperature below that of the dry bulb, the extent depending on the relative humidity, which is read from standard meteorological tables. A *hygrograph* (hygros, moist) keeps a permanent record of relative humidity by magnifying the changes in length of blond human hair that is particularly sensitive to water vapor in the air.

Pressure distribution is a primary characteristic of weather systems migrating over the earth's surface. The *mercury barometer* is the standard instrument for measuring air pressure. The mercury rises to a height such that its weight balances the weight of a column of air on an area of the well surface equal to the area of the tube. As the air pressure changes with changes in weather, the height of the mercury column changes. Air pressure is stated in length, or pressure units. At sea level the average height of the mercury column is 76 cm (29.92 in). The *bar* is the common pressure unit (10^6 dynes per square centimeter). Average sea-level pressure is 1.0132 bars or 1013.2 millibars.

The *aneroid barometer* is portable and easier to read than a mercury barometer (Fig. 23–8). It contains an evacuated chamber supported by ribbed walls. Variations in air pressure cause the chamber to expand or contract and are reflected by a dial indicator through gears and levers. The face of the dial is calibrated in centimeters, inches, or millibars. A *barograph* is a recording aneroid barometer that produces a continuous record of pressure.

One type of *rain gauge* consists of a cylinder 10 or

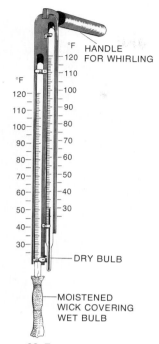

°F HANDLE FOR WHIRLING

DRY BULB

MOISTENED WICK COVERING WET BULB

Figure 23–7 A sling psychrometer is used for the measurement of relative humidity. After whirling, the thermometers are read. The difference in temperature between the wet bulb and dry bulb is proportional to the dryness of the air. (From Navarra, J. G., *Atmosphere, Weather and Climate*. Philadelphia: W. B. Saunders Company, 1979, p. 185.)

Figure 23–8   The aneroid barometer responds to variations in air pressure. (Photo courtesy of Lloyd Black.)

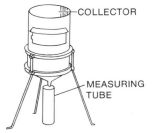

Figure 23–9   A rain gauge. (From Navarra, J. G., *Atmosphere, Weather and Climate.* Philadelphia: W. B. Saunders Company, 1979, p. 205.)

20 cm in diameter, leading into a funnel that collects the water in a long, narrow tube (Fig. 23–9). A calibrated stick indicates the depth of water that would be caught in a straight-sided, flat-bottomed pan. Snow is measured by sampling the snow depth, then converting the average into the equivalent of water, which is about one-tenth the depth of the snow.

A *ceilometer* measures ceiling, the height of the lowest cloud cover of the sky. The device projects a beam of light to the cloud base. The base of the cloud reflects the beam, which is then picked up by a detector that gives a reading of the ceiling or cloud height.

TIROS *weather satellites* (*T*elevision and *I*nfra *R*ed *O*bservation *S*atellites) were first placed in orbit in 1960 to analyze infrared energy and cloud covers. Over the years they produced vast quantities of data that were analyzed by computers. Nimbus satellites, first launched in 1964, photograph still larger areas of the earth's surface. They are larger and more complex weather satellites than TIROS, and can make nighttime photographs of cloud cover. Data recorded on magnetic tape for readout upon command are used in weather analysis and forecasting. Synchronous orbit satellites that keep pace in orbit with the earth's own rotation continuously monitor the atmosphere within their field of view.

23-8   THE HYDROLOGIC CYCLE

The average water-vapor content of the earth's atmosphere is rather constant. Evaporation from oceans, lakes, rivers, and moist soil replaces the moisture lost by rain, dew, snow, sleet and hail. The process of maintaining this steady state is known as the *hydrologic cycle* (Fig. 23–10).

Air is saturated with water vapor at a particular temperature when the maximum possible amount, called the *capacity*, is present. *Specific humidity* refers to the weight of water vapor (grams) per weight of air (kilograms). *Relative humidity* is the ratio of the actual amount of water vapor present in the air to the capacity at that temperature:

$$\text{relative humidity} = \frac{\text{specific humidity}}{\text{capacity}} \times 100\%$$ Equation 23–1

Saturation is achieved primarily by a lowering of the temperature of the air until its capacity is reached. The temperature at which saturation occurs is called the *dew point*. At or below the dew point water vapor is condensed to liquid water, provided that condensation particles or nuclei are present.

Dew is deposited as fine beads of moisture on grass and other matter close to the earth's surface. It forms mainly on clear, calm nights when the ground cools rapidly by radiation, and particularly in summer when there may be considerable water vapor in the air. When the dew point is below freezing, *frost* forms instead through sublimation. *Fog* develops when there is condensation throughout a layer of air near the land or water surface cooled below the dew point. A thick fog blanket called an *advection fog* often results from the movement of warm, moist air over a cooler surface (Fig. 23–11). The famous "pea soup" fogs of Britain occur during the cold season when relatively warm, moist air from the Atlantic is carried over the colder British Isles.

Precipitation occurs when large masses of moist air are cooled rapidly below the dew point. Condensation continues until water drops or ice particles are formed that are too large to remain suspended in air and fall earthward. *Drizzle* is precipitation composed of tiny droplets, each less than 0.5 mm in diameter. *Rain* consists of water droplets from 0.5 mm to 5 mm in

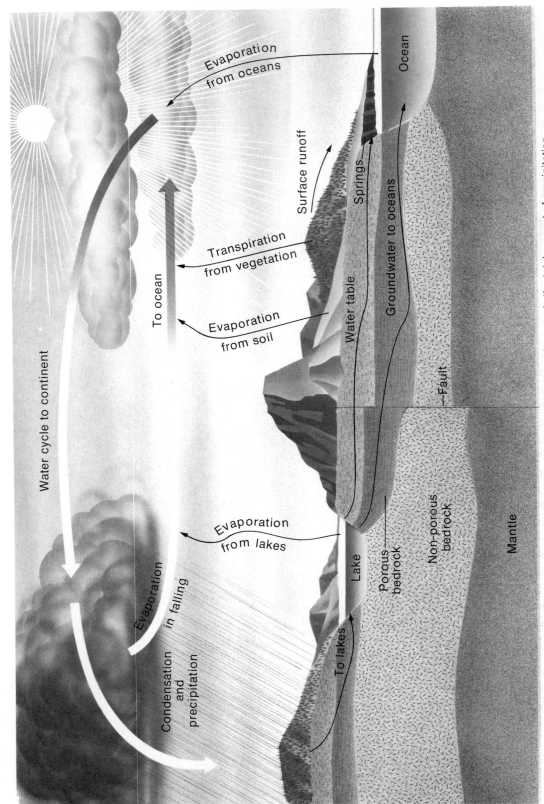

Figure 23-10 The hydrologic cycle. For the world as a whole, the total amount of evaporation equals the total amount of precipitation.

Water cycle to continent

Evaporation from oceans

Ocean

Surface runoff

Springs

To ocean

Transpiration from vegetation

Groundwater to oceans

Evaporation from soil

Water table

Fault

Condensation in falling

Evaporation in falling

Evaporation from lakes

Lake

Porous bedrock

Non-porous bedrock

Mantle

To lakes

Figure 23–11 The Golden Gate Bridge surrounded by an advection fog.

diameter. Above this size the drop breaks apart as it falls. *Snow* is a form of ice in hexagonal (six-sided) crystals. Raindrops falling through cold air may freeze, forming small pellets of ice called *sleet. Hail* consists of rounded pieces of ice, often in concentric layers like the layers of an onion. The ice layers may result from the repeated lifting of hailstones upward through warm, moist air layers. An *ice storm* or *glaze* is a coating of clear ice that forms on branches, wires, and other surfaces. It occurs when rain falls through a layer of cold air close to the ground, causing the droplets to freeze as they touch exposed surfaces. The weight of the ice may cause branches and wires to snap.

23–9 CLOUD TYPES

Clouds are composed of tiny water droplets or ice cyrstals. Those above 6 km high are usually composed of ice crystals because air temperatures there are below freezing. Below this region, most clouds are composed of water droplets. The brilliant snowy appearance of clouds in bright sunlight is due to their high *albedo,* that is, their capacity to reflect sunlight over the entire visible spectrum. Dense clouds appear gray or black because sunlight is absorbed rather than reflected.

Stratiform clouds are clouds formed into layers. *Cumuliform* clouds have a flat base and a massive, globular shape. Clouds are grouped into classes according to height and form, whether stratiform or cumuliform. An international system of classification lists four classes: high clouds, middle clouds, low clouds, and clouds of vertical development (Fig. 23–12).

High clouds above 6 km are cirrus, cirrostratus, and cirrocumulus, all composed of ice crystals (Table 23–2). *Cirrus* clouds are thin and featherlike and usually form streaks called mares' tails. *Cirrostratus* clouds are also thin and often cause a halo around the sun or moon. This is evidence that the particles consist of ice rather than water droplets. *Cirrocumulus* clouds are composed of small cumulus masses called a "mackerel sky."

Middle clouds extend from 2 km to 6 km in height. *Altostratus* clouds form a thick, gray blanket with a smooth underside that shuts out sunlight but causes the sun to be seen as a bright spot through the cloud. *Altocumulus* clouds are grayish cumuliform cloud masses that lie in a distinct layer and are associated with fair weather. Blue sky is visible in breaks between the masses.

Low clouds occur from ground level to a height of 2 km above the earth's surface. *Stratus* clouds form a low, uniform cloud sheet that completely covers the sky. *Nimbo-stratus* clouds are dense, dark gray stratus

Figure 23–12 Classification of cloud forms based on height and vertical development.

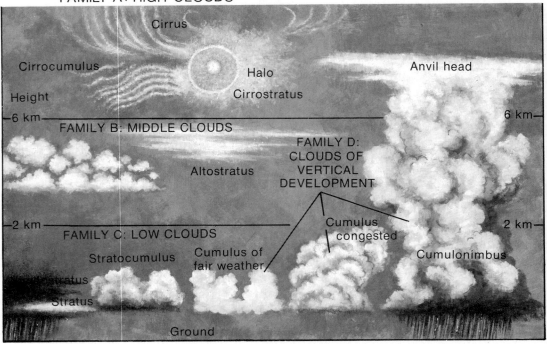

FAMILY A: HIGH CLOUDS

FAMILY B: MIDDLE CLOUDS

FAMILY D: CLOUDS OF VERTICAL DEVELOPMENT

FAMILY C: LOW CLOUDS

Cirrus

Cirrocumulus

Halo

Anvil head

Cirrostratus

Height

6 km

6 km

Altostratus

2 km

2 km

Stratocumulus

Cumulus of fair weather

Cumulus congested

Cumulonimbus

Nimbostratus

Stratus

Ground

Table 23–2 CLOUD TYPES AND WEATHER

ALTITUDE OVER MIDDLE LATITUDE (In Feet)	NAME, ABBREVIATION, AND SYMBOL	DESCRIPTION	COMPOSITION	POSSIBLE WEATHER CHANGES
High Clouds 18 000 to 45 000	Cirrus (Ci)	Marestails Wispy and feathery	Ice crystals	May indicate a storm, showery weather close by
	Cirrostratus (Cs)	High veil Halo cloud	Ice crystals	Storm may be approaching
	Cirrocumulus (Cc)	Mackerel sky	Ice crystals	Mixed signifi- cance, indica- tion of turbu- lence aloft, possible storm
Middle Clouds 6500 to 18 000	Altocumulus (Ac)	Widespread, cotton ball	Ice and water	Steady rain or snow
	Altostratus (As)	Thick to thin overcast; high, no halos	Water and ice	Impending rain or snow
Low-Family Clouds Sea Level to 6500	Stratocumulus (Sc)	Heavy rolls, low, wide- spread Wavy base of even height	Water	Rain may be possible
	Stratus (St)	Hazy cloud layer, like high fog Somewhat uni- form base	Water	May produce drizzle
	Nimbostratus (Ns)	Low, dark gray	Water, or ice crystals	Continuous rain or snow
Vertical Clouds Few hundred to 65 000	Cumulus (Cu)	Fluffy, billowy clouds Flat base, cotton ball top	Water	Fair weather
	Cumulonimbus (Cb)	Thunderhead Flat bottom and lofty top Anvil at top	Ice (upper levels) Water (lower levels)	Violent winds, rain, hail are all possible Thunderstorms

Source: Navarra, J. G., *Atmosphere, Weather and Climate*. Philadelphia: W. B. Saunders Company, 1979, p. 189.

clouds that produce rain or snow. *Stratocumulus* clouds are low cylindrical cloud masses, dark gray on the shaded side but white on the illuminated side. Blue sky is visible between the narrow breaks. They are best developed during a clearing period following a storm.

Clouds of vertical development are all of the cumuliform type. The smallest are the *cumulus of fair weather* clouds, found when the weather is fair and there is much sunshine. They are snow white and cottonlike, with flat bases and round tops. Congested cumulus clouds are larger and denser than small cumulus and have tops resembling heads of cauliflower. They may grow into gigantic *cumulonimbus* clouds, or thunderheads, that extend to heights of 20 km and are the source of violent thunderstorms with rain, hail, wind, thunder and lightning.

23–10   AIR MASSES

We can often detect the presence of air masses by our senses. For example, after suffering through a hot, oppressively sticky summer heat wave we appreciate the arrival of cool, dry air. A large, hot, humid air mass was simply replaced by a cool, dry air mass. A knowledge of air masses helps in understanding our day-to-day weather changes.

An *air mass* is a large body of air that may extend over a large part of a continent or ocean. It is fairly uniform in temperature and humidity at a given altitude level. Vertically, it includes the lower part of the troposphere from ground level to heights of 3 km to 6 km. It usually has different properties than those of adjacent air masses, and has distinct boundaries.

An air mass acquires its properties of temperature and humidity from its *source region*, the land or sea surface over which it originates (Fig. 23–13). Air-mass source regions usually coincide with large bodies of land or water. For example, air over a warm, tropical ocean will be warmed by radiation and will absorb water vapor by evaporation from the sea surface. The entire air mass will acquire a high moisture content. When an air mass breaks loose from its source region, it moves in a generally eastward direction, steered by strong winds in the upper troposphere. Most of the United States does not lie in any air-mass source region but is affected by the passage of air masses from source regions lying both to the north and south.

Cold northern air masses are called *polar air masses*, designated P on weather maps, or *arctic air masses*, A. Depending on whether they form over land or water, the air masses will be dry or humid. This leads

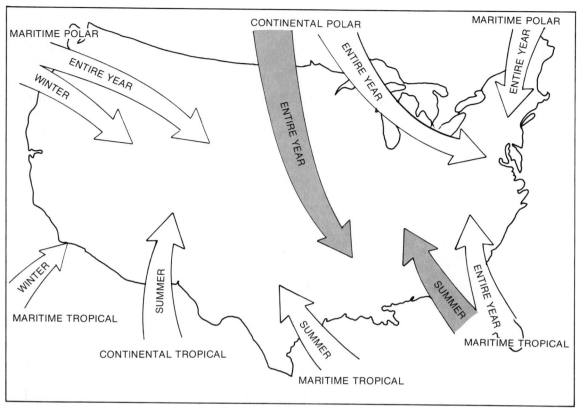

Figure 23–13   Air mass source regions. (From Navarra, J. G., *Atmosphere, Weather and Climate.* Philadelphia: W. B. Saunders Company, 1979, p. 252.)

to two subdivisions, *continental* and *maritime*. The polar and arctic continental air masses (cP and cA) are cold and dry (0.5 g H_2O/kg air); of the two, the arctic air mass is the colder. They originate in a vast source region from Hudson Bay to Alaska. Their influence is greatest in winter, when they bring cold waves to much of the country. Polar maritime air (mP) is of oceanic origin, the North Atlantic, and is cold and moist, bringing drizzle to the northeastern United States in the warmer seasons and often heavy snowfalls from storms called northeasters in the winter.

Warm air masses originating in low latitudes are called *tropical air masses*, T. Maritime tropical air masses (mT) originate in the Caribbean Sea, Gulf of Mexico, and neighboring parts of the Atlantic Ocean. They are warm and moist (15–17 g H_2O/kg air). Their encounter with cP, cA, and mP air masses from the north produces frequent weather disturbances in the central and eastern United States.

23–11   WEATHER FRONTS

Air masses of different properties do not mix easily and tend to be separated by distinct boundaries called *weather fronts.* The movement of one air mass into a region occupied by another is often marked by major changes of weather. Three types of fronts may develop:

   1. Cold Front.   A *cold front* is formed when a cold air mass invades a region occupied by a warmer air

Figure 23–14 A cold air mass invading a region occupied by a warmer air mass. (From Navarra, J. G., *Atmosphere, Weather and Climate.* Philadelphia: W. B. Saunders Company, 1979, p. 267.)

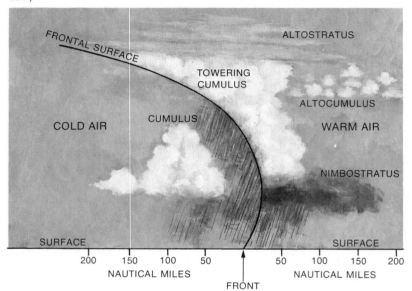

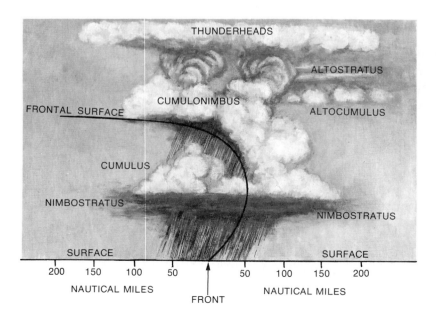

mass. The forward edge of the cold air, being denser, keeps contact with the ground displacing the warmer, lighter air along this line and forcing it to rise upward (Fig. 23–14). The front is the entire surface of contact between the two air masses. The cold front zone often develops a narrow band of cumulus and cumulonimbus clouds from the steeply rising warm air just ahead of it. Cloudiness is normally present, and precipitation if it occurs takes the form of showers and thundershowers. The passage of the front often brings a marked drop in temperature and humidity and a shift in wind direction.

2. Warm Front. A relatively warm air mass moving into a region occupied by colder air forms a *warm front.* The warm air rises over the denser, cold air that is close to the ground over a broad, gently sloping front and forms stratus-type clouds. A long period of cloudiness and precipitation may follow. If the cold air beneath a warm front is below freezing, rain falling through it may form sleet or freeze upon contact with the ground, producing glaze or an ice storm. As the warm front passes, the temperature rises gradually and the wind shifts to the south or west.

3. Occluded Front. A cold front may overtake a warm front and completely lift the warm air off the ground. The warm front is said to be *occluded* and is then located at higher altitudes, where it produces precipitation. A cold front travels fairly rapidly because cold air can more easily displace warm air than vice versa.

23–12 WEATHER MAPS

A weather map gives a broad view of weather conditions over a large region. Specialized maps show specific data such as temperature, pressure, and precipitation of a region (Fig. 23–15).

Frontal systems are represented by lines indicating the character of the air mass. A cold front is drawn with a series of triangles on the side of the line toward which the air is moving. A warm front is drawn with a series of half-circles. An occluded front is shown with alternating triangles and half-circles pointing in the same direction. These symbols and others are designed and set by international agreement.

Equal temperature regions are connected by lines

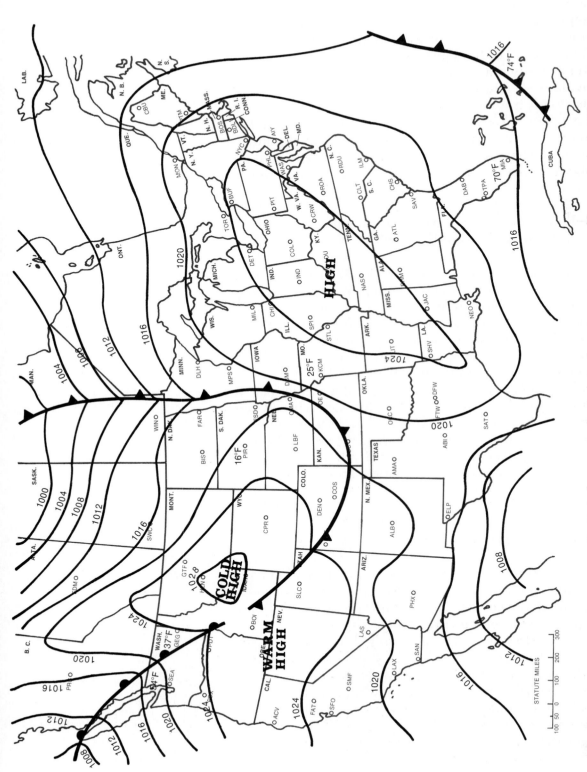

Figure 23–15   A typical daily weather map. (From Navarra, J. G., *Atmosphere, Weather and Climate*. Philadelphia: W. B. Saunders Company, 1979, p. 266.)

called *isotherms*. Regions of equal pressure are connected by lines called *isobars*. When isobars are close together, the horizontal change in pressure, or pressure gradient, is great. When relatively far apart, the pressure gradient is weak. Roughly circular isobar patterns enclosing highs and lows are particularly important features of a weather map. Estimates of the direction of movement of weather systems are possible. *Station models* containing data about cloud cover, temperatures, precipitation, and other variables for different stations throughout the country may also be shown.

23–13 VIOLENT WEATHER— THUNDERSTORMS AND TORNADOES

Thunderstorms are found along both cold and warm fronts. Those along a cold front are often particularly violent. When dissimilar air masses meet, large masses of warm, moist air may be forced to rise several kilometers in altitude (Fig. 23–16). As it rises the air expands, since there is less air pressing on it from above. As the air expands, it cools with no decrease in heat energy; the same quantity of heat is spread over a greater volume, and the temperature therefore drops. The rising convection column of warm, less dense air resembles the updraft in a chimney in a fireplace.

When the air arrives at a level at which the dew point is reached, a cumulus cloud begins to form, then builds rapidly into the cumulonimbus cloud from which heavy showers will fall. Condensation occurs at the upper levels with the release of latent heat and pro-

Figure 23–16 Stages of a thunderstorm. (Adapted from NOAA/PA 75009, 1977.)

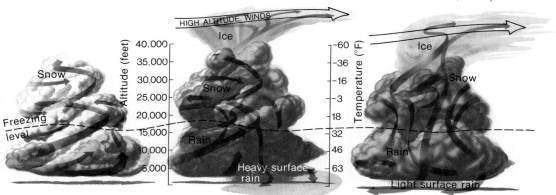

Figure 23-17   A tornado. (From Navarra, T. G., *Atmosphere, Weather and Climate.* Philadelphia: W. B. Saunders Company, 1979, p. 307.)

duces downdrafts. Violent updrafts and downdrafts sweep through the storm cells, rain falls on the ground level, and there is lightning and thunder. In the final stage of a thunderstorm, downdrafts occur over the entire cell, updrafts decrease, rain slows because the storm is cut off from its source of warm, moist air, then stops, and the cloud evaporates.

The most violent form of weather are tornadoes, related to very severe thunderstorms. They are produced when the greatest contrast exists in temperature and moisture between two air masses. Although they are most common in the central United States from early spring to late summer, they have been known to occur in nearly every state.

Dark tapering funnel clouds extend from a cumulonimbus cloud to the ground (Fig. 23–17). The winds composing the funnel have been estimated to have velocities as high as 500 mph. Devastation follows along a swath 300–400 m wide and several kilometers long when a funnel cloud touches the ground, both from the great wind velocity and from the sudden lowering of air pressure as the funnel passes over. Buildings may "explode" from the expansion of the air within, especially if the doors and windows are closed. Waterspouts are essentially tornadoes that occur over the ocean. "Dust devils" are swirling masses of air that are less extensive and violent than tornadoes.

23-14   WEATHER MODIFICATION

In their research into artificially produced rainfall in 1946, Vincent Schaefer (b. 1906) and Irving Langmuir (1881–1957) of the General Electric Research Laboratory induced precipitation by seeding a supercooled cloud with dry ice. The temperature in the vicinity of the dry ice fell to below −40°C, ice crystal nuclei were formed from the water vapor in the cloud, and rain fell.

It has since been found that silver iodide will produce the same reaction at a temperature of −6°C. It causes a rapid growth of ice crystals in the cloud that leads to precipitation. But weather modification is not yet an important process on a large scale. Much more research is needed. Besides the scientific questions, there are complex legal and moral questions that must be addressed.

EXERCISES

23–1.   How does the troposphere differ from the stratosphere?

23–2   Why are hydrogen and helium scarce in the earth's atmosphere?

23–3.   How is it possible for radio waves, which travel in straight lines, to follow the curvature of the earth?

23–4.   Explain how auroras form.

23–5.   Why were the radiation belts encompassing the earth discovered only recently?

23–6.   What important role does ozone play in the atmosphere?

23–7.   What is meant by the "hydrologic cycle"?

23-8.   Why does the relative humidity rise at night?

23-9.   Distinguish between condensation and precipitation.

23-10.   Give the common units in which air pressure is expressed.

23-11.   How are weather fronts related to air masses?

23-12.   List the principal air mass source regions that affect the United States.

23-13.   Compare the amount of snowfall to rainfall.

23-14.   How do fronts cause clouds and precipitation?

23-15.   How does a thunderstorm develop?

23-16.   Discuss five meteorological instruments.

23-17.   What is the current status of weather modification?

23-18.   Discuss the characteristics of an air mass.

23-19.   What makes a tornado so life-threatening?

23-20.   Discuss the major classes of clouds.

23-21.   *Multiple Choice*

A. Atmospheric pressure is usually recorded in
   (a) grams   (c) millibars
   (b) meters   (d) isobars

B. The relative humidity of the air is measured with a
   (a) barometer   (c) thermograph
   (b) sling psychrometer (d) ceilometer

C. The lowest region of the atmosphere is called the
   (a) troposphere
   (b) stratosphere
   (c) ionosphere
   (d) exosphere

D. Rain or snow is most likely to be produced by which of the following clouds?
   (a) cirrostratus
   (b) nimbostratus
   (c) stratocumulus
   (d) altocumulus

E. A boundary that separates air masses of different properties is known as a (an)
   (a) weather front
   (b) cold front
   (c) warm front
   (d) occluded front

Battan, Louis J., *Harvesting the Clouds: Advances in Weather Modification.* Garden City, N.Y.: Doubleday & Company, Inc., 1969.
> Can we reduce the damage done by hailstorms, harness tornadoes, divert hurricanes? Should we?

Lynch, David K., "Atmospheric Halos." *Scientific American,* April, 1978.
> Rings around the sun and moon are caused by ice crystals. Physicists still want to know precisely how.

Stewart, R. W., "The Atmosphere and the Ocean." *Scientific American,* September, 1969.
> The atmosphere drives the ocean circulations. The atmosphere in turn derives its energy from the ocean.

Woodwell, George M., "The Carbon Dioxide Question." *Scientific American,* January, 1978.
> Carbon dioxide may be moving toward a central role as a major threat to world order.

THE CHEMICAL BASIS
OF LIFE

24-1 INTRODUCTION

If life comes only from life, how did the first living forms arise? During the early part of the twentieth century, scientists were too convinced that the great chemist Louis Pasteur (1822–1895) had disproved *spontaneous generation*—the idea that life evolves from nonliving matter—to consider the origin of life a legitimate scientific question. Interest in the origin of life on earth and in the possible existence of life elsewhere in the universe was reawakened by experiments in the 1950's and the beginning of space exploration in the 1960's.

Any account of the origin of life must be regarded as hypothetical and speculative. Although various theories have been proposed, scientists must offer the best possible explanation based upon the available data from chemistry, physics, astronomy, geology, and biology and on inferences about the nature of the earth 3 or 4 billion years ago. It is necessary to demonstrate that the theories are either possible or probable.

24-2 CRITERIA FOR LIFE

Organisms vary considerably in structure but they are functionally similar. They grow, they have the ability to reproduce, and they transform and use energy for activities such as locomotion, circulation of materials, and movement of material into and out of cells. How may one look for life that may be very different from our own? How can we be sure that the observations reveal the existence of life rather than the presence of nonbiological processes?

The Space Science Board of the National Academy of Sciences conducted an Exobiology Study in 1964–65 to seek answers to such questions. The scientists agreed that anything organized to draw nourishment from the environment and reproduce itself should be considered "life." They found no reason to consider life unique to earth. They held that life does not manifest some mysterious force lying outside the realms of chemistry and physics. They considered the possibility that life elsewhere might not use water as its primary solvent or carbon atoms to form the backbones of its molecules, two essential features of life on earth.

24–3   THE OPARIN-HALDANE THEORY

Alexander Oparin (b. 1894) was the first to go into the problem of the origin of life in detail. He presented his ideas of *chemosynthesis* or the *heterotroph hypothesis* in a paper before the Botanical Society of Moscow in 1922. Two years later this was published as a monograph, but it had no effect on his contemporaries. Only when Oparin expanded the article into a book published in 1936 called *The Origin of Life* did his ideas begin to attract attention.

Thinking independently along similar lines, although he never read Oparin's paper, John B. S. Haldane (1892–1964), a biochemist, published a theory of the origin of life in an article in 1929. It, too, elicited no reaction at the time. Biochemists failed to appreciate that Oparin and Haldane were proposing not that life evolves from nonliving matter today but rather that life once evolved from nonliving matter under the conditions prevailing on the primitive earth.

According to their theory, before life began there must first have been a period of *chemical evolution.* During this period, the gases of the atmosphere and ocean gradually became more and more complex. The atmosphere contained hydrogen (H_2), water vapor (H_2O), ammonia (NH_3), methane (CH_4), hydrogen cyanide (HCN), and carbon dioxide (CO_2), but no free oxygen (O_2) (Fig. 24–1). Without oxygen, there would have been no ozone layer to block most of the ultraviolet radiation from the sun. The radiation reaching the surface of the earth could then have provided the

694   THE CHEMICAL BASIS OF LIFE

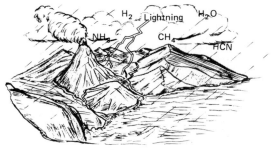

Formation of the primitive atmosphere consisting of H_2O, NH_3, CH_4, H_2, and HCN

Figure 24–1
The primitive atmosphere of the earth, consisting of free hydrogen and compounds of hydrogen, but no free oxygen. (From Tortora, G. J., and Becker, J. F., *Life Science,* 2nd ed. New York: Macmillan Publishing Co., 1978, p. 59.)

energy for the synthesis of many organic compounds nonbiologically from the gases present (Fig. 24–2). Without free oxygen to destroy them again, such compounds would have accumulated in the oceans until the primitive oceans reached the consistency of a warm, dilute broth nicknamed "Haldane soup."

The first living organisms would have been little more than a few chemical reactions wrapped in a film to keep them from being diluted and destroyed (Fig. 24–3). They would absorb compounds, grow, divide, and obtain energy from the organic molecules around them. Eventually, photosynthesis, a process by which carbohydrates are synthesized from carbon dioxide and water, would arise as an alternative energy source when the food supply ran low (Fig. 24–4). The oxygen released by photosynthesis would screen out the ultraviolet radiation with an ozone layer it would form in the atmosphere, and gradually the atmosphere would turn from reducing to oxidizing. Free oxygen would lead to the evolution of respiration and to a metabolism involving enzymes that protect oxygen-using organisms from the toxic side effects of the gas. In the process, the nonbiological synthesis of organic matter would end, and the stage would be set for true

Figure 24–2 The formation of simple organic monomers from atmospheric gases might have occurred under the influence of ultraviolet radiation and other forms of energy. (From Tortora, G. J., and Becker, J. F., *Life Science,* 2nd ed. New York: Macmillan Publishing Co., 1978, p. 59.)

H_2O
CH_4
NH_3
HCN
$\Bigg\}\longrightarrow\Bigg\{$
Monosaccharides
Glycerin
Fatty acids
Amino acids
Pyrimidines
Purines

Organic compound formation from the atmospheric gases in the ancient seas

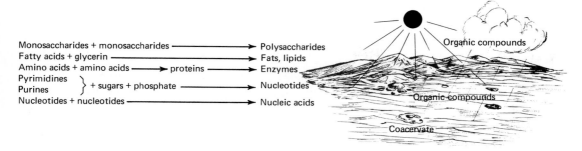

Formation of complex organic compounds and coacervates.

biological evolution. The appearance of life on the planet would change the planet and destroy the conditions that made it possible.

The Oparin-Haldane theory is the basis for ideas still held today. This is remarkable because in 1929 none of the biochemical details were known. Enzymes were a mystery, and the nature of the genetic code was unknown. Much of what has been learned since then amounts to a filling in of the details in their theory. Even if life did not arise on the earth at all but, as some have suggested, might have been deliberately seeded by intelligent beings living in other solar systems, life must have evolved on a planet not drastically different from the earth that is so hospitable to the kind of life found on it. The question then really is: How might life have evolved on an earthlike planet?

Figure 24–3 Formation of complex organic polymers and coacervates, structures from which life may have originated. (From Tortora, G. J., and Becker, J. F., *Life Science*, 2nd ed. New York: Macmillan Publishing Co., 1958, p. 59.)

24–4 EVOLUTION OF THE EARTH'S ATMOSPHERE

The most common atoms in the universe are the two simplest: hydrogen (92 percent) and helium (7 percent). Most of the atoms of the other common elements consist of carbon, nitrogen, oxygen, sulfur, phosphorus, neon, argon, silicon, and iron. When a planet forms out of dust and gas, it ought to be mostly hydrogen and

Figure 24–4 Photosynthesis enabled organisms to convert carbon dioxide into glucose, an energy-rich molecule.

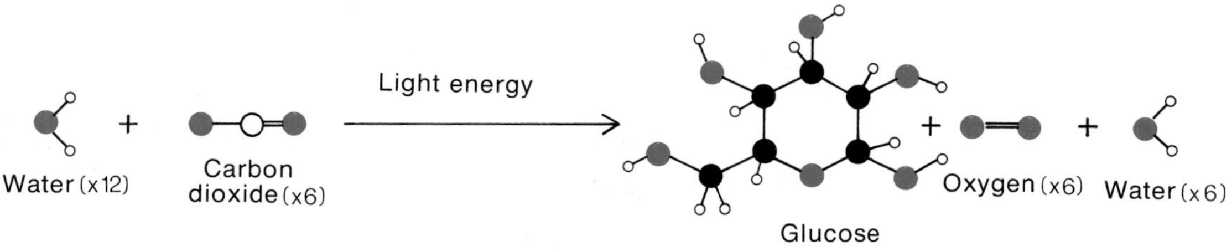

Water (x 12)   +   Carbon dioxide (x 6)   Light energy   →   Glucose   +   Oxygen (x 6)   +   Water (x 6)

helium, and these gases would make up most of the original atmosphere. The composition of the large Jovian planets—Jupiter, Saturn, Uranus, and Neptune—is close to that of the universe at large. The small terrestrial planets—Mercury, Venus, Earth, and Mars—have more of the heavier elements and are poorer in such gases as helium that could escape from their weak gravitational pull.

Helium atoms do not combine with other atoms, but hydrogen atoms do. They are so numerous that any atom that can combine with hydrogen will do so. Each carbon atom combines with four hydrogen atoms to form methane (CH_4); nitrogen forms ammonia (NH_3); oxygen forms water (H_2O); sulfur forms hydrogen sulfide (H_2S). These compounds would all be found in the primitive atmosphere of the earth, Atmosphere I, and the ocean would have contained much ammonia in solution.

Water vapor released from the interior of the earth by degassing processes such as volcanism would have been a component of a secondary atmosphere as well, Atmosphere II. The decomposition of metal carbides could give rise to methane (CH_4), carbon dioxide (CO_2), and carbon monoxide (CO); nitrides, ammonia (NH_3) and nitrogen (N_2); and sulfides, hydrogen sulfide (H_2S). Atmosphere II would be composed chiefly of nitrogen (N_2), carbon dioxide (CO_2), and water (H_2O) vapor, and the ocean would contain much carbon dioxide (CO_2) in solution.

The present atmosphere of the earth, Atmosphere III, consists mainly of nitrogen (N_2), oxygen (O_2), and water (H_2O) vapor, with only small quantities of gas dissolved in the ocean. The oxygen was put there only after life had developed by early organisms making glucose through photosynthesis and releasing oxygen as a by-product. Life, then, must have originated in either Atmosphere I or Atmosphere II.

24–5   LABORATORY EVIDENCE

Harold Urey (b. 1893), a chemist then at the University of Chicago, restated the Oparin-Haldane thesis in 1952 in his book *The Planets*. Like Oparin, he felt that life might have started in Atmosphere I and planned experiments to test this idea. In 1953 a key discovery in the search for the origin of life was made.

What if one were to start with a mixture of gases that represents Atmosphere I and add energy in a way that it might have been added on the primitive earth? Will more complicated molecules be formed out of the simple gases? And if so, will they be the kind that are found in living organisms?

Stanley Miller (b. 1930), a graduate student in his twenties working on his Ph.D., acted on Urey's suggestion. He set up an artificial atmosphere of hydrogen, methane, and ammonia in a large glass reaction vessel, and water in another vessel (Fig. 24–5). For an energy source, Miller considered working with ultraviolet radiation in the first experiments, but problems with windows in the reaction vessel caused him to use an electric spark discharge, a laboratory simulation of a small bolt of lightning.

In a typical experiment, the water was boiled and the steam passed into the gas mixture, forcing the mixtures through another tube back into the boiling water. The mixture was kept circulating past the electric spark. At the start, the water and gases were colorless, but after 24 hours the water had turned pink, growing a deeper and deeper red as the days passed. The progress of synthesis was monitored by taking samples out of the boiling flask for analysis.

It was surprising to find that several common amino acids (corresponding to the general formula H_2NCOOH), the kind that serve as protein building blocks, were synthesized, along with other molecules that are present in living matter, such as urea (NH_2CONH_2), formic acid ($HCOOH$), acetic acid (CH_3COOH), and many others. The compounds

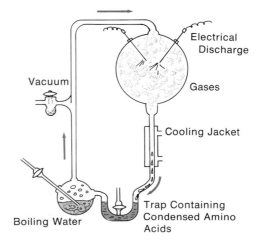

Figure 24–5 Apparatus used by Stanley L. Miller and Harold C. Urey to produce complex molecules from simple gases presumed to be present in the atmosphere of the primitive earth. (From Levin, H. L., *The Earth Through Time.* Philadelphia: W. B. Saunders Company, 1978, p. 251.)

Electrical Discharge

Vacuum

Gases

Cooling Jacket

Boiling Water

Trap Containing Condensed Amino Acids

formed quickly and in relatively large quantities from the small amount of gas used in the experiment. Miller and Urey observed that the concentration of ammonia fell steadily and that its nitrogen atoms appeared first in hydrogen cyanide (HCN). Amino acids were synthesized more slowly from hydrogen cyanide and compounds called aldehydes (RCHO) by a mechanism well-known to organic chemists.

Miller and others tried many variations of the experiment, substituting carbon monoxide or carbon dioxide for methane, nitrogen for ammonia and ultraviolet radiation for the spark discharge. The main requirement for success seemed to be the absence or near-absence of oxygen. By 1968, every amino acid important to protein structure had been formed in such experiments. There is little doubt that all the necessary chemicals of life could have been produced in the warm ocean of the early earth, full of ammonia, with winds of methane blowing over it under ultraviolet radiation. Enormous amounts of these compounds may indeed have made the ocean become a "Haldane soup." Thus, guesses about the origin of life starting with Oparin and Haldane are supported by laboratory experiments.

24-6   POLYMERS; MICROSPHERES; COACERVATE DROPS

In order to join together and form polymer chains of proteins, a molecule of water must be removed from two amino acid monomers at each link in the chain (Fig. 24-6). This can be done by heating amino acids to 140°C and higher. Volcanic activity on the primitive crust of the earth could provide such temperatures.

Sidney W. Fox (b. 1912), a biochemist at the University of Miami, discovered that the polymerization reaction also occurred at temperatures as low as 70°C if phosphoric acid was present. Fox and his associates created chains resembling proteins from a mixture of eighteen common amino acids. He called these structures *proteinoids,* and suggested that billions of years ago they could have been the transitional structures leading from amino acids to true proteins. Fox discovered proteinoids similar to the ones he synthesized in the laboratory among the lavas adjacent to the vents of Hawaiian volcanoes. Amino acids may have formed

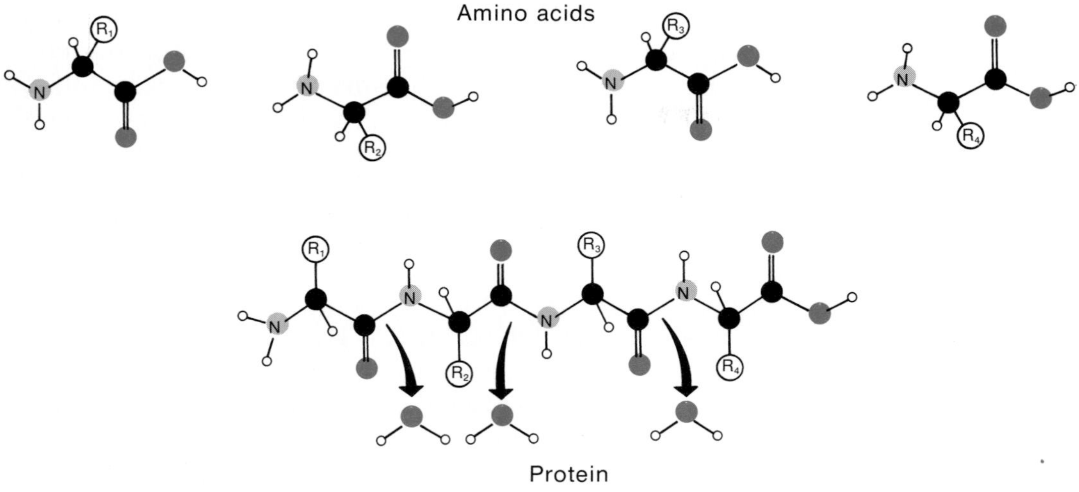

Amino acids

Protein

Figure 24–6 Amino acids link up to form a section of a protein chain through the removal of a molecule of water at each linkage point, creating a peptide bond between carbon and nitrogen atoms.

in the volcanic vapors and combined into proteinoids under the influence of the heat of escaping gases.

When hot aqueous solutions of proteinoids are cooled, the proteinoids form into tiny spheres that show many of the characteristics of living cells. These *microspheres,* 1 or 2 micrometers in diameter, are surrounded by a film-like outer wall and can swell and shrink. They exhibit budding as do yeast organisms and divide into "daughter" microspheres. Like some bacteria, they occasionally link up linearly to form filaments and exhibit a streaming movement of internal particles similar to that in living cells.

Oparin for many years studied the segregation of matter in solution into droplets that were possible precursors of life. He focused on the tendency of aqueous solutions of biological polymers—proteins, carbohydrates, nucleic acids—to separate spontaneously into *coacervates:* colloidal droplets rich in polymers, ranging in diameter from 1 micrometer to 500 micrometers, suspended in the surrounding watery medium. Many coacervates seem to be set off from the medium by a kind of membrane. Some coacervates are unstable, sink to the bottom of the medium, and coalesce into a separate layer. When coacervates become too big, they break up into several daughter droplets.

Oparin's and Fox's experiments show that separation into coacervates or microspheres is common for biological polymers in solution. Before living cells evolved, the primitive ocean may have been teeming with such droplets.

Droplets containing catalysts able to induce polymerizations that would increase their bulk or strengthen their membrane would survive longer than others. They would be particularly favored if they could synthesize molecules promoting the survival of the daughter droplets as well as the parents. This may not be life but it is getting close to life.

24–7   GEOLOGICAL EVIDENCE OF LIFE

Fossils—the remains of organisms that lived in the geologic past—preserved in sedimentary rocks are the basis for dividing geologic time into eons, periods, and epochs (Fig. 24–7). The fossil deposits form layers that can be identified in geological formations. The ages of the deposits can be calculated from the constant rate of decay of radioactive isotopes in the earth's crust. A date can be assigned to a rock unit and to nearby fossil-bearing strata by determining how much of an isotope has decayed since the minerals in the rock unit crystallized.

A dramatic boundary in the rock record separates the Cambrian period from all that came before. The eleven periods of geologic time since the start of the Cambrian are referred to as the Phanerozoic eon—the era of manifest life. The preceding era is simply called the Precambrian. It has been established that the Phanerozoic eon began about 570 million years ago. Radioisotope dating places the age of the earth and the rest of the solar system at 4.6 billion years. Thus, the Precambrian eon involves approximately 90 percent of the earth's entire history.

Cambrian strata contain many fossils of marine plants and animals such as seaweeds, sponges, mollusks, and the early arthropods called trilobites. It was thought for many years that the underlying Precambrian strata contained no fossils. Life seemed to come into existence abruptly during the Cambrian period. Yet it did not seem reasonable to assume that life began with organisms as complex as trilobites.

In the 1950's it became evident that many Precambrian rocks *are* fossil-bearing. These fossils had escaped detection earlier because they are the remains of microscopic forms of life. Precambrian fossils have even been found in some of the most ancient sedimentary deposits known.

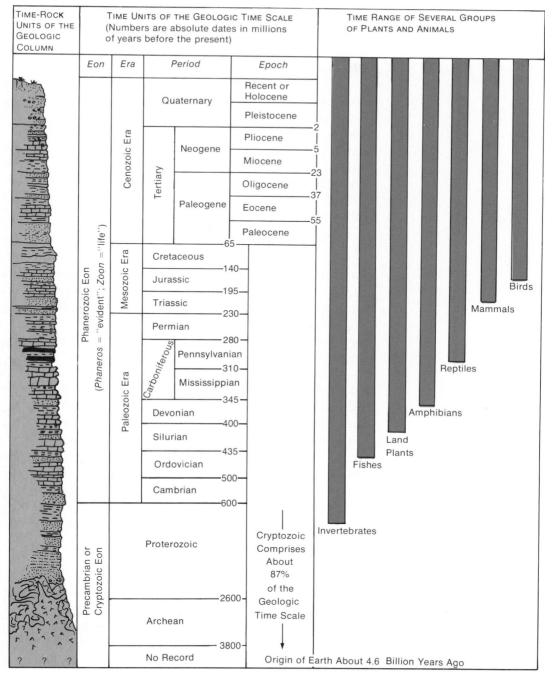

TIME-ROCK UNITS OF THE GEOLOGIC COLUMN	TIME UNITS OF THE GEOLOGIC TIME SCALE (Numbers are absolute dates in millions of years before the present)				TIME RANGE OF SEVERAL GROUPS OF PLANTS AND ANIMALS
	Eon	Era	Period	Epoch	

Figure 24-7 Geologic time scale. (From Levin, H. L., *The Earth Through Time.* Philadelphia: W. B. Saunders Company, 1978, p. 103.)

The paleobotanist Elso S. Barghoorn (b. 1915) of Harvard, working with the sedimentologist Stanley A. Tyler (b. 1911) of the University of Wisconsin, in 1954 discovered fossil microscopic plants in an outcropping of Precambrian rocks called the Gunflint Iron formation near Lake Superior in Ontario. Most of the Gun-flint fossils resemble modern bacteria and blue-green algae (cyanobacteria). The rocks have been dated to an age of about 2 billion years (Fig. 24–8).

Since then, microfossils have been identified in many places and in even older rocks. The oldest are the Fig Tree and Onverwacht deposits of South Africa, which are 3.2 and 3.4 billion years old, respectively. Both deposits contain microfossils that resemble bacteria. The world's oldest known multicellular organisms left their record in rocks that were to be called the Torrowangee Group of Australia about 1 billion years ago.

The search for geological evidence of ancient life is a laborious process. Thousands of rock specimens have to be sawed into thin slices and polished so that they can be studied under the light microscope and the electron microscope. But the fossils reveal a surprising amount of information about the size, shape, structure, and even the internal structure of the cells of the organisms.

The geological record of the evolution of life on earth shows that life on earth appeared within about a billion years after its formation. We have no direct geological evidence to tell us when the transition from nonliving to living occurred. For perhaps three billion years all life on earth was at or below the unicellular level of bacteria or unicellular blue-green algae. The step from nonbiological organic matter to life seems to have been easier than the step from one-celled bacteria to many-celled organisms, a step that took twice as long.

24–8 THE SEARCH FOR EXTRATERRESTRIAL LIFE

A driving force in the space effort has been the search for life elsewhere in the solar system and in the universe. *Exobiology* is the name given to this new discipline, the study of extraterrestrial life. The reasoning is that given the right conditions life may arise spontaneously and evolve.

These conditions for life as we know it include a surface temperature low enough to provide water in the liquid state and one that is suitable for the chemical reactions of life processes; a store of the chemical elements required for these processes, such as carbon and

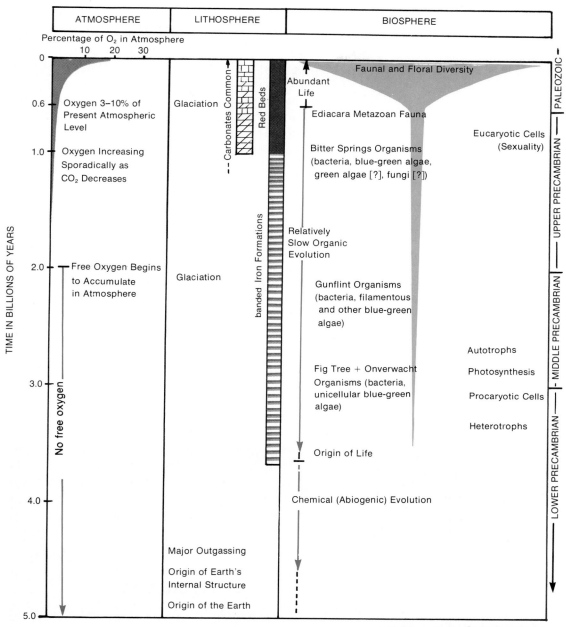

ATMOSPHERE	LITHOSPHERE	BIOSPHERE

Figure 24–8 Correlation of fossil deposits with events in the atmosphere and lithosphere. (From Levin, H. L., *The Earth Through Time*. Philadelphia: W. B. Saunders Company, 1978, p. 263.)

oxygen; and a source of radiant energy. Even if life is not found on other planets, various stages of chemical evolution may be found.

We learned from the Apollo program that life did not evolve on the moon. There is no life there today or any indication of life having been there in the past. Mercury and Venus seem most inhospitable to life as we know it. Some of the outer planets, such as Jupiter, and some large satellites, are candidates for life. Mars seems to present the only possibility of life among the terrestrial planets besides the earth.

If there is life on Mars, it must exist in a world drier than the driest deserts of earth. It must withstand ultraviolet radiation stronger than on any mountaintop of earth. And it must contend with temperatures that dip to hundreds of degrees below zero each night. Harsh as these conditions may be, laboratory experiments with earth organisms under simulated Martian conditions have shown that such adaptation is possible.

The Viking search for life on Mars involved two spacecraft. Viking 1 set down safely on Chryse Planitia in the northern hemisphere of Mars on July 20, 1976. Viking 2 landed at Utopia Planitia on September 3, 1976. The Viking landers scooped up soil from the Martian surface and performed experiments with the Automatic Biological Laboratory until June 1, 1977, when the laboratories were switched off from earth (Fig. 24–9). They had run out of the nutrient liquids and helium gas needed to conduct the experiments. The two spacecraft did not detect life in the samples of Martian soil they tested. But the findings may not negate the possibility of life on Mars. The Vikings only

Figure 24–9 Viking lander spacecraft on the surface of Mars searched for evidence of Martian life. (Courtesy of NASA.)

searched a few square meters of the planet. Perhaps there is life on Mars and we have simply missed it so far.

Within the universe there may be stars similar to our sun with a planet that has an environment like earth's. The Harvard astronomer, Harlow Shapley, suggested that one star in a million has the necessary conditions for the existence of life. In the Milky Way Galaxy alone there may be 100 000 planets similar to the earth among the billions of planets believed to exist there. But even our largest optical telescopes are not capable of detecting a planet in orbit around another star. Evidence supporting the hypothesis comes from a close neighbor of our sun, Barnard's star, 6 light years away. From a wiggle in its path, astronomers have concluded that two planets about the size of Jupiter and Saturn are orbiting the star and exerting a gravitational pull that affects its course. Smaller planets may also be orbiting Barnard's star.

Exobiologists are certain that the creation of life anywhere in the universe is more the rule than the exception. In any event, there is growing interest among scientists not only in the search for extra-terrestrial life but in the search for extraterrestrial intelligence (SETI). Space travel as we know it would be out of the question for contacting a civilization in another solar system because of the vast distances involved. Traveling at rocket speeds of 8 km/sec, a space ship would take 160 000 years or more to reach Proxima Centauri, the nearest star to the sun, only 4.3 light years away.

Our capability for detecting extraterrestrial intelligence lies rather in radio astronomy and involves radio telescopes, advanced computers for processing data, and human imagination, skill, and commitment. The first such effort was organized in 1959 by Frank Drake (b. 1930) at the National Radio Astronomy Observatory in Greenbank, West Virginia. Drake directed an antenna toward two nearby stars, Epsilon Eridani and Tau Ceti, and for two weeks listened for possible radio signals but heard none. Since then, hundreds of other stars have been examined with negative results, but the quest is just beginning. There are scientists who believe that it would be more difficult to understand a universe in which we are the only technological civilization than to imagine one brimming over with intelligent life.

CHECKLIST

Amino acid	Microsphere
Cambrian period	Phanerozoic eon
Chemical evolution	Precambrian eon
Coacervate	Protein
Exobiology	Proteinoid
"Haldane soup"	Spontaneous generation
Microfossil	Trilobite

EXERCISES

24–1.   What is meant by spontaneous generation?

24–2.   How is "Haldane soup" related to the problem of the origin of life?

24–3.   What did the Urey-Miller experiments prove?

24–4.   Why are amino acids important in the evolution of life?

24–5.   How did the earth's atmosphere probably evolve?

24–6.   What is the role of oxygen in the origin of life on earth?

24–7.   What energy sources were available on the primitive earth?

24–8.   What sorts of organisms were the principal inhabitants of the earth during most of its history?

24–9.   Explain the apparently sharp discontinuity between the fossil deposits of Precambrian and Cambrian rocks.

24–10.   How are fossils dated?

24–11.   List some of the attributes of life.

24–12.   How are coacervate drops related to the origin of life?

24–13.   What are the prospects for the existence of life elsewhere in the solar system?

24–14.   Discuss the value of supporting exobiology.

24–15.   How probable is life elsewhere in the Milky Way Galaxy?

24–16.   Do you think there is extraterrestrial intelligence?

24–17.   What would it matter to humanity if the existence of other technological civilizations was definitely proved or disproved?

24–18.   Is space travel to other solar systems a realistic goal?

24–19.   How might communication between worlds occur?

24–20.   Do you think it likely that human beings will ever colonize other planets?

24–21.   *Multiple Choice*

A. The most common element in the universe is
   (a) helium   (c) carbon
   (b) hydrogen   (d) oxygen

B. The formation of complicated molecules out of the simple gases presumed to exist on the primitive earth was first shown to be possible by
   (a) Oparin   (c) Urey and Miller
   (b) Haldane   (d) Fox

C. Solutions of proteins and other biological polymers separate into droplets called
   (a) "Haldane soup"
   (b) coacervates
   (c) Spontaneous generation
   (d) trilobites

D. The study of extraterrestrial life is
   (a) exobiology
   (b) paleontology
   (c) sedimentology
   (d) astrophysics

E. A key science in the search for extraterrestrial intelligence is
   (a) psychology
   (b) chemistry
   (c) radio astronomy
   (d) biology

SUGGESTIONS FOR FURTHER READING

Dickerson, Richard E., "Chemical Evolution and the Origin of Life." *Scientific American,* September, 1978.
> The earth formed 4.6 billion years ago. Within one billion years, one-celled organisms had evolved out of organic molecules produced nonbiologically.

Fox, Sidney W., Kaora Harada, Gottfried Krampitz, and George Mueller, "Chemical Origins of Cells." *Chemical and Engineering News,* June 22, 1970.
> Experiments have led toward definitions of primordial life. Such definitions may yield deeper insights into contemporary life.

Sagan, Carl, "The Quest for Intelligent Life Is Just Beginning." *Smithsonian,* May, 1978.
> Advances in the design of radiotelescopes and data analysis make possible a search for other civilizations in our galaxy.

Schopf, J. William, "The Evolution of the Earliest Cells." *Scientific American,* September, 1978.
> For three billion years the only living things were primitive microorganisms. These cells gave rise to biochemical systems and the oxygen-enriched atmosphere on which modern life depends.

APPENDIX A
MEASUREMENT

A–1 WHY MEASUREMENT?

Measurement is the process of determining "how many." To measure the length of a table, for example, you select a suitable reference standard—say, meter stick—and lay it end to end until the comparison is completed. Then you express the measurement as the number of units it takes to equal the length of the table, let's say, 1.38 meters.

Precise measurement lies at the very heart of science. Words alone often mean different things to different people. How big is "big?" . . . or "small?" How hot it "hot?" What does "heavy" mean? Even in everyday life concepts mean more when they are expressed in numbers. We speak of a $3'' \times 5''$ file card, a car that averages 16 miles (or 25.6 kilometers) to the gallon, a temperature of 72°F (or 22°C). Our purpose here is to look at standards of measurement and some ways of expressing measurements conveniently.

A–2 WHY STANDARDIZATION?

For most of the billions of measurements made each day, we use the foot-pound-quart system for length, mass, and capacity. This system is known as the English Customary system of weights and measures. Many of the standards were originally derived from a convenient though crude source—the human anatomy. Thus, the inch was the length of the thumb from tip to knuckle; the palm, the width of four fingers; the foot, four palms; a yard, the distance from the tip of the nose to the tip of the middle finger of the outstretched arm. The anatomically based units are not easy to work with. There are 12 inches in a foot; 3 feet in a yard; 5½ yards in a rod; 320 rods in a mile; and 5280 feet in a mile.

A second drawback of customary units is that the same name quite often stands for different quantities from one place to another, or from one context to another. The gallon used in the United States today was Queen Anne's (1665–1714) wine gallon. It was smaller than the ale gallon used in her day and different from the imperial gallon eventually adopted. A Canadian imperial gallon, for example, is about 25 percent larger than an American gallon. Or consider the different meanings of "ounce." An ounce for measuring fluids is not the same as an ounce for weighing. Moreover, the avoirdupois ounce for ordinary weighing is lighter than the troy ounce for weighing precious stones and metals, and the apothecaries' ounce for prescriptions and drugs.

As commerce and industry expanded and the world became increasingly interdependent for raw materials and finished products, it became evident that a common and rational system of weights and measures was needed. And with the emergence of experimental science it became clear that physical measurement was a basic element of studying natural phenomena. Well-defined units of measurement therefore were essential to the exchange and comparison of experimental results. By the second half of the 18th century, there was a movement in several nations to establish a system of measurement based upon easily reproducible standards. This drive culminated at the time of the French Revolution with the development of the metric system.

A–3   THE METRIC SYSTEM

When the metric system came into existence in the 1790's, standards were important for only a few kinds of measuring units, primarily length and mass. Today we must also have standards for the accurate measurement of temperature, color, electric current, sound and light intensity, and many other physical quantities. Practically all, however, can be derived from four fundamental standards: the *meter* for length, the *kilogram* for mass, the *second* for time, and the *kelvin* for temperature.

The standard for length, the meter, was originally defined as one ten-millionth of the distance between the equator and the North Pole—a quadrant of the earth's meridian. The section of the quadrant lying between Dunkirk, France, and a point near Barcelona, Spain, was carefully measured by a surveying team and extrapolated by astronomical observations to the entire quadrant (Fig. A–1). The unit of mass, the kilogram, was originally defined as the mass of a cubic decimeter (a liter) of water at the temperature of its maximum density, 4 degrees Celsius, and then redefined as the mass of a particular platinum-iridium cylinder, Prototype Kilogram No. 1, kept at Sèvres, France. It is equivalent to 2.20 pounds.

The orginators of the metric system saw no reason to

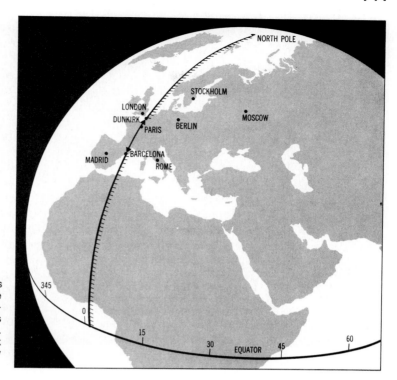

Figure A–1 The meter was originally intended to be one ten-millionth of the equator-to-pole quadrant of the earth's meridian passing close to Paris. The section between Dunkirk and Barcelona was accurately measured.

question the unit of time, the second, which had been in use for centuries. The second had long been defined as 1/86 400 of the mean solar day, based upon the rotating earth. The day is divided into 24 hours, each hour into 60 minutes, and each minute into 60 seconds.

When problems arose with the metric system as originally conceived, the French government arranged a conference in 1870 to work out standards for a unified measurement system, and in 1875 the Treaty of the Meter was signed in Paris by representatives from 17 nations.

The treaty set up an International Bureau of Weights and Measures to be the custodian of the standards for an international system of measurement with headquarters at Sèvres, a suburb of Paris. It also established a General Conference on Weights and Measures, which meets at least once every six years. Commissions were appointed to design prototype standards of length and mass—the meter and the kilogram—and copies were given to the nations adhering to the treaty (Fig. A–2).

Soon after the prototype copies were received in the United States, the Treasury Department redefined the yard and the pound (which had been based on copies of English standards) as appropriate fractions of the meter and the kilogram. Along with other customary units such as the gallon, they are based by law on metric standards.

The American copy of the international standard of

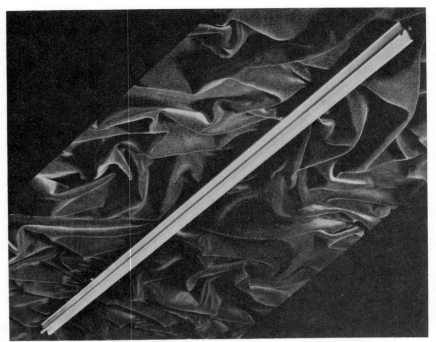

Figure A-2 The United States copy of the Standard Meter, Prototype Meter No. 27, was delivered from France in 1890. (Courtesy of National Bureau of Standards.)

mass, known as Prototype Kilogram No. 20, is kept in a vault at the Gaithersburg Laboratory of the National Bureau of Standards (Fig. A-3). It is removed not more than once a year for checking the values of other standards with a precision balance. Twice since 1889 it has been taken to France for comparison with the master kilogram. In practice, we use secondary working standards of length and mass, which have a high degree of accuracy.

Figure A-3 Prototype Kilogram No. 20, the national standard of mass for the United States. It is a platinum-iridium cylinder 39 millimeters in diameter and 39 millimeters high. (Courtesy of National Bureau of Standards.)

A–4 METRIFY OR PETRIFY?

The United States has been committed to the metric system ever since the Metric Conversion Act was signed into law in 1975. The law states that it is United States policy to coordinate and plan the increasing use of the metric system. A National Metric Board coordinates the voluntary conversion both in and out of government. The key word is "voluntary." No one is going to be put in jail for using feet and pounds. Living with two systems at once should pose no great problems.

The switch to metric has a strong economic dimension. For example, multinational corporations and exporters were influenced by a directive by the European Economic Community requiring exporters to the Common Market countries to label their products in metric units. Dozens of major corporations organized metric conversion programs. Their products have a better chance of success in foreign countries if designed to the metric standards observed in most countries. Much of the nation's heavy industry has already converted, particularly the auto industry. Thousands of suppliers must either go metric, too, or miss out on the contracts. Department store chains are manufacturing their products in metric measures. Football programs list the players' heights and weights in metric as well as customary units. Soft-drink companies have introduced half-liter and liter bottles. In the National Parks, new signs give road distances in metric equivalents. Grocery store items often include metric-equivalent units on the labels. The vitamin and mineral content of cereals are given in milligrams. Doctors prescribe metrically, and some record pa-

---

Table A–1 ENGLISH-METRIC CONVERSION FACTORS

LENGTH

1 inch (in.)	= 2.54 centimeters (cm)
1 foot (ft)	= 0.3048 meter (m)
1 yard (yd)	= 0.914 meter (m)
1 mile (mi)	= 1.609 kilometers (km)

MASS

1 grain (gr)	= 64.799 milligrams (mg)
1 ounce (oz)	= 28.35 grams (g)
1 pound (lb)	= 453.6 grams (g)
1 ton (tn)	= 907.2 kilograms (kg)

VOLUME

1 fluid ounce (fl. oz)	= 29.57 milliliters (ml)
1 quart (qt)	= 0.946 liter (l)
1 gallon (gal)	= 3.785 liters (l)

Table A–2   METRIC-ENGLISH
CONVERSION FACTORS

LENGTH

1 millimeter (mm)	= 0.03937 inch (in.)
1 meter (m)	= 1.09 yards (yd)
1 kilometer (km)	= 0.621 mile (mi)

MASS

1 gram (g)	= 0.03527 ounce (oz)
1 kilogram (kg)	= 2.2 pounds (lb)

VOLUME

1 liter (l)	= 1.06 quart (qt)
1 cubic meter (m^3)	= 1.308 cubic yards (yd^3)

---

tients' weights in kilograms, heights in centimeters or meters, and temperatures in degrees Celsius. Camera hobbyists work with film cut to metric dimension, such as 35 millimeters. The Olympics and other international athletic events have been using the metric system for years. In some baseball stadiums the dimensions of the playing field are expressed in meters as well as feet. The chemical, optical, pharmaceutical, film, and jewelry industries all have converted to the metric system.

Some common English-metric and metric-English conversion factors are given in Tables A–1 and A–2.

A–2   WHY SI UNITS?

To be useful, a system of units must change with the times. In 1960, the 11th General Conference on Weights and Measures extended and refined the metric system, so that it now goes far beyond the original standards of length and mass. The Conference also officially adopted the International System of Units (Système Internationale d'Unités), which is abbreviated SI in all languages. This system is built on seven fundamental physical quantities for length, mass, time, temperature, electric current, light intensity, and amount of substance.

The meter was redefined as the length of 1 650 763.73 wavelengths of a certain line in the emission spectrum of the krypton-86 atom.

Refinements based upon clocks controlled by certain properties of quartz crystals made it clear that the second as previously defined was not constant but varied, because the earth's rotation was more irregular than had been suspected. Astronomers were using a more reliable time scale, ephemeris time, based upon the motion of the earth around

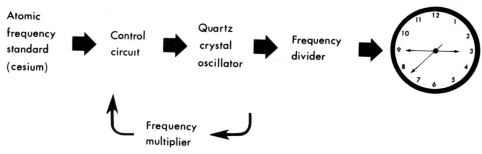

Figure A–4 Schematic diagram showing the essential parts of a cesium atomic clock. The cesium resonance frequency serves as a primary standard, not as a clock. (From *Chemical and Engineering News,* October 26, 1964, p. 143.)

the sun, and in 1960, the 11th General Conference adopted a new definition of the second.

In the belief that a second defined in terms of the frequency of vibration of an atom or a molecule would be even more accurate, the Conference urged that work proceed on an atomic definition. The most suitable candidate for timekeeping was found to be a vibration of the cesium-133 atom, which can be determined with great accuracy. In 1967 the 13th General Conference on Weights and Measures redefined the second as 9 192 631 770 cycles of that particular vibration. The accuracy of a cesium clock is such that in 6000 years it would not gain or lose more than a second (Fig. A–4). Exquisite timing accuracy is necessary in such fields as satellite tracking and astronomical observations. The standards of length and time have shifted, then, from the earth to the atom. Table A–3 gives the basic SI units and their symbols.

The basic SI units and the units derived from them now form a legal basis of measure. The customary system of units is defined in terms of SI units. Like most of the world, we are going directly to the metric system, rather than a system derived from it, for commerce, industry, the arts, medicine, agriculture, and engineering, as well as for science.

In practice, scientists use two subsystems of SI units: the centimeter-gram-second (cgs), and the meter-kilogram-second system (mks). Chemists and others often prefer the

Table A–3 SI BASE UNITS

PHYSICAL QUANTITY	NAME OF UNIT	SYMBOL
length	meter	m
mass	kilogram	kg
time	second	s
thermodynamic temperature	kelvin	K
electric current	ampere	A
luminous intensity	candela	cd
amount of substance	mole	mol

former as a matter of convenience, and physicists the latter. Both subsystems are used in this text.

A–6 MULTIPLES AND SUBMULTIPLES

Since many of the quantities measured in science—the speed of light, the distance of a star, the size of an atom, the quantity of electric charge—are either much larger or much smaller than 1, they are expressed in units appropriate to their dimensions. A big advantage of SI units is their decimal relationship. The prefixes that are used with the units show this. A centimeter is 1/100 of a meter, or 0.01 m; a kilometer is 1000 meters. A megameter is one million meters; a kilogram, one thousand grams; a nanosecond, one billionth of a second. The 11th General Conference has approved the use of the 16 prefixes listed in Table A-4, which range from the very large to the very small and are used with any of the fundamental SI units.

A–7 WHY EXPONENTIAL NOTATION?

Science and technology have so developed that it is possible to think of a clock having an accuracy of 1 part in

Table A–4 SI PREFIXES

PREFIX	SYMBOL	FACTOR	MEANING
exa-	E	10^{18}	1 000 000 000 000 000 000 one quintillion
peta-	P	10^{15}	1 000 000 000 000 000 one quadrillion
tera-	T	10^{12}	1 000 000 000 000 one trillion
giga-	G	10^{9}	1 000 000 000 one billion
mega-	M*	10^{6}	1 000 000 one million
kilo-	k*	10^{3}	1000 one thousand
hecto-	h	10^{2}	100 one hundred
deka-	da	10^{1}	10 ten
Unit-one			
deci-	d	10^{-1}	0.1 one-tenth
centi-	c*	10^{-2}	0.01 one-hundredth
milli-	m*	10^{-3}	0.001 one-thousandth
micro-	μ*	10^{-6}	0.000 001 one-millionth
nano-	n	10^{-9}	0.000 000 001 one-billionth
pico-	p	10^{-12}	0.000 000 000 001 one-trillionth
femto-	f	10^{-15}	0.000 000 000 000 001 one-quadrillionth
atto-	a	10^{-18}	0.000 000 000 000 000 001 one-quintillionth

*Most commonly used

Table A–5 DISTANCES IN THE UNIVERSE

meters

10^{28}	—— Visual limit (greatest distance from which light can reach us in the expected age of the universe)
10^{25}	——
10^{22}	—— Earth to nearest galaxy
10^{19}	——
	Earth to nearest star
10^{16}	——
	Distances between planets
10^{13}	——
	Earth to sun
10^{10}	——
10^{7}	——
10^{4}	—— Height of Mt. Everest
10^{1}	——
	Human height
10^{-2}	——
10^{-5}	—— Width of paper thickness
10^{-8}	—— Diameter of a virus / Diameter of a molecule of DNA
10^{-11}	—— Diameter of an atom
10^{-14}	—— Diameter of a nucleus

1 000 000 000 000. That is equivalent to 1 second in 30 000 years. We can deal with a few parts of mass in 1 000 000 000, which is equivalent to the mass of ink in one punctuation period compared to the mass of a whole book. We know that a concentration of phosphorus as small as 0.000 000 01 g per ml of lake water can cause the lake to begin to die from the overgrowth of algae. It is estimated that there are 602 300 000 000 000 000 000 000 carbon atoms in 12 grams of carbon.

Such numbers are awkward to write, since there are so many zeros to keep track of, making them difficult to work with. (Try calculating the number of carbon atoms in 1 gram of carbon without worrying about the zeros.) It's far better to express such numbers in exponential notation—also known as standard notation, scientific notation, or powers of ten. This compact way of writing numbers makes calculations easier and saves us a good deal of paper and ink in the process (Tables A–5, A–6, and A–7).

Note that the exponential form of a number involves the product of two numbers: a number between 1 and 10, called the *coefficient* (for example, 6.02), times a power of

---

Table A–6   RANGE OF TIME INTERVALS

seconds

10^{18}	——	Age of universe
10^{15}	——	Age of earth
10^{12}	——	Time of appearance of earliest men to present
10^{9}	——	Human life span
10^{6}	——	One year
		One day
10^{3}	——	Time required for light to travel from sun to earth
	——	
1	——	Time between heartbeats
10^{-3}	——	Period of a sound wave
10^{-6}	——	Period of a radio wave
		Time for electrons to go from source to screen in TV tube
10^{-9}	——	Time for light to penetrate a window
10^{-12}	——	
10^{-15}	——	
10^{-18}	——	Period of visible light wave
10^{-21}	——	Period of an X ray
10^{-24}	——	Time for light to cross a nucleus

10, the *exponential* (for example, 10^3). Let's focus on the exponential

$$\underbrace{10^3}_{\text{base}} \text{\{exponent}$$

We call 10 the base, and 3, written as a superscript following the 10, the exponent or power of the base 10. We read the number 10^3 as "10 cubed" or "10 to the third power." The exponent, 3, tells how many times the base, 10, is used as a factor, that is, the number of 10's multiplied together: $10 \times 10 \times 10$. It is also equal to the number of zeros between the 1 and the decimal point in the expanded form of the number. Thus, we could write 10^3 as 1 followed by three

---

Table A–7   RANGE OF MASSES

grams

10^{50}	——	The Milky Way
10^{40}	——	Sun
10^{30}	——	Earth
10^{20}	——	Moon
		A ship
10^{10}	——	A person
1	——	Postage stamp
10^{-10}	——	Red blood cell
10^{-20}	——	Oxygen molecule
10^{-30}	——	Electron

Table A–8 EXPONENTIALS

$10^6 = 10 \times 10 \times 10 \times 10 \times 10 \times 10$	$=$	$1\,000\,000$
$10^5 = 10 \times 10 \times 10 \times 10 \times 10$	$=$	$100\,000$
$10^4 = 10 \times 10 \times 10 \times 10$	$=$	$10\,000$
$10^3 = 10 \times 10 \times 10$	$=$	1000
$10^2 = 10 \times 10$	$=$	100
$10^1 = 10$	$=$	10
10^0	$=$	1
$10^{-1} = \dfrac{1}{10^1} = \dfrac{1}{10}$	$=$	0.1
$10^{-2} = \dfrac{1}{10^2} = \dfrac{1}{10 \times 10} = \dfrac{1}{100}$	$=$	0.01
$10^{-3} = \dfrac{1}{10^3} = \dfrac{1}{10 \times 10 \times 10} = \dfrac{1}{1000}$	$=$	0.001
$10^{-4} = \dfrac{1}{10^4} = \dfrac{1}{10 \times 10 \times 10 \times 10} = \dfrac{1}{10\,000}$	$=$	$0.000\,1$
$10^{-5} = \dfrac{1}{10^5} = \dfrac{1}{10 \times 10 \times 10 \times 10 \times 10} = \dfrac{1}{100\,000}$	$=$	$0.000\,01$
$10^{-6} = \dfrac{1}{10^6} = \dfrac{1}{10 \times 10 \times 10 \times 10 \times 10 \times 10} = \dfrac{1}{1\,000\,000}$	$=$	$0.000\,001$

zeros: 1000. Table A–8 gives some positive and negative exponentials. A negative exponent indicates a reciprocal:

$$10^{-3} = \frac{1}{10^3}$$

Suppose we wish to express the number 634 in exponential form. We know at the outset that there is a decimal point following the 4. We change 634 to a form consisting of a number between 1 and 10, multiplied by a power of 10, that is, 6.34×10^n. We get the number 6.34 by moving the decimal point two places to the left. The exponent, n, is equal to the number of places the decimal point has been moved, in this case, 2. If the decimal point must be moved to the left, the exponent is positive; if it must be moved to the right, the exponent is negative. Therefore we express the number 634 as 6.34×10^2. Since $10^2 = 100$, 6.34×10^2 really does equal 634.

For a number less than one, such as 0.000284, move the decimal point to the right to obtain the coefficient. Moving the decimal point four places to the right, we get the coefficient 2.84. The exponential, therefore, is 10^{-4}. So, $0.000284 = 2.84 \times 10^{-4}$. Since $10^{-4} = 0.0001$, $2.84 \times 10^{-4} = 2.84 \times 0.0001 = 0.000284$. The numbers below are expressed in exponential form. Check them.

$62.4 = 6.24 \times 10^1$	$0.00361 = 3.61 \times 10^{-3}$
$19\,357\,000 = 1.9357 \times 10^7$	$0.000096 = 9.6 \times 10^{-5}$
$2740 = 2.740 \times 10^3$	$0.459 = 4.59 \times 10^{-1}$
$58\,900 = 5.8900 \times 10^4$	$0.000\,005\,8 = 5.8 \times 10^{-6}$

A–8 MULTIPLICATION AND DIVISION OF EXPONENTIAL NUMBERS

One of the bonuses we get from exponential notation is that we can readily multiply and divide numbers expressed in this way. All we have to do is to apply a few rules.

(A) Multiplication. To multiply two exponential numbers, *add* the exponents. If one or both exponents happen to be negative, follow the same procedure and add them algebraically. The general rule is

$$10^a \times 10^b = 10^{(a + b)}$$

Therefore

$$10^2 \times 10^3 = 10^{(2 + 3)} = 10^5$$

This is because

$$10^2 = 10 \times 10, \text{ and } 10^3 = 10 \times 10 \times 10$$

Thus

$$10^2 \times 10^3 = (10 \times 10)(10 \times 10 \times 10) = 10^5 = 10^{(2 + 3)}$$

Similarly,

$$10^5 \times 10^{-2} = 10^{[5 + (-2)]} = 10^{(5 - 2)} = 10^3$$

and

$$10^{-3} \times 10^{-5} = 10^{[(-3) + (-5)]} = 10^{(-3-5)} = 10^{-8}$$

To multiply one exponential number by another, first multiply the coefficients together, then add the exponents algebraically. If the product of the coefficients is a number greater than 10 (for example, $8 \times 7 = 56$), it may be expressed in standard exponential notation ($56 = 5.6 \times 10^1$), and the exponentials combined.

EXAMPLE A–1

Multiply 30 000 × 200 000.

SOLUTION

1. *Change to exponential form:* $(3 \times 10^4) \times (2 \times 10^5)$

2. *Rearrange, separating the coefficients from the exponentials:* $(3 \times 2) \times (10^4 \times 10^5)$
3. *Multiply the coefficients together and add the exponents:* $= 6 \times 10^{(4 + 5)}$
$$= 6 \times 10^9$$

EXAMPLE A–2

Multiply 400×0.002.

SOLUTION $\quad 4.0 \times 10^2 + 2.0 \times 10^3 = 8 \times 10^{-1}$

Change to exponential form: $(4 \times 10^2) \times (2 \times 10^{-3})$
Rearrange: $(4 \times 2) \times (10^2 \times 10^{-3})$
Solve: $= 8 \times 10^{2 + (-3)}$
$$= 8 \times 10^{(2 - 3)}$$
$$= 8 \times 10^{-1}$$

EXAMPLE A–3

Multiply $600\,000 \times 7\,000\,000$.

SOLUTION $\quad 6 \times 10^5 \times 7 \times 10^6 = 4.2 \times 10^{12}$

Change to exponential form: $(6 \times 10^5) \times (7 \times 10^6)$
Rearrange: $(6 \times 7) \times (10^5 \times 10^6)$
Solve: $= 42 \times 10^{(5 + 6)}$
$$= 42 \times 10^{11}$$
$$= 4.2 \times 10^1 \times 10^{11}$$
$$= 4.2 \times 10^{12}$$

(B) Division. To divide, *subtract* the exponents. The general rule is

$$\frac{10^a}{10^b} = 10^{(a - b)}$$

Therefore

$$\frac{10^5}{10^2} = 10^{(5 - 2)} = 10^3$$

This is because

$$10^5 = 10 \times 10 \times 10 \times 10 \times 10, \text{ and } 10^2 = 10 \times 10$$

Thus

$$\frac{10^5}{10^2} = \frac{10 \times 10 \times 10 \times 10 \times 10}{10 \times 10} = 10^3 = 10^{(5-2)}$$

Similarly,

$$\frac{10^2}{10^6} = 10^{(2-6)} = 10^{-4}$$

and

$$\frac{10^3}{10^{-2}} = 10^{3-(-2)} = 10^{3+2} = 10^5$$

To divide one exponential number by another, separate the coefficients from the exponential terms, divide the coefficients, and subtract the exponents.

EXAMPLE A–4

Divide **60 000** by **0.003**

$6.0 \times 10^4 \div 3 \times 10^{-3} = 2 \times 10^7$

SOLUTION

1. *Change to exponential form:* $\dfrac{6 \times 10^4}{3 \times 10^{-3}}$

2. *Rearrange:* $\dfrac{6}{3} \times 10^{[4-(-3)]}$

3. *Divide the coefficients and subtract the exponents:* $= 2 \times 10^{(4+3)}$
$= 2 \times 10^7$

EXAMPLE A–5

Evaluate $\dfrac{9 \times 10^6}{4.5 \times 10^4}$ $= 2 \times 10^2$

SOLUTION

Rearrange: $\dfrac{9}{4.5} \times 10^{(6-4)}$

Solve: $= 2 \times 10^2$

EXERCISES.   Carry out the following operations.
1. $(4.00 \times 10^3) \times (2.00 \times 10^4)$ = 8×10^7
2. $(6.00 \times 10^5) \times (1.50 \times 10^{-3})$ = 9×10^2
3. $(5.00 \times 10^{-4}) \times (1.60 \times 10^{-5})$ = 8×10^{-9}
4. $(9.50 \times 10^6) \times (3.00 \times 10^{-2})$ 28×10^4
5. $35\,000 \times 200\,000$ $(3.5 \times 10^4) \times (2.0 \times 10^5)$ = 7×10^9
6. $6400/80\,000$
7. $20\,000/0.004$
8. $(7.20 \times 10^5)/(3.60 \times 10^{-3})$
9. $(4.80 \times 10^{-3})/(6.00 \times 10^6)$
10. $\dfrac{(4.20 \times 10^4) \times (2.00 \times 10^{-5})}{(7.00 \times 10^6)}$

$$\frac{6.4 \times 10^3}{8.0 \times 10^4} = 10^{-1}$$

A–9   DIMENSIONAL ANALYSIS

It doesn't mean much to say that the length of a swimming pool is 50. Fifty feet? Yards? Meters? Until we include the dimensional unit, the number has no meaning. Most physical quantities can be expressed in terms of length, mass, and time, or combinations of these. But lengths can be expressed in a variety of units. We often have to know how to convert from one unit to another. This is very easy to do in SI. Since all subunits are related by powers of 10, all we have to do is to move a decimal point or change an exponent. Thus, 1 meter = 10^2 centimeters = 10^3 millimeters. As long as customary units are also used, we will have to be "bilingual." We must be at home with both systems and know how to convert from one customary unit to another or from a customary unit to an SI unit.

Such conversions are easy to make with *dimensional analysis.* In this process the dimensional unit in which a measurement is expressed is treated as an algebraic term that may be multiplied or divided.

A conversion factor is the numerical relationship between any two units (Fig. A–5). In the customary system, 1 mile = 5280 feet; 1 foot = 12 inches; and so on. For SI units, 1 kilometer = 1000 meters; 1 meter = 100 centimeters; and so on. The fraction 5280 feet/mile actually equals 1, so that when we multiply a number by this conversion factor, we are multiplying by 1. Doing this does not change the value, only the units. To get the number of feet in 5 miles, use the conversion factor 5280 feet/mile. Start with the given 5 miles and multiply by the conversion factor. See how the miles cancel out, leaving the answer in the unit we want (feet).

$$5 \text{ miles} \times \frac{5280 \text{ feet}}{1 \text{ mile}} = 26\,400 \text{ feet}$$

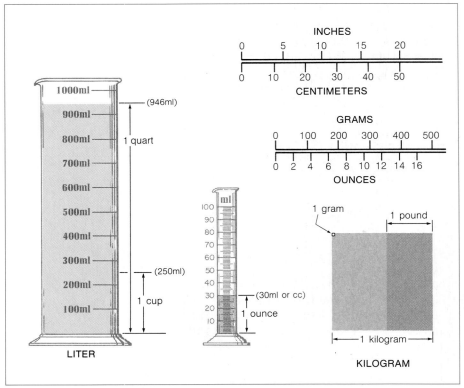

Figure A–5 Comparison of some metric-English units.

EXAMPLE A–6

Determine the equivalent in feet of a 50-meter Olympic swimming pool.

SOLUTION

The question is, 50 meters are equal to how many feet? Begin with the measurement of 50 meters. Then use all the appropriate conversion factors, one after another, canceling the units as you progress. In this calculation, you might go from meters to yards, and finally to feet. Then do the mathematical operations.

$$50 \text{ m} \times \frac{1.09 \text{ yd}}{\text{m}} \times \frac{3 \text{ ft}}{\text{yd}} = 163.5 \text{ ft}$$

EXAMPLE A–7

What is the number of centimeters in 1 kilometer?

SOLUTION

You know that 1 kilometer is equal to 1000 meters, and that 1 meter contains 100 centimeters. Using these equivalents as conversion factors, multiply the given unit by them. In the process, the kilometers and meters cancel out, and the desired unit, centimeters, is the answer.

$$1 \text{ km} \times \frac{10^3 \text{ m}}{\text{km}} \times \frac{10^2 \text{ cm}}{\text{m}} = 10^5 \text{ cm}$$

EXERCISES. Carry out the following conversions:
1. 5.00 cm to in.
2. 60 km to mi
3. 150 lb to kg
4. 760 mm to in.
5. 1 min to sec
6. 3000 mi to km
7. 1 year to sec
8. 100 yd to m
9. 55 mi to km
10. 1 oz to mg

cgs system
Conversion factor
Coefficient
Dimensional analysis
Exponential
Metric system

mks system
Scientific notation
SI units
Standard kilogram
Standard meter
U. S. customary system

CHECKLIST

A–1. Express the following in exponential notation: EXERCISES
(a) 440.74
(b) 1 200 000
(c) 0.000 263
(d) 0.00145
(e) 27 600
(f) 0.182
(g) 0.0000319
(h) 83.7
(i) 60 322 000
(j) 7840

A–2. Write out the following in ordinary notation:
(a) 8.92×10^2
(b) 5.63×10^{-4}
(c) 3.59×10^{-6}
(d) 7.482×10^3
(e) 6.21×10^5
(f) 8.02×10^{-7}
(g) 4.274×10^6
(h) 2.003×10^{-4}
(i) 9.70×10^8
(j) 3.065×10^6

A–3. Carry out the following conversions:

(a) 7.50 g to mg   (f) 100 mm to in.
(b) 8.00 in. to cm   (g) 186 000 mi to m
(c) 1050 g to oz   (h) 2000 lb to g
(d) 106 lb to kg   (i) 1 ft to cm
(e) 75 watts to kilowatts (j) 1 gal to ml

A–4. Carry out the indicated operations and express the answers in standard exponential notation:

(a) $(10^3)(10^2)(10^6)$

(b) $\dfrac{(1.8 \times 10^3)(2 \times 10^5)}{4 \times 10^9}$

(c) $\dfrac{(500)(6200)}{20\,000}$

(d) $(1.67 \times 10^{-27})(6.02 \times 10^{23})$

(e) $\dfrac{1.44 \times 10^{10}}{(3 \times 10^4)(4 \times 10^7)(6 \times 10^{-3})}$

(f) $(24) \times (3\,000\,000) \times (1/600\,000)$

(g) $\dfrac{(3 \times 10^8)(2 \times 10^{-3})}{(6 \times 10^{11})}$

(h) $(8 \times 10^5)/(4 \times 10^3)$

(i) $\dfrac{(2 \times 10^{-43})(6 \times 10^8)}{(8 \times 10^{-29})(4 \times 10^{-17})}$

(j) $\dfrac{(10^2)(10^{-3})(10^5)}{(10^6)(10^{-9})}$

A–5. The 10 000-meter run is one of the events in the Olympic games. What is the distance in miles?

A–6. (a) Give your height in customary units. (b) Express this in meters.

A–7. On certain flights, the maximum weight of luggage allowed a passenger is 20 kg. If you travel on one of these flights, what is the weight in pounds you could take with you?

A–8. To produce a ton of steel, 5×10^4 gal of water are required. Write this number in its expanded form and give it its appropriate name (for example, 10^2 is one hundred; 10^3 is one thousand; and so on).

A–9.   Convert the customary measurements in this saying to their SI equivalents: "An ounce of prevention is worth a pound of cure."

A–10.   A person with 6.0 quarts of blood has 1.2×10^3 mg cholesterol per liter of blood. What is the mass of cholesterol in the blood in (a) grams; (b) ounces?

A–11.   The ordinary chair seat is about 450 mm high. How many inches is this?

A–12.   How many liters of gasoline does a 21-gallon tank hold?
                               79 l

A–13.   A speed limit of 55 miles per hour is the same as how many kilometers per hour?

A–14.   According to a highway sign, you are 150 kilometers from your destination. How many miles is this? 93 mi

A–15.   The moon is about 240 000 miles from earth. How long would it take, in seconds, for a radio signal to travel from the moon to the earth at the rate of 3×10^8 meters per second?

A–16.   Multiple Choice:

A. The prefix "kilo" means
   (a) one-thousandth   (c) ten
   (b) one-tenth   (d) one thousand

B. Of the following, the smallest measurement is
   (a) 500 mm   (c) 0.005 km
   (b) 5.00 m   (d) 0.002 mi

C. The meter was originally based upon the dimensions of the
   (a) arm   (c) ocean
   (b) foot   (d) earth

D. The prefix that does not match the number is
   (a) centi-, 10^{-1} (c) milli-, 10^{-3}
   (b) kilo-, 10^3 (d) micro-, 10^{-6}

E. A yard equals
   (a) 1.1 m   (c) 39.37 m
   (b) 30.48 m   (d) 0.92 m

F. The number of grams in 2000 pounds is
   - (a) 9.909×10^5
   - (c) 436
   - (b) 1.60×10^4
   - (d) 6.24×10^{-2}

G. The product of $(1.30 \times 10^{-4}) \times (6.00 \times 10^5)$ is
   - (a) 7.80×10^{-9}
   - (c) 7.80×10^9
   - (b) 7.80×10^{-1}
   - (d) 7.80×10^1

H. A proper unit for designating volume is
   - (a) mg
   - (c) ml
   - (b) mm
   - (d) m

I. The result of dividing (8×10^3) by (2×10^{-5}) is
   - (a) 4×10^{15}
   - (c) 4×10^2
   - (b) 4×10^{-2}
   - (d) 4×10^8

J. The number 4.36×10^{-4} is
   - (a) 0.0436
   - (c) 0.000436
   - (b) 43 600
   - (d) 0.00000436

SUGGESTIONS FOR FURTHER READING

Adamson, Arthur W., "SI Units? A Camel is a Camel." *Journal of Chemical Education,* October, 1978.
> Argues that the SI system is not convenient to physical chemistry or superior in consistency.

Astin, Allen V., "Standards of Measurement." *Scientific American,* June, 1968.
> Discusses the standards of length, mass, time, and temperature.

Masterton, William L., and Emil J. Slowinski, *Mathematical Preparation for General Chemistry.* Philadelphia: W. B. Saunders Company, 1970.
> Good discussions of exponential numbers, significant figures, unit conversions, and other topics.

National Bureau of Standards, "Policy for NBS Usage of SI Units." *Journal of Chemical Education,* September, 1971.
> Encourages the use of SI units to facilitate communication.

Ritchie-Calder, Lord, "Conversion to the Metric System." *Scientific American,* July, 1970.
> Shows what a country may expect when it makes the transition to the metric system.

Socrates, G., "SI Units." *Journal of Chemical Education,* November, 1969.
> Discusses the development and use of SI units used in virtually every country.

APPENDIX B
ANSWERS TO
SELECTED EXERCISES

Appendix A

A–1.   (a) 4.4074×10^2
       (b) 1.2×10^6
       (c) 2.63×10^{-4}
       (d) 1.45×10^{-3}
       (e) 2.76×10^4
       (f) 1.82×10^{-1}
       (g) 3.19×10^{-5}
       (h) 8.37×10^1
       (i) 6.0322×10^7
       (j) 7.84×10^3

A–3.   (a) 7.50×10^3 mg
       (b) 20.32 cm
       (c) 37.03 oz
       (d) 48.08 kg
       (e) 0.075 kw
       (f) 39.37 in.
       (g) 2.99×10^8 m
       (h) 9.072×10^5 g
       (i) 30.48 cm
       (j) 3.785×10^3 ml

A–5.   6.21 mi

A–7.   44 lb

Chapter 2

2–7.   3.6 mi

2–9.   (a) 175 mi
       (b) 137 mi

2–11.   250 mi N; 135 mi W

2–13.   20 mi

2–15. $82 \dfrac{\text{ft}}{\text{sec}}$

2–17.   11 sec

2–19.   (a) $20 \dfrac{\text{mi}}{\text{hr}}$ or $29 \dfrac{\text{ft}}{\text{sec}}$
       (b) 145 ft

2–21. $29.4 \dfrac{\text{m}}{\text{sec}}$ or $96 \dfrac{\text{ft}}{\text{sec}}$

Chapter 4

4–7. $4 \dfrac{\text{m}}{\text{sec}^2}$

4–9. 3.75×10^4 dynes

4–11.   3600 N

4–13.   140 N

4–17. $F_{\text{earth-moon}} = 7.69 \times 10^{28}$ N
       $F_{\text{earth-sun}} = 4.93 \times 10^{33}$ N

Chapter 5

5–1. $400 \dfrac{\text{kg-m}}{\text{sec}}$

5–7.   50 joules

5–9. 1.82×10^{-18} joule

5–11.   (a) 58.8 joules
       (b) $44.27 \dfrac{\text{m}}{\text{sec}}$

5–15.   3750 joules

5–17. 1.25×10^6 joules
5–19. 49 joules

Chapter 6

6–1. (a) $-17.7°C$
      (b) 255.3 K
6–3. $-110.2°F$
6–5. $-297.4°F$
6–7. (a) $20°C$
      (b) 293 K
6–15. 150 N
6–17. 1.68×10^5 cal
6–19. $10.64°C$
6–21. $36.67°C$
6–31. 2.76×10^4 cal
6–35. 2.1×10^3 cal
6–37. 0.25 ft

Chapter 7

7–5. 1072 cal
7–7. 1.7×10^4 m
7–11. 37.5%
7–13. 45%

Chapter 8

8–7. 282.3 ft
8–9. 0.76 cm
8–11. $30 \frac{cm}{sec}$
8–13. 300 m
8–17. $8 \frac{beats}{sec}$
8–21. 0.67 m
8–23. 0.39 m
8–29. 4 m; 2 m; 1.33 m
8–31. 2640 Hz
8–33. 100 cm; 50 cm; 33.3 cm

Chapter 9

9–11. (a) 2.4×10^4 mi
      (b) 3.84×10^{14} km

9–13. 5.1×10^{14} Hz
9–19. Eye: 1 "octave"
      Ear: 10 "octaves"
9–21. 60 cm

Chapter 10

10–13. 8.1×10^{-8} N
10–25. 18 ohms
10–27. (a) 120 ohms
       (b) 0.92 ampere
       (c) 0.50 ampere; 0.25 ampere;
          0.17 ampere
10–29. 156.25 ohms

Chapter 11

11–13. 11 000 Å
11–15. 8.82×10^{-35} m
11–21. 2.56 sec

Chapter 12

12–1. (a) 9p   9e   10n
      (b) 16p 16e 16n
      (c) 33p 33e 42n
      (d) 38p 38e 50n
      (e) 80p 80e 121n
12–3. 24.310 amu
12–9. (a) 9
      (b) 18
12–11. (a) $1s^2 2s^2 2p^1$
       (b) $1s^2 2s^2 2p^5$
       (c) $1s^2 2s^2 2p^6 3s^2 3p^6 4s^2 3d^1$
       (d) $1s^2 2s^2 2p^6 3s^2 3p^6 4s^2 3d^{10} 4p^3$
       (e) $[Ar]4s^2 3d^{10} 4p^6 5s^2 4d^{10} 5p^6 6s^2$

Chapter 13

13–5. (a) $Na^+ \; \ddot{\underset{..}{F}}:^-$

                :$\ddot{Cl}$:
       (b) :$\ddot{Cl}$: C :$\ddot{Cl}$:
                :$\ddot{Cl}$:

(c) $:\overset{\displaystyle ..}{\underset{\displaystyle ..}{F}}:\overset{..}{\underset{..}{B}}:\overset{\displaystyle ..}{\underset{\displaystyle ..}{F}}:$ with $:\overset{..}{\underset{..}{F}}:$ above

(d) $:N::N:$

(e) $:C::O:$

13–11.

(a) $H:\overset{\displaystyle H}{\underset{\displaystyle H}{C}}:H$

(b) $H:\overset{\displaystyle H}{\underset{\displaystyle }{N}}:H$

(c) $H:\overset{..}{\underset{..}{O}}:$

(d) $H:\overset{..}{\underset{..}{F}}:$

(e) $:\overset{..}{\underset{..}{Ne}}:$

13–13.   C—N   B—Cl   P—F   H—S   MgO

Chapter 14

14–1.
(a) Ca_3P_2
(b) $Ba(NO_3)_2$
(c) $MgBr_2$
(d) $Sn(NO_2)_2$
(e) $Fe_2(Cr_2O_7)_3$
(f) $CuSO_4$
(g) $Pb(ClO_4)_2$
(h) K_3PO_4
(i) Hg_3N_2
(j) $SnCl_2$

14–3.
(a) Calcium carbonate
(b) Barium sulfate
(c) Sodium sulfite
(d) Strontium phosphate
(e) Magnesium nitride
(f) Iron (II) acetate
(g) Zinc nitrite
(h) Iron (III) perchlorate
(i) Boron trifluoride
(j) Aluminum phosphide

14–5.   227 amu

14–7. 5.57×10^{24} atoms

14–9.   12.31% C
7.18% H
15.90% P
7.18% N
16.41% S
41.03% O

14–11.   66.4%

14–13.   (b) $CO(NH_2)_2$

14–15. C_4H_{10}

14–17. CH_4N_2O

Chapter 15

15–1.   (a) $K = \dfrac{[NO]^2}{[N_2][O_2]}$

(b) $K = \dfrac{[CO][H_2O]}{[CO_2][H_2]}$

(c) $K = \dfrac{[NO]^4[H_2O]^6}{[NH_3]^4[O_2]^5}$

15–9.   Shift to the right

15–17. $6.75 \times 10^{-3} \left(\dfrac{\text{mole}}{\text{liter}}\right)^{-1}$

Chapter 16

16–7. 1.89×10^{-6} g

16–9. $-3.72°C$

16–13.   (a) 0.1 mole
(b) 5.85 g

16–15.   Less than 7.0

Chapter 17

17–3.   (a) $CH_3CHCH\!=\!CHCH_2CH_2CH_3$ with Br, Br below

(b) $CH_2C\!\equiv\!CCHCHCH_3$ with Cl, Cl Cl below

(c) $CH_3\overset{\displaystyle CH_3}{\underset{\displaystyle OH}{C}} \; \underset{\displaystyle Br}{CHCH_2CH_3}$

(d) $CH_3\!-\!O\!-\!CH_2CH_3$

(e) $CH_3CH_2\underset{\displaystyle Br}{CH}\;CH_2COOH$

17–5.   (a) $CH_3CH\!=\!CHCH_3$
(b) CH_3CH_2OH
(c) $CH_3CH_2\!-\!O\!-\!CH_2CH_3$

(d) $CH_3\overset{\displaystyle O}{C}\!-\!O\!-\!CH_3$

(e)

$$CH_3\overset{\displaystyle O}{\overset{\|}{C}}\!-\!H$$

17–13. (a) $HC\!\equiv\!CH$

(b) $CH_3CH_2CH_2\!-\!O\!-\!CH_2CH_2CH_3$

(c) $CH_3\overset{\displaystyle CH_3}{\underset{\displaystyle CH_3Cl}{\overset{|}{\underset{|}{C}}}}\ \ CHCH_3$

(d)

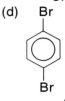

(e)

$$CH_3\overset{\displaystyle O}{\overset{\|}{C}}\!-\!O\!-\!CH_3$$

17–19. 1.65×10^5 tons

17–21. (a) trichloromethane

(b) ethoxyethane
(c) 2,4,6-trinitrotoluene
(d) ethyne
(e) 2-propanol

Chapter 18

18–1. (a) ^4_2He
(b) $^{39}_{19}K$
(c) $^0_0\gamma$
(d) $^{32}_{16}S$
(e) $^{14}_6C$

18–5. $^{246}_{96}Cm + ^{12}_6C \rightarrow ^{254}_{102}No + 4\,^1_0n$

Chapter 19

19–15. 3.08×10^5 years

GLOSSARY

Absolute Zero of Temperature   The theoretical lower limit to temperature: $-273°C$.

Acceleration   The rate at which the velocity of an object changes.

Accelerators   Machines designed to impart high energies to electrons, protons, and other particles for the purpose of serving as probes in the investigation of properties of matter.

Acid   A substance that, when added to water, tastes sour, turns the color of litmus from blue to red, neutralizes bases, and reacts with certain metals to give free hydrogen.

Action-at-a-Distance   The action of forces—gravitational, electric, magnetic—between pairs of bodies that are not in contact.

Action-Reaction Pair   Newton's Third Law of Motion states that to every action force there is an equal and opposite reaction force; the two forces are an action-reaction pair.

Addition Reaction   A typical reaction of unsaturated hydrocarbons—alkenes and alkynes. The breaking of relatively weak pi bonds is followed by the formation of bonds with a substance that in effect adds to the hydrocarbon.

Aether   In Aristotle's physics, the sun, planets, and stars were said to be composed of this perfect material unlike the four elements found on earth.

Air Mass   A large body of air that is relatively homogeneous in temperature and moisture content.

Albedo   The ratio of the amount of the sun's radiation reflected by a body to the amount striking it.

Alcohol   A hydrocarbon derivative in which one or more hydrogen atoms are replaced by —OH groups.

Algal Bloom   An overgrowth of tiny green plants, algae, that produces a foul-smelling scum on the surface of natural waters.

Alkanes   Saturated hydrocarbons. Organic compounds containing hydrogen and carbon that have two kinds of single bonds, carbon-to-hydrogen and carbon-to-carbon.

Alkenes   Unsaturated hydrocarbons containing at least one double bond between carbon atoms.

Alkyl Group   A radical derived from an alkane by the removal of one hydrogen atom.

Alloy   A solid solution of two or more metals.

Amber   A fossil resin that becomes strongly electric when rubbed. The Greek word for amber is "electron."

Amino Acids   Nitrogenous hydrocarbons that serve as building blocks of proteins and are thus essential to living things.

Ampere   The unit of electric current equal to the rate of flow past a given point of 1 coulomb per second.

Amplitude   Maximum displacement of a medium through which a wave is passing from its equilibrium condition.

Analysis Reaction   A chemical reaction in which a compound is broken down into simpler substances.

Anemometer   An instrument for measuring wind speed.

Angle of Declination   The angle between magnetic north and geographic north for any locality.

Angle of Dip   The angle, measured from the horizontal, taken by a magnetized needle that is suspended at its center and free to rotate in a vertical north-south plane.

Angular Momentum   A measure of the rotation of an object around a point.

Angular Momentum Quantum Number   Designates a subshell within a main energy level: s, p, d, f.

Anion   A negatively charged ion.

Applied Science   The application of science to the solution of practical problems.

Aromatic Hydrocarbons   Hydrocarbons that are chemically similar to benzene.

Arrhenius Theory of Ionization   When certain substances (electrolytes) dissolve in water, they dissociate into charged particles called ions.

Asteroids   "Minor planets," most of whose orbits around the sun lie in a belt between the orbits of Mars and Jupiter.

Astronomical Unit   The average distance between the earth and the sun; approximately 93,000,000 miles.

Atomic Number   The number of protons in the nucleus of an atom.

Atomic Weight   A weighted average of the naturally occurring isotopes of an element compared to carbon-12.

Audiogram   A profile of an individual's threshold of hearing for a range of sound frequencies.

Aufbau Principle   The building-up principle of the periodic system of the elements, based on four quantum numbers assigned to every electron in an atom.

Aurora   Display of bands of light in the sky caused by charged particles from the sun interacting with the earth's magnetic field and upper atmosphere.

Avogadro's Number The number of units in one gram-atom, one gram-molecular weight, and one gram-formula weight: 6.02×10^{23}.

Balmer Formula A formula that yielded the prominent lines in the visible region of the hydrogen spectrum with an amazing degree of accuracy.

Band Theory The band theory of solids assumes that isolated energy levels of atoms are broadened into energy bands that belong to the crystal as a whole.

Bar A unit of pressure equal to 1000 millibars or 29.53 inches of mercury.

Barograph A recording barometer, usually of the aneroid type.

Barometer An instrument for measuring atmospheric pressure.

Base A substance that, when added to water, has a bitter taste and soapy feel, turns the color of litmus from red to blue, and neutralizes acids.

Basic Science The pursuit of pure science that aims at extending human knowledge.

Bathythermograph A device that measures and records the change in temperature with depth of water.

Beats A periodic fluctuation in loudness caused when two wavetrains of sound of slightly different frequencies interfere.

"Big Bang" Theory The theory that the universe started from a highly compressed and extremely hot state that exploded, and that it has been expanding ever since.

"Big Science" The conduct of science by interdisciplinary teams of scientists.

Biodegradable A substance that is capable of being broken down biologically into harmless end products such as carbon dioxide and water by microorganisms in the soil or in sewage treatment plants.

Binary Compound A substance consisting of two different elements in chemical combination.

Binding Energy The energy that must be supplied to separate the particles in a nucleus.

Black-Body Radiation The radiation distribution of a perfect emitter of radiation.

Black Hole Thought to be the final stage in the evolution of certain stars. Composed of the densest matter known in the universe, the gravitational field would be so great that neither radiation nor matter could escape.

Boiling Point The temperature at which the vapor pressure of a liquid is equal to atmospheric pressure.

Boiling-Point Elevation Constant The change in the boiling point of a 1-molal solution of a nonelectrolyte as compared to the pure solvent.

Bond Angle The angle formed between three atoms in a molecule.

Bone Conduction Hearing through vibrations of bones in the skull that bypass the middle ear and are transmitted directly to the inner ear.

Breeder Reactor A nuclear reactor in which the relatively abundant, non-fissionable uranium-238 is transformed into fissionable plutonium-239.

British Thermal Unit (Btu) The amount of heat that is required to raise the temperature of one pound of water one Fahrenheit degree.

Brownian Motion The zig-zag motion of tiny particles suspended in a fluid due to the impacts of molecules of the fluid on these particles.

Calorie ("small") The amount of heat that is required to raise the temperature of one gram of water one Celsius degree (from 14.5°C to 15.5°C).

Calorimetry The measurement of the quantity of heat transferred from one substance to another.

Cambrian Period Designates the lowest systems of Paleozoic rocks.

Carbon-14 Dating A method based on the half-life of carbon-14 by which archaeological discoveries can be dated.

Carnot Engine An ideal engine—completely insulated, frictionless, and leak-proof.

Catalyst A substance that increases the rate of a chemical reaction without itself being used up or formed during the reaction.

Catalytic Hydrogenation The addition of hydrogen, in the presence of a catalyst, to the unsaturated bonds of vegetable oils, converting them into semisolid shortenings or oleomargarine.

Cation A positively charged ion.

Ceilometer An instrument for measuring cloud height.

Celsius Scale A temperature scale on which the ice point is 0° and the steam point is 100°.

Centrifugal Force A "center-fleeing" force; an outward force that is the reaction to centripetal force.

Centripetal Force A "center-seeking" force that causes an object to seek a circular path.

Cepheid Variable A class of pulsating stars that act as a "celestial lighthouse" and as a means of measuring distances deep into the universe.

CGS System A subsystem of SI units, the centimeter-gram-second system.

Chemical Evolution The transformation of simple gases in the primitive atmosphere and ocean to complex compounds that are presumed to sustain life.

Chemical Formula A combination of chemical symbols that tells what elements are present and how many atoms of each are in a unit of a compound.

Chemical Kinetics The study of the rate and mechanism of chemical reactions.

Chondrite A stony meteorite.

Chondrule A small crystalline spherical mass in the core of a stony meteorite.

Chromosphere A relatively thin atmospheric layer of the sun lying between the photosphere and the corona.

Close Packing A space-filling arrangement in metals in which as many atoms as possible are packed into a given volume.

Coacervates Colloidal droplets rich in polymers that separate spontaneously from aqueous solutions of proteins, carbohydrates, and nucleic acids.

Cold Front The leading edge of a relatively cold air mass that replaces warmer air.

Colligative Properties Properties of a solution that depend only on the number of particles present and not on their chemical nature. Examples are vapor pressure, boiling point, and freezing point.

Collision Theory The theory that chemical reactions occur when molecules encounter each other, and the energy of the molecules exceeds a certain minimum threshold, the energy of activation.

Colloid A mixture containing groupings of molecules intermediate in size between molecules and the particles in a suspension. Milk and blood are examples.

Comet Members of the solar system that orbit the sun in highly elongated orbits and when near the sun show a coma and a tail.

Complementary Colors Any two colors that give white when added, such as blue and yellow.

Components The parts into which a vector can be separated.

Compton Effect X-ray bombardment of various targets produces primary scattering and secondary scattering, the latter from the collision of an X-ray photon with an electron of the target material, as Arthur Holly Compton demonstrated.

Concentrated Solutions that contain a relatively large amount of solute.

Condensation The change in state from a vapor to a liquid.

Conduction The transfer of heat through the collisions of neighboring molecules in a material without the conducting material itself moving.

Conductor Any material that can readily transmit electricity or heat.

Conjugated System A molecule containing alternating single and double bonds between the carbon atoms.

Conservation of Energy Energy cannot be created or destroyed, but it can be transformed from one kind into another.

Conservation of Mechanical Energy In the conversion of kinetic energy into potential energy, and vice versa, the sum of kinetic energy and potential energy is a constant, if friction is disregarded.

Constellation Grouping of the brightest stars in configurations to which the names of mythological figures are attached and which designate regions in the sky.

Constructive Interference An effect caused when the crests of two waves having the same amplitude arrive at a given point and the two amplitudes combine.

Continental Drift The theory that the continents were once part of a single supercontinent that began splitting apart millions of years ago, and that the pieces—continents—drifted to their present locations.

Continental Shelf The gently sloping surface around the margin of a continent or island.

Continental Slope The steeply sloping surface reaching from the continental shelf to the ocean bottom.

Convection The transfer of heat through fluids by currents that cause the mixing of the warmer with the cooler regions of a liquid or gas.

Conversion Factor The numerical relationship between any two units.

Coordinate Covalent Bond A covalent bond in which both electrons of the bond come from the same atom.

Corona The outermost layer of the sun's atmosphere seen most spectacularly during a total solar eclipse.

Cosmology The study of the structure and development of the universe.

Coulomb's Law The force between two electric charges is proportional to the magnitude of each of the charges and inversely proportional to the square of the distance between them.

Crest The highest point of a wave.

Critical Mass The minimum amount of fissionable material that will sustain a nuclear chain reaction.

Crookes Tube Cathode-ray tube designed by Sir William Crookes that showed that cathode rays travel in straight lines.

Cumulus A cauliflower-shaped cloud type.

Cyclotron A nuclear accelerator introduced by Ernest O. Lawrence that is based on the concept of accelerating a particle many times in succession through a relatively low potential difference in such a way that it picks up energy at each acceleration.

DeBroglie Wavelength The wavelength associated with a moving particle, such as an electron or proton.

Decibel (dB) A unit of sound intensity.

Declination The celestial equivalent of latitude measured in degrees north or south of the celestial equator.

Degassing The expulsion of gases during the crystallization of minerals from cooling magma.

Destructive Interference A crest of one wave and a trough of a second wave encounter and cancel each other, giving a combined amplitude of zero.

Detergent Any cleansing agent, such as soap. In popular usage, the word refers to synthetic detergents, or syndets.

Detergent Builder A component of detergents that softens water and prevents dirt that has already been washed out from being redeposited.

Dew Point The temperature at which condensation will occur in a parcel of air.

Diesel Engine A high-compression engine in which air is the working fluid. The fuel is injected near the end of the compression stroke and is ignited by the heat of compression.

Diffraction The bending of waves around obstacles or through openings.

Dilute Solutions that contain a relatively small amount of solute.

Dimensional Analysis A process of converting from one unit to another by treating the dimensional unit in which a measurement is expressed as an algebraic term that may be multiplied or divided.

Dipole A molecule in which the centers of positive and negative charge do not coincide.

Dipole Moment A measure of the electrical character of a molecule, it is defined as the product of the effective charge and the distance between the centers of opposite charge.

Dispersed Phase The colloidal particles in a colloid.

Dispersing Medium The medium in which the colloidal particles in a colloid are dispersed.

Displacement Distance traveled in a specified direction; a vector quantity.

Doppler Effect The change in the frequency (or wavelength) of waves as sensed by an observer when there is relative motion between a source of waves and the observer.

Double Bond The atoms are joined by sharing two pairs of electrons.

Double Refraction A property of calcite crystals and other materials of splitting an incident beam of light into two beams and causing a double image.

Drizzle Droplets of precipitation less than 0.5 mm in diameter.

Dropsonde A radiosonde that is dropped by parachute from an aircraft to take measurements of the atmosphere below.

Dynamic Equilibrium If substances A and B react to form C and D, and C and D in turn react to form A and B, a condition of dynamic equilibrium exists when the two reactions proceed at the same rate.

Dynamo An electric generator.

Dyne The cgs unit of force; the force that accelerates a mass of 1 gram by 1 centimeter/sec^2 in the direction of the force.

Ecliptic The plane of the ecliptic is the plane of the earth's orbit around the sun.

Electric Intensity The strength of an electric field at a point, defined as the electric force that would be exerted on a very small positive test charge placed at that point.

Electrolytes Solutions that conduct electricity. Also, solutes whose solutions are conducting.

Electromagnetic Wave Electric and magnetic fields that propagate through space at the speed of light at right angles to each other and to the direction of propagation.

Electronegativity A measure of the ability of a bonded atom in a molecule to attract electrons.

"Electron-Sea" Model The picture of a metal crystal as an array of positive metal ions suspended in a "sea of electrons," delocalized valence electrons.

Electrophorus A device that produces large electrostatic charges.

Electroscope A device that may be used to determine the presence and kind of electric charge on an object, and to some extent the quantity of charge.

Electrostatic Generator Any device that produces electrical charges.

Electrostatic Precipitator A device that removes particulate industrial wastes—fly, ash, dust, and fumes—from air.

Electrovalent Bond A chemical bond formed by the transfer of electrons between atoms.

Element A pure substance that cannot be decomposed into simpler substances by ordinary chemical or physical means.

Ellipse A conic section on which the sum of the distances of any point from two fixed points, the foci, is constant.

Empirical Formula The simplest formula for a compound, giving the relative numbers of each kind of atom present in the compound.

Emulsification The dispersal of oil or grease into tiny droplets.

Endothermic Reaction A chemical reaction in which heat is absorbed.

Energy The capacity of an object to do work.

Energy of Activation The minimum energy that molecules must have if collisions between them are to lead to reaction.

English System The traditional system of units of measurement, based upon feet, pounds, and quarts.

Entropy A measure of the disorder of a system.

Enzyme Biological catalysts. High molecular weight protein molecules that are endowed with highly specific catalytic activity toward some chemical reaction.

Equilibrium Constant A numerical value that applies to a reversible chemical reaction when it is in equilibrium at a specific temperature.

Erg The cgs unit for work: 1 dyne · cm.

Ester A class of organic compounds formed when an alcohol reacts with an acid.

Ether A class of organic compounds corresponding to the general formula R—O—R—.

Ether Wind The concept that if the earth moved through a medium called the ether, it should be possible to detect an "ether wind" opposite to its motion.

Evaporation The change in state from a liquid to a vapor.

Exobiology The study of extraterrestrial life.

Exosphere The outermost region of the atmosphere, above an average altitude of 500 km.

Exothermic Reaction. A chemical reaction in which heat is evolved.

Expanding Universe A theory based upon Hubble's law and Einstein's general theory of relativity that the expansion of the universe began with the "big bang" and is continuing.

Experiment A process undertaken to demonstrate a known truth or to discover an unknown effect or principle.

Exponential A number expressed in scientific notation, or powers of ten.

Fahrenheit Scale A temperature scale on which the ice point is 32° and the steam point is 212°.

Field A region of space in which certain physical effects exist and can be detected, such as gravitational fields, electrostatic fields, and magnetic fields.

First Law of Thermodynamics The law of conservation of energy: energy cannot be created or destroyed, but may be converted from one form to another.

Fission The splitting of an atomic nucleus into two smaller nuclei with the release of energy and several neutrons.

Fixed Points Reference temperatures on a thermometer, usually the ice point and steam point of water.

Fluorescence Light is produced by exciting a phosphor, a substance that absorbs electromagnetic energy and then re-emits it as visible light.

Fog Visible, minute water droplets suspended in the atmosphere near the Earth's surface, reducing visibility below one kilometer (0.62 mile).

Force A push or a pull that imparts an acceleration to an object.

Formula Weight The sum of the atomic weights of all the atoms in the chemical formula of a substance.

Frame of Reference A coordinate system with respect to which the position and motion of an object may be described.

Free Radical An atom or molecule containing an unpaired electron, for which reason it is quite reactive.

Freezing Point The temperature at which a liquid changes to a solid at standard atmospheric pressure.

Freezing-Point Depression Constant The change in the freezing point of a 1-molal solution of a nonelectrolyte as compared to the pure solvent.

Frequency The number of waves passing a given point per unit time.

Fuel Melt The danger that excessive heat within the

core of a nuclear reactor could cause the core to melt and discharge large amounts of radioactivity into the environment.

Functional Group An atom or group of atoms that imparts common chemical and physical properties to all members of a family containing it, such as the —OH group in alcohols.

Fundamental The lowest frequency of which a vibrating string or air column is capable.

Fusion A process in which light nuclei are fused together into a heavier nucleus and energy is released.

Galaxy A basic building block of the universe that includes stars, star clusters, clouds of gas and dust, and interstellar molecules.

Geiger-Mueller Counter A device that detects radioactivity.

Geocentric Hypothesis The idea that the earth is at the center of the universe.

Globular Cluster A spherical group of thousands of stars located above and below the plane of the galaxy and in relative motion with it.

Gondwanaland In continental drift theory, a temporary large land mass in the Southern Hemisphere.

Gram-Atomic Weight The amount of an element in grams that is numerically equal to its atomic weight.

Gram-Formula Weight The amount of a substance in grams that is numerically equal to the sum of the atomic weights appearing in its formula.

Gram-Molecular Weight The amount of a substance in grams that is numerically equal to the molecular weight of the substance.

Ground State In the ground state of an atom, the electrons occupy the lowest energy levels available and the atom is stable.

Group A vertical column in the periodic table occupied by elements with related chemical and physical properties that make up a chemical family.

Guyot A flat-topped volcanic feature on the ocean floor.

Hail Precipitation in the form of balls or lumps of ice.

"Haldane Soup" The condition of the primitive ocean before the appearance of life—a warm dilute system of organic compounds.

Half-Life The time required for the radioactivity of a sample to drop to one half of any value.

Hard Water Water containing a high concentration of calcium and magnesium ions. Soap does not lather in hard water and may form a scum.

Harmonic The frequencies above the fundamental frequency of which a vibrating system is capable.

Heat The transfer of molecular energy from one body to another because of a temperature difference.

Heat Budget The equilibrium between the heat absorbed by the earth in one year and the amount of heat radiated back into space in one year.

Heat of Combustion The heat produced per unit mass of substance burned in oxygen. Used to determine the heat value of fuels and foods.

Heisenberg Uncertainty Principle According to this principle, if we can measure the energy of an electron precisely, we cannot simultaneously determine the exact position of the electron.

Heliocentric Hypothesis The idea that the sun is at the center of the universe.

Hertzprung-Russell Diagram A graph of brightness vs. temperature or spectral class for a group of stars.

Homologous Series A group of related organic compounds, such as the alkanes, in which the molecular formulas of the members differ from one another by a constant increment.

Horsepower Work performed at the rate of 33 000 foot-pounds per minute or 550 foot-pounds per second. 1 H.P. = 746 watts.

Hubble's Law Other galaxies are moving away from ours at speeds that are proportional to the distances of those galaxies.

Humanities A group of studies that includes the languages and literature, the fine arts, philosophy, and religion.

Hund's Rule Electrons do not enter into joint occupancy of an orbital until all of the available orbitals of a given subshell are half occupied.

Hydrocarbon The simplest organic compounds, containing only hydrogen and carbon.

Hydrogen Bonding An intermolecular force between a hydrogen atom of one molecule and an oxygen or nitrogen atom in a nearby molecule.

Hydrogen 21-cm Waves Radio waves emitted by hydrogen atoms in space.

Hydrophilic Water soluble.

Hydrophobic Repelled by water.

Hydrosphere The earth's water shell, including the oceans, lakes, streams, glaciers, ground water, and similar features.

Igneous Rock Rock formed by the cooling and solidification of magma.

Impulse The product of force and time. A measure of the change in momentum of an object.

Induction A method of producing electrostatic charges in which there is no direct contact between the charging body and the object.

Inertia The resistance of an object to changes in its state of motion.

Infrared The region of the electromagnetic spectrum with wavelengths longer than 8000 Å.

Inhibitor A negative catalyst, one that slows down a chemical reaction.

Insulator Any material through which electricity or heat does not readily flow.

Intensity The amount of energy carried by a wave.

Interdisciplinary Involving more than one field of knowledge, such as biochemistry.

Interference The superposition of waves, producing regions of reinforcement and regions of cancellation.

Internal Combusion Combustion takes place inside the working fluid without any need for a firebox and a boiler.

Ionic Bond See Electrovalent Bond.

Ionization Energy The energy necessary to remove an electron from an atom in the gaseous state.

Ionosphere A layer of the atmosphere extending from about 50 km beyond the earth's surface that has a large concentration of free electrons and positively charged ions. The ionosphere reflects radio waves to the earth's surface.

Ion-Product of Water The dissociation constant of water.

Island Universe Galaxies lying far beyond the Milky Way.

Isobar A line on a chart connecting points of equal pressure.

Isomerism The existence of more than one compound with a given molecular formula due to differences in structure or to the spatial orientation of the atoms.

Isotherm   A line connecting points of equal or constant temperature.

Isotopes   Different varieties of the atoms of an element. Their nuclei have the same atomic number but different mass numbers.

Joule   The SI unit for work: 1 newton·meter.

Kelvin Scale   A temperature scale based upon the absolute zero of temperature. The size of a degree is the same as that of a Celsius degree.

Kepler's Laws   The three laws of planetary motion discovered by Johannes Kepler.

"Kernel"   The nucleus of an atom and all of the electrons in the inner filled shells of the atom.

Kilocalorie   "Giant" calorie. The amount of heat that is required to raise the temperature of one kilogram of water one Celsius degree. Used to rate the heat value of foods and fuels.

Kinematics   The scientific description of motion, without regard to the cause.

Kinetic Energy   The energy a moving object has by virtue of its motion: K.E. = $\frac{1}{2}mv^2$.

Kinetic Theory   The theory that matter is made up of tiny particles called molecules, and these molecules are in a state of constant motion.

Lapse rate   A temperature decrease with height.

Laser   An acronym for "light amplification by stimulated emission of radiation." An optical device that emits coherent waves that are highly monochromatic and can be propagated in the form of well-collimated beams.

Latent Heat of Fusion   The quantity of heat that must be added to one gram of solid at its melting point to change it to liquid at the same temperature and pressure.

Laurasia   In continental drift theory, a temporary large land mass in the Northern Hemisphere.

Lava   Molten material that flows out of volcanic vents as hot streams or sheets and solidifies as rock.

Law of Charles and Gay-Lussac   The volume of a gas varies directly with its absolute temperature, if the pressure is held constant.

Law of Combining Volumes   When gases combine to form new compounds, the volumes of the reactants and products are in the ratio of small whole numbers.

Law of Conservation of Mass   Atoms are neither created nor destroyed in chemical reactions; they are merely rearranged.

Law of Mass Action   The rate of a chemical reaction is usually proportional to some power of the concentrations of the reactants.

Law of Reflection   For all waves, the angle of incidence is equal to the angle of reflection.

Law of Refraction   The ratio of the sine of the angle of incidence to the sine of the angle of refraction is a constant.

Le Châtelier's Principle   When a stress is imposed on a system at equilibrium, the equilibrium shifts in such a way as to minimize the effect of the stress.

Lens Equation   An equation that relates the object distance, image distance, and focal length of a lens.

Lewis Structure   A representation of atoms, ions, and molecules in which the valence electrons are shown as dots surrounding the symbol of an element.

Light-Year   The distance that light travels in a year, 6×10^{12} miles. A useful standard for expressing stellar distances.

Limiting Nutrient   The element in shortest supply with respect to the nutritional requirements of an organism.

Linear Expansion   The change in length of a solid upon heating.

Lines of Force   Lines that are used to represent the direction of an electrical or magnetic field.

"Little Science"   The conduct of science by individuals working on problems which they themselves select within a narrow field of knowledge.

Local Group   A cluster of twenty or so galaxies that includes the Milky Way.

Lodestone   A natural magnetic stone.

Lone-Pair Electrons   A pair of electrons not involved in a chemical bond.

Longitudinal Wave   A wave consisting of alternating compressions and rarefactions, such as a sound wave, in which the particles of the medium oscillate in the direction of the disturbance.

Loudness   A physiological sensation that is experienced when sound waves strike the ear drum.

Luminiferous Ether   A hypothetical medium that carries the disturbance called light.

Lunar Eclipse   An eclipse of the moon; this occurs when the moon passes into the earth's shadow.

Macromolecule   Giant molecules, or polymers.

Macroscopic   Large-scale observable properties of matter.

Magnetic Domain   Clusters of atoms that have a common orientation. If aligned, the substance is magnetic.

Magnetic Field Reversal   The earth's magnetic field has two stable states and has alternated between them over the ages.

Magnetic Quantum Number   Designates a sub-subshell or orbital.

Main Sequence   A long, narrow, diagonal band on the Hertzprung-Russell diagram on which over 90 percent of all stars near the sun fall.

Manganese Nodules   Lumps containing oxides of manganese, iron, nickel, and copper found scattered over the ocean floor.

Mantle   A layer of dense rocky material about 2900 km thick that lies between the earth's crust and core and occupies about 80 percent of the volume of the earth.

Maria   The name originally given to the dark areas of the moon because they were thought to be seas (*mare*, sea).

Mascon   Regions of dense material under the surface of the moon; mass concentrations.

Mass   A measure of an object's inertia; also, the quantity of matter in an object.

Mass Defect   The difference between the mass of a nucleus and the sum of the masses of its protons and neutrons.

Mass Spectrometer   An instrument that separates the isotopes of an element.

Maxwell's "Demon"   An imaginary being that could establish a temperature difference from a source of uniform temperature by separating fast from slow molecules that could then drive a heat engine.

Mechanical Equivalent of Heat Mechanical work is converted into heat by the ratio
Work/Heat = 4.18 joules/calorie.

Metallic Bond The chemical bond that joins the atoms of metals.

Metamorphic Rock Rocks that have been changed by heat and pressure in the earth's crust so that their original characteristics have been lost.

Metathesis Reaction A chemical reaction between two compounds in which there is a double replacement of atoms or radicals in the compounds.

Meteorites Small asteroids that have orbits overlapping that of the earth and impact the earth.

Meteorology The study of the atmosphere and its processes.

Metric System A decimal system of weights and measures, based upon the meter as the unit of length and the kilogram as the unit of mass.

Micelle Oil or grease droplets surrounded by soap molecules.

Microfossils Remains of microscopic forms of life.

Microspheres Tiny spheres that separate from aqueous solutions of proteinoids and show many of the characteristics of living cells.

Mineral A naturally occurring solid substance with a definite chemical composition and crystal structure.

MKS System A subsystem of SI units, the meter-kilogram-second system.

Model A representation of a structure or process that cannot be observed directly.

Moderator A material in a nuclear reactor that slows neutrons but does not absorb them.

Mohorovičić Discontinuity (MOHO) The sharp boundary between the earth's crust and underlying mantle.

Molality The number of moles of a solute in 1 kilogram of solvent. A measure of the concentration of a solution.

Molarity The number of moles of a solute per liter of solution. A measure of the concentration of a solution.

Mole A unit numerically equal to Avogadro's number: 6.02×10^{23}. This unit is used interchangeably with the terms gram-atom, gram-molecular weight, and gram-formula weight.

Molecular Formula The molecular or true formula represents the total number of atoms of each element present in one molecule of a compound.

Molecular Geometry The arrangement of the atoms of a molecule in space.

Molecular Weight The sum of the atomic weights of all the atoms in a molecule.

Molecule The smallest particle of a substance that possesses the properties of that substance.

Momentum The product of the mass of a moving object and its velocity.

Monochromatic A beam of radiation having a single well-defined frequency.

Monomer Any substance that can serve as a building unit of a polymer.

Multiple Bonds A double or triple covalent chemical bond.

Nansen Bottle A device used by oceanographers to obtain samples of ocean water.

Nebula A cloud of interstellar gas and dust.

Nebular Theory The theory that the sun and planets formed out of a cloud of gas and dust.

Negative The kind of electric charge that hard rubber acquires when rubbed with fur.

Neutralization A reaction between acids and bases in which hydrogen ions from the acid are combined with hydroxide ions from the base to form water molecules.

Neutron A nuclear particle with roughly the same mass as a proton but lacking an electrical charge.

Neutron Star An extremely dense star composed of neutrons, believed to be closely related to pulsars.

Newton The mks unit of force; the force that accelerates a mass of 1 kilogram by 1 meter/sec^2 in the direction of the force.

Newton's First Law An object will maintain its state of rest or motion unless acted upon by a force.

Newton's Law of Universal Gravitation The gravitational force between two objects is directly proportional to the product of their masses and inversely proportional to the square of the distance separating them.

Newton's Second Law An object will accelerate in the direction of a force applied to it.

Newton's Third Law For every action force, there is an equal and opposite reaction force.

Noble Gas One of the family of elements helium, neon, argon, krypton, and xenon. All except helium have eight electrons in their outermost shell.

Noise Sounds consisting of unrelated frequencies; any sound regarded as a nuisance.

Non-Electrolyte Substances whose solutions do not conduct an electric current.

Nova A star that increases in brightness explosively.

Nuclear Atom In Rutherford's model of the atom, the positive charge of an atom and most of its mass is confined to a small sphere, the nucleus, and the negative electrons are distributed at some distance from the nucleus.

Nuclear Chain Reaction The neutrons released when uranium-235 fissions induce fission reactions in other uranium-235 nuclei, thus sustaining the process.

Nuclide Any nuclear species that has a specific atomic number and a specific mass number.

Objective Based on observable phenomena; uninfluenced by emotion or personal prejudice.

Oceanic Trench A long, narrow, deep depression on the ocean floor.

Oceanography The scientific study of the sea.

Octane Number A measure of the anti-knock properties of a gasoline used as an automotive fuel.

Octet Rule Atoms enter into combination to obtain a structure that has eight electrons in the outermost shell.

Ohm's Law The current through a conductor is directly proportional to the potential difference across the conductor and inversely proportional to its resistance.

Oil-Drop Experiment An experiment designed by Robert A. Millikan that enabled him to measure the charge on the electron.

Olefin An alkene.

Orbit The path in space of a celestial object.

Orbital A sub-subshell of an atom that can accommodate two electrons of opposite spin.

Organic Chemistry The chemistry of carbon compounds.

Ossicles Three small bones in the middle ear.

Otto Engine A reciprocating, piston, internal-combustion engine that employs a 4-stroke cycle first demonstrated by Nicolaus August Otto: intake, compression, power, exhaust.

Overtones Multiples of the fundamental frequency of a vibrating string or air column.

Oxidation Number The positive or negative charge assigned to the atoms of an element in the formula of a compound. It is a measure of the combining capacity of an element.

Ozone A form of oxygen, O_3.

Pangaea In continental drift theory, the original supercontinent.

Paraffin An alkane, or saturated hydrocarbon.

Parallax The apparent displacement of an object against the background when viewed from two different positions.

Parallel Circuit A divided electrical circuit that provides more than one path for the current. The current passes through each resistance independently.

Parsec Parallax-second. The distance of an object having a parallax of 1 second of arc. One parsec = 3.26 light-years.

Partial Component tones of definite pitch present in a sound produced by a musical instrument.

Pauli Exclusion Principle No two electrons of the same atom may have the same set of four quantum numbers.

Percentage Composition The weight-percent of each element present in a compound.

Period A horizontal row running from left to right across the periodic table.

Period The time to complete one cycle of a wave.

Periodicity The recurrence at definite intervals of the chemical and physical properties of the elements when arranged in a table according to increasing atomic number.

Perturbation A discrepancy between the predicted and actual orbits of a planet or satellite.

pH A measure of the acidity of a solution on a scale of 1 to 14, 7 being neutral, values lower than 7 acidic, and values greater than 7 basic.

Phanerozoic Era The periods of geologic time since the start of the Cambrian 570 million years ago.

Phosphorescence Similar to fluorescence, except that the emission of light may persist for minutes or even hours.

Photochemistry A chemical reaction induced by light.

Photoelectric Effect The emission of electrons from a metal surface when light is shined on it.

Photon A packet of light energy, hf. A light quantum.

Photosphere The visible surface of the sun.

Pi Bond A covalent chemical bond formed by the lateral overlap of p orbitals on adjacent carbon atoms.

Planet A relatively large body revolving around the sun.

Plasma A hot gas in which atoms are reduced to nuclei and free electrons.

Plate Tectonics The theory that the earth's surface consists of a number of rigid plates that move with respect to one another.

Polar Covalent A covalent bond in which the electron pair is not shared equally by the joined atoms.

Polariscope An instrument that enables an engineer to infer the stresses in structural materials from colored patterns produced by the interference effects of polarized light.

Polarized Light Light waves that oscillate in just one direction at right angles to the beam's path.

Polaroid A material invented by Edwin H. Land that polarizes light.

Polymer A large molecule formed from the combination of many small molecules.

Positive The kind of electric charge that glass acquires when rubbed with silk.

Potential Energy The energy possessed by a body by virtue of its position or condition.

Power The rate at which work is done. Power = work/time.

Precambrian Eon The first 90 percent of the earth's history; the first 4 billion years.

Precipitation Moisture in any form falling from the atmosphere, such as drizzle, rain, snow, and hail.

Pressure Force per unit area.

Primary Carbon A carbon atom that is bonded to one other carbon atom.

Primary Colors Three pure colored lights—red, green, and blue—that form white when combined.

Primeval Fireball Radiation in the microwave band that is coming from all directions in space; interpreted to be the remnant signals of the "big bang."

Principal Quantum Number Determines the main energy level in which an electron is located.

Prism A triangular piece of glass that separates a beam of light into its component colors.

Protein A long-chain polymer synthesized from amino acids and present in all living cells.

Proteinoid Long-chain molecule resembling a protein synthesized from amino acids.

Proton A nuclear particle of an atom carrying a unit positive charge.

Psychrometer An instrument that measures humidity.

Pulsar A celestial object that emits bursts of radio pulses with periods ranging from 0.033 to 3.75 seconds; believed to be a rapidly rotating neutron star.

Quality Timbre or tone color of a sound.

Quantum The unit of radiant energy. According to Max Planck, radiant energy comes in discrete amounts, or packets, called quanta.

Quasar A quasi-stellar radio source. Starlike objects that have large red shifts and extremely high energy output.

Radiation A method of heat transfer. Any object above absolute zero radiates energy to its surroundings as electromagnetic waves.

Radical A group of atoms having either a positive or negative charge that occurs as a unit in compounds but is not stable by itself.

Radioactivity A process in which an element is transmuted into another element through the emission of nuclear particles and radiations.

Radiosonde A balloon-borne instrument package that measures and transmits meteorological data.

Radio Source A location in space from which radio waves are received, including the sun, planets, nebulae, stars, and galaxies.

Radio Telescope An antenna and reflecting disk for the detection and study of radio sources.

Rain Gauge An instrument that measures rainfall.

Raoult's Law The lowering of the vapor pressure of a solution is proportional to the molal concentration of the solute.

Reaction Mechanism A single step or sequence of steps by which reactants are changed to products in a chemical reaction.

Red Giant A relatively bright and relatively cool star.

Red Shift The shift of spectral lines toward the longer, red-wavelength region of the spectrum, interpreted as a Doppler effect in receding objects.

Red Spot A conspicuous feature of the planet Jupiter.

Reflection The change in direction that waves undergo when they encounter a barrier.

Refraction The bending of waves when they pass from one medium into another.

Regelation The process of melting a solid such as ice under pressure, then refreezing when the pressure is removed.

Relative Humidity The ratio of water vapor in the air compared to the amount held if the air were saturated at the same temperature.

Relativity In Einstein's special theory of relativity, the idea that there is no preferred reference frame.

Replicable The repetition of an experiment or observation with similar results.

Resonance The effect produced when a system that can vibrate at a certain frequency is acted upon by a periodic disturbance that has the same frequency.

Resultant The geometrical sum of two or more vectors.

Retrograde Motion The apparent backward motion of a planet caused by the relative motion of the planet and the Earth in their orbits.

Reversible Reaction The products of a chemical reaction react with one another to reform the reactants.

Revolution The orbiting of one celestial body around another.

Right Ascension Celestial longitude, measured in hours of time eastward along the celestial equator.

Rock Heterogeneous aggregates of minerals in varying proportions that form an appreciable part of the earth's crust.

Rotary Engine A rotating combustion engine invented by Felix Wankel in which there are no valves, the rotor itself opening and closing the fuel-air input port and the exhaust port.

Rotation The turning of a body around its own axis.

Salinity Salt content, as of sea water.

Satellite An object that orbits around an astronomical body.

Saturated Hydrocarbon An alkane.

Scalar A quantity such as mass or volume that is described by magnitude or size only without regard to direction.

Scattering The reflection of light by particles in many directions.

Science An approach to solving problems based on logic and experimentation.

Scientific Law A statement of an unchanging relationship.

Scientific Notation A number expressed as the product of two numbers: a number between 1 and 10 times a power of 10.

Scientism The belief that only scientific knowledge is worthwhile.

Sea-Floor Spreading A hypothetical mechanism for continental drift, according to which molten rock from the mantle rises up along the mid-oceanic ridge, then spreads outward and hardens into new ocean floor.

Seamount An undersea volcano.

Secondary Carbon A carbon atom that is bonded to two other carbon atoms.

Second Law of Thermodynamics One formulation of this law is that the energy in the world available for work is continually decreasing. Another formulation is that the entropy of the world is continually increasing.

Sediment Finely divided particles of organic or inorganic origin that accumulate in loose form.

Sedimentary Rocks Rocks formed from the material caused by the weathering and erosion of older rocks and carried elsewhere by water or wind.

Seismic Wave Waves in the earth produced by earthquakes or underground explosions.

Seismograph An apparatus that registers the shocks and motions of earthquakes and underground explosions.

Selective Absorption A subtractive process in which certain wavelengths of light are removed from the incident light and absorbed, while the remaining wavelengths are reflected.

Selenography The study of the same aspects of the moon that geography studies of the earth.

Semiconductor Materials such as silicon and germanium, which ordinarily are not good conductors or good insulators, can be made to conduct well when made with certain impurities or by raising the temperature.

Series Circuit An electrical circuit in which the same current flows through several conductors in turn.

Shell An orbit or main energy level of an atom.

Shock Wave A pulse of pressure that moves through a medium at a speed faster than the medium can transmit sound waves.

Sigma Bond A covalent bond that is cylindrically symmetrical around the line joining the nuclei.

Significant Figures Numbers that can be read directly from a measuring instrument, plus one estimated, or uncertain, figure.

Simple Replacement Reaction A chemical reaction in which an atom or radical replaces another atom or radical in a compound.

Sine Wave The path traced by the simplest type of regular wavetrain.

SI Units The International System of Units (Système Internationale d'Unités), an extension and refinement of the metric system.

Sleet Precipitation in the form of ice pellets, or a mixture of rain and snow.

Soap A mixture of the sodium salts of various organic acids obtained from natural fats or oils that serves as a cleansing agent.

Solar Eclipse An eclipse of the sun; this occurs

when the moon partially or totally covers the sun's disk.

Solar Flare Sudden increases in the light intensity of the chromosphere of the sun that release great amounts of energy.

Solubility The limit at which no additional solute goes into solution, often expressed as grams of solute per 100 grams of water.

Solute A substance dissolved in a solvent.

Solution A homogeneous mixture of two or more non-reacting substances.

Solvation A process in which the molecules or ions of a solute become surrounded by a number of water molecules in the solution.

Solvent The medium in which a substance is dissolved.

Sonic Boom Explosive sound associated with aircraft shock waves.

Specific Heat The amount of heat that is required to raise the temperature of one gram of a substance one Celsius degree.

Specific Humidity The mass of water vapor in the air compared to the total mass.

Speed A measure of how fast an object travels over a period of time without regard to direction; distance divided by time.

Spin Quantum Number One of two equivalent electron spin orientations.

Spontaneous Generation The idea that life evolved from non-living matter.

Standard Kilogram The mass of a platinum-iridium cylinder kept at the International Bureau of Weights and Measures near Paris.

Standard Meter The distance between two scratches on a bar of platinum-iridium alloy, kept at the International Bureau of Weights and Measures, calibrated in terms of the wavelength of light.

Steam Turbine An engine that converts the heat energy of steam into the kinetic energy of rotary motion.

Stimulated Emission The basis of the laser principle. Atoms in the excited state are stimulated to emit photons with wavelengths that are precisely in phase with light of the right frequency passing through them.

Stock System A system of chemical nomenclature in which the oxidation number of a metal is indicated by Roman numerals enclosed in parentheses following the English name of the metal; for example, Tin (II) chloride.

Stratosphere The layer of the earth's atmosphere from an altitude of 11 km to 50 km.

Sublimation The change in state directly from the solid phase to the gas phase, without passing through the liquid phase.

Submarine Canyon A V-shaped rocky canyon cut into the continental shelf or slope.

Substitution Reaction A typical reaction of alkanes, in which one or more hydrogen atoms are replaced by other atoms.

Sunspot Regions of the photosphere of the sun where the gases are cooler than those of the surrounding areas, causing them to appear dark by contrast.

Sunspot Cycle An 11-year cycle during which sunspots increase in number toward a maximum and then fall off to a minimum.

Superconductivity The disappearance of all resistance to the flow of electric current.

Supernova A major star explosion, flaring up to millions of times brighter than its original state.

Superposition Principle See Interference.

Suspension A mixture in which particles large enough to be visible with a microscope settle to the bottom of a medium.

Symbol A representation of an element.

Synthesis Reaction A chemical reaction in which atoms or molecules combine to form larger molecules.

Tablemount A flat-topped seamount.

Technology The application of science to the development of new products or processes.

Technology Assessment The effort to predict some of the consequences of introducing new technologies.

Telescope An instrument for detecting and observing distant objects.

Temperature A measure of the intensity of heat, of the kinetic energy of molecules.

Ternary Compound A compound composed of a metal in combination with a radical consisting of another element and oxygen; in all, three kinds of atoms are present.

Tertiary Carbon A carbon atom that is bonded to three other carbon atoms.

Tetrahedral Configuration The valence bonds of a carbon atom are directed toward the four corners of a regular tetrahedron.

Theory A statement that is intended to explain a scientific law.

Thermal Efficiency The efficiency of an ideal (Carnot) engine operating between temperatures T_1 and T_2 is given by $T_1 - T_2/T_1$.

Thermal Energy The sum of the kinetic energy, rotational energy, and vibrational energy of molecules.

Thermal Pollution The discharge of waste heat into natural bodies of water.

Thermocline A zone in the ocean that is characterized by a rapid decrease in temperature. It extends from a depth of 250 m to 1400 m.

Thermography The detection and recording of variations in body temperature on thermograms.

Thermometric Property A property of matter that varies with temperature in a measurable way, such as expansion or electrical resistance.

Thermonuclear Reaction Nuclear fusion reactions that occur at temperatures of millions of degrees.

Thermosphere A region in the atmosphere beginning at an altitude of about 85 km and extending to an average altitude of about 500 km.

Threshold Frequency In the photoelectric effect, light below the threshold frequency for a given material will not eject photoelectrons.

Threshold of Hearing The lowest intensity at which sound of a given frequency can be heard by an individual.

Thunderstorm A local storm produced by a cumulonimbus cloud accompanied by heavy wind, rain, lightning, and thunder.

Titius-Bode Law A formula that gives the radii of the orbits of the seven innnermost planets and the radius of the asteroid belt.

Tornado A violent, destructive storm nearly always

observed as a "funnel cloud," with wind speeds estimated at 100 to more than 300 mph.

Trajectory The path that an object such as a projectile describes in space.

Transition State Theory Colliding molecules may stick together to form an activated complex. The activated complex then either breaks apart to reform the reactants, or it proceeds to produce the products.

Transmutation The change of a nucleus of one element into a nucleus of another element either spontaneously, as in radioactivity, or by artificially induced means.

Trans-Uranium Elements Elements beyond uranium in the periodic table.

Transverse Wave A wave in which the disturbance travels in one direction while the particles of the medium oscillate in a direction perpendicular to it.

Trilobites Paleozoic marine arthropods that were especially abundant during the Early Paleozoic.

Troposphere The lowest region of the earth's atmosphere, with an average height of 11 km. It is a region of cloud formations, warm and cold fronts, and storms.

Trough The lowest point of a wave.

Turbidity Current A dense fast-flowing body of sediment-laden water.

Turbojet Engine A gas turbine or "jet engine"; a non-reciprocating internal combustion engine that transforms heat energy directly into work.

Ultraviolet The region of the electromagnetic spectrum consisting of wavelengths shorter than 4000Å.

Ultraviolet Catastrophe The hotter a body, the more it radiates at shorter wavelengths. Classical theory predicts that most of the radiation from a very hot body will be in the ultraviolet or X-ray range. This prediction fails.

Uncertainty A measure of doubt associated with any physical measurement.

Unsaturated Hydrocarbon A hydrocarbon containing one or more double or triple bonds.

U.S. Customary System See English System.

Valence Shell Electrons in the valence shell of an atom occupy the highest-energy orbitals of the atom.

Vapor Pressure The vapor pressure of a liquid is the pressure at which the substance in the gaseous state is in equilibrium with the liquid in a confined space at a given temperature.

Vector A quantity that has both magnitude and direction, such as velocity or force.

Velocity The speed of an object in a specified direction; a vector quantity.

Violent Motion In Aristotle's physics, motion contrary to the nature of an object.

Vital Force A hypothetical force that was supposed to regulate substances derived from plant or animal sources.

Volt The unit of electrical potential. 1 volt = 1 joule/coulomb.

Voltaic Pile The first electric battery. A series of alternating copper and zinc disks separated by pads moistened with salt water.

VSEPR Theory The valence shell electron pair repulsion theory, proposed by R. J. Gillespie, says that the number of valence-shell electron pairs surrounding an atom determines the arrangement of the bonds around the atom.

Wankel Engine See Rotary Engine.

Warm Front A front that moves in such a way that warmer air replaces colder air.

Watt The SI unit of power. Work performed at the rate of 1 joule/second.

Wave Equation The velocity of a wave is equal to its frequency multiplied by its wavelength.

Wavelength The distance between successive crests or troughs of the same wave.

Wave Mechanics A system based upon Erwin Schrödinger's wave equation, in which the wave nature of electrons is assumed and the solutions describe the arrangement of electrons in atoms.

Wave-Particle Duality The idea that matter and radiation manifest both wave and particle properties.

Weather Short-term variations of the atmosphere, usually thought of in terms of temperature, humidity, precipitation, wind, cloudiness, and visibility.

Weight The force of gravitational attraction upon an object.

White Dwarf The final stage in the evolution of certain stars in which the density of matter is approximately a million times greater than the sun's.

Work The product of force and the distance through which the force acts.

Xerography The dry-copy imaging process underlying the Xerox and other copying machines.

X Ray High energy photons with wavelengths from approximately 0.1Å to 50Å.

Young-Helmholtz Theory The three-receptor theory of color vision that certain cells in the retina sense red light, others green, and still others blue.

Zodiac The band of twelve constellations through which the sun, moon, and planets move in the course of the year.

INDEX

Page numbers in *italics* refer to illustrations; those followed by (t) refer to tables.